THE
TELECOMMUNICATIONS
HANDBOOK

EDITORS-IN-CHIEF

Kornel Terplan
Patricia Morreale

 CRC PRESS

 IEEE PRESS

A CRC Handbook Published in Cooperation with IEEE Press

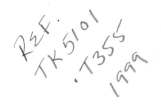

Library of Congress Cataloging-in-Publication Data

The telecommunications handbook / edited by Kornel Terplan, Patricia Morreale.
     p. cm.
    Includes bibliographical references and index.
    ISBN 0-8493-3137-4 (hc. : alk. paper)
    1. Telecommunication--Handbooks, manuals, etc.   I. Terplan, Kornel. II. Morreale, Patricia.
TK5101.T355 1999
384—dc21

                                              99-044580
                                              CIP

No claim to original U.S. Government works
International Standard Book Number 0-8493-3137-4
Library of Congress Card Number 99-044580
Printed in the United States of America 1 2 3 4 5 6 7 8 9 0
Printed on acid-free paper

Acknowledgments

The Editors-in-Chief would like to thank all contributors for their excellent, high quality work. Special thanks are due to our Associate Editors: Floris van den Broek, Rolf Stadler, and Dan Minoli. Without their help, we would not have been able to submit this manuscript on time.

We are particularly grateful to Dawn Mesa who has supported our editorial work by providing significant administrative help from CRC Press. We would also like to thank Suzanne Lassandro for leading the production of this handbook, in addition to Ramila Saldanha and Sameer Kaltra who assisted the co-editors in the compilation of the manuscript.

Foreword

Understanding telecommunications, in all its forms, has never been more critical. Numerous sources promise to provide the information and insight needed. However, a coherent, concise presentation of the essentials, which could serve as a starting point for providing the reader with a foundation of knowledge to build on has been lacking. Furthermore, much of the available information on telecommunications has been provided by vendor and manufacturer sources that are eager to advocate a proprietary viewpoint. *The Telecommunications Handbook* serves to address this need.

By gathering a group of talented contributors from all aspects of the telecommunications industry — including policy makers, regulators, manufacturers, operating companies, and academia— a total picture can be drawn, presenting the current operating environment and emerging technologies against the historical background of the telecommunications industry.

The organization of this handbook is designed to provide the reader with a reference for analyzing today's telecommunications revolution. Section 1 starts with a review of telecommunications services and their history, an enterprise model is developed, and growth strategies for telecommunications organizations are presented. A discussion of regulation, in the U.S. as well as elsewhere in the world, is also presented. Wireless telecommunications regulation receives particular attention as the international growth of wireless has, to some extent, advanced wireless communications in the U.S.

A discussion of the emerging standardization and expectations for future standards is provided.

For the reader who would like a review of basic telecommunications principles, telecommunication architecture is assessed, with a discussion of signal processing and coding schemes for use in telecommunications systems. Section 2 provides an examination of telecommunications services on broadband networks, and an overview of mobile and wireless telecommunications networks. Section 3 presents the integration of communication technologies, as found in such advanced services as FDDI/CDDI and real-time communications, multiple access communications networks, DSL, SONET, SDN, ATM, and multimedia. Finally, these technologies are presented in an advanced look at wireless and mobile networks. Video communications is discussed in Section 4.

This handbook is made possible by outstanding contributions from an international gathering of telecommunications experts. By providing a comprehensive starting point for interested readers, we hope to explain the fundamentals of telecommunications systems, current and future, and provide direction for further inquiry through our references and sources.

We wish our readers well as they investigate one of the most complex, challenging, and interesting topics in the scientific and engineering community today.

Patricia Morreale
Advanced Telecommunications Institute
Stevens Institute of Technology
Hoboken, NJ

Editors-in-Chief

Kornel Terplan, Ph.D., is a telecommunications expert with more than 25 years of highly successful multinational consulting experience. His book, *Communication Network Management,* published by Prentice-Hall (now in its second edition), and his book, *Effective Management of Local Area Networks,* published by McGraw-Hill (now in its second edition), are viewed as the state-of-the-art compendium throughout the community of international corporate users. He has provided consulting, training, and product development services to over 75 national and multinational corporations on four continents, following a scholarly career that combined some 140 articles, 19 books, and 115 papers with editorial board services.

Over the last 10 years, he has designed five network management–related seminars and given some 55 seminar presentations in 15 countries. He received his doctoral degree at the University of Dresden and completed advanced studies, researched, and lectured at Berkeley, Stanford University, University of California at Los Angeles, and Rensselaer Polytechnic Institute.

His consulting work concentrates on network management products and services, operations support systems for the telecommunications industry, outsourcing, central administration of a very large number of LANs, strategy of network management integration, implementation of network design and planning guidelines, products comparison, selection, and benchmarking systems, and network management solutions.

His most important clients include AT&T, AT&T Solutions, Georgia Pacific Corporation, GTE, Walt Disney World, Boole and Babbage, Solomon Brothers, Kaiser Permanente, BMW, Siemens AG, France Telecom, Bank of Ireland, Dresdner Bank, Commerzbank, German Telecom, Unisource, Hungarian Telecommunication Company, Union Bank of Switzerland, Creditanstalt Austria, and the state of Washington.

He is Industry Professor at Brooklyn Polytechnic University and at Stevens Institute of Technology in Hoboken, NJ.

Patricia Morreale, Ph.D., is Director of the Advanced Telecommunications Institute (ATI) and an associate professor in the School of Applied Sciences and Liberal Arts at Stevens Institute of Technology. Since joining Stevens in 1995, she established the Multimedia Laboratory at ATI and continued the work of the Interoperable Networks Lab in network management and performance, wireless system design and mobile agents.

Dr. Morreale holds a B.S. in computer science from Northwestern University, an M.S. in computer science from University of Missouri, and a Ph.D. in computer science from Illinois Institute of Technology. She holds a patent in the design of real-time database systems, and has published numerous articles.

Before joining academia, Dr. Morreale was in industry, working with Sears Communications Network (SCN, now Advantis-IBM) and, earlier, the McDonnell Douglas Corporation, in telecommunications management and network design. She has been a consultant on a number of government and industrial projects.

Dr. Morreale's research has been funded by the National Science Foundation (NSF), U.S. Navy, U.S. Air Force, Allied Signal, AT&T, Lucent, Panasonic, Bell Atlantic and the New Jersey Commission on Science and Technology (NJCST). She is a member of Association for Computing Machinery and a senior member of the Institute of Electrical and Electronic Engineers.

Contributors

Raymond U. Akwule
George Mason University
Fairfax, Virginia

David Allen
Information Economics and Policy Magazine
Concord, Massachussets

Willem F. Korthals Altes
Law School Professor
Amsterdam, The Netherlands

Chris B. Autry
Lucent Technologies
Atlanta, Georgia

Biao Chen
University of Texas at Dallas
Richardson, Texas

Chuck Cosson
AirTouch Communications
San Francisco, California

Rolf de Vegt
Renaissance Worldwide
San Francisco, California

Wouter Franx
Lucent Technologies
Leidschendam, The Netherlands

Rob Frieden
Pennsylvania State University
University Park, Pennsylvania

Zygmunt Haas
Cornell University
Ithaca, New York

Hisaya Hadama
NTT Telecommunications Network
  Laboratory Group
Tokyo, Japan

Al Hammond
Santa Clara University
Santa Clara, California

Marc Hendrickx
International Network Services
Amsterdam, The Netherlands

Jean-Pierre Hubaux
EPFL-ICA
Lausanne, Switzerland

John (Toby) Jessup
US West Communications
Seattle, Washington

Maarten Looijen
Delft University of Technology
Delft, The Netherlands

Nicholas Malcolm
Hewlett-Packard (Canada) Ltd.
Burnaby, British Columbia, Canada

Dan Minoli
Teleport Communications Group
Staten Island, New York

Emma Minoli
Red Hill Consulting
Red Bank, New Jersey

Steven Minzer
Lucent Technologies
Whippany, New Jersey

Patricia Morreale
Stevens Institute of Technology
Hoboken, New Jersey

Charles M. Oliver
Kelley, Drye & Warren LLP
Washington, D.C.

Henry L. Owen
Georgia Institute of Technology
Atlanta, Georgia

Pamela Riley
AirTouch Communications
Washington, D.C.

Izhak Rubin
University of California
Los Angeles, California

Larry Sookchand
Teleport Communications Group
Staten Island, New York

Kornel Terplan
Consultant and Industry Professor
Hackensack, New Jersey

Floris G. H. van den Broek
Level 3 Communications
Amsterdam, The Netherlands

Roger Wery
Renaissance Worldwide
San Francisco, California

Wei Zhao
Texas A&M University
College Station, Texas

Simon Znaty
ENST-Bretagne
Cesson-Sevigne, France

Michele Zorzi
University of California
San Diego, California

Contents

Section 1 Telecommunications Services, Regulation, and Standardization

Section Editor: Floris G. H. van den Broek

Section 2 Basic Communication Principles

Section Editor: Rolf Stadler

Section 3 Communication Technologies

Section Editor: Patricia Morreale

Section 4 Video Communications

Section Editor: Dan Minoli

1

Telecommunications Services, Regulation, and Standardization

Kornel Terplan
Consultant and Industry Professor

Floris G. H. van den Broek
Level 3 Communications

Marc Hendrickx
International Network Services

Rolf de Vegt
Renaissance Worldwide

Roger Wery
Renaissance Worldwide

Willem F. Korthals Altes
Law School Professor

Charles M. Oliver
Kelley, Drye & Warren LLP

Raymond U. Akwule
George Mason University

Rob Frieden
Pennsylvania State University

Pamela Riley
AirTouch Communications

Chuck Cosson
AirTouch Communications

Al Hammond
Santa Clara University

Wouter Franx
Lucent Technologies

Maarten Looijen
Delft University of Technology

David Allen
Information Economics and Policy Magazine

1.1 Trends in Telecommunications

Kornel Terplan

Changes in telecommunications are impacting all types of user group, which include business users, traveling users, small and home offices, and residential users. The acceptance rate of telecommunications and information services is accelerating significantly. Voice services needed approximately 50 years to reach a very high teledensity; television needed just 15 years to change the culture and lives of many families; the Internet and its related services have been penetrating and changing business practices and private communications over the last 2 to 3 years.

Trends in the telecommunication industry are analyzed from the following perspectives:

- Growth of the global telecommunications market
- Increasing network complexity

0-8493-3137-4/00/$0.00+$.50
© 2000 by CRC Press LLC

- Deregulation and privatization
- Communication convergence
- Customer orientation

Growth of the Global Telecommunications Market

Explosive expansion driven by internal growth and acquisition is forcing telecommunications providers to increase the productivity of their current support systems. Growth and acquisition mean that the number of subscribers grows for existing services, new services are provisioned on existing intrastructures, and completely new services on new infrastructures are deployed or acquired. Several support systems vendors have worked to capitalize on this opportunity with solutions that reduce complexity. These support systems vendors do not usually replace existing systems, but add functionality to accommodate new services, such as:

- Internet, intranets, and extranet
- Special data services on top of voice networks
- Wireless services and fixed wireless services
- Cable and video services
- Voice services on top of data networks

Adding functionalities that interoperate with each other opens new business opportunities for support systems vendors. The coming years will experience a bitter competition between circuit- and packet-switched services. Tradition, stability, and quality of existing services will compete against new technologies with easier maintenance and reduced operating expenses. The transition from circuit-switched to packet-switched technologies may take decades.

Increasing Network Complexity

As a result of customer expectations, the time-to-market of new services is extremely short. Incumbent and new telecommunications service providers do not have the time to build all new infrastructure, but combine existing and new infrastructures, such as copper, fiber, and wireless. They deploy emerging services on the basis of a mixture of infrastructures as an overlay. New services use emerged and emerging technologies, such as:

- Emerged technologies: voice networks, ISDN, circuit switching, packet switching, message switching, frame relay, Fast Ethernet, Fast Token Ring, and FDDI/CDDI.
- Emerging technologies: ATM, mobile and wireless, SMDS, Sonet/SDH, cable, xDSL, and B-ISDN.

Each of these technologies has its own support system solutions. The only elements in Public Switched Telephone Networks (PSTN) that should be managed are the switches themselves. On average, the ratio of managed elements to subscriber lines is around 1:10,000. The advent of distributed, software-based switching and transmission has created a large number of additional managed elements, about one for each 500 subscriber lines. Moreover, multiple elements per subscriber in digital loop carrier systems, digital cellular networks, or hybrid fiber/coax systems may cause an explosion in terms of managed elements. As a result, the size of configuration databases and event messages generated by more intelligent network elements have grown exponentially over the last 20 years.

Growth in the number of network elements has been accompanied by an increase in the complexity of items to be managed. Sonet/SDH, ATM, and digital wireless are highly complex, with a high degree of interdependence among network elements. This in turn makes service activation and fault isolation a challenge, especially as the number of service providers increases. As networks shift from lower-speed, dedicated-rate, and inflexible services to mobile, fully configurable, bandwidth-on-demand, and high-speed services, support systems must adapt to this new situation.

When services are offered in combination, support systems should be modified, re-engineered, and connected to each other. This opens new business opportunities for support systems vendors.

The introduction of standards for support systems is accelerating the demand for third-party support systems. Legacy systems are primarily proprietary systems not integrated across functional areas. Service providers depend upon custom development by internal development staff and outside integrators to connect various support systems. The introduction of technology standards, such as Telecommunication Management Network (TMN), Distributed Communication Object Model (DCOM), Common Object Request Broker Architecture (COBRA), Telecommunications Information Networking Architecture (TINA), and Web-Based Enterprise Management (WBEM) have begun to gain critical support by new support systems vendors.

The implementation of standard gateways enables interaction between newer client/server solutions and existing legacy systems, easing interoperability among all support systems. In particular, TMN may help to streamline support system processes and to position support systems.

Deregulation and Privatization

Telecommunications service competition began in the 1980s in the U.S., led by MCI with three operating support systems playing a key role. The AT&T divestiture in 1984 marked a major breakthrough. The second significant milestone was the Telecom Act of 1996. As telecom deregulation continues, with Regional Bell Operating Companies (RBOCs) actively pursuing the long-distance market and long-distance carriers moving into local services, major support systems re-engineering efforts are expected.

Under the pressure of the European Commission (EC), Europe is in the process of deregulation and privatization. It is a much slower process than in the U.S., because multiple countries are involved with their own agenda. Interoperability of support systems is more difficult than in the U.S.; but at the same time, it offers opportunities for support systems vendors. It is assumed that Asia/Pacific, South America, Eastern Europe, and Africa will follow this deregulation and privatization trend.

Competition is everywhere — long distance, local exchange, ISP, cable, and wireless. In many cases, support systems are the differentiators. The best opportunities are seen with Competitive Local Exchange Carriers (CLECs). Support systems requirements vary quite substantially from carrier to carrier. As a result, CLECs support system strategies are ranging from internal development, to outsourcing, to systems integrators, and to third-party software/service providers. CLECs could be small or medium sized, with or without facilities. In all cases, they must interoperate with Incumbent Local Exchange Carriers (ILECs) by opening the support systems to permit access by CLECs in various phases of provisioning and order processing and service activation. Key issues are:

- Local Number Portability (LNP): This allows customers to retain their telephone numbers even if they change service providers. It is not only the telephone number that is important; customers also typically want to retain access to advanced features they have come to expect from an intelligent network.

- Extranets connecting support systems of ILECs and CLECs: ILECs are required to provide access to information on five classes of support systems. They are preordering, ordering, provisioning, repair, and maintenance.

- Directory services: Real-time service processing requires additional customer-related data. The expanded directory role includes end-user authorization and authentication. It also includes the real-time allocation of network resources according to a user's class of service and other policy-based variables. Directory Enabled Networks (DEN) promise to increase momentum for directory services by bringing physical infrastructures under the directory umbrella and tackling the standardization of directory information.

- Fraud management: Offering multiple services that are accessible by user-friendly interfaces increases the risks of penetration. Service providers agree that up to 5% of their revenues is lost as a result of fraud. Real-time surveillance systems combined with customer analysis features of billing systems may help reduce fraud risks to a reasonable minimum.

- Usage-based billing: Time- or pulse-based billing has been very successful for decades in voice-related environments. As data-related services dominate the communications infrastructures, customers are expected to be billed on the basis of resources they have actually used. In data environments, resources are shared between multiple applications and multiple customers.

Incumbent service providers have turned to advanced support systems to differentiate their long-distance or local-exchange services from each other. After a substantial investment in custom systems over the last few years, many incumbents have begun to focus on upgrading their support systems with best-of-breed technologies. Many of them try to augment older systems to add more flexibility while sustaining traditional levels of performance and reliability. This creates additional complexity and requires that new management solutions designed for advanced equipment also work with older technologies.

As a result, "umbrella-types" of support systems are in demand, opening new opportunities for support system vendors with integration capabilities. To remain competitive, incumbent carriers need to deliver an increasingly larger number of new products and services. This has created a mixture of equipment, software, and services within many carriers.

Innovation and re-engineering on behalf of the incumbent carriers include:

- Better customer care: Based on Call Detail Record (CDRs) and other resource utilization-related data, unsophisticated customer analysis can be accomplished. It includes discovering trends in customer behavior, traffic patterns, reasons for frauds, and service-related items.
- Convergent billing: The customer may expect to receive one bill for all services, such as voice, data, video, and Internet. The minimal requirement is to receive multiple bills with electronic staples.
- Rapid provisioning of new services: Based on additional support systems, provisioning can be expedited by better interfaces and more accurate data.
- Service differentiation: Still using the same infrastructures, new services can be created and deployed. Carefully defining the value-added nature of new services will lead to customer differentiation.
- Offering new services, such as Internet access, xDSL, VPN, and VoIP: Incumbent service providers are expected to react rapidly to new communication needs, including a offering Internet access for reasonable payment, and deployment of xDSL, VPNS, and VoIP.

In each of these cases, either the deployment of new support systems or the customization of existing support systems is required. In both cases, additional market opportunities open for support systems vendors.

Communication Convergence

Advanced technology coupled with deregulation is driving communications convergence. Customers prefer to get all types of services, such as long distance and local voice, data/Internet, cable/video, wireless access, from the same service provider. Voice is expected to support both local and long distance, requiring it to play an LEC and IEX role at the same time. Data are gaining importance for both local and long distance, and usually includes Internet access. Data are supposed to reach voice volumes within 5 years, requiring total rebuilding of circuit-switching technology. Cable is expected to accommodate voice and data in addition to video. Wireless includes all kinds of mobile services and satellites that support voice, video, and data.

Deregulation was intended to encourage competition through proliferation of new entrants. Looking to gain share, carriers are entering each other's market, blurring traditional lines between services, geographic coverage, and communication platforms. Aggressive new carriers have moved rapidly to establish nationwide service networks, consolidating local, long-distance, Internet, wireless, and cable services under one umbrella. Incumbent carriers are trailing this way of convergence. The U.S. shows an excellent example of this convergence; the big eight carriers cover most end markets. These carriers are AT&T, Bell Atlantic, BellSouth, GTE, SBS, Sprint, US West, and WorldCom. But they still leave room for hundreds of point products, mostly best-of-breed telecommunications products and services. In particular, the number of Internet service providers (ISPs) is growing significantly. Also, other providers with or without facilities are gaining momentum in city areas or by offering services to special business communities.

Communication convergence necessitates the deployment of next generation support systems. Relying on advanced technologies, client/server, or Web-based support systems enable convergence carriers to offer their customers higher total value through new, innovative products and services; superior customer service; and customized pricing and billing. At the same time, support systems guarantee profitability by increasing effectiveness of processes by automating all routine processes and by supervising quality of services metrics.

Customer Orientation

Competition is driving telecommunications service providers to emphasize customer management. Driven by global competition, carriers are likely to focus on improving the total value of their services — quality, support, and price — as a means to retain customers. Many of these improvements will come from advanced support systems. Besides improving the customer interface — e.g., offering Web access — granular data available with new support systems can be utilized to retain key customers and reduce the amount of customer churn. Over the longer range, further differentiation is expected. High-margin customers may receive special treatment, average customers just average services — similarly to other industries.

Customer Network Management (CNM) incorporates a class of support systems that enable end users to view, troubleshoot, reconfigure, and generate reports securely on their subscribed telecommunication services. CNM provides strategic links to the customer and allows service providers to differentiate their offerings further. Support systems vendors are expected to offer the following:

- Performance: Extraction of the information from the network without slowing overall network processing.
- Customization: Packaging information so that customers can receive an appropriate level of detail, in a way they can understand.
- Security: Delivery of the information to the customer in a cost-effective and secure manner so that customers see only relevant information about their portion of the network.

It is expected that Web technology will be used to deliver this service. CNM represents a modest source of incremental growth for support systems suppliers.

Certain support services can also be outsourced. The customers may not know where the support services come from. Today's outsourced solutions are service bureaus. They may outsource all or part of the carrier support systems. In the latter case, the vendor relies upon remote access to the carrier's existing solution to deliver incremental functionality. For most emerging carriers, the benefits of outsourcing outweigh the negatives.

The Telecommunications Handbook addresses most of these trends and goes into depth on the following subjects:

- Evolution and future of voice-related services
- Evolution and future of data-related services
- Challenges in the wireless environment
- Use of the Internet for business applications
- Architecting and deploying intranets and extranets

- Administration and management of telecommunications systems, networks, and services
- Estimating future telecommunications technologies and their impact on user communities

1.1.1   Telecommunications Services

Floris G. H. van den Broek

1.1.1.1   History of Telecommunications Services Offerings

Telecommunications, "communicating over a distance," has existed for thousands of years, in the form of smoke signals by the Indians or light communication by various groups, which later developed into lighthouses to communicate with ships several hundred years ago, resulting in the *optical telegraph*. The real step forward in bridging a distance of more than a few miles was the invention of electricity-based telecommunications services by Samuel Morse in 1838. He invented the telegraph, a device that uses electrical signals, transported over a copper wire. He developed the Morse alphabet as a standard way of expressing our language in electric signals. Later, the telegraph was fitted to use radio waves for transmission. The telephone made it possible to communicate with normal voice and was invented by Alexander Graham Bell in 1879. The principle of operation of the telephone changed little between 1879 and today, but the technology for transmitting the electromagnetic signals from the telephone has changed drastically. Initial communications were based on overhead wires, that all connected to a manually operated exchange, introduced in most countries around 1900. The first electromechanical switches appeared in 1930s. They replaced the telephone switch operators and allowed dialing a number by the subscriber himself. However, it took until the 1960s in most Western countries for all the manually operated switches to be replaced by automatic switches, and some countries in the world still rely on manually switched networks today. Since 1966, when the first telecommunications satellite was launched, telecommunications between continents grew enormously and international transmission of television images was made possible.

1.1.1.2   Introduction of Multiple Suppliers

Historically, the telecommunications services offering in countries was a task of the government. The government operated a ministry or public body that would be responsible for offering telecommunications services. The services were seen as utilities to be provided to everyone at the same price, usually regardless of where the user resided. There were also technical reasons to keep the complete network in hands of one party. The technical solutions had not been found to account for use of a "bottleneck resource," such as the infrastructure, by more than one party.

The offering of services consisted of a limited number of services that the government deemed necessary and beneficial for the users in the country. The government also controlled the quality of the services offered. If services were offered by a state-owned company, requirements were placed on the company. These one-supplier environments still created problems. In some countries, where that situation still exists, the examination of a telecommunications services offering in that country is then an easy task, as only one information source is needed.

In the early 1980s, some governments started changing that situation and multiple suppliers emerged in some countries in the 1980s and 1990s. In the U.S., (one of the first countries to open for multiple suppliers) a series of lawsuits led to the decision of Judge Greene to open the U.S. market for long-distance telecommunications services on January 1, 1984. In February 1996, the market for local telecommunications services was opened. In other countries, the dates for introduction of multiple suppliers vary, and in many countries there is still only one supplier. More information on this is available in Section 1.2 on Regulation.

1.1.1.3   Categorizing Services

Telecommunications services can be characterized in different ways. Distinctions that are often used are public vs. private services, data vs. voice services and wireline vs. wireless services.

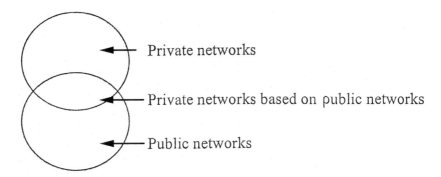

FIGURE 1.1 Private networks, public networks, and their primary management organization.

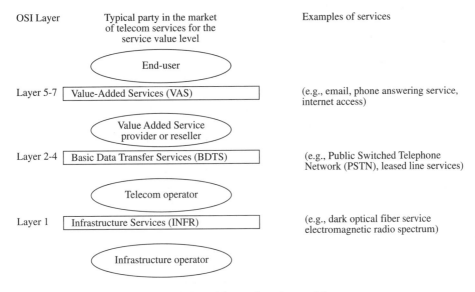

FIGURE 1.2 The service value model.

Figure 1.1 shows an overview of the relationship between public networks and private networks. Both public and private network services are commonly used by international organizations. The difference in essence stems from the ownership of and access to the network, which can be public (a body owns it and everyone has access to the same network) and private (only the party that leases or buys the network may access it). In practice, there are also services which are private, but are implemented on public facilities. In that case, the operator usually implements the technological measures on a public network to give the virtual private network user access to a seemingly private network.

Service Value Levels

Telecommunications services may be categorized in several levels, which we have called "service value levels." However, multiple ways of denoting such levels exist. The main purpose for categorizing services in such a way is that regulation in many countries uses a similar distinction to categorize services and the rule that the provision of a certain category of services is open to competition. The service value model is depicted with examples in Figure 1.2.

The service value model is related in this figure to another, more standardized model, which is the open systems interconnection (OSI) model. The OSI model is used more in technical descriptions and, in this introductory section, we will not address its details. The categorization in service value levels is adequate for most practical cases and is intuitively easy to understand.

In reality, there may be multiple parties (and supplier–customer relationships) at each layer of the service value model. Also, some regulatory environments in countries use the term *enhanced services,* rather than value-added services. What exactly is meant by enhanced services depends on the country. Since the regulatory environments for enhanced services are often more relaxed than for basic data transfer services, there is an incentive for suppliers of telecommunications services to try to categorize as many services as possible in the enhanced services category. That way the regulation for providing that service would be much less strict. An example of such a loophole in the legislation is the use of different protocols, such as Internet Protocol, to send voice over a private line. The resulting voice-connection service looks exactly the same to the end users, but the transmission protocol is different. Some regulatory environments see the translation of a circuit-oriented connection into packets, giving each of the packets its own path for transport, as an enhancement and therefore call it (e.g., in the U.S.) "enhanced service."

1.1.1.4   Examples of Services

The service value model already shows several examples of services in the rightmost column. Some services are addressed here in more detail as they are often used in industry. First of all, the standard telephone service is addressed. A more precise name for that service is Public Switched Telephone Network (PSTN). The PSTN is the most extensive in the world and allows people with a telephone set to speak with each other. In recent years, the PSTN has also been used for transmission of faxes (images from paper converted into data) and data (via modems). But, in essence, the PSTN service still transmits sounds between PSTN connections in the world. These connections are highly standardized and various kinds of telephone sets can be used in almost all parts of the world. In the service value model, we regard PSTN as a basic data transfer service.

Integrated Services Digital Network (ISDN) is a service that is less ubiquitous, and one that has become popular in recent years. The most important difference is that the circuit between the parties that are connected is digital rather than analog. It can therefore transmit more than just sounds and is much more efficient in transporting data signals. In one view it replaces the existing PSTN service, but in other views it adds a series of services to PSTN, which are quite different from anything that PSTN can offer. The new services are based on transport of data. Not only can data be transported very fast (basic rate ISDN is 144 kb/s), but data can also be sent while the ISDN circuits are in use, and therefore additional end-user services can be performed, such as call waiting, calling line identification, and transmittal of packet-oriented data (such as Internet traffic). Most regulatory environments still do not characterize ISDN as a value-added service and therefore it fits in the "basic data transfer service" category of our model.

An example of a value-added service would be the virtual private network service, which connects users in different locations with each other, so that they seem to work in a "closed user group." Working in a closed user group, they can call each other with short extension numbers and they can use different features of telephones, that are usually available for the users of phone extensions in businesses connected to a private branch exchange (PBX). There is also a "data version" of the virtual private network, which, just like that described in the voice example, forms a closed user group exchanging data.

Other examples of basic data transfer services are the services meant for the transport of pure data, such as leased lines (supplying a permanent circuit) or packet-switched service (supplying transport of packets of information).

An example of an infrastructure service is dark fiber service. This service is not yet widely offered, but gives operators the capability of completely controlling the transmission technology on both ends of a fiber and also reaches efficiencies by using types of equipment that make maximum use of the fibers.

Figure 1.3 shows an example of actual supplier–customer relationships, mapped on the service value model. Chosen is an example of a multiple country situation, to make it clear that operators can operate across borders. Infrastructure operators are most often active across international borders, as they have, for a long time, formed an actual bottleneck for communication and had to supply service across borders

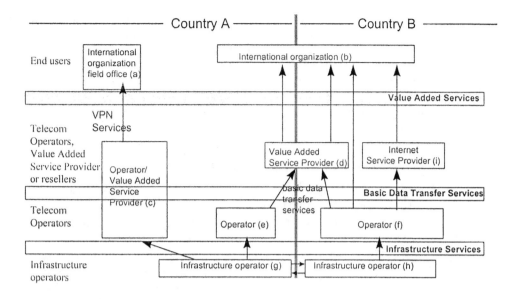

FIGURE 1.3 Example of supplier-customer relationships mapped on the service value model.

as a mandate by their governments. Infrastructure operators are at the bottom of the service value model for the actual owner and manager of the border, crossing links. Some value-added service providers try to become "international" by offering "managed services," e.g., at the value-added services level, such as the value-added service provider (d) that operates in both country A and country B and therefore maintains relationships with both infrastructure operators (g) and (h). The service value model here helps to understand the relationships between the parties that enable the operation of the international network. In Figure 1.3, both the end-customer organization (the multinational organization, b) and value-added service provider operate across international borders.

Standardization of Services

Telecommunications services can be categorized by their use (e.g., video services, voice services, and data services). The function of the service is sometimes even used to describe the service, but that can lead to various misinterpretations. *Telephone service,* for instance, is a term that is often used, but when looking at different countries and networks, many services can be identified that could be named telephone service. This includes telephone service with a certain technical interface between the telephone (an often-used type of terminal) and the central office location of the provider of the services. Telephone services could be provided using, for instance, an ISDN network, or a TCP/IP* network. Some services are standardized and named such that a wide audience knows what the service is exactly and may use it with standardized equipment and peripherals that connect well with the networks. More on standardization follows in Section 1.3 of the Handbook.

1.1.1.5 Quality of Service

Quality of service is an important aspect of a telecommunications service. Quality of service [Cole, 1991; Frieden, 1996] was measured the very first time that networks were made available, but as there was often only a single supplier in the country, quality of service was measured and reported to the government organization responsible for the network, and that was it. Of course, there were certain quality of service levels that were required in many countries, because of the importance of the network for, e.g., emergency calls. During the 1980s and 1990s, a more detailed system for measuring quality of service was established

---

*TCP/IP = Transaction Control Protocol/Internet Protocol.

in many countries. Quality of service is now measured with "performance indicators," that each cover a particular aspect of quality of service. For example, a few common performance indicators for PSTN services are system response time or speed of dial tone, the number of trouble reports compared with the number of users per month, and call completion. The indicators differ according to the type of service. The overviews of Frieden [1996] and Cole [1991] show that, 3 to 4 years after a change in the regulatory environment that resulted in more competitors for each of the telecommunications services, there was a measurable improvement in quality of service.* Today, quality of service is an important aspect of a telecommunications service and is used in marketing the service, just like price and other aspects of the service.

1.1.1.6   Current Developments in Telecommunications Services

Currently, telecommunications services are evolving rapidly. The demand for data-oriented services, such as TCP/IP, is showing a fast growth, and voice services (also growing at a steady pace) is becoming more and more ubiquitous. The demand for data services is also shifting from circuit-oriented services, such as leased lines, to packet-oriented services, such as TCP/IP or frame relay service, which make better use of the available network capacity and provide for the transmission of big chunks of data in relatively short time frames. As competition among providers of telecommunications services intensifies, customers demand more quality of service and lower prices. To supply these, service providers use a standardized basic data transfer service, such as TCP/IP, and establish only one worldwide network with this basic data transfer service and then provide various value-added services using the common basic data transfer service. This avoids use of multiple networks, which each has to be managed individually and can therefore be more costly than the management of one integrated network.

Sections 1.1.2 through 1.1.4 give an overview of how end-user telecommunications services are created by an operator and how that operator can create these services most efficiently and effectively to fulfill the needs of the users.

References

Cole, B.G., *After the Breakup, Assessing the New Post-AT&T Divestiture Era,* Columbia University Press, New York, 1991.
Frieden, R., *International Telecommunications Handbook,* Artech House, Norwood, MA, 1996.

1.1.2   Telecommunications Service Offerings for Business Use — What Does It Take to be a Credible Service Provider?

Marc Hendrickx

1.1.2.1   Introduction

This section is intended to provide a brief introductory overview of some of the key factors to be taken into consideration when assuming the role of a telecommunications service provider in today's world. The section puts particular emphasis on corporate data network services for the larger multinational organizations, such as private multinational companies.

At first, the business drivers of prospective customers are analyzed, which are subsequently translated into customer networking needs. Subsequently, we infer strategic principles for a telecommunication services provider aiming to satisfy the identified customer needs. The principles finally are linked to guidelines to construct a suitable offer to the customer and corresponding service provider operating model.

---

*The performance indicator "quality of service" in the study of Cole [1991] is expressed in "transmission quality, average PSTN dial tone delay, percentage of service orders completed on-time, percentage of calls completed, customer perception surveys" p. 261.

1.1.2.2   Telecommunication Services as an Organizational Resource

1.1.2.2.1   The Professional Telecommunication End-User Perspective

Although all professional organizations, profit or not-for-profit, aim for creation of sufficient *stakeholder* value, in order to maintain their success, they particularly need to be able to continuously distinguish themselves toward their customers. The customer in this respect may materialize into someone with whom the sales transaction is formally closed (the buyer) or into someone who consumes a service from the product bought (the end user). The position is taken here that the customer's end user is the final decision maker on whether the service derived from the product delivers the required benefit. Professional organizations that have taken distinctive positions in the last decade have narrowed their focus to delivering superior value toward *end users* according to three well-known value disciplines: operational excellence, customer intimacy, and product leadership [Treacy and Wiersema, 1993]. Superior customer value is achieved by organizations that are able to push the boundaries of one value discipline while meeting the standards of their professional sector in the other two. Key to this is the effective alignment of the organization's entire operating model to serve the chosen value discipline strategy. What are the primary actions that need to be taken to this extent and how can telecommunications be an effective enabler? The following items have been observed during recent surveys by OVUM* among their panel members:

- Improving business processes
- Improving customer service
- Cost reduction and control
- Creating a competitive advantage

Improving Business Processes

Changing working practices (increases in teleworking and remote working) and buying practices (consumers are more willing to make purchases remotely) force organizations to generate value quickly and be accessible all day, every day. This on its turn has led to an increased usage of business process re-engineering to reduce the time from request to order fulfillment and the use of telecommunications to reduce organizational complexity. The deployment of new processes essentially identifies the need for improved communications along the workflow. Telecommunications enables faster transportation of documents, allows on-line access to business critical information irrespective of storage place, and brings the work to the person. It thus facilitates teams of people that are either geographically dispersed, or are working as virtual teams in so-called knowledge communities.

Improving Customer Service

Through telecommunications, customers can be given workstation access to their supplier which may provide them with an increased level of convenience and service. The idea here is to reduce customer hassle to do business with suppliers. The workstation may range from a Touch-Tone™ telephone set to a dedicated computer terminal. Examples are electronic information dissemination (e.g., product portfolio information, order tracking, invoice queries) and electronic transactions (e.g., cash management, ordering of goods, telephone subscriber line removal).

Cost Reduction and Control

Traditional management information systems use an organization's accounting system that generates monthly reports on paper. Nowadays, organization's decision makers want information on a more regular basis and on-demand. To this extent, corporate information warehouses are being built that capture information as often as needed at its sources (production lines, points of sale, etc.). Another area where telecommunications is seen to enable cost control and reduction is inventory management. In fact,

---

*OVUM, company literature, 1996.

organizations that practice state-of-the-art, just-in-time operations rely heavily on telecommunications to obtain information to monitor and match demand for goods.

Competitive Edge

Telecommunications has been a critical enabler for many organizations to gain a competitive edge among their competitors. An example we already have seen above is electronic supplier access, which may at the same time secure an existing customer base and open inroads to new customers. Another example is the ability to differentiate a commodity product through speed of service, information by-products, and ease of access. But telecommunications can also facilitate new products or services through leverage on an existing installed technical facility. A large computer manufacturer that offers remote assistance through a call center may diversify using the same customer information and technical facility to start a telemarketing business.

1.1.2.2.2   The Corporate Telecommunications Manager Perspective

So far we have identified the generic business drivers from the end-user community, but now we need to perform the translation into corporate telecommunications network drivers. Within a customer's organization, this translation role is often taken up by the corporate telecommunications manager. A recent survey of the Yankee Group[*] identified the number one driver as geographic expansion, as organizations move into new markets and countries to gain new customers and suppliers. This is followed by a need for cost management (business driver = cost reduction and control) and then generic issues that will ensure better management of the business leading to improved levels of productivity and customer satisfaction (business driver = organizational processes).

Today's corporate networks consist of many individual networks and different technologies, often with their own specialized staff and operating procedures. About 50% of existing traffic in corporate WANs is telephony, modem data, facsimile, and video applications. The remainder decomposes into 35 to 40% LAN Data and 10 to 15% IBM/SNA. What has resulted is an overall structure that is both inefficient and expensive to operate and that creates barriers to change. The traditional organizational network architecture has become a performance and cost hassle or even a real bottleneck. In addition, a corporate telecom manager faces a strong demand from real-time desktop multimedia applications, leading to an increased demand for differentiated network performance and increased traffic load.

Issues that are of particular concern to today's telecom managers:

- *Network Management*, a pivotal function and specifically within that availability and reliability. If the network is business critical it has to be there when it is needed and working to full capability.
- *Cost Management*, essential in order to ensure that costs are in line with the benefits received. Today many organizations do not have a good measure of their telecommunications costs and the service performance delivered.
- *Skill Management*, the need for suppliers to be able to provide complex networks and the ability to change network configurations in line with business needs, is linked to the supplier having the right skilled people supported by processes and tools.

1.1.2.3   Telecommunications Services as a Business

Recalling the primary network drivers for our potential customers, identified above, how should a telecommunications service provider respond through incorporation of these drivers into its value proposition? Figure 1.4 captures the strategic responses to each of the drivers, which will be worked out in the next sections.

1.1.2.3.1   Business Process Improvement

The implementation of business process improvements is leading to more and more virtual teams being created on a short-term basis, and often in different physical locations. We expect the organization to

---

[*]Yankee Group, company literature, 1996.

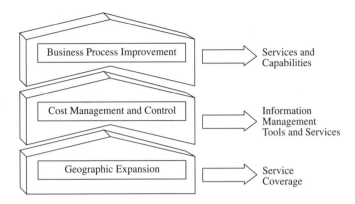

FIGURE 1.4 Threefold challenge to a telecommunication service provider.

invest heavily in voice-, call-, and information-processing systems to manage its workflow queues, originating from internal (field sales, production staff, etc.) and external calls (customers and suppliers) in the form of voice mail, E-mail, and fax. This development reinforces the importance of improved communications. A number of services and capabilities are developed specifically to aid either the remote worker or those on the move:

- Linking Remote Workers
  — Internet services, virtual network services, frame relay, ISDN
  — Audio and video conferencing
- Linking Workers "on the move"
  — Remote and mobile access to their corporate networks
  — Dial IP
- Sales/customer service via call centers

But at the same time there still remains continual demand for services that provide wide-area inter-connection between all kinds of "legacy" computer applications that operate along with proprietary protocols, such as IBM/SNA, DECnet, and AppleTalk. Hence, many of the larger corporations request total value-added (network) solutions that are able to handle these types of traffic.

1.1.2.3.2 Cost Management and Control

Organizations expect their telecommunications service providers to contribute significantly to the ongoing enhancement of overall operating efficiency. This is where we see the bigger providers offering global purchasing schemes, tuning of maintenance levels, rapid change management services (within a day), and various kinds of operational lease arrangements. In addition, organizations want a more accurate quantitative view on the wide-area network cost drivers themselves. A number of information management tools and services are offered by today's bigger telecommunications service providers to facilitate customers in their process of cost monitoring and network service productivity tracking. Some essential tracking tools include:

- Monitoring of network status, performance, and quality through on-site equipment and through extensive reports on, for instance, network service levels and implementation progress;
- Cost accounting through billing reports;
- Service trends and usage monitoring through service utilization reports.

More recent developments in this area focus on customization to specific business requirements, such as customer-defined service level agreements and enhanced interfaces to the service provider through Web-based reporting tools.

1.1.2.3.3 Geographic Expansion

An organization that develops itself internationally and creates in-country footholds outside its home country typically extends its telecommunications network requirements correspondingly. Major factors determining the network (or traffic) evolution in such a case are the type of business growth (e.g., organic or acquisition), size of the foreign-country business, and autonomy of the foreign operation. The most common customer network or traffic evolution scenarios that follow are "few-legs-abroad" and "sub-hub-abroad," the former consisting of a star-shaped network (or traffic) extension whereas the latter comes with one or more new networks deployed in the foreign country.

A corporate customer expects a telecommunications service provider to have its network points of presence nearby its own sites to maximize service availability and to minimize cost. However, from a service provider's point-of-view, economies of scale will generally decrease the closer its service access points move to the customer premises. Physical network access (e.g., by dedicated fiber, leased line, or ISDN) therefore often constitutes a major cost element to both service provider and customer, specifically in nonregulated or partly deregulated areas. But the advent of competing metropolitan area networks in major cities and business areas around the world will force access cost to go down.

Apart from the physical network interfaces, customers will also have requirements for *organizational* interfaces with their service provider(s), such as for service ordering, implementation, billing, and inquiry. This network of *management services* may evolve quite differently. During the first stages of internationalization through autonomous organizational growth, an organization may wish to keep all of its providers' interfaces at its headquarters. In case of internationalization through successive acquisitions, some of those interfaces may be required in more than one country, e.g., for account management and billing.

1.1.2.4 Organizing for Added Customer Value

In developing its strategy, the telecommunications service provider first needs to decide:

- Which solution packages it wants to create;
- Which services it wants to create itself and which it needs to buy;
- Which value discipline it wants to use for competitive differentiation.

1.1.2.4.1 Building a Telecommunications Solution

The first fundamental choice is strongly dependent on the type(s) of customer network requirements the telecommunications service provider wants to address. The principal choices lie along three axes from which solution packages can be built (Figure 1.5): service elements, process elements, and network elements. The basic criterion is to what extent each of these elements is marketed to the direct benefit of the customer.

Three of the most common types of solution packages have been shaded in the future: infrastructure services, network services, and value-added services. for instance, a public telecom operator (PTO) selling a leased line service typically delivers an infrastructure service if provisioning is limited to the electrical connection itself and some kind of help desk service that logs faults reported by the customer and ensures fault repair.

The PTO could add more *management* value through proactive monitoring of more service parameters, such as usage and performance. It could provide these measurements in the form of regular reports to the customer. It could also care for installation and maintenance of network equipment connected to the leased line on the customer premises, such as bridges or routers. In this way the PTO tasks itself to increasingly manage the wide-area network of the customer and hence starts delivering a network service or even a value-added service. The rather blurry line between network services and value-added Service is crossed if the PTO starts *operating* on customers' proprietary systems and data. Value-added services are "visible" not only to the IT manager, but also to the end users of a telecommunications service provider's customer. An example of the latter is an E-mail directory service or a fax store-and-forward service.

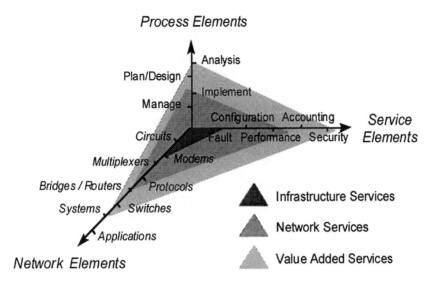

FIGURE 1.5 Telecommunications solution elements and some key service families.

1.1.2.4.2 Telecommunications Solutions on the Market

The next choice is to determine the service portfolio depth and broadness. The matrix in Figure 1.6 gives an impression of the wide range of telecommunications services and service packages offered by telecommunications service providers today. As we move from the bottom to the top of the services matrix, we see service providers adding more value to the information transported for their customers. Another characteristic of the matrix is that each service typically supports a service in the next layer.

The bottom layer consists of the so-called infrastructure services such as the cables of a metropolitan cable TV network or frequency slots of a satellite transponder. The network services layer still shows signs of the old telecommunications market paradigm: the data communication world (left-hand side) vs. the telephony world (right-hand side). Yet, we also see signs of services integration on this layer with the advent of ATM services, Voice-Over Frame (VOF), and Voice-Over IP (VOIP) as key examples. The latter service is an important building block for Internet telephony service. Fast-growing services in this layer are frame relay and IP, whereas demand for X.25 packet switched declines, although it still accounts for a large installed base of wide-area networks.

As stated before, network services are not generally visible to end users, but value-added services generally are. Examples are directory services offering navigation capabilities to various kinds of distributed applications or global freephone services that offer number transparency in all countries. Internet VAS include Web site mirroring, digital signature services, encryption services for secure file transfer, or Internet telephony. Approaching the top of the figure, we move into horizontal applications like workflow management services (e.g., SAP, Baan) or industry segment–specific solutions like broker voice services combined with dealing-room applications oriented toward financial trade institutions.

With the organization's demand for multiservice flexibility and cost control, service-independent network services such as ATM or IP rapidly seize the market (Figure 1.7). These cell or packet-oriented services are seen to be a more effective means of information transport, be it telephony, E-mail, SNA, or video. Techniques such as IP tunneling or priority-class PVCs arrange for special treatment of each class of traffic. The explosive growth of IP-based public services (the Internet) fuels the demand from corporate end users for similar services in their private environment. The Gartner Group[*] predicts that Internet connectivity will penetrate corporate user sites from 12% now to 40% in 2005 for the European Union, while the U.S. will see this penetration ratio grow from 25 to 70%. Sales staff in far-off places with no

---

[*]Gartner Group, company literature, 1997.

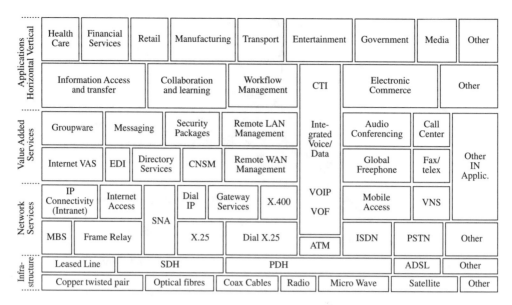

FIGURE 1.6 Today's telecommunications services landscape.

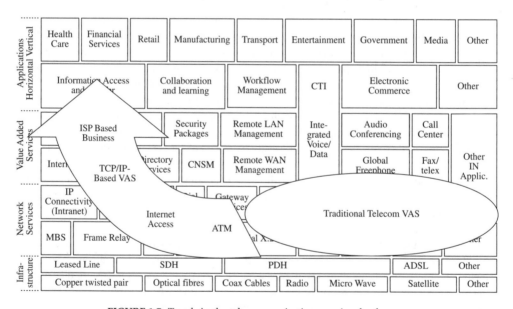

FIGURE 1.7 Trends in the telecommunications services landscape.

access to their corporate WAN in the neighborhood seek to use local ISPs (Internet service providers) as alternative means of access. As network sections that are publicly accessible are traversed, security becomes a major issue. But ease of network and service interconnectivity through a *de facto* standard suite of communication protocols drives the increasing amount of *inter-* and *intra*corporate data traffic to be carried by the TCP/IP service.

1.1.2.4.3 *Building the Value Proposition*

Delivering telecommunication services means that the customer's end user is a co-producer. Without his or her ongoing minute-by-minute cooperation, no service is delivered or produced and no value is created by the service provider's activity. This perhaps simple and straightforward statement has important

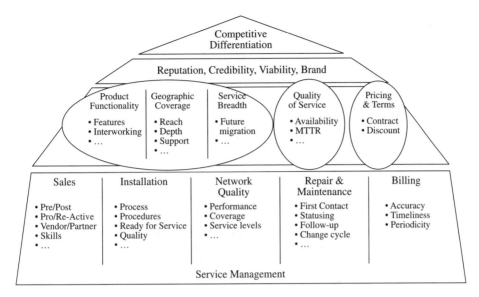

FIGURE 1.8 The telecommunications service differentiation pyramid.

consequences in our present effort to construct the right value proposition to the customer, in which the aspects of service management play an important role.

Let us first consider the basic elements of the telecommunications service value proposition, as the source of added customer value and thus of competitive differentiation in the telecommunications marketplace (Figure 1.8). Three groups of attributes, shaded in the figure, are seen to feed the proposition: the service *product* attributes and the service *production* attributes on the one hand, and the commercial attributes on the other. Taking a frame relay service as an example, the corresponding service product attributes will be commonly expressed in terms of committed information rate,* bursting capabilities,** geographic coverage of the service, and so on. Our marketing task here lies in the translation of the customer's requirements into a set of attributes that generates the benefits expected.

Service production attributes relate to the way the service is perceived by the customer throughout production. We first need to determine the points where customers will touch and interwork with our organizational processes and where they either see their performance expectations confirmed or denied. Typical touch points are the account manager, the implementation, and the help desk teams. Next we define the real-time measures of service quality for each customer touch point, in terms of *perceivable* values, such as ready-for-service or time-to-repair. These are the so-called Direct Measures of Quality (DMOQs). Finally we translate the DMOQs into internal equivalents (MOQs), alongside the required processes to deliver the attribute. This identification and translation task is at the heart of telecommunications service marketing, ranging from sales, implementation, repair and maintenance, and billing, to customer inquiry.

1.1.2.4.4 Building the Value Chain
The final step is to define the way the value proposition is delivered to the customer. At this point, we need to start talking about telecommunications business assets that need to be organized in the most productive way to deliver to the customer the agreed value proposition. Figure 1.9 aims to identify the most essential tangible and intangible assets of a telecommunications value chain, which has been simplified for the

---

Committed information rate refers to a minimum data throughput parameter, agreed and guaranteed on a per-customer basis.

**Bursting* allows a customer to exceed the committed information rate, although without throughput guarantees.

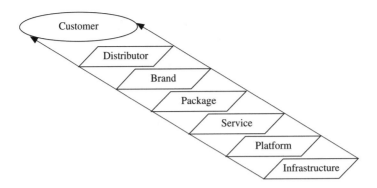

FIGURE 1.9 Telecommunications asset staircase.

purpose of this section. Each asset type can be financially valued and legally owned. And each asset builds upon a previous one to add value. Each asset is indispensable for making the value chain work.

The *infrastructure* assets typically are network utilities such as switching, transport, and signaling that create the basic telecommunication network, although without customer interfaces. The *platform* asset generates the customer service and customer intelligence (e.g., directory, ordering, customer care, and billing applications) through which a marketable end-user service can be created (e.g., leased lines, ATM/CBR, frame relay, or EDI). The recipe (or market formula) according to which this end-user service is designed, operated, and sold to the customer is contained within the *service* asset, which is intangible: it is intellectual property.

One or more services can be put into a commercial (such as price bundling, service level guarantee) or a technical (such as an integrated voice/data access service) *package* oriented toward selected customer segments. A *brand* is added to facilitate package distinction and represent commercial value in the market, before the package is delivered to a distributor function for sales to the targeted customer. In this model, the distributor takes the form of a sales agent, adding value as a customer interface from sales to delivery. But this value adding may be as limited as an automated order desk and as extensive as a fully fledged *value-adding reseller* that will nest another asset staircase. An example of the latter is a systems integrator that includes a network service like LAN-interconnect into a complete enterprise-wide ERP solution.

The next step is to distribute the business roles with respect to each of the assets among the telecommunications business partners: financing, ownership, marketing, and operation (Figure 1.10). The financing role includes all activities to fund the owner, the marketer, and the operator, whereas "ownership" refers to legal ownership of the particular asset. These first two roles enable value creation by the marketer and the operator. The marketer defines the way (the formula) a particular asset is marketed and operated by the operator, the latter, for instance, being a distributor or service provider. All four roles contribute to value creation in the chain, their actual distribution among the parties depending on the trade-off between flexibility on the one hand and commitment on the other.

Apart from pure business reasons, the government may impose additional criteria to the distribution of roles. Take the antitrust regulation as an example. A single partner such as incumbent operator KPN Telecom of the Netherlands is not allowed to own a vast majority of the country's fixed telecommunications infrastructure assets. Through this ruling, the Dutch regulator tries to prevent KPN Telecom from effectively playing all four roles with respect to the assets. Therefore, KPN Telecom decided to divest its cable TV network assets. Hypothetically, it could lease back the assets and could limit itself to the marketing and operator roles, provided that the new owner — a competitive consortium between France Telecom and Rabobank — would be prepared to restrict itself to a financial investor's role.

1.1.2.4.5 *Aligning the Value Chain*

Finally, sustainable success with the chosen value proposition on the market is strongly dependent on the effective alignment of corporate goals between the various financiers, owners, marketers, and operators identified in the previous paragraph. None of the roles need necessarily be combined into a single

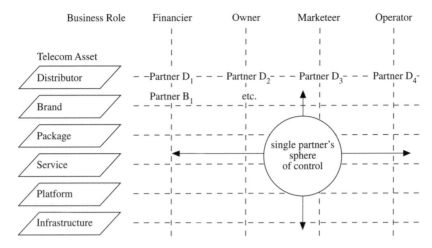

FIGURE 1.10 A telecommunications asset governance model.

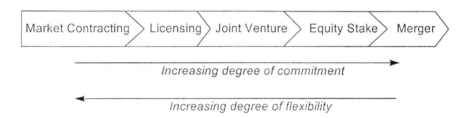

Increasing degree of commitment

Increasing degree of flexibility

FIGURE 1.11 Facilitating strategic alignment.

business partner, many examples of which we see in the current marketplace. For instance, a legal owner of a frame relay service platform may wish to define the way the platform is operated and marketed, but may outsource actual operations to a third party. Still, strategic alignment on market formula and speed of decision making between the partners can be optimized if the number of decision-making parties is limited or if only a few parties have a dominating influence (the so-called channel captain position) [Kotler, 1997].

Strategic alignment can be facilitated or even enforced through various forms of business relationships, ranging from market contracting to legal merger (Figure 1.11) [Elixmann and Hermann, 1997].

Market contracting implies the lowest degree of commitment and the highest degree of flexibility between the partners, with obvious possibilities for opportunistic behavior. Legal merger on the other side limits opportunistic behavior through mutual capital investment, transfer of resources, and formalized cross-managerial communication on a strategic level. Nowadays, we see the major global telecommunications alliances being governed by equity stake joint ventures or mergers, resulting in a jointly owned "service firm" that acts as the nucleus in the telecommunications service provider's operation.

This service firm fulfills at least the owner and marketer roles with respect to the platform, service, package, and brand assets (Figure 1.12). The infrastructure assets often are financed, owned, marketed, and operated by third parties, that is, PTOs or new operators, depending on the state of deregulation in a particular country. But there definitely is a trend toward more control of the infrastructure assets through ownership by the service firm or its shareholders.

The distributors are contracted by the service firm, but may often also be owned by the same shareholders that own the service firm. These shareholders typically act as strategic investors in addition to their default financial investor role. This means that they influence, for instance, the market strategy of the service firm and its distributors.

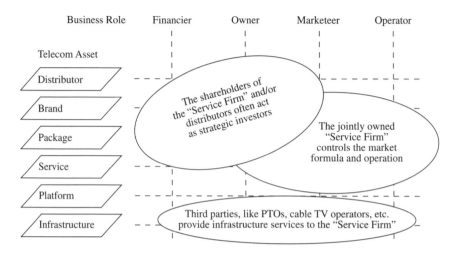

FIGURE 1.12 Typical telecommunications asset governance among leading players.

1.1.2.5 Summary

In our attempt to define some of the key factors that shape a telecommunications service provider addressing multinational companies as its prospective customers, we have first analyzed the main business drivers of the customers. We have seen that process re-engineering, improving customer service, cost control, and creating a competitive advantage are the top four drivers on the customer's mind. Telecommunications services prove to be a critical enabler with respect to the identified drivers, for which the corporate telecommunications manager in the customer's organization plays a pivotal role. He or she has to derive the needs for the specific telecommunications services that support the business goals. In that respect, the manager represents the end-user community within the corporate customer's environment. Besides that, he or she is concerned with maintaining the right skills, cost, and performance to deliver the telecommunications services, regardless of whether these are in- or outsourced.

Subsequently, we have inferred strategic principles for a telecommunications services provider aiming to satisfy the identified customer needs. Business process re-engineering is enabled by a state-of-the-art and robust telecommunications service provider's service portfolio that matches the customer's geographic reach. Cost management and control are facilitated by information tools and management services that a service provider should deliver along with its core services.

The principles finally are linked to guidelines to construct a suitable offer to the customer and corresponding service provider operating model. At first, telecommunications service providers need to make three decisions: (1) about the type of customers and corresponding network requirements they want to address, (2) a make-or-buy decision per network solution component offered, and (3) which value discipline to put emphasis on to keep a competitive advantage in the chosen customer segment.

Building the value proposition for the customer and establishing a suitable price means realizing, first, that the customer (the end user) is a coproducer of the telecommunications service. Hence, we need to take account of both service *product* and *production* attributes, the former relating to feature functionality, the latter referring to the way the service is perceived by the customer during operation. In this respect, the make-or-buy decision and subsequent value system are important as the partners in the value chain to the customer need to have a minute-by-minute alignment on the way the service is to be produced.

To facilitate the construction of such a value system, we have proposed a simple governance matrix based on two entries: telecommunications service assets and business partner roles. Six types of strategic assets have been identified: infrastructure, platform, service, package, brand, and distributor. Business partners governing these assets can assume any combination of four roles: financier, owner, marketer, and operator. The partners need to decide among themselves the distribution of their roles over the assets. Strategic alignment along the telecommunications value system can be facilitated through corporate governance arrangements, such as equity stakes.

The "channel captain" is typically taken up by a service firm jointly owned by a set of governing shareholders and acts as the nucleus in a telecommunications service provider's system. The service firm often combines the owner and marketer roles with respect to the platform, service, package, and brand assets.

References

Elixmann D. and Hermann, H., *Strategic Alliances in the Telecommunications Services Sector.* Wissenschaftliches Institut für Kommunikationsdienste GmbH, Bad Honnef, Germany, 1997.
Keen, P.G.W. and Cummins, J.M., *Networks in Action*, 1st ed., Wadsworth, Belmont, CA, 1994.
Kotler, P., *Marketing Management*, 9th ed., Prentice-Hall, Englewood Cliffs, NJ, 1997.
Treacy, M. and Wiersema, F., Customer intimacy and other value disciplines, *Harvard Business Review*, Jan.–Feb.: 84–93, 1993.

1.1.3   An Enterprise Model for Organizing Telecommunications Companies

Rolf de Vegt

1.1.3.1   Context and Objectives

The objectives of this section are to

- Provide high-level insight into the different activities that occur inside telecommunications services providers and the dependencies between the different activities;
- Provide insight into the value added and costs associated with each of the major activities;
- Highlight trends and issues in the "customer-facing" activities;
- Describe major organization structural dilemmas that executives of telecommunications service providers face.

This section will deal predominantly with trends, issues, and dilemmas of telecommunications companies in the developed markets. But, before diving into the activity models and economic models, let us have a look at the type of services a telecommunications services company offers.

Telecommunications companies in developed markets have gone beyond offering basic point-to-point voice communications service. A clarifying framework to think about the myriad of services that are, or will be, enabled by telecommunications companies is represented in Figure 1.13. A useful categorization of services is by the type of media (voice, image, data, video) and nature of communication, i.e., store and forward, point-to-point and workgroup. A third dimension that can be added to the categorization of services is the mobility aspect. This dimension defines what types of access devices are used for the communication, e.g., phone, PC desktop, pager, personal digital assistant (PDA), or public pay device.

1.1.3.2   The Enterprise Model

Telecommunications services companies perform an intricate combination of activities to deliver their end product — high-quality network services — to their customers. The enterprise map and the enterprise economic model provide a high-level overview of the groupings of activities involved and the economic importance of each of these activities.

1.1.3.2.1   The Enterprise Map for a Telecommunications Company

To understand better the inner workings of a telecommunications company, one can use a variety of activity modeling techniques to describe the operations of a telecommunications provider. The functional decomposition technique is often used in the information strategy planning discipline and provides a thorough top-down insight into the inner workings of a particular organizational entity. In a functional decomposition, the activities are broken down along the lines of primary functions and support functions,

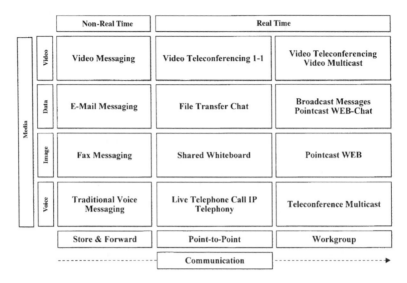

FIGURE 1.13 Telecommunications services framework.

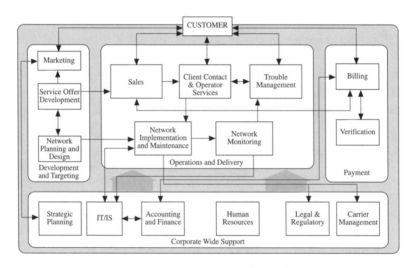

FIGURE 1.14 Generic enterprise map for a national carrier.

where the primary functions are unique to the particular business and directly contribute to the production of the end product or service. For example, order handling and production scheduling are typical primary functions in a manufacturing company. Secondary functions can also be labeled support functions and typically include functions like human resources and strategic planning.

A variation of this functional decomposition technique is the enterprise modeling technique, whereby the main activities are placed in a diagram that depicts the major dependencies and information flows between the main activities and the customer.

Apart from the insight that can be gained from identifying the major activities in the telecommunications company and their interdependencies, the enterprise modeling technique becomes more powerful when coupled with what we will call the "economic x-ray" of the company. In this economic x-ray, the "value-add" of each main activity will be determined and graphically represented.

Figure 1.14 represents the enterprise map of a telecommunications company. This is a particular example for a facilities-based national carrier. Facilities based means that the provider "owns" the majority of the network (switching, transmission, and operations support systems) it uses to deliver its services.

How to read the enterprise map

The enterprise map has been read along two dimensions, vertically and horizontally. Starting vertically on the top the diagram you will find the "customer." At this generic level of modeling, this can range from a residential customer with just one phone line to a small business customer to a large multinational business customer. Just below the customer, but within the boundaries of the telecommunications service provider, are the main customers facing activities. The supporting activities are located on the bottom of the diagram.

The horizontal dimension is structured along the life cycle of products and services. Reading the diagram from left to right, it starts with the "development and targeting" of products and services activities, followed by the "delivery and maintenance" of those products and services. Finally on the right-hand side, the customer pays for the products and services.

Benefits and limitations

The benefits of using a representation like the enterprise map are

- It structures the thinking and strengthens the comprehensive picture of the organization.
- It highlights the main relationships and information flows between main activities.
- It creates a common language and basis for discussion.
- It can be used to identify key areas for businesses process improvement and potential leverage of IT solutions.

The limitations of the model are that it is a very high level representation and will often require further decomposition of the business functions identified in the enterprise map.

Furthermore, it is not an organizational chart and some fundamental organizational problems in telecommunications companies cannot be highlighted in the model. For example, a key organizational issue such as how to best structure the relationships between market facing units (e.g., large business customers business unit) and shared resource units (e.g., network operations business unit) will be glossed over.

Activity	Definition
Marketing	Packaging and promotion of offerings to the marketplace market planning, product management, market management, and advertising
Service offer development	Creation of new offers to be made available to the market and product development
Network planning and design	Analysis of physical needs; projections on capacity and configuration requirements; establishment of standards; network design and configuration
Sales	Pre-sale analysis, proposal delivery, sales closure, sales support, and client management
Client contact	Handling of customer inquiries; receiving and entering problem notifications; handling service order and change requests; opening and closing trouble tickets
Operator services	The handling of directory inquiry services
Trouble management	Determination and diagnosis of trouble tickets; delivery of corrective action; repair and inspection; tracking and logging of problems and escalation needs
Network implementation and maintenance	Deployment of infrastructure, customer provisioning, implementation of changes, and inspection of facilities
Network monitoring	Tracking and management of network performance; proactive problem detection; network optimization and tuning; and report on usage statistics
Billing	Usage and outage collection; calculation of rebates and physical billing; collections; billing design and development

Activity	Definition
Verification	Verification and management of billing errors
Strategic planning	Definition of strategic objectives and subsequent planning and budgeting
Accounting and finance	Analysis of financial data for asset management, costing of services, and investment decisions
IS	Systems development and maintenance, systems administration and application management
Legal and regulatory management	Regulatory strategy development and ongoing management of relations with government bodies that regulate the telecommunications industry
Carrier management	Contracting and day-to-day management of the interactions with other telecommunication service providers
Human resources	The definition of human resources policies and procedures, salary and benefits administration, union relations, training and development programs

1.1.3.2.2 The Enterprise Economic Model for a Telecommunications Company

Figure 1.15 depicts the economic importance of each of the major activities in the telecommunications provider. The diagram is developed through using an "activity-based costing" approach, whereby the relative costs of each of the major activities are assessed.

The major sources for this type of analysis are

- Company annual reports;
- Industry analyst reports;
- Company management reports;
- Headcount information (i.e., an approximation of the number of full-time-equivalent staff dedicated to a particular function);
- Interviews with people working in the finance function of telecommunications service providers.

Like the enterprise map in Figure 1.14, the economic model in Figure 1.15 is specific for the category of facilities-based interexchange carrier. The access to local infrastructure cost element is a typical cost element for facilities-based interexchange carriers since this represents the fees that the national carrier pays to the local operator for origination and/or termination. This particular cost element is equivalent to the cost of goods sold in a manufacturing company and does not contribute to the internal value added of the telecommunications company. For the example in Figure 1.15, these costs represent 32% of the total expenses of the company.

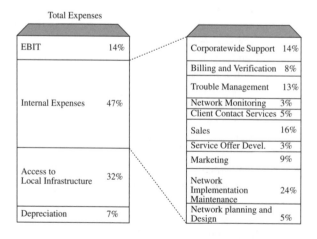

FIGURE 1.15 Enterprise economic model of a facilities-based interexchange carrier.

To illustrate the point that the economic models differ significantly depending on the type of telecommunications services provider, Figure 1.16 contains the enterprise economic model for a typical reseller. This diagram depicts that the costs for access to infrastructure are much higher, since the reseller is paying another company for the transportation capacity and depending on the type of reseller (switched vs. switchless), also for the switching capability. This leads to a very different picture of the internal value-added of the company. In particular, the sales and the marketing functions are relatively more important in the reseller model, whereas the network implementation and maintenance function is obviously relatively less important in the reseller model.

This section highlighted the enterprise economic models for two types of telecommunications providers. There are nine different common types of providers, which does not include emerging operators like satellite-based operators and other emerging types of operators.

Different Types of Providers

Type of Provider	Definition	Example
Facilities-based national carrier	Long-distance carrier that owns call switching and transmission lines nationally. Carrier has switching offices in all service areas of the country and provides originating service nationwide.	AT&T, MCI, Sprint
Facilities-based reseller	A long-distance company that does not own its own transmission lines. It buys lines from other carriers and resells them to its subscribers. Owns its own switches and a mix of leased and owned lines.	LCI, Winstar
Pure reseller	A company which purchases a big block of long-distance calling minutes for resale in smaller blocks to its customers.	Phoenix Network, Matrix Telecom
Local exchange carrier	The local phone company which can be either a Bell operating company or an independent which traditionally had the exclusive franchised right and responsibility to provide local transmission and switching services. Prior to divestiture, the LECs were called telephone companies or telcos. With the advent of deregulation and competition, LECs are now known as ILECs (incumbent local exchange carriers).	Bell Atlantic, Ameritech, GTE
Competitive access provider	An alternative, competitive local exchange carrier.	Colt, Teleport
Wholesaler	A network operator that sells unbranded bandwidth to other telecommunications providers and has no direct contact with the end customer.	Qwest
Wireless service provider	A carrier authorized to provide wireless communications exchange services (for example, cellular carriers and paging services carriers).	Airtouch, Libertel, Orange
PTT-PTOs	Post Telephone & Telegraph–Public Telephone Operator. Usually controlled by their governments, provide telephone and telecommunications services in most countries outside North America.	Deutche Telecom, France Telecom
Competitive local exchange carrier (CLEC)	A term coined for the deregulated, competitive telecommunications environment envisioned by the Telecommunications Act of 1996. The CLECs intend to compete on a selective basis for local exchange service, as well as long distance, Internet access, and entertainment (e.g., cable TV and video on demand). They will build or rebuild their own local loops, wired or wireless. They will also lease local loops from the incumbent LECs at wholesale rates for resale to end users.	Intermedia

Note: Data are taken from *Newtons Telecom Dictionary.*

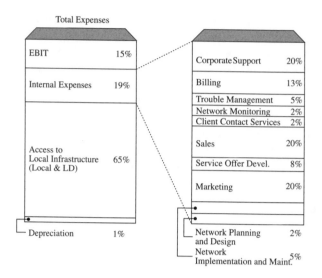

FIGURE 1.16 Reseller economic model.

Transaction Marketing	Relationship Marketing
• One transaction at a time	• Building relationships with customers
• No immediate contacts with end customers	• Interface with customers expanded throughout the organization
• Short-term focus on profits from today's exchanges	• Longer term focus on creating results in the long run
• One way flow of communication	• Two way flow of communication
• Economies of scale matter	• Economies of scope/custom wallet share/matter
• Focused on optimizing 4P's for a discrete offering	• Continually adapt the offerings based on changing and emerging customer needs
• Results measured by total revenues	• Focus on marketing production measuring marketing effectiveness in terms of return
	• Highly applicable for service firms

FIGURE 1.17 Distinction transaction marketing vs. relationship marketing.

1.1.3.3 Trends and Issues in Customer-Interfacing Functions in the Enterprise Map

1.1.3.3.1 Introduction

New entrants in a deregulating communications market typically start to compete by offering lower prices than the incumbent players. Shortly after market entry, however, new entrants often realize that they can differentiate themselves by offering superior customer service to potential customers. Either in response to new entrants in the marketplace or as preparation for pending competition, incumbent players have also made significant changes in the way they approach their customers. This section outlines the most significant trends and issues in the customer-facing activities as listed in the enterprise map.

1.1.3.3.2 Marketing

This function is often "underexposed" in the traditional PTOs, but is key in a competitive marketplace. Even a significant number of new entrants make the mistake of gearing up the marketing function once the network is up and running and the salespeople are hired.

A distinction needs to be made in the marketing function between business-to-business marketing and mass marketing. Business-to-business marketing approaches are applied to medium to large business and government customers. A major consideration in business-to-business marketing is whether to adopt an industry marketing approach in addition to product marketing or whether to maintain a product marketing orientation solely. Industry marketing implies the selection of the most attractive industries (e.g., banking) and the development of industry-specific integrated business solutions aimed at the companies in the target segments.

For the mass marketing of telecommunications services, lessons derived the marketing of other mass market industries are trickling down into the telecommunications industry. A key trend in mass marketing of telecommunications services is the migration from product marketing to relationship marketing.

Making the transformation from product marketing to relationship marketing involves significant investments in the following:

- Transformation and integration of existing processes and systems from a strictly functional orientation to a customer-centric orientation. This means that, whereas a customer used to have to deal with different departments for each type of request, now the customer can deal with one knowledgeable customer representative who can assist in most requests.
- Creation of customer data warehouses or "datamarts" (databases in which the data about customers from multiple systems are integrated). Historically, customer data are collected and stored in multiple systems, e.g., billing systems; revenue reporting systems; installation and repair systems; and product-specific systems (e.g., separate systems for "wireless" service, "Internet" service and "local access service"). Hence, it was very difficult to get a holistic view of a particular customer.
- Acquisition of new skill sets in the customer-facing operations. For example, a customer service representative has to be able to address multiple types of issues and requests. Another area where different skill sets are required is the marketing function where there is now a need for data mining and statistical modeling skills.

1.1.3.3.3 Service Offer Development

Service offer development is defined as "creation of new offers to be made available to the market," and in most telecommunications companies this refers to the product or service development process.

Most telecommunications services companies do not have a hugely impressive track record in the area of innovation and product development. For example, most established North American telecommunications companies were caught off guard by the Internet "tidal wave" and were (some still are) very slow to respond. The innovation cycle is historically slow and predominantly driven by the switch manufactures, like Nortel, Lucent, Siemens, or Ericsson, when they offer new features with new releases of switch software.

Rapid service offer development requires a project-management approach whereby multiple players in the organization (e.g., marketing, network planning, IT/IS, sales, regulatory, and billing) work jointly on a particular project (Figure 1.18). Traditional telecommunications companies are strongly functionally organized (stovepiped). Effective project management-driven service offer development in such a setting is difficult and makes telecommunications companies vulnerable. All the traditional telecommunications services providers have undertaken major process improvement initiatives, however, to streamline their service offer development processes.

1.1.3.3.4 Sales

The sales function encompasses the following activities: presale analysis, proposal delivery, sales closure, sales support, and client management.

Similar to the situation in the marketing process, the volume, type, and service requirements for telecommunications services and the nature of the sales activities vary strongly by the type of customer that is served.

Typically, telecommunications companies distinguish among four broad categories of customers:

1. Very large business and government customers
2. Large-sized business and government customers
3. Medium and small-sized business and government customers
4. Residential/consumer customers

The major distinction between these groups is the *size* of the customers. The number of lines per customer is still often used as a criterion to categorize a specific customer, although some companies are

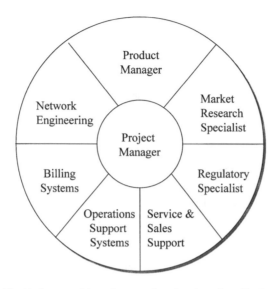

FIGURE 1.18 The ideal composition of a cross-functional service offer development team.

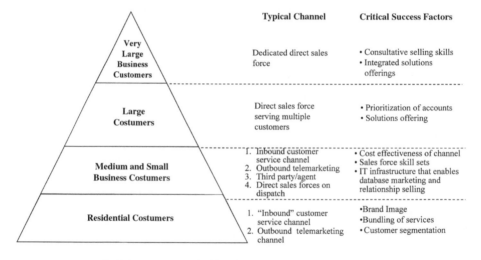

FIGURE 1.19 Typical channels per major customer segment.

beginning to adopt more-sophisticated criteria such as industry, company growth, company size, and type of primary process for segmenting their customer base.

Telecommunications companies typically utilize the following channels for the sales of their services:

- Direct sales force
- Inbound call center customer service representatives
- Outbound telemarketing call centers
- Agents/third party sales force

Figure 1.19 lists the typical sales channels for each of the major customer segments and also identifies the major critical success factors for selling to each of these segments. Critical success factors are defined as key elements that deserve focused management attention, in order for the entire activity to be successful.

The critical success factors per major customer segment indicate that each of these major segments requires a unique approach for managing the sales channel. To serve customers effectively in the very large business customer segment, a dedicated sales team for each customer is a requirement. These sales teams ideally consist of an account executive, with overall responsibility for the account, and the primary relationship manager, supported by technical specialists. These specialists often cover areas like voice communications, data networking, billing, and project management. Another success factor for serving the very large business customer segment is the ability to offer integrated solutions, i.e., offer customers the ability to conduct one-stop shopping for their communications needs. This means that the telecommunications service provider has to be able to offer bundled network services, networking equipment, professional services, and support services.

For the large business customer segment, the critical success factors are the ability to offer integrated solutions (similar to the needs for very large business customers), and ability to prioritize accounts for sales force attention. This refers to the ability to figure out which accounts yield the highest return on the investment for the time spent by the sales representative. This segment is typically served by a direct sales force, with account managers who are calling on multiple customers, and who do not have enough time to cover all their customers extensively.

The critical success factors for selling to the small and medium business market segment are cost effectiveness of the channel, and an IT infrastructure that enables database marketing and relationship selling.

The cost effectiveness of the channel refers to the fact that the revenues and margins generated from a small to medium-sized customer typically are not large enough to allow in-person visits by sales representatives. Service providers are addressing this issue by building up "telesales" offices, using outside channel partners like agents or value-added resellers and careful segmentation of the customers based on their potential value to the service provider. The IT infrastructure success factor refers to the need for a holistic customer view, i.e., having a single repository of all the products, services, usage, billing information, customer interactions, customer demographics, and needs. This single repository can then be used for customer segmentation and prioritization, relevant targeted offers and campaigns, etc.

To market and sell to the residential customer segment effectively, there are three critical success factors:

- Brand image
- Bundling
- Customer segmentation

The fact that brand image is very important is attested by the huge amounts of money spent on brand advertising campaigns targeting residential customers in competitive marketplaces. The name of the service provider and a positive connotation with that name in the minds of the customers, together with the customers' perception of price, are the key decision factors in the purchasing decisions of residential customers. A further illustration of the importance of the brand are that, after 14 years of deregulation, AT&T still has approximately 60% of the long distance market for consumers, even though it does not consistently offer the lowest rates.

Bundling of multiple types of telecommunications services into one integrated offer for the customer (like wireline, wireless, Internet access, etc.) is a relatively new phenomenon. There are three major reasons it is important for a telecommunications services provider to offer bundled services:

- Important customer segments want the convenience of one bill and integrated services.
- The providers capture a larger "wallet share" of the customers' overall spending on telecommunications services.
- It is harder for customers to switch providers if they receive multiple services (for example, they have to change their E-mail address if they want to change their Internet access provider); thus, bundling improves the "lock in" of the customer.

Customer segmentation is the activity of identifying groups of customers with similar needs. There are too many residential customers to treat each of them on an individual basis (to customize pricing, billing options, features, etc.). So it is important to identify groupings of customers who can be targeted

Communications companies must invest in processes, systems, and organizational capabilities to offer one-stop service for all customer transactions. This means information capture, sharing, and coordination of activities across the customer contact spectrum.

	Request for Information	Request for Service	Request for Discontinuation	Repair/Problem Resolution	Billing/Remittance	Outbound Marketing
Opportunities	• Suggest new programs and services • Compare offerings to competition • Request customer satisfaction information	• Offer additional services • Discover reason for switching • Offer more appropriate services	• Offer bonuses for staying • Inform customer if more appropriate program exists • Gather information on reasons for leaving	• Suggest new, different, additional services • Demonstrate attention to quality	• Include marketing materials with statements • Design customer-specific billing format, schedules, etc.	• Keep current customers informed of new offerings • Offer specialized programs and services

Key Success Factors	• One-call resolution:
	  - No transfers
	  - Seamless transfers if necessary
	• Information technology systems that facilitate information sharing, customer knowledge and real-time access
	• Highly-trained agents with generalist knowledge and specialized skills
	• Associate rewards and incentives based on customer satisfaction
	• Detailed market research and segmentation allows for real-time, accurate targeting of customer requirements and possible upgrades					

Customer Contac Spectrum

FIGURE 1.20 Customer interaction points.

with mass customized price plans, features, and promotions that are relevant for the members of these groups. Furthermore, segmentation is used to estimate the current and future profitability of customers. This in turn will determine if the service provider is interested in retaining or winning back the customer, and how much effort the service provider can afford to spend on a customer.

1.1.3.3.5 Client Contact/Operator Services

A key distinction needs to be made here between (1) the operator service function, accessible for the general public and (2) the function handling customer inquiries, receiving and entering problem notifications, handling service order and change requests, and opening and closing trouble tickets. Operator services are typically handled in a dedicated call center environment and have limited or no relationship with the general customer service function

Handling the primary interactions with the customer in a high-quality manner is essential in a competitive environment and massive process improvement initiatives have been executed in telecommunications services companies who prepare for competition. The quality of the connection in most services nowadays provides no source of differentiation in the eyes of most customers. Therefore, brand, price, and a superior customer experience are the key means of competition between telecommunication companies. Leading telecommunications companies look at each of the customer interaction points as an opportunity to increase loyalty and potentially upsell additional services.

Figure 1.20 provides an overview of each of the customer interaction points, the potential sales and customer satisfaction opportunities, and the potential and the "critical success factors" (the things that must go "right" at each of the customer interaction points).

1.1.3.3.6 Billing

The billing function is one of the most critical customer-facing activities. For many customers, the only direct interaction they have with their telecommunications service provider is through the monthly bill.

Most of the incumbent players face some significant challenges in the area of billing since most billing systems in place today were designed to meet the internal organizational needs of a monopoly. Billing was considered a "backroom operation" and software changes have been "incrementally" layered on the systems that are often over 20 years old. All this is to say that the current billing systems in place often cannot support the competitive environment that the industry is facing with multiple new products, services, and pricing structures.

A good illustration of the challenges associated with billing is the flat-fee pricing for Internet access service in the U.S. A number of the providers had to resort to this relatively unsophisticated pricing model because they do not have the capability to bill for the service on a usage basis. The needs regarding billing differ by type of segment.

Summary of Billing Customer Needs

Business Customers	Consumers
• Integrated bills (voice, data, leased lines, etc.)	• Simple bills, easy to understand
• Consolidated across multiple locations	• One bill for all services
• Timely (up to real-time billing)	• Flexible timing in billing (get billed when I want it)
• Available on multiple media (CD ROM, EDI, paper)	
• Facilitate internal billing and/or external rebilling (e.g., for law firms)	
• Flexible reporting structures	

Most major telecommunications companies that are entering a competitive environment are making significant investments in revamping their billing environment. The requirements for the new billing environment can be summarized as follows:

- Ability to support new network technologies (CCS7, AIN, broadband services, VATV, PCS, Internet access)
- Ability to simplify and consolidate billing from multiple products, services, and vendors
- Ability to facilitate customized bill formatting
- Ability to support rapid introduction of services, pricing plans, plan changes, and regulatory changes
- Ability to flexibly enable multiple bill calculations, taxing, and discounting
- Ability to enable rapid software/program changes (table driven systems)
- Ability to enable sophisticated billing data mining

New entrants have the option to outsource the billing operations to a third-party provider, which may be an attractive alternative in the initial stages of the life cycle of the new company. Another relevant development is the emergence of billing mediation tools. Mediation software achieves a virtual integration of disparate billing systems by correlating events from multiple systems to a single "billing event" that can be reported to the customer.

1.1.3.4 Fundamental Organizational Dilemmas for Telecommunications Services Provider

Sections 1.1.3.2 and 1.1.3.3 described the typical enterprise activities for a telecommunications services provider and their financial/economic significance. This section focuses on some of the fundamental organizational dilemmas. The dilemmas highlighted are:

1. Balancing the requirement for flexibility and acknowledgment of unique customer-need sets for key customer segments on the one hand and the optimization of the shared asset base (network and operations support systems) on the other hand.
2. Significance of wholesale of the network, i.e., how much opportunity do we allow for other parties to resell our network and compete with our own retail operations.

1.1.3.4.1 Marketplace Flexibility vs. Optimal Allocation of Shared Resources

Any telecommunications services provider that services multiple segments of customers faces the dilemma of how to deploy capital and other shared resources most effectively. Figure 1.21 is a stylistic representation of the typical organizational structure of a telecommunications services company. Typically there are mul-

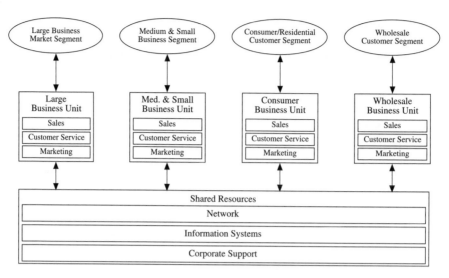

FIGURE 1.21 Typical organizational model.

tiple market-facing units that each focus on separate segments of the customer base. Examples of such market-facing units are large business, consumer, or wholesale/other carriers. The core activities performed in the market-facing units are sales, marketing, customer service, and service offer development.

Unlike companies in a number of other industries, telecommunications service providers cannot allocate specific portions of their primary asset base (network and operations support systems) to particular strategic business units. The network (switching and transportation) and the operations support systems (billing, service order processing, etc.) are shared resources. Each of the organizational units is focused on a particular market segment (e.g., large business or consumer markets) to increase or maintain market share and increase revenues. This leads to requests from multiple market units to the shared resource units for such resources as investments in network infrastructure, changes in billing format, or the development or new services. The financial, technical, and human resource constraints in the shared resource units (network operations, engineering, IS) lead to resource allocation dilemmas. Most telecommunications companies, however, are not structured to deal with those dilemmas in a streamlined manner since the autonomy and flexibility of market-facing units is often valued more highly by top management than the optimization of the allocation of shared resources.

To address this dilemma, some companies are now investing in implementing a process solution to this organizational problem, by introducing cross-company project portfolio management processes, capital management processes, and cross-business unit prioritization forums that decide on the allocation of the shared resource. In the U.S., some RBOCs have created separate organizational units that coordinate and prioritize product development efforts and network investments as well as serve as an intermediary between the market-facing organizational units and the shared resource units.

New entrants can address this dilemma by selecting a particular market segment and focusing the majority of the resources on that specific segment. For example, MCI focused on two segments, large business and consumer, but did not spend much attention on the mid-sized and small business market, while Worldcom is neglecting the consumer market all together.

1.1.3.4.2 The Retail–Wholesale Dilemma

The term *retail* refers to the situation where the telecommunications service provider sells its services and products directly to the end customer. *Wholesale* describes the situation where the provider sells services and products to other players who in turn sell the product to the end customer. The key organizational dilemma for many telecommunications services stems from the conflicting strategic requirements of (1) maximizing network utilization, and (2) selling communications services directly to the end customer, leveraging the marketing and sales channels that are part of or controlled by the company so that a significant profit margin can be derived from the sale of telecommunications services.

In most cases, the facilities-based telecommunications service providers offer their network capacity both on a retail and a wholesale basis. The local exchange carriers (LECs) in the U.S. are forced by the Telecommunications Act of 1996 to offer local access services on a wholesale basis. This means that the LECs have to give access to the network elements between the central office and the customer terminal to competitors on a wholesale basis.

The internal dilemma focuses on the question about how aggressive by the company needs to push its wholesale offers, and how accommodating the company will be to offer wholesale network services.

Arguments Pro and Con Emphasizing Wholesale Offers

Pro	Con
• Generates revenues from otherwise underutilized network capacity	• Potentially cannibalizes the companies own higher margin retail revenue stream
• Forces the retail product units to be as efficient as potential competitors	• Distorts the market for network capacity by creating an oversupply
• Expands the entire market by spurring innovation by small, lean companies who can now offer products and services for niche markets and applications that are neglected by the retail market units due to lack of scale of the market niche or application	• Creates too many internal political battles that defocus the organization from the task at hand

This debate will continue within all major telecommunications carriers in a deregulating market. It is, however, a typical symptom of the transitory state toward full competition that the developed markets are in right now. Under full competition the convincing argument will be, "If we don't offer the capacity ourselves, someone else will." This will most likely lead to the following results: (1) the facilities-based carriers will offer their excess network capacity on a wholesale basis and (2) the role of the market-facing elements of the telecommunication services provider will be changed to become the value-added integrator of telecommunications solutions, for which they will also be utilizing networks from other facilities-based operators.

1.1.3.5 Conclusion

The telecommunications services marketplace is becoming increasingly diverse, complex, and competitive in terms of (1) services offered, e.g., media (voice, image, data, video) and type of communications flow (store and forward, point-to-point, or workgroup) and (2) enterprise models and economic models (e.g., resellers versus facilities-based carriers).

Furthermore the technological developments, the hyper growth in IP-based networking in particular, and increasingly higher customer expectations are forcing the established telecommunications service providers to become more flexible and customer focused. The relative inertia of the established players is leaving room for aggressive newcomers in this industry where, in principle, "bigger is better," based on the scale advantages that can be achieved.

Senior management of telecommunication service providers have to address the following dilemmas in their decision making:

- Focusing on certain target segments to achieve agility vs. targeting all segments for regulatory and or scale purposes
- Building, maintaining, and upgrading their own network elements vs. reselling somebody else's network
- Centralizing prioritization of the allocation of shared resources (network and IT/IS) to achieve corporate optimization vs. decentralizing decision making to achieve flexibility and market-place agility
- Building and protecting in-house retail operations vs. aggressively pursuing the wholesale market

Tools like enterprise models and economic models can be used to gain valuable insights to facilitate this decision-making process.

1.1.4 Growth Strategies for Telecommunications Operators

Roger Wery

1.1.4.1 Introduction: Turbulent Telecom Waters — New Players, New Technology, New Regulations, and New Customers

One can hardly pick up a newspaper or listen to the evening news today without hearing about the myriad changes going on in the telecommunications industry. Once a monopolistic and slowly changing industry, today's telecommunications industry proves to be dynamic and unpredictable. Telecommunications providers now find themselves in a challenging environment that increasingly requires their executives to equip themselves with more robust strategic insights and capabilities. They must use these strategies and capabilities to anticipate, or at least develop, the agility and velocity required to win in the new telecommunications era. Although the $600 billion and growing global telecommunications industry of today represents significant opportunities for a number of traditional and nontraditional competitors, (Figure 1.22) only those with sound, realistic, and disciplined growth strategies will overcome the challenges to earn and maintain a place in the telecommunications marketplace of tomorrow.

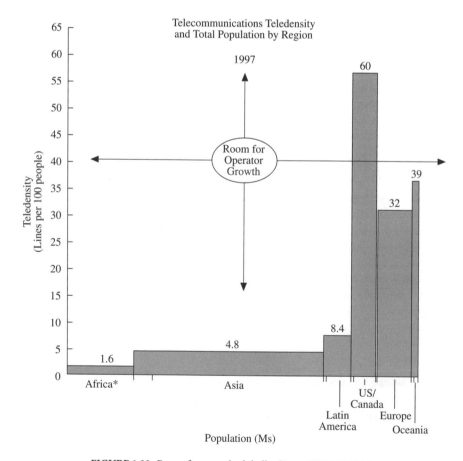

FIGURE 1.22 Room for growth globally. (From ITU 1996/97.)
*The discussion in text does not focus on Africa.

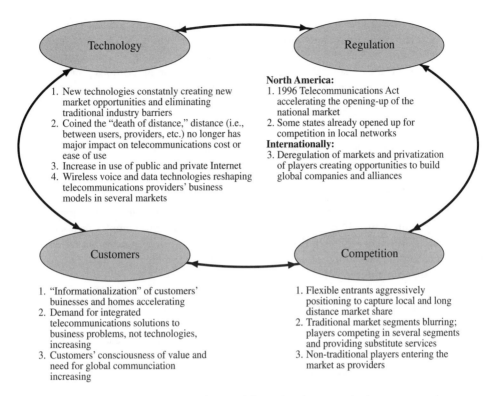

FIGURE 1.23 Four drivers of change in telecommunications.

Telecommunications providers must determine their optimal strategic direction through careful consideration of the four main drivers of change in the telecommunications market: technology, regulation, customers, and competition (Figure 1.23), for example, regulatory bodies in developed and developing markets are significantly redefining the business model of existing telecommunications service providers.*
New technologies allow for the provision of new or improved forms of telecommunications products and services. Customers, especially large business customers, are increasingly telecom savvy and expect more innovation and better customer service from their telecommunications providers. Given the impact of these drivers, local market conditions, local regulatory environments, and targeted customer characteristics all play a significant role in the development of a successful telecommunications growth strategy.

The nature of the telecommunications provider will also greatly influence its optimal growth strategy. Generally, there are two distinct, but complementary types of telecommunications players:

1. "*Incumbents*": Large, established telecommunications providers; these are previous monopolies that have progressively opened several lines of business to some kind of local competition.
2. "*New Entrants*": New telecommunications ventures; these are either true entrepreneurial ventures or new entities formed by foreign established telecommunications providers in association with large, indigenous investors or industrial companies.

To understand the types of growth strategies that will determine tomorrow's successful telecommunications providers, we will look at the fundamental elements of an effective telecommunications strategy.

---

*An example of this redefinition of the business models is the migration from guaranteed rate of return on capital pricing to market-based pricing of telecommunications services.

Then we will review the profile of both *incumbents* and *new entrants,* outlining the types of strategies and approaches that have and have not worked for them, and examining the problems facing both. To conclude, we will identify the business processes and core capabilities required in today's successful growth strategies. These are the growth strategies that will allow savvy telecommunications providers to achieve a sustainable competitive position in the marketplace tomorrow.

1.1.4.2 Elements of Telecommunications Strategies

There are many elements to creating an effective telecommunications strategy. In the last 10 years since markets began deregulating, telecommunications providers have had to act progressively to reinvent their basic strategy. Their new strategy must enable them to survive, continue to grow, and provide value to shareholders and to customers.

For decades, there was no need for true strategic planning, strong marketing and sales, or world-class capital management in the telecommunications environment. Telecommunications monopolies functioned primarily as finely tuned "operations" houses, managing efficiently complex regional or national networks and infrastructure with virtually no competition. Network operations was the dominating function in the telecommunications industry. The regulatory environment and the lack of competition for customers supported this type of business model for years.

Although the creation of shareholder value resulted from an accounting process in yesterday's market, today's shareholders are increasingly demanding. They scrutinize their telecommunications investments and ask providers to implement changes that will minimize the future risks and volatility of their investment. Telecommunications providers must now create strategies that enable them to meet the expectations of their shareholders.

In addition to shareholders, successful telecommunications strategies must also enable companies to meet the expectations of their customers. Telecommunications providers can no longer afford to take their customers for granted. In yesterday's market, customers had limited options in choosing a telecommunications provider, if any. Today's customers can often choose from an array of providers, and they do so based on value and price. Value and price are now the avenues through which telecommunications providers meet the expectations of their customers.

Customer value in today's telecommunications environment means that a customer perceives the benefits of using a provider's products and services as exceeding the cost of those products and services (Figure 1.24). Knowing what customers value enables providers to create a telecommunications strategy that will bring measurable value to customers.

The type of strategy a provider pursues will vary by the type of customer segment or segments it has decided to pursue. The telecommunications market is not homogeneous; specific submarkets, or market segments of customers, exist. Each segment has distinct characteristics and needs based on, for example, the type of products and services they frequently use, or the region in which they live. An effective telecommunications provider has divided its customer base into these customer segments and determined which segments it wants to dominate, based on how it can best use its core competencies and areas of strength to meet the expectations of those segments. Then, based on these selected customer segments, the provider can determine if it should pursue one of two generic types of strategies: a *price* strategy or a *value* differentiation strategy (since customers choose their telecommunications provider based on *value* and *price*). In other words, a provider *does not* have the privilege to select which generic strategy it will develop and implement by segment — its customers do.

Using these targeted customer segments, the telecommunications provider determines whether to pursue a price strategy or a value differentiation strategy through a two-step process: first it measures the price sensitivity of a segment; then it measures the customer segment sensitivity to the attributes of its products and services (Figure 1.25).

The whole challenge and value of formulating a distinctive strategy lies in how a telecommunications provider defines which strategic segment it plans to target, and how it intends to compete in this segment in order to provide value or a good price to a target customer while also generating value for shareholders.

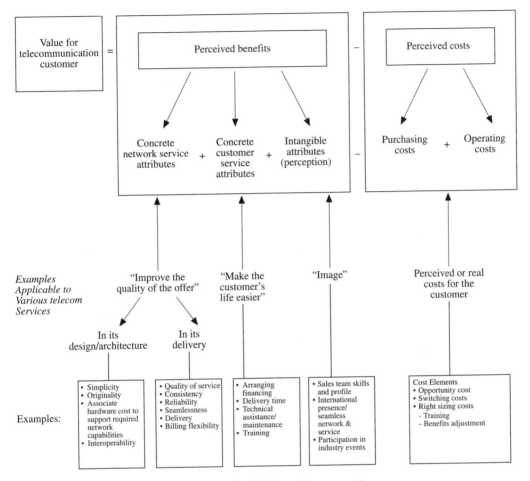

FIGURE 1.24 The definition of customer value.

1.1.4.3 Profiles of Growth Strategies

Despite the favorable outlook for the global telecommunications industry, the uncertainty and constant state of flux of the industry requires providers to become superior growth strategists. Current economic, legislative, competitive, and technology developments, as well as global telecommunications industry trends, are imperatives forcing providers to streamline their operations and grow rapidly in new areas while protecting their core capabilities (Figure 1.26).

1.1.4.3.1 Growth Options for Incumbents

The strong motivation behind any incumbent's growth strategy is a defensive one — protect core market share and retain revenue at current levels. Regulatory changes have progressively opened their markets to other incumbents and new entrants. Unprecedented mergers and acquisitions are swallowing incumbents whole. They must now grow in ways they never could before to retain what was once a given (Figure 1.27).

The intensity and speed of the local regulatory environment are the main catalysts driving an incumbent's telecommunications strategy. However, because this environment varies significantly by country, strategy formulation specific to a country requires a thorough analysis of that country's regulatory environment. For example, in some countries incumbents are already in the throes of competition, while

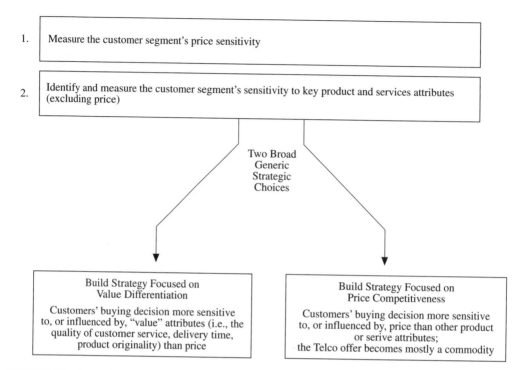

FIGURE 1.25 Determining a generic telecommunications strategy. (*Note:* In some segments, *valve* and *price* components may coexist, but one will always dominate.)

Economic	Legistative	Competitive Substitution Threats	Global Telecommunications Industry trends	Technology

1. Favorable global macro-economic conditions

2. Global growth in capital investments

3. Uncertain impact of recent regional economic and liquidity crises

3. Volatility of legislative environment

4. Non traditional entrants and non-traditional offerings

5. Uneven price structure at the global account level

6. Declining overall profitability/risk of intermediation /commoditization

7. Growing role of Internet and Intranets

8. Convergence of voice and data communications into integrated enterprise networks

9. Network security initiatives

10. Capacity expansion

11. Globalization and alliances among network service providers

12. Integrated view of the network (delivery of whole network solutions)

13. Roll-out of productivity improvements

14. Rapid displacements among technology curves

15. Progressive migration toward rapid transmission technology (various technology competing)

FIGURE 1.26 Current telecommunications trends.

in others the incumbents still find themselves in a monopoly situation (Figure 1.28). The duration of each deregulation phase can also vary significantly by country — anywhere from 2 to 10 years. For example, in Central and Latin America, deregulation has generally been phased in at different stages in different countries (Figure 1.29).

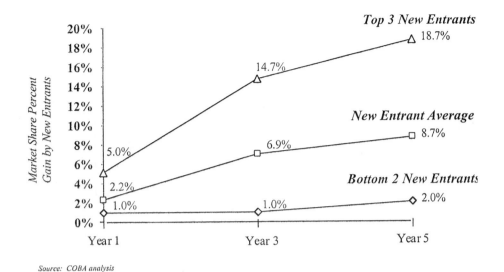

Source: COBA analysis

FIGURE 1.27 Market share gain of new entrants against incumbents.

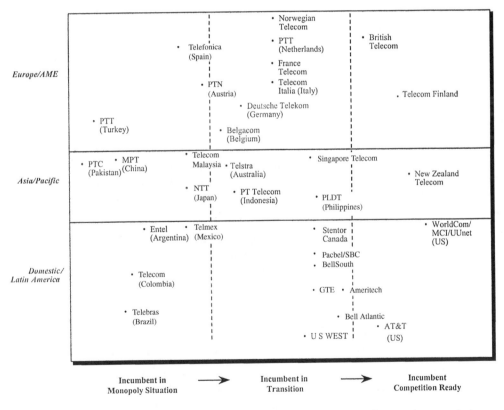

FIGURE 1.28 Incumbents are at different stages of the global telecommunications regulatory evolution.

These varying environments illustrate the instability facing incumbents. The new competitive telecommunications environment requires established providers to make strategic decisions under uncertainty, which were not required in the monopolistic era. To deal with this uncertainty, incumbents need to address key strategic questions that help them plan for growth (Figure 1.30).

Positioning on the Spectrum of Deregulation

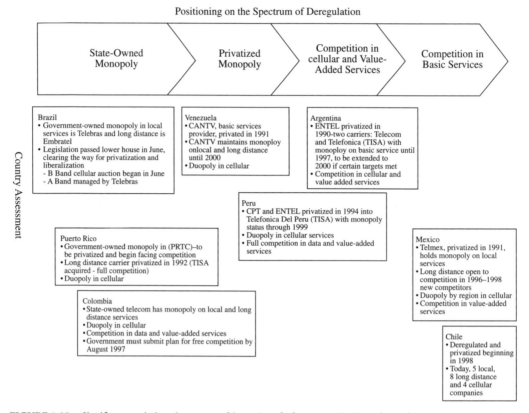

FIGURE 1.29 Significant variations in scope and intensity of telecommunications deregulation exist in Central and Latin America. (From the U.S. Dept. of Commerce.)

Common Incumbent Growth Strategies — As previously regulated markets open up to competition, incumbents are inevitably under attack. The most successful incumbents are the ones adopting strategies that enable them to limit the overall competitor's gain in market share, while growing new market opportunities. All actual growth strategies developed by incumbents are a combination of four major elements (Figure 1.31):

1. **Geographical expansion**: By increasing its geographic span, can increase its customer base and potentially decrease the competitor's customer base.
2. **Customer expansion**: By gaining more customers, providers increase their revenue-generating opportunities.
3. **Product and Service extension through new technology platforms**: The emergence of new technologies can result in new products and services that increase sales.
4. **Product and Services diversification** (in general into higher value-added services): Specialized products and services can better meet the needs of specific customer segments and increase sales.

Because incumbents have generated deep pockets after years of functioning as a monopoly, they can usually afford to experiment simultaneously with multiple growth initiatives, and even to make multiple, capital intensive "growth bets." Incumbents generally fund the implementation of a selected portfolio of growth strategies, but often their deployment of these growth strategies lacks velocity due to the sheer size, management structure, and complexity of the incumbent itself.

Most integrated, successful growth options for incumbents include geographical, technological, and business dimensions (Figure 1.32). In identifying potential growth options, incumbents assess each dimension individually on the basis of value creation potential, risk, and resource requirements.

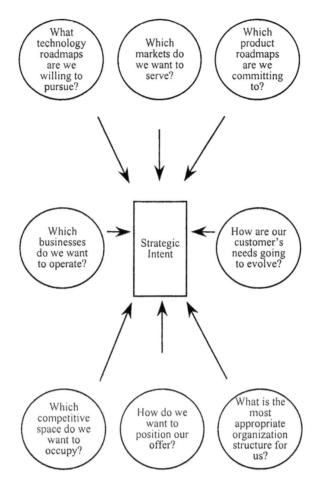

FIGURE 1.30 Incumbents strategic questions focusing on growth.

Frequent Incumbent Strategic Problems — As previously described, most incumbents have enough capital to fund ambitious growth programs; however, they often have vulnerabilities that threaten their ability to make decisions rapidly and implement programs successfully. Some of the most common areas of vulnerability that incumbents encounter in making their growth strategy work are as follows:

1. No clearly communicated innovation vision and strategy.
2. Limited experience or knowledge of new markets (i.e., value-added services and solutions).
3. *Ad hoc* valuation and prioritization of growth opportunities and lack of analysis of business implications.
4. Highly complex and convoluted operations and billing systems that require an excessive amount of resources to enable the introduction of a new solution.
5. Geographically dispersed innovation and growth-focused team members which result in high coordination costs and extended cycle times.
6. Company culture generally not supportive of innovative growth strategies:
   - Desire for the "100% solution," not change or advancement in incremental steps;
   - Lack of trust in the organization's own capabilities (the "not invented here" syndrome);
   - Process averse ("following a formal process slows you down");
   - Risk averse ("it is more important *not* to make mistakes then stick your neck out").

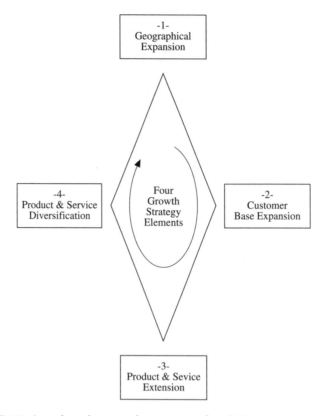

FIGURE 1.31 Incumbents base growth strategies on four distinct components/elements.

7. Overall insufficient skill sets in:
   • Marketing and finance;
   • Project management, technology, and operations which results in the involvement of an excessive number of people on each project.

These types of vulnerabilities tend to impact negatively an incumbent's core business growth opportunities. To manage these growth opportunities proactively, several incumbents have adopted the successful strategy of creating autonomous organizations to manage their growth initiatives. Often these new organizations demonstrate a better ability to acquire and retain required skills, are easier to manage from a financial performance point of view, and are overall more nimble than their parent organizations.

1.1.4.3.2 Growth Options for New Entrants

At the other end of the telecommunications provider spectrum, we find a slew of new entrants vying with the incumbents for a share of the growing global industry. However, unlike the incumbents, these new entrants are formulating ambitious growth strategies with deliberate *offensive* views that penetrate a market that appears simultaneously attractive and not tapped into, as well as suboptimally served by local incumbents.

Just as for incumbents, the market share of new entrants in both dollars and in services is controlled by regulation. In 1997, roughly a third of the global telecommunications revenue streams are liberalized — that is, deregulation made them accessible to non-PTT (Post Telegraph and Telecommunications, or former incumbent monopolies) players. In fact, liberalized revenue streams could grow to 80% by the end of the millennium, provided that the levels of deregulation continue to grow at current rates. However, these growth projections represent maximum market *potential;* and new entrants will, in reality, only achieve a fraction of those results in the years to come.

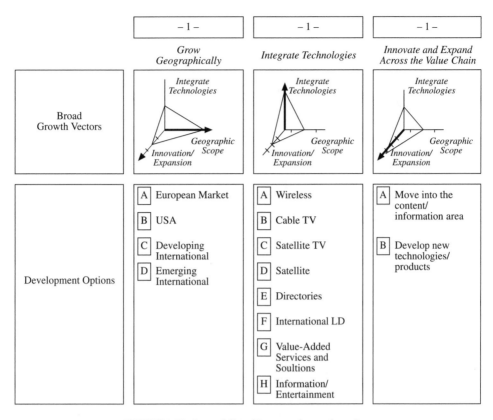

FIGURE 1.32 A portfolio of integrated growth option.

New entrant growth strategies are generally more focused and implemented more rapidly than those of incumbents, assuming a permissive regulatory environment. On the other hand, the impact of a failed telecom venture often threatens the survival of this type of entrepreneurial organization. For this reason, new entrants have to overcome a number of strategic challenges to compete successfully in the industry. This may be particularly challenging, as sometimes these new entrants are not fully aware of what these challenges might be.

All new entrants are not alike and different new entrants have different challenges. Studies identify six different categories of new entrants (Figure 1.33):

1. Foreign PTT (Post Telegraph and Telecommunications, or former monopolies)
2. Pure start-up
3. Managed network
4. CATV (cable TV) operator
5. Utility company
6. Industrial company

To deal with their differing challenges, these different types of new entrants have varying strategies and differentiate themselves from the competition in many different ways. However, most new entrants follow a fairly typical market entry and market penetration strategy. This strategy usually plans for them to provide data products and services to niche segments in the beginning, and then moves them to providing data and voice products and services to the broad public market in the future (Figure 1.34).

To service these markets, new entrants generally adopt one of three business models. Two of these models are facilities-based and involve the ownership of their own network and infrastructure and usually result from either "greenfield" deployment of an infrastructure or adaptation of an inherited running

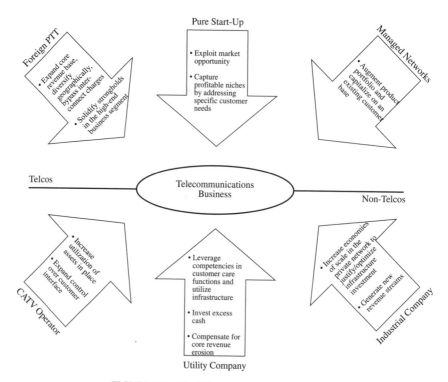

FIGURE 1.33 Six different types of new entrants.

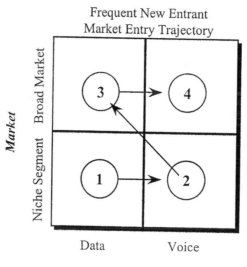

FIGURE 1.34 New entrants start with a niche strategy, and eventually migrate to a broader market.

network. Nonfacilities-based models mainly resell services from the local PTT through minimal inter-connect infrastructure deployment.

In the medium term, however, most new entrant strategy deployment plans rely on leased lines from the local incumbent monopoly and withstand the following consequences on the economics of their business: (1) minimal variations in network and traffic management highly impact the new entrant's bottom line since their profitability is greatly impacted by network efficiency due to high fixed costs or (2) access charges required by incumbent monopolies represent a substantial enough expense. New entrant start-up strategies often include the investment of significant resources in lobbying in an effort to accelerate deregulation or reduce interconnect charges.

Common New Entrant Growth Strategies

Successful new entrants generally employ five types of winning strategies:

1. **Organize and market the business around and for the customer.** The only optimal organizational structure is one that directly supports the processes that serve the needs of the customer. An organization that seeks highly differentiated offerings and positioning centered around superior quality and flexibility will be compelling and will build a customer base. For example, a residential customer strategy will increase the probability of success if initial marketing efforts focus on disloyal, unsatisfied customers. This illustrates how marketing and sales functions are crucial to growing the top line sufficiently and gaining economies of scale that will offset the nonnegligible fixed costs to the business.

2. **Develop simple, efficient processes and superior customer care capabilities.** Processes enable operational efficiencies and relatively lower cost structure than incumbents. The customer care strategy should guarantee that customer care capabilities exceed those of the incumbent monopoly. Incorporating state-of-the-art customer care, processes, and systems from the outset is crucial. Technology choices and network deployment activities are critical, but only insofar as they are effective, stable, realistic, and methodically subordinated to market strategy.

3. **Define, by target market, a compelling value proposition that extends the company's reach far beyond an initial price leadership strategy.** The sales force delivers clear and concise product and service offering messages that are highly targeted and consistently communicated. This practice will lead customers to buy a whole new concept rather than a mere discount in the long term. It is also crucial for new entrants, as part of their strategy deployment, to carefully synchronize customer acquisition cycles with operational implementation cycles. Doing so avoids either the sale of services that are not supportable or the deployment of infrastructure that does not bolster the top line.

4. **Understand the capital-intensive nature of the business while carefully managing a focused and orderly investment plan.** Make steep investments in the business, investing only in the basic infrastructure required to maintain adequate levels of asset utilization and maximize return on investment.

5. **Maintain strong pressure on regulatory bodies.** New entrants must accelerate changes in the legal environment to broaden their customer base and boost their top line. They must also secure favorable interconnect rates to optimize the short-term bottom line.

Frequent New Entrant Strategic Problems

Few new entrants are experiencing the success that they expected at the speed they had expected, and most are finding the level of competition in the telecommunications arena is tougher than anticipated. Often the difficulties new entrants encounter are the direct consequence of management misdirection which results from broad misconceptions of the business. Two common misconceptions may be summarized as follows:

Strategic Misconception 1: "If we build it, they will come." Many new entrant executives operate under the assumption that merely their status as a new entrant will guarantee them market share. It is not uncommon to see business plans based on the assumption that the new entrant will capture 10 to 15% of market share from the incumbent monopolies in time frames ranging from 2 to 5 years. The reality is that while the idea of having a choice may be appealing to customers, they may still be accustomed to thinking of telecommunications as a monopoly business. As a result, new entrants often experience the following reactions from the market:

- *Overcautiousness* with respect to the legitimacy of the new entrant; this deters customers from subscribing to the new player's offerings.
- *Psychological captivity* of customers to the incumbent PTT.
- *Customer overexpectations* of the new entrant which inevitably lead to dissatisfaction and high churn.
- *Short-term trial runs* that generate unstable market presence with the lowest possible market profitability.

Strategic Misconception 2: "The local incumbent monopoly is a defenseless giant." While it is indeed true that David defeated Goliath, the story never suggests that Goliath failed to defend himself. Incumbents are, as previously discussed, not ready to sit by the sidelines and watch their core business be chipped away. They have been anticipating liberalization for the last few years, and, although they have been lobbying vehemently to slow the pace of deregulation, they have also been preparing themselves, both strategically and operationally, to meet competition head-on. Some actions include:

- *Accumulating financial reserves:* PTTs have been stockpiling financial resources which will afford them a clear edge on investing in a very capital-intensive industry.
- *Solidifying customer share of mind:* incumbents have fewer operational concerns than their emerging competition and are now aggressively focusing their marketing and promotions strategies on reliability, customer care, and price.
- *Preemptive strategic alliances:* globalization is driving the need for alliances for the purpose of extending the market reach and competitiveness. Large, established players have consistently been securing the most attractive and viable alliances, often before new entrants even realize their need for international partners.

These common misconceptions inevitably lead executives of new entrants to make managerial choices and decisions that are not necessarily optimal or sound. Some of the most typical issues at some of the more successful new entrants are:

1. **Erratic strategies:** New entrants often mistake lack of focus and confusion for nimbleness and marketing aggressiveness. They randomly pursue any shadow of an opportunity regardless of whether it fits within their stated strategy or not. Eventually, strategic goals and intent are negated by a long series of opportunistic, contradictory tactical moves. For example, CEOs of companies that provide data services to the top 200 corporations of their country find themselves feverishly pursuing international dial-around offerings for the consumer market. That consumer market sales tactic leads them to develop a costly and inefficient direct sales force, when in fact the overall strategy, which focuses on the small business market segments, would have required the development of indirect channels.

2. **Purposeless network deployment:** Several new entrants are basing network deployment decisions on erratic strategies as described above, and on very loosely identified service definitions. This approach yields equally organic networks, which are a long string of patches neither suitable for the corporation's long-term needs nor efficient and reliable. One company that sought to offer voice and fax services to the public at large had, as a result of patches, invested millions in building a network of PBXs instead of the suitable public switches.

3. **Neglect of marketing and customer care functions:** As exemplified by purposeless network development, several new entrants are erroneously staking their success on their technology strengths rather than on their marketing strengths. In many instances, new entrants subordinate marketing efforts to technological initiatives and customer care functions so that marketing initiatives are consistently neglected. Yet, these marketing initiatives are often key to the company's differentiation and competitive immunity in the next few years. A company's value proposition is often undefined and in most cases limits that company to a *20% discount over the incumbent monopoly*. The company often mistakenly perceives this discount, which is the cost of entry into a former

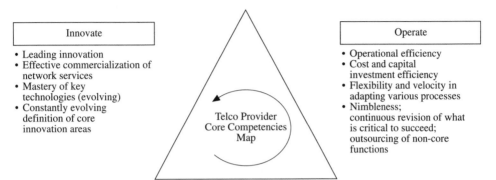

FIGURE 1.35 Three core competencies for telecommunications providers. (From COBA Methology, based on Tracy and Wiersema, *The Discipline of Market Leaders.*)

monopoly business, as a sustainable differentiating factor. A vivid example of poor service differentiation happened when a new entrant based its provisioning time on the local incumbent monopoly — one of the industrialized world's worst — on the grounds that customers in their market were not accustomed to world-class-level service.

4. **Rudimentary customer segmentation schemes:** Some approaches trivialize end-user complexities down to a mere number of lines or revenue levels. These approaches have clear shortcomings which incumbents have finally figured out on their own. There is no reason why those approaches would be effective for new entrants; however, many of them are applying them. For example, one company whose marketing executives sought to offer services to the 1000 largest corporations of the country, found two main segments: 1 to 500 and 501 to 1000.

5. **Recreation of the incumbent monopoly**: Several new entrants are, consciously or not, recreating an incumbent monopoly. Examples include a company that secured a telecommunications license from the local regulator, but was already selecting a location to build the company corporate headquarters with an IS staff of 60. Or the company that did not have a single customer in its second year of operation but had reorganized five times in that period and created four levels of managerial decision-making committees.

1.1.4.4 Conclusion: Core Competencies Required to Win in a Growth Strategy

Current changes and developments make the telecommunications industry dynamic and exciting. The advent of new technologies and the progressive deregulation of several markets will make it increasingly competitive. However, we have seen that both groups of carriers will face tough challenges while earning or maintaining their rightful position in the industry.

Industry benchmarking and customer needs analyses identified a number of core competencies, or capabilities, required of telecommunications providers to maintain a sustainable competitive advantage. These core competencies can be grouped in three major categories: innovate, operate, and captivate (Figure 1.35).

To continue to create value for shareholders, providers will have to maintain at least two of these core competencies categories at a *world-class level* and the third at least at *competitive parity*. Generally the incumbents and new entrants are strong in different core competencies (Figure 1.36).

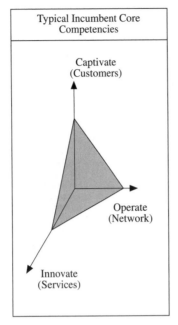 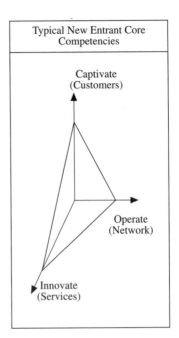

FIGURE 1.36 Incumbent and new entrant core competencies differ.

An incumbent's competitiveness will usually be anchored in the *Operate* cluster of capabilities. It may also possess some potentially formidable assets in the area of *Captivate,* which are usually underutilized. The ability of incumbents to *Innovate* usually lags behind the best in class.

Both types of telecommunications providers, however, are launching numerous strategic initiatives to remedy their relative areas of vulnerability. Incumbents are investing in processes and systems to liberate the power of captivating their customers. (This can be done by database marketing, relationship marketing, segmentation, industry marketing, training and recruitment of new sales forces, knowledge enabled services, and telemarketing centers). These processes and systems also improve their ability to increase the volume, velocity, and variety of their innovation and commercialization engine (a provider's process for developing customer offers) while continuously streamlining their operations.

To build their fledgling competencies, new entrants will focus on a few attractive customer segments, form alliances with other competitors (usually international telecommunications providers), build open, modular infrastructure and systems, and hire a balanced mix of executives from the industry and other competitive intensive industries.

In conclusion, only telecommunications providers who realize the importance of a sound, disciplined strategy will eventually prove their success through healthy financial statements. However, all providers will eventually realize three things:

1. **Ongoing growth strategy planning needs to be a highly adaptive process**. Executives cannot simply define options and recommendations and start implementing strategies for the next 3 years. Telecommunications growth strategies are about highly volatile hypotheses and knowledge. Unveiling visions and strategic intents is not sufficient and top management may need to treat their growth strategy formulation like a sequence of experiments. Ongoing analysis of industry structures and market discontinuities is critical. There is no certainty when it comes to strategy, and providers need to become nimble in order to achieve usable patterns of competitiveness. The best way to figure out what to do is maybe to formulate hypotheses on possible futures and test them in a controlled environment.

2. **Profitability takes time**. In most cases, realistic business plans for growth opportunities should not project excessive market share, or expect profitability prior to the third of fifth year.

3. **Not everyone will make it.** The basic constraints of market size (both revenue and users) and growth rates limit the number of new players that a country can viably accommodate. Some smaller countries are seeing the emergence of three or more players in addition to the incumbent monopoly. Only a few of the very best players will survive.

1.2 Regulation

1.2.1 Regulation Instruments from a Legal Perspective

Willem F. Korthals Altes

1.2.1.1 Introduction

Regulation of telecommunications is inevitable. Ownership and control of the telecommunications infrastructure; the right to lay cables across public and private land; the use of communication channels; and the protection of privacy are among the issues which require rules of law. This is true not only in a completely monopolized environment, but also, and perhaps even more so, in a situation in which competition is allowed to thrive.

Since airwaves by their very nature do not stop at national borders, telecommunications is one of the earliest areas in which regulation at the international level was applied. The International Telecommunications Union (ITU) and its predecessors were among the first international organizations. While the ITU is primarily based on technical cooperation, current practice, in particular in the European Union, shows that considerations of an economic nature can be a driving force in creating a more contemporary type of regulatory environment.

This chapter purports to provide a general overview of the regulation of telecommunications, including the types of instruments which are and may be used to regulate telecommunications, both at the national and at the international level, and a discussion of the rationale of such regulation and the way in which it is enforced.

1.2.1.2 Distinctions between Telecommunications and Media

Traditionally, most countries have always had separate regulation for telecommunications on the one hand and media on the other. A notable exception is the U.S., which combined the two in the Communications Act of 1934. The U.S. thereby also created an agency, the FCC, which was made responsible for the proper exercise of all activities under the Communications Act. In most European countries, the legislatures not only enacted separate laws, but also decided that different sections of the government should deal with the implementation and other aspects of the legislation. In the Netherlands, for instance, the Ministry of Transportation is responsible for telecommunications, while the Ministry of Culture deals with the media.

The rationale of the European approach may be that telecommunications has always been considered to be a purely technical matter, whereas the media concern content and culture. In addition in Europe, unlike in the U.S., those who provide the content of radio and television programs, whether public or commercial, do not own or fully control the broadcast facilities (transmitters, etc.). These facilities are provided for by either the PTT or a consortium controlled at least in part by the PTT. Telecommunications and media are also separated at the level of the European Union. Following the tradition of the various Member States, the European Commission, the European Union's lawmaking and executive body, has different Directorates General dealing with these activities. Since this handbook deals with telecommunications and not with media, the remainder of this section will be limited to the former. It is important to keep these distinctions in mind, however, in particular when convergence is at stake.

1.2.1.3 Why Regulation of Telecommunications?

First, the question why regulation of telecommunications is necessary has to be answered. An important reason is the fact that telecommunications activities require the use of either public or private domain, or both. This applies to airwaves as well as to the use of public and private grounds for cables and other elements of infrastructure. Although one might say that, due to their intangible nature, airwaves are not subject to public ownership, it is easy to counter this by stating that, by occupying any segment of the spectrum, one makes it impossible for others to use the same part of the spectrum. Since the spectrum is not finite, there is an element of scarcity which necessitates rules preventing chaos.

Another reason for regulating telecommunications is that telecommunications concerns a public service of vital importance. People can communicate through telecommunications channels and such channels are used to provide the public with a still increasing amount of information and vital services. Finally, telecommunications activities have a strong impact on the individual's privacy. Individuals use telecommunications channels to transmit a large amount of information which is not meant to be made public. Rules are therefore needed to regulate (or prevent) access to such information by:

- Those who provide the facilities
- Those who have the capability of breaking into the system
- Those who claim access to information for specific reasons (for instance, the government or the judicial authorities)

1.2.1.4 What Should Be Regulated in Telecommunications?

The next issue to be discussed is what has to be regulated to serve the interests listed above and at what level regulation should take place. It is necessary to have rules dealing with the following: the procedures to be applied in building, maintaining, and securing the telecommunications infrastructure; the allocation of and control over the communication channels (whether terrestrial or cable); terminal equipment; services; quality standards and quality control; and fees and costs.

Procedures are vital, in particular because cables, whether or not underground, cross both public and private property. Regulation is needed to give the organization responsible for building the telecommunications infrastructure the authority to construct and maintain the necessary installations and cables. If private property has to be taken, the law should provide for proper procedures, since the taking of property is a far-reaching breach of individual rights. Although the individual's position is not at stake in the case of public grounds, the use of public property similarly requires rules of law.

There should also be rules about standards for safety and reliability of the infrastructure, as well as rules about the consequences of damage caused by malfunctioning or dysfunctioning. Finally, there has to be regulation to allocate the responsibility for securing the network against improper use and other types of events causing damage.

Apart from international agreements (to be discussed later), national law needs to contain precise rules on the issue of who can use the frequencies allocated to the state and for what purposes they are to be used. The same goes for the channels and wires which form part of the infrastructure. The need for regulation follows from the inherent scarcity of both airwaves and channels. Regulation in this area includes rules providing for sanctions against those who violate the rights of the lawful users of the frequencies and channels.

Evidently, the law has to create a provision telling who may or should produce and sell terminal equipment. Although the use or taking of private ground is not at stake in the case of the production and application of terminal equipment, some additional regulation may be required. In today's fairly typical case of a monopolized (albeit privatized) telecommunications infrastructure and a market situation for terminal equipment, there have to be rules on the applicability of equipment to the infrastructure. Without such regulation, the provider of the infrastructure may directly or indirectly favor one or more producers of terminal equipment. Countries dealing with this issue often appoint or create independent organizations for setting and applying the standards of applicability.

Regulation of this type can be absent both in the case of the traditional total monopoly and in the case of a completely demonopolized environment. In the former, monopolist providers will make sure that their equipment can be connected to their infrastructure. In the latter, market forces will ensure that equipment is suitable.

As in the case of terminal equipment, there has to be some regulation telling who should or may provide services, whether basic or enhanced. Should there also be additional rules like in the case of infrastructure and terminal equipment?

Neither the issue of use of private ground nor the issue of applicability of equipment to the infrastructure is at stake in the case of services. Service providers are dependent on the infrastructure and the terminal equipment for the kind of services they can offer. In a demonopolized environment, this does not necessitate specific regulation, other than, perhaps, general legislation preventing unfair competition.

As is the case with the telecommunications infrastructure, certain types of services may require regulation protecting the privacy or other vital interests of the users in that it should be absolutely certain that no one without authorization can have access to the information transmitted. In general, this raises an issue fundamental to telecommunications, i.e., the issue of whether the service provider or, as the case may be, the provider of the infrastructure should have anything to do with the content of what is transmitted. Whatever choice is made, the law should make sure that it is clear whether or not either of the two may exercise some sort of content control. A contemporary example of this issue is whether providers of Internet facilities should act in preventing the transmission of child pornography.

There should be regulation on the issue of who sets the standards for quality and who enforces them. The kind of regulation depends on the extent to which the telecommunications activities are demonopolized and/or privatized. In the traditional monopolized environment, quality standards and quality control were by and large the domain of the PTT. Because of the lack of competition, the PTT did not have to worry about innovation, about providing high-quality services as quickly as possible, or in general, about meeting consumer demand. The law provided for few incentives, other than the requirement that the PTT would provide a universal service, accessible at a reasonable rate to all citizens. Every country has its proverbial widow living in a remote area who should not be deprived of the facilities offered to inhabitants of big cities, including affordable rates.

Telecommunications in traditional European-style PTT was definitely the domain of the PTT management, not only technically, but also legally. Until about 1980, there were virtually no lawyers outside the PTT with any knowledge of telecommunications law.

What about quality control and quality standards in a competitive environment? Is it necessary to have rules of law or is it sufficient to leave quality standards and quality control to the market forces? In general, there is no reason why there should be specific laws in this area. There are also no specific laws providing rules on the quality of other products, such as furniture, paper clips, and cookies. Such rules will be part of general laws on product liability, etc. Each lawmaker will have to decide for his or her own country whether telecommunications activities should be treated differently in this respect.

As in the case of quality standards and quality control, the kind of regulation on fees and costs will depend on the extent of demonopolization. Nevertheless, here too, some rules have to be made about the way in which fees are set and costs are accounted for. This is true in particular because telecommunications concern a public service. It is quite conceivable that, in a completely monopolized environment, the government has a strong say in the level of the fees charged to the users, whereas setting the fees can be left to market forces to the extent that telecommunications activities are left to the market forces. It is arguable that regulation is not required when all activities are solely subject to market forces. But even there, the legislature may have to create impediments against cross subsidies.

1.2.1.5 The Level of Regulation

There are basically three levels at which regulation can take place:

- The constitutional level
- The level of the national legislature
- The level of the executive power

In telecommunications, additional rules have often been made (and, in some cases, still are made) by the management of the organization engaging in the telecommunications activities — in Europe, the PTT.

Rules will be enacted at the constitutional level if they touch on vital rights and duties. For example, most national regulations providing for telecommunications secrecy will be laid down in a body of law with the weight of a constitution. Acts of the national legislature will be used for the general provisions regulating most of what is listed above. Such acts of the national legislature should also contain criminal law provisions about the consequences of violating certain rules. Provisions of these national acts will subsequently mandate the executive to carry out more practical aspects. In general, regulating telecommunications is considered to be an activity of a technical nature with a low political profile.

1.2.1.6   What Is Regulated?

1.2.1.6.1   Traditional Telecommunications Regulation
In a completely monopolized environment, regulating telecommunications can be relatively simple. All the legislature essentially has to do is create a basic act providing the general framework for the organization destined to exercise all telecommunication activities. Such an act will typically contain a mandate to this organization (in European tradition, the PTT) and provisions dealing with the PTT's overall responsibility for quality, quality control, and price-setting. Whatever additional regulation may be needed will be made and enforced by the PTT management. This has traditionally been the approach in Europe.

Traditional "PTT-regulation" also did not make a distinction between infrastructure, terminal equipment, and services. All these activities were in the hands of one *monopolized* organization. Therefore, there was no need to separate them in the law. Producing and selling equipment or services and engaging in infrastructure-like activities by others than the PTT were simply subject to criminal prosecution.

1.2.1.6.2   Contemporary Regulation in Telecommunications
Since the advent of privatization and separation of activities in the 1980s, new legal systems had to be created. Legislation in a more or less privatized and demonopolized environment inevitably deals with the following issues:

1. The law will determine who is responsible for constructing and maintaining the telecommunications infrastructure, who may manufacture and sell equipment, and who will provide the services.
2. There will be rules about the types of technical installations and terminal equipment needed for the different kinds of communication.
3. There will be regulation dealing with the way in which fees are set and the extent to which providers of infrastructure and services can be held responsible for malfunctioning.
4. The law will provide for a system of supervision, i.e., an agency or other type of independent body with the task of overseeing the way in which the various actors carry out their activities.

Of course, many rules will be accompanied by provisions dealing with the various types of procedure needed to carry out the law, as well as criminal law provisions. As has been discussed before, the answer to the question of whether there will also be specific rules about quality standards and quality control depends on the extent to which telecommunications activities have been privatized and demonopolized. It is conceivable that such elements are fully covered by general laws on consumer protection, product liability, etc. It is also quite likely that many rules will be more clear-cut and more extensive than in the traditional monopolized environment in the interest of a fair competition. In addition, such rules will be enacted at a higher level (i.e., acts of the national legislature rather than ministerial decrees) because of their significance for all parties in the market.

1.2.1.7   Regulation at the International Level

1.2.1.7.1   The International Telegraph Union
The need for regulation at the international level was already recognized in the earliest stages of telecommunications practice. In the middle of the 19th century, the telegraph was developed as the first means of telecommunications. Regulation could be purely domestic as long as communication by telegraph stayed within the national boundaries. If messages had to be sent across the border, they were first wired

to some border town and then carried across the frontier to be put on wire again after reaching the first town in the neighboring country.

Governments soon realized that such an interruption would defeat the purpose of sending messages across as quickly as technology allowed. Therefore, a number of sovereign European states got together in Paris in 1865 and created the International Telegraph Union (ITU). From the outset, cooperation and rule-making in the ITU was of a purely technical nature, without interference by politics. The only notable exception was the exclusion of South Africa in 1965 because of its apartheid policies. Secondly, the ITU members decided that only countries with government-run telecommunications organizations could become members. Britain, for instance, was not admitted until it nationalized its telecommunications organizations. On the other hand, the ITU had as its first members principalities which later became part of Germany.

1.2.1.7.2 The International Telecommunications Union

The nature of the ITU never really changed. Although the ITU always retained its nonpolitical character, its activities and organization were affected by major events, such as World War I and the *Titanic* disaster. Of course, the ITU followed technical developments. The telephone was added to its activities and so were, later on, radio and satellite. In 1932, at the conference of Madrid, the ITU was renamed International Telecommunications Union. The Madrid conference also created the International Telecommunications Convention (ITC).

The ITC is, in fact, the ITU constitution. It establishes its structure; sets forth its purposes; defines its membership; fixes its relationship with the United Nations; and sets forth primary regulations dealing with telecommunications in general and radiocommunication in particular. Basically, the ITU has two main purposes (as set out in the ITC): (1) the maintenance of an efficient, worldwide telecommunications network, and (2) the constant upgrading of the technologies and procedures in that network.

The Member States convene once every 5 to 8 years at a Plenipotentiary Conference. Most of the ITU day-to-day activities are carried out by its agencies, such as the International Frequency Regulations Board (IFRB), which plays an important role in the allocation of frequencies.

The need for regulation at the international level follows from the same elements as the need for national rules. An organized partition of the spectrum, technical standardization, the use of non-national territory (the oceans), the use of the geostationary orbit (for satellite communication), and international cooperation in general are all part of what is regulated within the ITU. This does not mean that the ITC and its offspring contain detailed provisions on every aspect of telecommunications. Much is left to the Member States, both with regard to purely domestic matters are with regard to some matters that require bilateral agreements between individual Member States. The ITU lacks an enforcement mechanism, so conflicts are solved by way of negotiation.

1.2.1.7.3 Monopolism vs. Market Forces: The BT Case

The character of the ITU as a champion of monopolist, government-run telecommunications organizations was challenged in a major case which came before the European Court of Justice (ECJ) in 1985: the *British Telecom* case. This case dealt with the issue of whether British Telecom violated the antitrust provisions of the Treaty of the European Communities by setting price schemes which effectively made competition impossible for companies which at that time were allowed to carry out certain telecommunications activities in Britain. After the European Commission decided that the price schemes violated Article 86 of the EC Treaty, the U.K. government took no further action. By that time, Britain had already demonopolized and privatized a number of telecommunications activities and had decided to continue in this direction. The British government accepted the Commission's decision.

It was the Italian government that appealed to the ECJ. In the U.S., a state is, in principle, exempt from the purview of the antitrust laws on the basis of the so-called *state action doctrine* (*Parker v. Brown*, 317 U.S. 341 (1943)). That is not the case in the European Union. Under European Union law, anticompetition rules are applicable not only to activities carried out by private companies, but also to legislative measures of Member States (*Leclerc v. Au Blé Vert*, Case 229/83, and *Cullet v. Leclerc*, Case 231/83, both decided on January 29, 1985). Therefore, telecommunications activities, even if run by the government, could be subject to anticompetition scrutiny. Nevertheless, certain services can be exempted from such scrutiny, if one can claim the public body exception of Article 90(2) of the EC Treaty.

Italy, like most European countries, still had a complete PTT monopoly and was not planning to give it up. Italy argued that telecommunications was a public service and was therefore not subject to the anticompetition provisions of the EC Treaty. The ECJ rejected this argument. The Court held that BT activity, notwithstanding its status of a public undertaking, was a commercial activity since it consisted of offering public telecommunications services to the user against payment. Such an activity falls under Article 86 of the EC Treaty. While the *Leclerc* cases dealt with books and fuel, the Court thus extended the anticompetition rules to activities which were still widely considered as public services.

The Court's decision did not imply that Member States were totally prohibited from creating monopolies. In 1974, the ECJ had already held that:

> Nothing in the Treaty prevents Member States, *for considerations of public interest of a non-economic nature* [italics added] from removing radio and television transmissions, including cable transmissions, from the field of competition by conferring on one or more establishments an exclusive right to conduct them. (*Sacchi v. Italy,* Case 155/73, 1974 ECR 409)

By explicitly talking about considerations of a *noneconomic nature,* the Court drew a sharp line. It is conceivable that such considerations could play a role in television or radio, for instance when it comes to protecting national culture or language. But such elements are unlikely to come into play in telecommunications.

1.2.1.7.4 *The BT Case and Its Aftermath: A Green Paper*

The impact of the ECJ decision in the *British Telecom* case was enormous. It in fact meant that telecommunications could no longer be seen as an activity of a purely technical nature free from market forces. The Italian government argued that ITU rules, accepted by all Member States of the EC in the framework of the ITU, prohibited activities by anyone other than the monopolist PTT. Although the Court did not say so in a direct way, its decision implied that Article 86 of the EC Treaty prevailed over ITU law. While the U.S. had already divested the AT&T monopoly at the national level in 1982, the BT judgment applied to more than one country.

The European Commission used the momentum of the BT case by issuing a Green Paper in 1987. The essence of this paper was that in the course of the years a substantial percentage of telecommunications activities in the European Union Member States would no longer be carried out by a monopolist PTT, but by private companies competing in an open market. Since 1987, the European Commission has guided the Member States in achieving this purpose by issuing directives telling the Member States how to adapt their national laws and by creating institutions such as ETSI, which deals with standards.

Although much of what happens in telecommunications is still highly technical, the European Union drive for demonopolization and privatization has created a strong economic force next to the ITU. In the U.S. and Europe, law and economics are much more important than has ever been the case in telecommunications. While in the traditional monopolized environment legal experts in telecommunications hardly existed outside the PTT, many major law firms now advertise their expertise in this area.

1.2.1.7.5 *GATS*

A recent addition to regulation of telecommunications at the international level is the *General Agreement on Trade in Services* (GATS), concluded under the auspices of the World Trade Union (WTO). The GATS deals with a variety of services including telecommunications. Regulation in the framework of the WTO is created by a body consisting of representatives of the Member States at the executive level. Dispute resolution takes place primarily through negotiations and is conducted by the organization's General Council, which then acts as a Dispute Settlement Body. Private parties cannot institute suits. They have to address their national government, which then has to decide whether to enter into a dispute settlement proceedings.

The GATS purports to create an open universal market in telecommunications. The agreement deals with the way in which member states should treat each other. The GATS requires a most-favored nation clause between the member states in addition to transparency of rule-making at the national level. On February 5, 1998, the Fourth Protocol dealing with basic telecommunications services entered into force. Under the GATS, the member states agreed to create a liberalized nonmonopolistic telecommunications market by the end of 2004.

1.2.1.7.6 Supranational and International Regulation

A distinction can be made between supranational regulation and international regulation. Supranational regulation is at stake when an organization creates rules which those subject to the organization's jurisdiction have to apply as they are made by the organization. Such rules are created by international organizations with strong decision-making and rule-making power, such as the European Union. Supranational regulation is fairly rare, because it implies that states have to give up some of their sovereignty. Even the European Union applies such procedures in specific situations only.

Most regulation at the international level concerns international regulation, i.e., rules made by an international organization which have to be implemented in national law by each of the organization's member states. In telecommunications, international regulation is the rule. Due to the technical and nonpolitical nature of most decisions made by the ITU and its agencies, ITU rules are generally implemented as adopted at the organization's level. The same goes for the GATS.

1.2.1.8   Final Remarks

It is obvious that telecommunications activities are of vital importance to the lives of an increasing number of people in this world. Technology makes it possible for almost anyone to use telecommunications services in one way or another. Regulation is needed to provide for the proper national and international environment. Whereas originally regulation, both at the traditional national and at the international level, supported a monopolistic environment, developments in the major industrial countries show that such a policy is no longer viable. The U.S. and, subsequently, the European countries are now on the way to an open market and this is enhanced by the creation of such bodies of law as the General Agreement on Trade in Services, which has a universal scope.

1.2.2   A Model for Assessing Regulation in a Country

Floris G.H. van den Broek

Regulation, or rather the regulatory environment in a country, can be assessed and expressed in a model.[*] Literature [OECD, 1997] confirms that the regulatory environment plays a role in the management of the network. The regulatory environment determines, for instance, what telecommunications services an organization that manages a network or a telecommunications operator may provide. The model presented here consists of a series of three elements: the legal framework for competition, the regulatory body, and competition active. The three elements each model an aspect of the regulatory environment in the country. Modeling each of the elements is done with a quantification, such that each of the elements can be scored and added together for a total score of the regulatory environment in the country. Other research is done on the validity of this model and the practical use in management of international networks [van den Broek, 1999].

---

[*]In this section we present a model for measuring the regulatory environment in a *country*, as is common in most literature. So, even though particular reference may be made to supranational regulation (such as European Union directives), the entity for examination of the regulatory environment will be the *country*. This section is an adaptation of a chapter in the book *Management of International Networks* by Floris van den Broek, also published by CRC Press in 1999 [van den Broek, 1999].

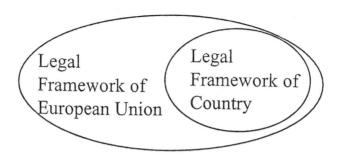

FIGURE 1.37 In a country, laws from different legal frameworks may be applicable.

1.2.2.1 Legal Framework for Competition

The first element we chose for modeling the regulatory environment is the legal framework for competition. For the purpose of this handbook, the legal framework for competition of a country consists of a series of laws regarding telecommunications competition. The applicable telecommunications laws in a country can be assessed in terms of the possibility that they create for telecommunications service suppliers to compete when offering their services. Laws may be enacted at the country level (e.g., by the country government) or at a higher level, the so-called *supranational* level (see Section 1.2.1). As we have taken the *country* as the unit of analysis, we look at applicable laws in the country. However, the applicable laws in a country may have been made by *supranational* organizations, such as the European Union. Laws from the supranational organizations may or may not have direct effect on the citizens and companies in a country. For instance, in case of the European Union, laws or other regulations only have direct effect if they are self-executing, which is usually defined in the laws or regulations themselves. The World Trade Organization (WTO) is another example of a supranational organization. The WTO, however, does not make laws, but forms agreements that are ratified by country governments that then implement those agreements in laws. When we keep the unit of analysis the *country*, a mix of different laws can be applicable. In practice, national governments take care that laws at a country level do not contradict laws that have influence on that country at a supranational level. In Figure 1.37 an example of different legal frameworks is shown. The example assumes that the country is in the European Union.

The legal framework for competition as we describe it is represented by a listing of telecommunications services categorized in service value levels (see Section 1.1.1.3.1) in the country and symbols showing for each service value level that the laws rule that the provision of the services of that service value level is open to more than one supplier.

Service value levels are part of the service value model, a model that we developed to categorize telecommunications services according to their place in the value chain for the provision of telecommunications services. Laws sometimes differentiate between types of services. The legal framework for competition, however, does not address such differences between types of services. It only addresses differences between service value levels of the services. The political, social, and economic aspects that may influence the regulatory environment are not taken into account separately here, but we recognize that they may influence the legal framework for competition.

Four categories are used to show the content of the laws with respect to allowing suppliers to offer the telecommunications service. The categories used are monopoly (M), duopoly (D), partial competition (PC), and competition (C).[*]

M means that there is only one supplier allowed that supplies telecommunications services of that service value level, D means that there are two suppliers allowed that compete in supplying services of that service value level, PC means that there is only competition in certain areas, and C means that there are several suppliers of similar services and that there is competition in the whole country. Current

---

[*]These categories are also used by the Organization for Economic Cooperation and Development (OECD) [OECD, 1997].

Legal Framework for Competition	Score Starting Value
M	0 points
D	1 point
PC	2 points
C	3 points

	Legal Framework for Competition						
	Value-Added Services		Basic Data Transfer Services		Infrastructure Services		
	Code	Score	Code	Score	Code	Score	Total
X	C	3	M	0	M	0	3
Y	C	3	C	3	C	3	9
Z	C	3	C	3	C	3	9

FIGURE 1.38 Starting values for the legal framework for competition.

FIGURE 1.39 Example of a legal framework for competition for sample countries X, Y, and Z.

literature [OECD, 1995] considers the order of liberalization from less liberalized to more liberalized as: M, D, PC, C. To use the regulatory environment in a quantitative way, we would like to score the categories, with a higher score, meaning a legal framework for competition that allows more freedom for parties to offer telecommunications services. We have chosen starting values for the scores as an assumption, using our experience and literature: M, D, PC, and C are awarded 0, 1, 2, and 3 points, respectively, as starting values. This results in the starting values as shown in Figure 1.38.

These starting value scores are used to quantify the legal framework for competition and do statistical analysis in future case studies. The legal framework for competition is a *prescriptive* element, which means that it quantifies a situation that is prescribed. The use of laws, for instance, can prescribe a behavior, but does not give any feedback if the behavior actually occurs. Figure 1.39 gives an example of scores for the legal framework for competition for sample countries X, Y, and Z.

1.2.2.2 Regulatory Body

The second element of the regulatory environment is the *regulatory body*. Regulatory bodies are organizations that are responsible for implementation of part of the regulation and, in some cases, are responsible for development of regulation. Regulatory bodies exist in several countries. Usually, regulatory bodies are founded as soon as more than one telecom operator emerges, but even in a monopoly environment, a regulatory body can exist and control the offering of telecommunications services by the monopoly as well as handle the control of tariffs and quality of the telecommunications services.

The regulations that the regulatory body is allowed to establish are within boundaries set by the laws and associated legal instruments, defined here in the legal framework for competition. There is no standard arrangement to determine if a regulation should be made by the regulatory body or by laws. Regulations from regulatory bodies are, however, more flexible than laws, since laws usually have to be made in a formal process involving a majority vote by the representatives of the population in a congress or senate, whereas regulatory body regulations usually do not have to pass these barriers. The laws of a country, however, usually must provide a legal basis for the regulatory body to issue regulation in the first place [Melody, 1997].

1.2.2.2.1 Purpose of a Regulatory Body

In practice, we observe that regulatory bodies exist to promote *good competition* in the telecommunications services market, such that the consumers in that country can benefit from low prices and more choice of telecommunications services. Therefore, we describe good competition as competition where *unequal advantages of competition are neutralized.*

Two examples of the role of the regulatory body

The regulatory body can play an important role in the regulatory environment and can make or break the effects of the legal framework for competition. For instance, in the U.K. during a period (1984 to 1992), the regulatory body in the U.K. (OFTEL) ruled that Mercury had to pay access charges to British Telecom, for access to the local telephone lines, which were mostly owned by BT. The access charges were higher than British Telecom's cost and that were only on a per-minute-use basis and not on a per-call basis [Cave, 1995]. The given cost structure made it impossible for Mercury to compete effectively. When this was changed by the regulatory body in 1992, Mercury gained a better cost basis and could start to capture significant market share. Another example is the regulatory body in New Zealand (the Ministry of Commerce) that established very few regulations and had little enforcement power after the market opening in 1987, which resulted in years of battles between the dominant telecommunications operator Telecommunications Corporation of New Zealand (TCNZ) and new Telecom operators on interconnection between networks. The first new telecommunications operator, Clear Communications, was connected with TCNZ more than 3 years after the first negotiations for interconnection started [Yankee Group, 1997]. Technically, such a process would only need to require 3 to 6 months. In a later stage, the regulatory body was assigned more enforcement power and established more regulations in order to speed up negotiations for interconnection and access to other TCNZ infrastructures.

Regulatory bodies in various countries
Some examples of regulatory bodies in various countries are shown below [Oliver, 1996; Clifford Chance, 1997; Melody, 1997; Noam, 1997].

> *AUSTEL,* the regulatory body of Australia, formed in 1989, that regulates the telecommunications environment. On July 1, 1997, the PSTN telecommunications market in Australia was fully opened and went from a *duopoly* state to a *competition* state.
>
> *Ofkom,* the regulatory body of Switzerland, was founded in 1997 and developed regulation for interconnection of PSTN services, which opened for competition January 1, 1998.
>
> *CRTC* (Canadian Radio-Television and Telecommunications Commission), formed in 1996, chose a liberalization of national long-distance communications first. International traffic opened for competition later, in October 1998. There are special rules for traffic to and from the U.S.
>
> *FCC* (Federal Communication Commission), the regulatory body of the U.S., plays a key role in supervising competition in the U.S. For instance, it carries the responsibility for implementing competition in local PSTN services, as described in, e.g., the Telecommunications Act of 1996. Competition in long-distance PSTN services was introduced January 1, 1984 and in local PSTN service in 1996.
>
> *MF PTE* (Ministere Française de Poste, Télécommunications et Espace), which in France performs the regulatory body tasks until a regulatory body is founded. Competition in PSTN services was introduced on January 1, 1998.
>
> *OFTEL* (OFfice of TELecommunications), the regulatory body of the U.K., founded 1982. OFTEL is in principle independent from the government. Opened the U.K. for one new telecommunications operator in voice services (PSTN) in 1982, resulting in a duopoly (legal framework for competition score D) and for competition by an unlimited number of suppliers in 1994.
>
> *OPTA* (Onafhankelijke Post en Telecommunicatie Autoriteit), the regulatory body of the Netherlands, founded in 1997, is charged with implementing the higher level directions, given by the office of the Ministry or HDTP (Hoofd Directie Telecommunicatie en Post, General Directorate on Telecommunications and Postal services). It opened the market for voice (PSTN) services in July 1997 [Tempelman, 1997].

1.2.2.2.2 Categories of the Regulatory Body
Six categories of regulations have been identified for examining and quantifying the influence of the regulatory body in a country. Ideas for these categories came from practice and various literature, such as the description of the Meta Telecom Maturity model [Johnson, 1997]. The categories were chosen to be as much as possible independent from each other, but not all categories are expected to be completely independent, which should be taken into account when the scores are used in an analysis. The starting values for the scores are mentioned in the categories. The scores for each category will be added and result in a total score for the element regulatory body. The categories are numbered 1 to 6.

1. *Regulations concerning regulatory body independence and enforcement power*. This is the only regulatory category that is not made by the regulatory body, but made by the government for the regulatory body to be more effective. A regulatory body needs to be independent of all telecommunications operators, responsible for enacting telecommunications regulations that assure competition and quality of service in order to be effective. The regulatory body should have the power to enforce the rules that it establishes. Starting values were chosen as follows for question a and b:
   a. Who is the regulatory body? One of the following answers is possible:
      • The (dominant) telecommunications operator is the regulatory body: score 0 points.
      • The (dominant) telecommunications operator is privatized (independent of government) and the regulatory body is a government department: score 1 point.
      • The (dominant) telecommunications operator and the regulatory body are both independent of government: score 2 points.
   b. Does the regulatory body have *enforcement power*? One of the following answers is possible:
      • Penalties or sanctions can be imposed for violation of regulatory body regulations*: score 1 point.
      • Penalties or sanctions cannot be imposed for violation of regulatory body regulations: score 0 points.

For each of the two questions above, the scores are added to form the score of the category. This means that category 1 of the regulatory body may have scores ranging from 0 to 3 points.

2. *Regulations concerning the licensing process*. If there is a need for parties to obtain a *license* before they may offer telecommunications services in a country, the regulatory body can influence competition by influencing the process that regulates the awarding of telecommunications licenses to parties. The regulatory body can determine that the licensing process fulfills the following criteria:
   • Reasonably *short*. As a guideline we establish that the average length of the total licensing process is less than 6 months. This is the total licensing process including a public appeal process. This criterion gets 1 point if applicable and 0 points if not applicable.
   • *Open* to the public in its criteria and process (transparent). This is reflected in requirements of openness for, e.g., hearing sessions or publications in newspapers. This criterion gets 1 point if applicable and 0 points if not applicable.
   • *Nondiscriminatory*. The regulatory body treats all applicants in the same way. This criterion gets 1 point if applicable and 0 points if not applicable.

For each of the three criteria, the starting values chosen are 1 point and scores are added to get a score for this category ranging from 0 to 3. This means that, if the licensing process is reasonably short, open, and nondiscriminatory, the category scores 3 points and it scores less in other cases. Literature [Cave, 1997] suggests active use of the licensing process by the regulatory body influences competition considerably, such as timing of competition on a per service basis. This can be done by making the possibility to offer services dependent on other parties in the industry. The U.S. Telecommunications Act of 1996 allows local telecommunications operators (Regional Bell Operating Companies or RBOCs in the U.S.) to offer long-distance voice services, but only *after* the geographic market area of the RBOC has already encountered competition of new suppliers.

3. *Regulations concerning equal access to infrastructure*. With these regulations, the dominant telecommunications operator gives equal access to some of its facilities, including switches, cables, ducts, antennas to telecommunications operators (including to the service providing organization within

---

*It is noted that procedures can be lengthy before penalties can actually be imposed.

its own company*). Equal access can be provided in various areas. We use three criteria, which are each scored with starting values of 1 point if they are applicable and 0 points if not applicable:

a. Right to build a link/node infrastructure oneself. This includes the *right-of-way,* which means the right of a party (e.g., new telecommunications operator), to lay cable in areas of land that the party does not own. Score is 1 point if applicable; 0 points if not applicable.

b. Access to link/node infrastructure (equipment) of the (dominant)** telecommunications operator is mandatory and based on internationally accepted technical standard interfaces. Score is 1 point if applicable; 0 points if not applicable.

c. The dominant infrastructure service provider is obliged to accept co-location of equipment at its sites. Score is 1 point if applicable; 0 points if not applicable.

Under (a) "Right to build link/node infrastructure oneself," the criteria are used similarly for all different kinds of infrastructure, such as equipment at links/nodes and cables, but also the use of common media, such as frequency allocations for wireless and satellite applications. For fast provision of services, often wireless links are used in both networks for *mobile* operators as well as *fixed infrastructure* operators. Therefore access to frequency spectrum can be important, even when there is only competition allowed in wireline networks.***

Adding the starting values of the three criteria in this category results in scoring for this category to range from 0 to 3 points.

4. *Regulations concerning price of interconnection.* The network for infrastructure services of the dominant telecommunications operator, or the entity that owns the infrastructure that provides the direct access to the locations of the subscribers in the country, should be open for connection (also called interconnection) with other telecommunications operators. The regulations concerning interconnection are identified in this category by the conditions a, b, and c:

a. The price and conditions for access to infrastructure are cost-justified: 1 point if applicable or 0 points if not applicable.

b. The price and conditions for access to infrastructure are known to the public: 1 point if applicable or 0 points if not applicable.

c. The price and conditions for access to infrastructure are exactly the same for all competing telecommunications operators, including the competing part of the dominant telecommunications operator that uses the network infrastructure. This criterion is referred to as non-discriminatory: 1 point if applicable or 0 points if not applicable.

For each of the three regulations, the starting values chosen are 1 point and scores are added for the category, leading to a 3-point score if all regulations are present or less if some of them are not present and a total range of scores of 0 to 3 points.

---

*In order to carry out this regulation, often the dominant operator is split in terms of legal structure and accounting measures in a "service provider" and an "infrastructure operator." This way "cross subsidization" can be avoided between the "infrastructure operator," that operates in a monopoly environment as the only provider of infrastructure and the "service provider," which operates in a competitive environment.

**Essential in practice is to have access to an operator of facilities that are a bottleneck, meaning facilities that are only in possession of that particular operator *and* that are needed to realize the intended provisioning of the service.

***Wireless and wireline (or fixed) services of operators are usually defined on basis of what the end user sees. The end user with a wireless handset that communicates directly with the operator (using, e.g., GSM) considers himself on a wireless service (e.g., by a wireless operator). The end user with a wireline handset considers herself connected to wireline service of a wireline operator. The actual transmission of the signal from user A to user B can use various media to complete its path, e.g., including satellite, terrestrial wireless, fiber cable, or copper cable. Access to all infrastructure (to both wireless frequency spectrums, as well as equipment and cables of the fixed network) is essential for both kinds of operators to establish service.

The importance of these interconnection regulations is shown in various practical cases (see the BT/Mercury example shown earlier). Cost-justification (condition a) is an extensive subject that requires detailed accounting rules. Especially the allocation of fixed costs of a network (which are very high in this capital-intensive type of operation) is a detailed accounting process and is often challenged in court cases between existing telecommunications operators and new telecommunications operators. An example of such a challenge in New Zealand is listed in Section 1.2.2.2.1.

5. *Regulations concerning fair competition.* There are measures that can promote the fairness of competition, by establishing regulations aimed at particular parties. Depending on the nature of the competition, some of these measures can be used temporarily or permanently. The two kinds of regulations addressed here are *price regulations,* listed under a, and *universal service,* listed under b. Opinions vary widely on the effectiveness of price regulations [van Cuilenburg and Slaa, 1995; Cole, 1991]. We have chosen starting values that increase when fewer regulations are imposed. Universal service is a regulatory criterion that has been estimated in literature to promote fair competition as it equals the obligations for telecommunications operators that are offering services (see Section 1.2.8). The two kinds of regulations concerning fair competition are

   a. Price regulations. Either one of the following three situations may exist:
      - Prices for services are established with government approval: score is 0 points
      - Price regulations applicable (for instance minimum or maximum prices to be charged for certain services): score is 1 point
      - No price regulations: score is 2 points

   b. Universal service. Universal service means that telecommunications operators have to offer their services in all areas of a country at the same price and conditions or have to pay into a "Universal service fund" to subsidize the provision of service to certain nonprofitable customers that they do not provide service to. Universal service regulation is made because telecommunications operators would otherwise not offer service to low-revenue customers or customers in areas where there is a high cost of provisioning the service, resulting in less profit. Starting values are as follows: 0 points if Universal service does not exist, 1 point if Universal service exists in some form in a country.

A description of universal service in more formal wording is: *Universal service is the offering of a service on a non-discriminatory basis, with the same quality and price, independent of where the user is requesting the service.* Today, universal service exists in most countries for basic telephone service (public switched telephone network), but only exists in some places for other telecommunications services. As universal service often plays a role in lawsuits between telecommunications operators, we provide a more detailed example of how universal service can be implemented.

Universal service example

An example of an implementation of universal service can be found in the U.S. Universal service in the U.S. is regulated in various laws, including antidiscrimination laws, as well as the Telecom Act of 1996 in section 254, which provides for special funds, financed by all telecommunications operators for maintaining universal service for various types of services. Also, there are special programs with government subsidies that can give access to basic services. Increasingly, access to PSTN (in the law described as basic telephony) is not considered sufficient anymore in U.S. politics. Access to advanced telecommunications services, such as the Internet, is seen as another service that all people need to be able to have [Oliver, 1997]. As regulations for universal service for the access to basic telephony, several programs exist, such as the *Lifeline* and *Link-up* programs, which are actually subsidies to the users of telecommunications services. For access to advanced telecommunications services, a multitude of regulations exist, such as antidiscrimination laws, as well as specific programs to ensure universal service for certain groups in the population, such as members of schools and libraries. Figure 1.40 gives an overview of current universal service regulations in the U.S.

Subject	Applicable Regulation
Access to basic telephony	Lifeline and Link-up programs (subsidies for low-income households)
Access to advanced telecommunications services	Antidiscrimination laws
	Regulation based on Section 254 and 706 of the Telecom Act of 1996, "Universal Service in all regions"
	Regulations that arrange targeted subsidies for:
	Schools and libraries
	Rural health care providers
	Community-based organizations

FIGURE 1.40 Example: Universal service regulations in the U.S.

The scores of category 5, fair competition regulations, will be tracked both as an aggregate sum to represent the whole category, but also individually in order to be able to do more detailed statistical analysis. With the starting values chosen, the scores in the category range from 0 to 3.

    6. *Regulations concerning number portability.* This category handles regulations on number portability or subscriber identification and numbering. An important aspect of subscriber identification and numbering is *number portability,* the possibility to keep one's current phone number, domain name, or TCP/IP address while changing telecommunications operators or service providers.

Number portability exists in different forms. Meant here is to assess the *local number portability,* which concerns the possibility for subscribers to keep their own number or identification. Number portability has been shown to be important criterion for subscribers as very few users would like to give up their number or identification, in order to get a lower price or better service from a different telecommunications operator (see Sections 1.1 and 1.2.9). Number portability scores 3 points if it is implemented and 0 points if not implemented. The number portability requirement is fulfilled only when the *numbering plan* is administered by an independent body or department, in order to assure fair treatment of the new telecommunications operators that request identification numbers.

1.2.2.2.3 Total Overview of the Scoring for the Element Regulatory Body

For each of the six categories of regulatory body, we are using the starting values as mentioned in the categories. The starting values were chosen to cover all categories with about the same weight and such that a higher score is expected, according to literature [Van Cuilenburg and Slaa, 1995] and experience, to result in *more, lower priced, or faster provisioned* services. The weighting may be changed after the statistical analysis that we plan to do in case studies. The answers are scored and added per country in the rightmost column in Figure 1.41. The total score is the total of the category scores and is expected to give an indication of the total regulations of the regulatory body. This means that the higher the score, the more the regulatory body is expected to promote *good competition* as referred to earlier in this section. Each of the categories has a score ranging from 0 to 3, so the total score for the regulatory body should thus range between 0 and 18.

    When the regulatory body of multiple countries is described, the format of Figure 1.41 can be difficult as only few columns fit on a page. Therefore, the figure can also be depicted in a way, mirrored across the diagonal, resulting in a figure in condensed horizontal format as depicted in Figure 1.42. The condensed horizontal format does not allow room for the scores of each of the individual bullet items (criteria) within the categories. An exception is made for category number 5, fair competition regulations, since it is built upon two very different criteria, *price regulations* and *universal service.* Splitting the scores of category 5 makes it easier to do analysis of potential relationships between these criteria when applying the model in practice.

Regulatory Body	Country X	
	Status	Category
Categories (point scores in parentheses)	(starting value)	(starting values)
1. Independence/Enforcement power		
1a. Regulatory body independence. One of the following:		
=> dominant telecommunications operator is regulatory body (0 points)		
=> regulatory body is government department and telecommunications operator privatized (1 point)		
=> independent telecommunications operator and regulatory body (2 points)	Independent (2)	
1b. Does regulatory body have enforcement power? (Y = 1, N = 0)	Y (1)	3
2. Licensing process		
Licensing process is:		
• short (Y = 1, N = 0).	Y (1)	
• transparent (Y = 1, N = 0).	N (0)	
• nondiscriminatory (Y = 1, N = 0).	N (0)	1
3. Equal access to network infrastructure		
A new telecommunications operator is allowed to:		
• build network infrastructure (Y = 1, N = 0).	Y (1)	
• have access to network of a dominant (bottleneck) operator (Y = 1, N = 0).	Y (1)	
• require mandatory co-location from a dominant operator (Y = 1, N = 0).	Y (1)	3
4. Price of interconnection		
Access charges are:		
• cost-justified (Y = 1, N = 0)	Y (1)	
• published (Y = 1, N = 0)	Y (1)	
• nondiscriminatory (Y = 1, N = 0)	Y (1)	3
5. Fair competition		
Price regulations set the following restrictions:		
= Services pricing only with government approval (0 points)		
= Price caps (minimum/maximum prices) (1 point)		
= No price regulations (2 points)	No reg (2)	
Universal Service or fund contribution is mandatory (Y = 1, N = 0).	N (0)	2
6. Number portability		
Number portability mandatory (Y = 3, N = 0)	Y (3)	3
TOTAL Regulatory Body Score		15

FIGURE 1.41 Regulatory body scores for a sample country X (Y = yes, N = no).

	Regulatory Body							
	1. Independence/ Enforcement Power	2. Licensing Process	3. Equal Access to Infrastructure	4. Price of Interconnection	5. Fair Competition Regulations		6. Number Portability	TOTAL Regulatory Body Score
					Price Regulations	Universal Service		
Country X	3	1	3	3	2	0	3	15
Country Y	2	3	2	3	1	1	0	12
Country Z	2	3	2	2	0	0	0	9

FIGURE 1.42 Regulatory body scores in a figure in condensed horizontal format.

Competition Active	Code	Starting Value
No competitor	0	0 points
One competitor	1	1 point
Two competitors	2	2 points
More than two competitors	>2	3 points

FIGURE 1.43 Starting values for competition active.

	Competition Active						
	Value-Added Services		Basic Data Transfer Services		Infrastructure Services		Total Competition Active
	Number	Score	Number	Score	Number	Score	
Country X	> 2	3	1	2	> 2	3	8
Country Y	1	2	1	2	> 2	3	7
Country Z	> 2	3	> 2	3	1	2	8

FIGURE 1.44 Example of scores for the element competition active for countries X, Y, and Z.

1.2.2.3 Competition Active

The third element of the *regulatory environment* is *competition active*. Competition active, or the number of competitors for a telecommunications service in a country, differs from the other two elements in the sense that it is *descriptive* rather than *prescriptive*. This element is part of the regulatory environment merely in order to confirm that the two other elements actually result in competition or not. A high number of (legally operating) competitors in a country shows that the regulatory environment has created *low* barriers for entry, and is a sign of a competitive market. The number of competitors offering a service at a level in the service value model are counted, which forms the basis for the score. A competitor is only counted as such, if its market share is above 5%. The market share number turns out to be essential as it often happens that many competitors are starting in a certain market, but in some cases competition is so tough that the new entrants get almost no market share. The starting values for the scores are chosen so that the number of points is in line with the experience of expected competitiveness. *More than two competitors* denotes a competitive market and is therefore awarded 3 points with lower scores for fewer competitors. This results in the scoring as shown in Figure 1.43.

The competitors are counted for each of the three layers in the service value level and the scores for all of the three layers in the service value model is then added by country. Figure 1.44 shows an example for the scoring of the element competition active.

1.2.2.4 Overview of the Regulatory Environment

When the three elements described as part of the regulatory environment are combined, they form a total model of the regulatory environment. An example is shown with scores for a sample country X in Figure 1.45.

Each of the three elements fulfills a *role* in the regulatory environment model, which is probably best explained by viewing each of the three elements as fulfilling a role in a stage of competition active in a country. When a country opens markets of certain services for competition, usually first the *laws* are made to allow this. Then more detailed regulations are made (usually by a *regulatory body*) and finally *competitors* should start showing up that would like to take advantage of the newly created opportunities. The information needed to fill in the boxes in the regulatory environment model can be obtained from various sources. Examples are geographical area studies, such as those done by Noam [1997, 1994, 1992]; high-level overviews that can be found in handbooks (see Section 1.1) [Frieden, 1996]; or consultant reports and reports of official organizations [OECD, 1997]. The scores are added up on a per-country basis in order to arrive at a total score of the regulatory environment. In the example in Figure 1.45 the *Country X* would get the addition of all the *Country X* columns, 3 + 3 + 15 = 21 points. The resulting

Legal Framework for Competition						
Value-Added Services		Basic Data Transfer Services		Infrastructure Services		Total Legal Framework Score
Code	Score	Code	Score	Code	Score	
Country X	C	3	M	0	M	0	3
Country Y	C	3	C	3	C	3	9
Country Z	C	3	C	3	C	3	9

	Regulatory Body						
	1. Independence/ Enforcement Power	2. Openness and Length of Licensing Process	3. Equal Access to Infrastructure	4. Price of Interconnection	5. Fair Competition Regulations	6. Number Portability	TOTAL Regulatory Body Score
					Price Regulations / Universal Service		
Country X	3	1	3	3	2 / 0	3	15
Country Y	2	3	2	3	1 / 1	0	12
Country Z	2	3	2	2	0 / 0	0	9

Competition Active						
Value-Added Services		Basic Data Transfer Services		Infrastructure Services		TOTAL
Number	Score	Number	Score	Number	Score	
Country X	> 2	3	1	2	> 2	3	8
Country Y	1	2	1	2	> 2	3	7
Country Z	> 2	3	> 2	3	1	2	8

FIGURE 1.45 Example of scores in the regulatory environment model.

total score of the regulatory environment per country can also be used as input parameter for the *cost-effective management model.*

1.2.2.5 New Developments in Regulatory Environments

Several member countries of the World Trade Organization (WTO) signed an agreement in February 1997 called "General Agreements on Trade in Services Concerning Basic Telecommunications," often referred to as the WTO Basic Telecom Services Agreement [Oliver, 1997] in order to commit each other to opening their telecommunications markets for competition. Opening the markets for competition would happen according to a time schedule that was proposed by each of the countries that signed the agreement. Each country had a so-called *offer,* which includes a series of plans including a date to open the market, categorized mostly by a service value level or a similar description. The four main categories of services that are covered by the WTO Basic Telecom Services agreement are domestic public switched telephone network (PSTN), domestic long-distance PSTN, international PSTN, and satellite services. The countries that signed the WTO Basic Telecom Services Agreement represent 79% of the world economy.* Figure 1.46 shows an example of "offers" of a few countries that were committed in the WTO Basic Telecom Agreement.

---

*This is in terms of gross national product; however, in terms of world population, they only represent 19% of the population.

Situation Regarding a Service	Greece	Spain	Venezuela
Competition in domestic PSTN	In year 2003	In 12/1998	In 11/2000
Competition in international PSTN	Now	In 12/1998	In 11/2000
Possibility to own infrastructure	In year 2003	In 12/1998	In 11/2000
Allows bypass into country	In year 2003	Now	Not planned

FIGURE 1.46 Example of *offers* of a few countries in the WTO Basic Telecom Services Agreement.

As shown, most offers contain commitments regarding competition for particular services. Countries state a date in their offer by which they commit competition to be allowed for that particular service. The WTO Basic Telecom Services Agreement then states requirements for when opening each of the services actually happens. For example, when a country declares that competition exists in PSTN services, the interconnection of competitive telecommunications operators to the PSTN network has to be non-discriminatory and cost-based. There are numerous developments in the regulatory environment of many countries and both country organizations as well as supranational organizations such as WTO and European Union are developing regulations to implement in their countries or areas.

1.2.2.6   Summary

The regulatory environment has been described by modeling it in a way that determines three elements: *legal framework, regulatory body*, and *competition*. Each of the elements assesses the situation in a country and defined starting values are assigned to them to express the situation in a country in a quantitative set of scores. The combination of the scores filled in the description of the elements is called the *regulatory environment model*. There are numerous developments in the regulatory environment of many countries and both country organizations and supranational organizations such as WTO and European Union are developing regulations to implement them in their countries or other areas.

References

Cave, M., Alternative telecommunications infrastructures: their competition policy and market structure implications, OECD Competition Consumer Policy Division, 1995.

Clifford Chance, Telecommunications Regulations in the Netherlands, Internal Research, October 1997.

Van Cuilenburg, J. and P. Slaa, Competition and innovation in telecommunications, *Telecommun. Policy,* 19(8), 647–663, Nov. 1995.

Cole, B.G., *After the Breakup, Assessing the New Post-AT&T Divestiture Era,* Columbia University Press, New York 1991.

Frieden, R., *International Telecommunications Handbook,* Artech House, Boston, 1996.

Johnson, J. T., International telecom, know how to negotiate, *Data Commun.,* June 1997.

Melody, W. H., Ed., *Telecom Reform; Principles, Policies and Regulatory Practices,* Denmark Technical University, Lyngby, Denmark, 1997.

Noam, E., Ed., *Telecommunications in Western Asia and the Middle East,* Oxford University Press, New York, 1997.

Oliver, C., The Telecommunications Act of 1996, 100 A.B.A. Sec. Science and Technology, *Bull. Law Sci. Technol.,* 5, Dec. 1996.

Oliver, C., The information superhighway: trolls at the tollgate, *Federal Communications Law Journal,* December 1997.

OECD, *Communications Outlook,* Organisation for Economic Cooperation and Development, Paris, 1993, 1995, and 1997.

Tempelman, J., Dutch telecommunications, *Telecommun. Policy,* 21(8), 733–742, 1997.

Terplan, K., *Communication Networks Management,* 2nd ed., Prentice-Hall, Englewood Cliffs, NJ, 1992.

van den Broek, F., *Management of International Networks,* CRC Press, Boca Raton, FL, 1999.

Yankee Group, Regulatory Audit, Telecommunications Services Markets in Latin America, 1997.

1.2.3 The World Trade Organization Agreement on Basic Telecommunications Services and Related Regulations in the U.S.

Charles M. Oliver

1.2.3.1 Introduction[1]

On February 15, 1997, delegations from 69 countries concluded an historic series of negotiations by accepting the World Trade Organization (WTO) Agreement on Basic Telecommunications Services.[2] The Agreement was scheduled to enter into force on January 1, 1998, if all of the governments involved had ratified it by November 30, 1997.[3] In August 1997, the U.S. Federal Communications Commission (FCC or the Commission) adopted its *Benchmark Rates Order*,[4] limiting the international settlement rates that U.S. carriers will be permitted to pay foreign carriers that terminate international traffic originating in the U.S. In November 1997, the Commission adopted the *International Satellite Service Order* and the *Foreign Participation Order* liberalizing entry standards for foreign satellite and other telecommunications services.[5] These decisions completed the basic regulatory framework for Commission treatment of foreign telecommunications service providers under the Basic Telecom Agreement.

This section addresses the Basic Telecom Agreement and governmental decisions in the U.S. that fall within the framework of WTO requirements.[6] Those decisions can be grouped into three categories. First, the FCC has adopted rule-making orders that respond specifically and explicitly to WTO commitments. Second, U.S. courts and regulatory authorities are implicitly addressing WTO commitments in decisions implementing the broader, pro-competitive requirements of the Telecommunications Act of 1996 (the '96 Act).[7] Because the Basic Telecom Agreement was modeled, in large part, upon the principles embodied in the '96 Act, adoption of the '96 Act can be regarded as a form of implicit implementation before the agreement was signed. A corollary is that, if administrative decisions impede implementation of the '96 Act or if courts discover limitations in the '96 Act that were not recognized by U.S. negotiators when they signed the Basic Telecom Agreement, other countries might have a basis for arguing that the U.S. is failing to live up to its international commitments.

It is important to recognize, however, that the '96 Act is only one important step in a process that has been ongoing for a quarter-century, as the U.S. opened successive market segments to increased competition. The Basic Telecom Agreement was inspired in significant part by pro-competitive regulatory initiatives that preceded the '96 Act and, to some extent, continue to develop independently of the '96 Act's requirements. The marketplace consequences of the '96 Act were only beginning to develop in February 1997, when the Basic Telecom Agreement was signed, but the benefits of competition engendered by earlier regulatory decisions were readily apparent. This section addresses those developments as well.

For WTO member countries, the Commission eliminated existing requirements that foreign carriers seeking entry into the U.S. demonstrate that their home markets afford opportunities for entry similar to those permitted in the U.S. The so-called Effective Competitive Opportunities (ECO) test continues to apply to non-WTO member countries. While generally loosening restrictions on foreign entry, the Commission adopted several regulatory measures designed to prevent anticompetitive conduct by foreign entities and their affiliates.

The Commission takes the position that WTO obligations both permit and require participating governments to prevent anticompetitive conduct in ways that do not discriminate against companies on the basis of foreign ownership. The agency compared its foreign entry requirements to safeguards it has applied to domestic Incumbent Local Exchange Carriers (ILECs) when they enter competitive telecommunications markets like interexchange service, mobile telephony, and enhanced services.

1.2.3.2 U.S. Regulation of Foreign Carrier Entry before the WTO Basic Telecom Agreement

For many years, the FCC engaged in *ad hoc* reviews of foreign entry applications. The *Foreign Carrier Entry Order*, adopted in November 1995, provided a more structured framework and reflected an inten-

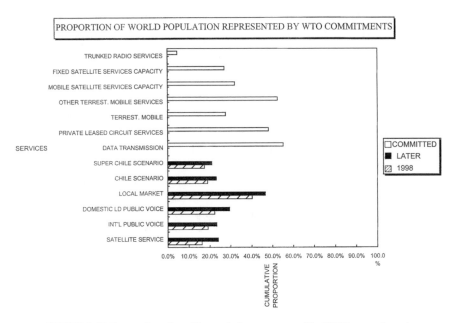

FIGURE 1.47 Proportion of world population represented by WTO commitments.

sified determination to pry open foreign markets.[8] The rules adopted in that order dealt both with applications for facilities authorizations under Section 214 of the Communications Act of 1934 (the Act) and with applications for common-carrier radio licenses under Title III of the Act. The Commission applied an ECO test to applications for international facilities-based, switched resale, and non-intercon- nected private line resale under Section 214 in circumstances where an applicant sought authority to provide the service between the U.S. and a destination market in which an affiliated foreign carrier had market power. In general, for purposes of applying the ECO test under Section 214, the FCC considered an applicant to be affiliated with a foreign carrier when the foreign carrier owned more than 25% of the applicant or controlled the applicant by other means. In the Title III context, the Commission applied the ECO test to common-carrier radio applicants or licensees that owned the applicant or sought to expand their ownership interest in the applicant beyond 25%. The ECO test looked at the *de jure* ability and the practical ability of U.S. carriers to enter the home market of the foreign carrier.

In May 1996, the Commission proposed to adopt rules applying the ECO test to non-U.S. licensed satellite operators seeking to serve the U.S.[9] but placed that proceeding on hold pending completion of the WTO basic telecommunications negotiations.

1.2.3.3 World Trade Organization Agreement on Basic Telecommunications Services

When the WTO was created in 1994, the U.S. and other WTO members committed to allow market access for a broad range of services, including value-added telecommunications services. The members agreed to extend the negotiations for a limited number of service sectors, including basic telecommunications.

The *Foreign Participation Order* states that, under the Basic Telecom Agreement, 44 WTO members representing 99% of WTO members' total basic telecommunications service revenues will permit foreign ownership or control of all telecommunications services and facilities, but that is a highly simplified characterization of the Agreement.

The accompanying charts illustrate the scope of the Basic Telecom Agreement, along with its limita- tions. As shown in Figure 1.47, national delegations representing 82% of world telecommunications revenues have committed to open four of the biggest bottleneck monopolies — local public telephone services, domestic long-distance voice services, international public voice services, and satellite services — by 1998. The same nations represent 79% of the world economy (cumulative GNPs). The WTO

	Satellite Service	International Public Voice	Domestic LD Public Voice	Local Market	Chile Scenario	Super-Chile Scenario
1998	16.1%	19.0%	22.3%	40.1%	18.9%	17.4%
Later	24.0%	23.3%	29.4%	46.4%	23.1%	21.0%

	Data Transmission	Private Leased Circuit Services	Terresrial Mobile	Other Terrestrial Mobile Services	Mobile Satelite Services	Fixed Satellite Services Capacity	Trunked Radio Services
Committed	55.1%	48.0%	27.6%	52.4%	32.0%	27.1%	4.6%

FIGURE 1.48 Population represented by WTO tel commitments.

commitments are far more modest when characterized as a percentage of world population. Figure 1.48 shows that only 19% of the world's population is represented by countries that have agreed to open all four major bottleneck monopoly services by 1998. That figure barely rises in subsequent years, to 23%.

1.2.3.3.1 *Substance of Promises Made in the Basic Telecom Agreement*

The WTO negotiations revolved around a generic statement of commitments, referred to as the "Reference Paper," which most countries folded into their national schedules of specific commitments,[10] and two official notes by the WTO Chairman interpreting the Reference Paper.[11] Schedules representing all but four of the governments involved in the Agreement include the Reference Paper with few, if any, modifications, but nearly all of those schedules contain separate lists of limitations defining where and how a government will not conform to the Reference Paper. The Reference Paper supplied a default set of options from which countries were free to depart during the negotiating process, provided that they defined the exceptions.

The first chairman's note was prepared in response to concerns expressed by satellite service and other suppliers. It stipulates that, unless otherwise noted, any basic telecommunications service listed in a government's schedule of commitments may be provided through any means of technology, including cable, wireless, or satellites. The second chairman's note responds to requests for clarification from WTO member governments by stating that any market-opening commitments are implicitly subject to the availability of radio spectrum.

The Reference Paper reads like a capsule summary of the U.S. Telecommunications Act of 1996. Its most far-reaching paragraph provides that interconnection with a major supplier will be ensured "at any technically feasible point in the network," under nondiscriminatory terms and conditions, in a timely fashion under terms, conditions (including technical standards), and cost-oriented rates that are transparent, reasonable, and sufficiently unbundled that the competitive entrant will not need to pay for network components or facilities that it does not require. In the U.S., similar language is interpreted as requiring incumbent local exchange telephone companies to provide unbundled local loops at cost-based rates and to allow interconnection with the competitive supplier's network at the telephone company's switching office.

If the U.S. experience is a guide, ferocious battles will be fought over the prices that telephone companies will be allowed to charge for access to their local loops. The significance and complexity of this issue is being heightened by the advent of digital subscriber line (DSL) equipment, which is capable of pumping data at very high speeds over local loops, at rates of 6 megabits per second or more over distances up to 10,000 meters. Depending upon implementation, the confluence of DSL technologies with unbundled local loops could lead to a significant expansion in the market for fiber-delivered data services and for international satellite services, delivered via earth stations situated at or near telephone company switching offices.

The General Agreement on Trade in Services provides that each member must promptly and at least annually inform the WTO Council for Trade in Services of the introduction of any new, or any changes to existing laws, regulations, or administrative guidelines which significantly affect trade in services covered by its specific commitments rendered in the WTO framework.[12] Each member is also required to establish one or more inquiry points to provide specific information to other members, upon request, on all relevant matters.[13]

1.2.3.3.2 *Significance of Promises Made in the Basic Telecom Agreement*

Over the past quarter century, the U.S. has introduced competition to successive layers of its telecommunications infrastructure, the most recent major step being the congressional decision in 1996 to open urban telephone markets to local competition. This process of progressive liberalization has brought significant benefits to consumers, but it has come slowly, and it has revealed that, if some, but not all, bottlenecks in the transmission path are cleared of monopoly control, whoever controls the remaining bottleneck will discourage competition in the hope of collecting monopoly rents. To borrow an analogy that was once applied to Gorbachev's way of reforming the Soviet Union, the U.S. has made a gradual transition from driving on the left-hand side of the road to driving on the right-hand side of the road. The transition has been difficult and painful, and may not have worked at all if regulatory traffic cops had not been present to micromanage the process.

At least one other country, a small one, has moved more rapidly toward competition. Chile's transition has been both quick and successful. In that country, international rates fell by 46% from 1989 to 1994 as a result of limited competition, and domestic long-distance rates fell by 38% over the same period. The government then took liberalization a step further by authorizing several companies to lease satellite capacity directly from INTELSAT and acquire and operate their own earth stations. And, in October 1994, a new law went into effect enabling end users to access the long-distance telephone carriers of their choice. Prices plummeted still further and, by December 1994, the volume of international calls to and from Chile increased by 35% from the pre-October level. By June 1995, prices for international calls had stabilized at a level that was approximately 60% lower than they had been before October 1994, and even further below the prices that had prevailed in 1989.[14]

Prices fell and traffic volume rose in Chile because that country cleared several of the bottlenecks that had previously constrained supply, including both domestic long-distance and international service monopolies. Yet Chile still has not committed itself to allow competition in the provision of local exchange services. Figures 1.47 and 1.48 portray both a Chile scenario and a "Super-Chile" scenario. Under the WTO Basic Telecom Agreement, countries representing 82% of world telecommunications revenues and 79% of the world economy have committed themselves to implement the Super-Chile scenario by 1998.

1.2.3.3.3 *Enforcement of the Basic Telecom Agreement*

The WTO Basic Telecom Agreement is meaningful only to the extent that it can be effectively enforced. Enforcement mechanisms will come into play at two and, in some cases, three levels. The least contentious scenario would involve a national government that is voluntarily enforcing commitments under the Agreement in a way that does not antagonize any other participating countries. The Reference Paper provides that the committing government will maintain or establish a regulatory body that is separate from, and not accountable to, any supplier of basic telecommunications services and that its decisions and procedures will be impartial with respect to all market participants. The regulator must make publicly available all licensing criteria and the period of time normally required to reach a decision concerning an application for a license. Service suppliers requesting interconnection with a major supplier will have recourse to an independent domestic body for resolution of disputes within a reasonable period of time.

Such language is, of course, subject to a wide range of interpretations. The real meaning of the Reference Paper will ultimately be defined in the WTO dispute resolution process, though for some countries there could be an intermediate step: the European Council has indicated that it is prepared to prosecute non-complying countries through its own processes. For purposes of enforcing a plurilateral agreement like the Agreement on Basic Telecommunications Services, the WTO will establish a Dispute Settlement Body (DSB) consisting of only those members that are parties to that agreement, i.e., only those members that are parties to the telecom agreement will be empowered to participate in decisions or actions taken by the DSB with respect to that agreement.[15]

WTO litigation will proceed as follows: if consultations between the disagreeing national governments fail to settle a dispute within 60 days after the date of receipt of a request for consultations, the complaining party may request the establishment of a WTO panel.[16] Panels will usually consist of three persons unless the parties to the dispute mutually agree to a five-person panel.[17] To assist in the selection of panelists, the WTO Secretariat will maintain an indicative list of governmental and non-governmental individuals possessing requisite qualifications.[18] The Secretariat will propose nominations for the panel to the parties to the dispute; if the parties do not agree on the panelists within 20 days, the WTO Director-General, in consultation with the Chairman of the DSB and the chairman of the relevant DSB committee, will determine the composition of the panel.[19] Panelists are supposed to serve in their individual capacities and not as representatives of governments or any other organization.[20] The panel will render a written decision after oral and written presentations by the parties.

Panel decisions can be appealed to a three-person tribunal drawn from a standing Appellate Body established by the DSB, consisting of seven persons. The DSB will appoint persons to serve on the Appellate Body for 4-year terms.[21] Appellate tribunals will limit their decisions to issues of law covered in lower panel reports.[22] For all practical purposes, it appears that appellate tribunal decisions will be the end of the line, because their reports "shall be adopted by the DSB and unconditionally accepted by the parties to the dispute unless the DSB decides by consensus not to adopt the Appellate Body report."[23] This process represents a significant strengthening of authority compared with the situation prior to the formation of the WTO in 1995, when decisions under the General Agreement on Tariffs and Trade did not become legally effective until a positive consensus was achieved among signatories. DSB consensus to override the Appellate Body will presumably be unusual, since the winning party in every dispute will be a member of the DSB.

The WTO Agreement provides that losers under its dispute settlement process shall ensure that their laws and regulations conform to the decision.[24] So far, litigants have generally been willing to comply at this point. Those that do not can be required to adopt alternative compensatory measures satisfactory to the complainant or face suspension of the complainant's reciprocal obligations to the defendant under the Agreement on Basic Telecommunications Services.[25]

The WTO dispute resolution process will not be freely accessible to any company that considers itself an injured party. Under U.S. law, no person other than the U.S. government itself will have any cause of action or defense before the WTO.[26] This is consistent with the WTO Dispute Annex, which itself provides an opportunity for action or defense only by WTO members, i.e., governments. Similar provisions are found in the laws of other nations. The implication is that national governments or other governmental members of the WTO will serve as gatekeepers and will exercise their authority to choose which cases to litigate.[27]

The U.S. has been the most litigious member of the WTO, and that is not likely to change with the entry into force of the Basic Telecom Agreement. The U.S. is the largest telecommunications market and is home to several large companies with global aspirations. Despite that, USTR officials have indicated that they will be highly selective in pursuing telecom grievances, because they want to present the WTO with cases that the U.S. will win and will establish significant precedents. Under the process that USTR envisions, a fully articulated interpretation of the Basic Telecom Agreement will emerge slowly through many years of litigation.

As discussed below, the FCC will seek to avoid committing any fouls under WTO requirements, but it is not willing to place the nation's destiny entirely in the hands of the WTO. The Commission has preserved its ability to wield a variety of mechanisms that could make life difficult for countries that fail to honor their WTO commitments or honor them in a half-hearted way.

1.2.3.4   Explicit Implementation of the Basic Telecom Agreement by the U.S. Federal Communications Commission

1.2.3.4.1   The FCC Stated and Unstated Goals

If the FCC discussion of goals and purposes in its international orders were taken at face value, one would infer that encouraging foreign governments to open their markets has now receded to a tertiary status on the agency's international agenda. The Commission has stated that the primary purpose of its new rules is to promote effective competition in the U.S. telecommunications services market by inviting foreign entrants in, and the secondary purpose is to prevent anticompetitive conduct in the provision of international services or facilities.

In fact, however, the Commission continues to be intensely interested in the openness of foreign markets open for U.S.-based companies, but the Basic Telecom Agreement now precludes the U.S. government from overtly applying ECO-like reciprocity tests outside of the WTO dispute resolution process. The WTO does allow governments to prevent anticompetitive conduct, however, and the Commission has made clear that it will continue to exercise its powers under that new banner. If defending the American consumer at home happens to further the strategic interests of American companies abroad, the FCC will not be discomfited.

1.2.3.4.2   Open Entry Policies toward WTO Member Countries

For applicants from WTO member countries seeking to enter the U.S. market, the *Foreign Participation Order* eliminates the requirement of demonstrating that the foreign markets involved offer effective competitive opportunities and replaces it with a rebuttable presumption in favor of approval. In those circumstances, the ECO test will no longer be applied to applications for Section 214 service authorizations, cable landing licenses, or permission to increase foreign indirect ownership of non-broadcast radio licensees above 25%. The *International Satellite Service Order* applies a similar presumption in favor of approval for applicants seeking authorization to provide domestic or international telecommunications service to the U.S. through satellites licensed by WTO member countries. The presumption in favor of entry will apply both to private companies and to affiliates of intergovernmental satellite organizations (IGOs) licensed by WTO members. The ECO test will continue to apply to non-WTO countries, and it will be applied to services not covered by the Basic Telecom Agreement — direct-to-home (DTH), direct broadcast satellite, and digital audio radio services.

The Commission also addressed the unique circumstances of COMSAT, the U.S. signatory to INTEL-SAT and Inmarsat and the exclusive provider of services through those entities in the U.S. As IGOs that are not members of the WTO, INTELSAT and Inmarsat do not have any direct rights under the Basic Telecom Agreement. As part of its market-opening efforts, however, the Commission decided that COM-SAT should be permitted to obtain authorization to provide U.S. domestic service via INTELSAT or Inmarsat satellites, if COMSAT waives any IGO-derived immunity from suit and demonstrates that the services proposed will enhance competition in the U.S. market. The Commission did not discuss how it would respond if COMSAT were to propose beaming DTH signals from INTELSAT satellites into the U.S. market. In the past, the U.S. executive branch has unsuccessfully opposed provision of DTH services by INTELSAT to other countries on the ground that such services are *ultra vires* with respect to the INTELSAT Agreement. An INTELSAT DTH proposal directed toward the U.S. would probably raise both eyebrows and blood pressure levels in Washington, even though the FCC has not specifically ruled out an authorization of that kind.

The Commission revised a special rule that had applied to carriers seeking to connect international private lines to the public-switched network and provide services to the public. Previously, the agency had required carriers seeking to provide such service to demonstrate that the foreign country on the other end of the private line allows resale opportunities equivalent to those permitted in the U.S. The Commission said that it will no longer require equivalency demonstrations in such applications involving WTO member countries, if at least 50% of the U.S.-billed traffic on the routes in question are at or below the relevant benchmarks adopted in the *Benchmark Rates Order*. Otherwise, equivalency demonstrations will continue to be required before the agency will grant authorization to interconnect international private lines to public switched telephone networks.

1.2.3.4.3 The Benchmark Rates Order

The *Benchmark Rates Order* limits the amounts that U.S. carriers will be permitted to pay foreign carriers that terminate international traffic originating in the U.S. The timing of the decision, 3 months before rules were adopted opening the U.S. market to foreign competitors, provides some insight into the Commission's sense of priorities. The prospect of increased foreign entry strengthened the Commission's determination to deter foreign carriers from exploiting monopoly advantages, even though other countries were challenging the agency's proposals as a form of extraterritorial regulation.

The FCC asserted that its action was necessary because the settlement rates that U.S. carriers have been paying to foreign carriers are, in most cases, substantially above the costs foreign carriers incur to terminate that traffic. The Commission said that this long-standing concern had become more pressing as it moved toward implementation of the market-opening commitments in the Basic Telecom Agreement. If a foreign carrier were to enter the U.S. market for the purpose of originating calls bound toward its home market, inflated settlement payments made to its home-based affiliate would represent not merely an internal transfer of funds for the foreign carrier, but would represent a real out-of-pocket cost to unaffiliated carriers, thereby placing them at a competitive disadvantage.

If it is successfully implemented, the *Benchmark Rates Order* will reduce the rates paid by American consumers when dialing foreign points and ameliorate the U.S. balance of payments deficit. Policy makers have been wringing their hands over this issue for many years, however, and it is more than coincidental that they decided to do something about it not long after the Basic Telecom Agreement was adopted. If foreign carriers gained the right to enter the largest telecommunications market while retaining the ability to collect monopoly rents from American competitors, U.S. telecommunications carriers would be strategically disadvantaged.

The first target date for U.S. carriers to negotiate rates at or below the settlement rate benchmarks in the *Benchmark Rates Order* was on January 1, 1999, for carriers in upper income countries ($0.15 per minute). Subsequent deadlines at 1 year intervals will apply to carriers in upper middle income countries ($0.19 per minute), followed in successive years by lower middle income countries (also at $0.19 per minute), low income countries ($0.23 per minute), and countries with fewer than one telephone line per 100 inhabitants ($0.23 per minute). These deadlines will be accelerated, however, for carriers seeking authorizations to provide facilities-based switched or private line service to foreign affiliates: such authorizations will be conditioned upon the foreign carrier offering U.S.-licensed international carriers a settlement rate at or below the relevant benchmark adopted in the *Benchmark Rates Order*.

The *Benchmark Rates Order* generated heated criticism from other countries, but in August 1998, the FCC reported that carriers in 21 countries representing 30% of total U.S. net settlement minutes were conducting settlements at or below the relevant FCC benchmark, and that carriers in another 12 countries representing 22% of U.S. settlement minutes had negotiated agreements with U.S. carriers that would bring them into compliance.

1.2.3.4.4 Other Safeguards

Other international safeguards appear to represent either a relaxation of past constraints or an application of separation requirements comparable to those applied to domestic carriers. In the past, the Commission has generally prohibited U.S. carriers from entering into exclusive arrangements with foreign carriers. The *Foreign Participation Order* narrows the "No Special Concessions" rule so that it only prohibits exclusive arrangements with foreign carriers that possess sufficient market power on the foreign end of a U.S. international route to affect competition adversely in the U.S. international services market. The Commission adopted a rebuttable presumption that a carrier with less than a 50% market share in each relevant foreign market lacks such market power.

The *Foreign Participation Order* modified the Commission's tariffing requirement to remove a 14-day advance notice requirement and accept international tariff filings on 1 day's advance notice with a presumption of lawfulness. The Commission adopted a limited structural separation requirement and required foreign-affiliated dominant carriers to file regular reports on traffic, revenue, provisioning, maintenance, and circuit status. Significantly, however, the agency declined to adopt proposals it had floated in the notice stage of the proceeding to limit the ability of U.S. carriers to enter into exclusive arrangements with their foreign affiliates for the joint marketing of basic telecommunications services, the steering of customers by the foreign affiliate to the U.S. carrier, or the use of foreign market telephone customer information. (Approval by U.S. customers is required before using information about them obtained from foreign affiliates.)

The Commission reserved its authority to apply special safeguards on the basis of *ad hoc* determinations. The agency said it retains authority to bar market entry in highly unusual circumstances, where it is apparent that participation by a foreign entity will adversely affect competition in the U.S. market or where the Executive Branch of the U.S. government expresses concern over national security issues.

1.2.3.4.5 *Implications of the FCCs WTO Implementation Decisions*

By adopting the *International Satellite Service Order* and the *Foreign Participation Order*, the FCC has signaled its willingness to deliver on U.S. commitments under the Basic Telecom Agreement. It appears likely that all or nearly all applications from WTO member countries for permission to enter or invest in the U.S. telecommunications market will be granted. This represents an opportunity for foreign companies, for American companies seeking capital and strategic partners, and for American consumers, who will benefit from intensified competition.

The controversies that survive will center around what conditions are attached to foreign entry applications. It is inevitable that at least some foreign entities will argue that they are victims of xenophobia or nationalistic protectionism, though the FCC will argue that it has a long-standing history of applying similar safeguards to dominant domestic carriers.

The *Benchmark Rates Order* could be a continuing source of controversy, because other countries, rightly or wrongly, tend to see it as a form of extraterritorial regulation. For many of them, the first impulse will be to attack it head-on as another expression of American hegemony. Those attacks will probably be unsuccessful, if only because the Administrative Procedures Act does not recognize hegemony as a basis for reversible error. In the American court system, the *Benchmark Rates Order* is more vulnerable on other grounds under domestic law. It is also possible that competitive and other pressures will resolve the controversy by means other than litigation: as indicated above, many non-U.S. carriers have already brought their settlement rates into compliance well ahead of the deadlines set forth in the order.

1.2.3.5 Telecommunications Regulation in the U.S. under the Telecommunications Act of 1996

A leader of the Massachusetts Bay Colony proclaimed that the Pilgrims had come to America not just to make a better life for themselves, but to build a "City on a Hill, a new Jerusalem," that would serve as a light and an example for all mankind. The original idea was that American influence would be expressed not by military or economic power, but by example. The *Benchmark Rates Order* falls outside of that tradition, but the '96 Act is squarely within it. Along with European Community regulatory changes that had already been targeted for implementation in 1998, the '96 Act served as an important model for commitments made under the Basic Telecom Agreement, which was adopted voluntarily by 69 nations.

Continuing implementation of the '96 Act by federal and state regulatory agencies and the courts is being closely watched by other countries, though that process provides, at best, a faltering guide. Other countries should carefully study the '96 Act and its implementation, both to draw inspiration from its lofty goals and to avoid the kinds of quagmires discovered (and in some cases commended) by American lawyers.

1.2.3.5.1 *The Status Quo Ante — Before the Telecommunications Act of 1996*

After starting with the FCC's *Carterfone* decision in 1968,[28] competition in the U.S. telecommunications sector by 1996 had spread to include customer premises equipment, value-added resale services, and long-distance services, including satellite. The last bastion of monopoly was the local telephone exchange, and it was heavily regulated. Prices in local exchanges, including so-called exchange access charges applied to long-distance carriers for the privilege of traversing local exchanges, were heavily politicized.

The core of the Telecommunications Act of 1996 is a *quid pro quo:* the RBOCs will be allowed to get into the long-distance[29] and manufacturing businesses,[30] in return for which they must open their markets to local competition.[31] The titanic lobbying struggle that preceded this legislation never questioned the basic terms of that bargain. The battle was fought over the terms and conditions that would be imposed on the BOCs as a condition precedent to their liberation.[32]

The '96 Act provides only part of the answer. The rest of the answer is being decided in the courts. On November 25, 1997, the FCC Office of General Counsel listed 112 pending court cases involving challenges to state government decisions affecting interconnection agreements.[33]

1.2.3.5.2 *Local Competition Requirements*

Procedural complexities aside, the open competition requirements contained in the legislation bear a remarkable resemblance to wish lists that competitive access providers have been circulating for several years:

- LECs are directed to unbundle their networks and allow interconnection at any technically feasible point, with competitive providers allowed to pick and choose what portions of those networks they will use, paying just, reasonable, and nondiscriminatory prices for use of those selected piece parts.[34] For BOCs, this means, at a minimum, that customers and competitors must be able to obtain on a separate stand-alone basis local loop transmission from the customer's premises to the nearest telephone switching office; local switching unbundled from transmission services; or trunk lines running between telephone company offices, unbundled from switching[35] The BOCs are also required to provide nondiscriminatory access to databases and associated signaling necessary for call routing and completion.[36]
- Competitive providers are given the right to locate their equipment on the premises of incumbent LECs, unless the LEC can demonstrate that doing so would be impractical.[37]
- The LECs are required to offer for resale at wholesale rates any telecommunications service that the carrier provides at retail to subscribers who are not telecommunications carriers, with wholesale rates excluding the portion of retail rates attributable to any marketing, billing, collection, or other costs that will be avoided by the LEC.[38] State commissions may, however, restrict resale by categories, to avoid anomalies like resale of circuits bought at residential rates to business customers.[39]
- The LEC is required to provide reasonable notice of changes in the information necessary for the transmission and routing of services using the LEC facilities.[40]

These duties will apply to all local exchange carriers, except the carriers with less than 2% of U.S. access lines and rural LECs may qualify for exemptions. Rural telephone companies are automatically exempt until they receive bona fide requests for interconnection and the relevant state commissions determine that complying with the requests would not be unduly economically burdensome. Any carrier other than the BOCs, GTE, and Sprint may petition state commissions to suspend application of the requirements on grounds of economic or technical infeasibility.[41]

The following requirements will apply to all LECs, including rural LECs, except that companies other than the BOCs, GTE, and Sprint may petition state commissions to suspend the requirements:

- Number portability — the duty to provide, to the extent technically feasible, the ability to switch carriers without changing telephone numbers. The BOCs are required to provide interim number portability through remote call forwarding and comparable arrangements.[42]
- Dialing parity — the ability to have nondiscriminatory access to telephone numbers, operator services, directory assistance, and directory listings, with no unreasonable dialing delays.[43]
- Access to rights-of-way — the duty to afford access to poles, ducts, conduits, and rights-of-way to competing providers, on the basis of specified rate structures.[44]

- Reciprocal compensation — the duty to establish reciprocal compensation arrangements with competing local carriers for the transport and termination of telecommunications.[45]

Finally, all telecommunications carriers without exception are required to interconnect directly or indirectly with the facilities and equipment of other carriers, and they are enjoined not to install network features, functions, or capabilities that do not comply with industry guidelines and standards.[46] Telecommunications carriers are defined as providers of the means of transmission, between or among points specified by the user, of information of the user's choosing, without change in the form or content of the information sent and received (i.e., enhanced service providers are not subject to these requirements).[47]

1.2.3.5.3   Universal Service Fund Provisions of the '96 Act
The authors of the '96 Act believed that increased competition would lead to service innovations and improvements in urban areas, but powerful senators from states with low population densities correctly perceived that competition would also erode the ILECs ability to sustain hidden subsidies for service to rural areas. Congress responded by setting in motion a process to ensure that service to high-cost areas would continue to receive subsidies in the newly competitive environment. It also expanded the scope of services being funded.

The provision of subsidies is not, in itself, a violation of the WTO Basic Telecom Agreement. Indeed, the Agreement specifically provides that "[a]ny Member has the right to define the kind of universal service obligation it wishes to maintain. Such obligations will not be regarded as anti-competitive *per se*, provided they are administered in a transparent, non-discriminatory and competitively neutral manner and are not more burdensome than necessary for the kind of universal service defined by the Member."[48]

1.2.3.5.4   FCC Implementation Proceedings
The FCC conducted a trilogy of major rule-makings to implement the major telecommunicans provisions of the '96 Act — dealing with local competition, exchange access charges, and universal service — plus many smaller proceedings.

The local competition proceeding required the BOCs to make parts of their network available separately, on an "unbundled" basis, and required them to price the unbundled elements on a forward-looking incremental cost basis. The pricing aspects of the decision embroiled the Commission in litigation leading to a review by the U.S. Supreme Court, centered around the question of whether the FCC or the states have authority over the prices established for unbundled network elements.[49]

In the universal service proceeding, the FCC established new subsidies for schools, libraries, and rural health services and maintained existing subsidies for low income individuals and for rural telephone companies.[50] The Commission decided to continue studying proposed cost models for service to rural areas, leaving observers to wonder which end of the $5- to $14-billion-per-year range of estimates for that purpose was most likely to prevail. The agency decided that it would generate the required funds by collecting a percentage of gross retail revenues from most telecommunications providers. The providers were allowed flexibility to decide on their own how to pass the levy through to end users.[51]

In May 1997, the FCC released its first report and order in the access charge proceeding.[52] Interexchange carriers were disappointed to learn that the Commission had not adopted a flash-cut reduction in access charges consistent with the forward-looking cost models that the agency had been promoting in its local competition and universal service proceedings. The agency chose instead to accelerate the annual rate at which ILECs will be required to reduce their access charges. Beyond that, the Commission indicated that it would rely on the availability of unbundled interconnection as an alternative to drive access prices downward.

1.2.3.6   Regulation of Information Service Providers before and after the '96 Act
As noted above, the WTO adopted a framework for provision of value-added network services in 1994, as an annex to the General Agreement on Trade in Services.[53] The U.S. was already in compliance but has continually revisited its regulatory regime for data services. In 1980, the Federal Communications Commission (FCC) established a definitional boundary between two kinds of telecommunications ser-

vices: basic and enhanced. It defined basic service as the common-carrier offering of telecommunications services in which information received is identical to the information transmitted by the customer. Enhanced services were defined as services offered over common-carrier transmission facilities which employ computer processing applications that act on the format, content, code, protocol, or similar aspects of subscriber-provided information to produce additional, different, or restructured information, or involve subscriber interaction with stored information.[54]

Subsequent regulatory decisions by the FCC have been premised on the assumption that enhanced services can be provided on a competitive and essentially unregulated basis, but that the underlying basic services are obtainable only from facilities-based carriers that have monopoly control over key transmission pathways. The Commission assumes that, when monopoly basic service providers are themselves involved in the provision of enhanced services, they have incentives to discriminate against competing enhanced service providers and, therefore, should be subject to regulatory safeguards.

In the late 1980s and early 1990s, the FCC established rules to protect competing enhanced service providers in two kinds of situations:

- *Comparably Efficient Interconnection (CEI)* rules protect enhanced service providers from discrimination by basic carriers that provide identical, or nearly identical, enhanced services.
- *Open Network Architecture (ONA)* rules address the needs of companies providing enhanced services that are not similar to those being offered by the basic carrier, and thus may require that basic services be offered in new configurations.

1.2.3.6.1 Comparably Efficient Interconnection Requirements

The CEI requirements are the foundation layer of the Commission's enhanced service rules, both in the sense that they were the first to be applied and in the sense that they address the most obvious cases of discrimination — those in which large, monopoly local exchange carriers provide enhanced services over their own basic service facilities, and then discriminate against competitors seeking to offer similar enhanced services. In their CEI plans, BOCs were required to describe: (1) the enhanced service or services to be offered, (2) how the underlying basic services would be made available for use by competing enhanced service providers (ESPs), and (3) how the BOC would comply with the other nonstructural safeguards *Computer III* imposed. Such other safeguards governed: timely disclosure to competing ESPs of network information, including technical interfaces; access to and use of customer proprietary network information (CPNI); and quarterly reporting to help ensure that BOC provision of basic services to competing ESPs was nondiscriminatory in terms of quality, installation, and maintenance.[55]

1.2.3.6.2 Open Network Architecture Rules

The Open Network Architecture rules go beyond the CEI rules by requiring the BOCs and GTE to submit comprehensive plans, updated annually, showing how they will deploy new services and network functionalities requested by enhanced service providers. The ONA rules also give ESPs the right to request a new ONA basic service and receive a response from the BOC within 120 days either offering the service or explaining in specific terms why it is declining to do so. The ESP may then file a complaint with the FCC it finds the response unsatisfactory. Once the Commission approved a carrier's ONA plan, the carrier was to be permitted to provide integrated enhanced services without prior Commission approval of service-specific CEI plans. Court decisions prior to the adoption of the '96 Act prevented the Commission from lifting the requirement to file CEI plans, but the '96 Act will probably provide the Commission with sufficient legal authority to move forward with its original plan. In the meantime, the BOCs and GTE are required to file both CEI plans and ONA plans.

1.2.3.6.3 Impact of the '96 Act on Regulation of Information Service Providers

The '96 Act introduces a new terminology. Instead of "basic" and "enhanced" services, it refers to "telecommunications" and "information services."[56] The statute defines information service as "the offering of a capability for generating, acquiring, storing, transforming, processing, retrieving, utilizing, or making available information via telecommunications, and includes electronic publishing, but does not include any use of any such capability for the management, control, or operation of a telecommunications system or the management of a telecommunications service."[57] "Telecommunications" is defined as "the transmission, between or among points specified by the user, of information of the user's choosing, without change in the form or content of the information as sent and received."[58] The Commission has concluded that the '96 Act's definition of information service includes all services that the agency had previously classified as enhanced services under its Computer II and Computer III rules, but also includes some other services as well.[59] Enhanced services are defined in the Commission's rules as "services, offered over common carrier transmission facilities used in interstate communications, which employ computer processing applications that act on the format, content, code, protocol or similar aspects of the subscriber's transmitted information; provide the subscriber additional, different, or restructured information; or involve subscriber interaction with stored information."[60] Information services could also include value-added communications transmitted over the facilities of entities other than interstate common carriers.

Before the '96 Act was adopted, an antitrust decree restricted the BOCs from providing long-distance services between metropolitan areas (interLATA service). After satisfying certain conditions specified in the '96 Act, the BOCs will be permitted to provide interLATA services, including interLATA information services. However, the '96 Act requires that any BOC interLATA information services be offered only through fully separated subsidiaries at least until February 8, 2000 — 4 years after the Act's adoption date. For electronic publishing, the separated subsidiary requirement applies to intraLATA as well as interLATA services. The FCC may choose to extend the separated subsidiary requirement beyond 2000, but it will have discretion to eliminate the requirement after the sunset date. At the beginning of 1998, the FCC continued to apply its Computer II and III rules but was reviewing proposals to modify them in light of other, potentially redundant requirements under the '96 Act.

The boundary between basic and enhanced services has been the subject of titanic regulatory struggles in the past, because it defined the limits of utility-style regulation. The boundary between telecommunications and information services promises to be the scene of even more ferocious battles in the future because, under the '96 Act, information services are exempt from contributing to universal service subsidies. With the advent of Internet telephony, classification as an information service provider can also provide a means to avoid paying exchange access charges or international settlements. These considerations generate powerful incentives for service providers either to characterize their offerings as information services or to characterize competitors as telecommunications providers.

Notes

1. This section is derived in part from articles by the author published earlier under the following titles: The World Trade Organization Agreement on Basic Telecommunications Services and Implementation by the U.S. Federal Communications Commission, 15 A.B.A. *F. Comm. L. Comm. Lawyer* No. 4 (1998) p. 13 (reprinted by permission of the American Bar Association); The information superhighway: trolls at the tollgate, *Fed. Comm. L.J.* 53 (1997); The Telecommunications Act of 1996, A.B.A. *Sec. Sci. Tech., Bull. La Sci. Tech.,* Dec. 1996, p. 5 (reprinted by permission of the American Bar Association); The Telecommunications Act of 1996: What It Does to Telecommunications Service Providers, 49 *International Engineering Consortium Annu. Rev. Comm.* 359 (1996); and Domestic Spectrum Regulation in the United States, in *Worldwide Wireless Commun.* 135 (Frank S. Barnes et al. eds., 1995) (reprinted by permission of Cohn and Marks).

2. Fourth Protocol to the General Agreement on Trade in Services, WTO Doc. S/L/20 (11 April 1996) (Fourth Protocol) and Annex: Schedules of Specific Commitments and Lists of Exemptions from Article II of the General Agreement on Trade in Services Concerning Basic Telecommunications (hereinafter collectively referred to as the Basic Telecom Agreement).

3. Fourth Protocol at 1. By mid-January, 1998, several signatory governments had not yet ratified the Basic Telecom Agreement, and the WTO had extended the deadline for ratifications to the end of

July 1998. U.S. Regulatory Scene: WTO Accord Still Adrift; FCC Reassures Industry, *Telecommunications Reports,* January 16, 1998 (available online at http/www.newsnet.telebase.co, Dialog record no. 03907666).

4. In the Matter of International Settlement Rates, IB Docket No. 96-261, *Report and Order* (FCC 97-280, released Aug. 18, 1997).

5. In the Matter of Amendment of the Commission's Regulatory Policies to Allow Non-U.S. Licensed Space Stations to Provide Domestic and International Satellite Service in the U.S., *Report and Order* (FCC 97-399, released November 26, 1997) ("International Satellite Service Order"); In the Matter of Rules and Policies on Foreign Participation in the U.S. Telecommunications Market, *Report and Order and Order on Reconsideration* (FCC 97-398, released November 26, 1997) ("Foreign Participation Order"). In early 1998, FCC staff indicated that the effective date of these orders could be suspended if the entry into force of the Basic Telecom Agreement were significantly delayed. *WTO Accord Still Adrift, supra* note 3.

6. Prior to the Basic Telecom Agreement, WTO members had agreed to a more limited set of telecommunications commitments, mainly affecting provision of customer premises equipment, value-added network services, and intracorporate communications. These requirements remain in effect. General Agreement on Trade in Services (April 1994), Annex on Telecommunications.

7. Telecommunications Act of 1996, Pub. L. No. 104-104 (codified in scattered sections of 47 U.S.C. §§151 *et seq.*)

8. Market Entry and Regulation of Foreign-Affiliated Entities, *Report and Order,* 11 FCC Rcd 3873 (1995).

9. In the Matter of Amendment of the Commission's Regulatory Policies to Allow Non-U.S. Licensed Space Stations to Provide Domestic and International Satellite Service in the U.S., *Notice of Proposed Rulemaking,* 11 FCC Rcd 18178 (1996).

10. See, e.g., United States of America Schedule of Specific Commitments, WTO Doc. GATS/SC/90/Supp. 2 (April 11, 1997), pp. 4-6.

11. Note by the Chairman, *Notes for Scheduling Basic Telecom Services Commitments,* WTO Doc. S/GBT/W/2/Rev. 1 (16 January 1997) ("the first Chairman's note"); Chairman's Note, *Market Access Limitations on Spectrum Availability,* WTO Doc. S/GBT/W/3 (3 February 1997) ("the second Chairman's note").

12. General Agreement on Trade in Services, 33 I.L.M. 1167 (1994), part II, art. III, ¶3.

13. Ibid., part II, art. III, ¶4.

14. See B. Petrazzini, *Global Telecom Talks: A Trillion Dollar Deal,* Washington, D.C.: Institute for International Economics, 1996, p. 33; J. Friedland, Chile's ENTEL seeks long-distance affair, *the Wall Street J.,* June 5, 1995, p. A10; K. Lynch, Chile gives the industry a lesson in competition, *Commun. Week Int.,* December 12, 1994, p. 1; K. Lynch, Latin Overhaul, *Commun. Week Int.,* February 7, 1994, p. 4; Price-cutting war arrested by heavy losses: chilean phone firms think again, Latin American Newsletters, Ltd., May 4, 1995 (available on NEXIS); W. R. Long, Investors dial up chilean market: there is fierce competition since the government ended the long-distance monopoly, *Los Angeles Times,* Home Edition, December 6, 1994, p. 3.

15. Marrakesh Agreement Establishing the World Trade Organization, Annex 2: Understanding on Rules and Procedures Governing the Settlement of Disputes ("Dispute Annex") art. 2, ¶1.

16. Ibid., art. 4, ¶7.

17. Ibid., art 8, ¶5.

18. Ibid., art. 8, ¶4.

19. Ibid., art. 8, ¶7.

20. Ibid., art. 8, ¶9.

21. Some of the initial appointees will begin with 2-year terms. Ibid., art. 17, ¶¶1–2.

22. Ibid., art. 17, ¶5.

23. Ibid., art. 17, ¶14. Lower panel reports must likewise be adopted by the DSB if they are not appealed to the Appellate Body, unless the DSB decides by consensus not to adopt the lower panel decision.

24. Ibid., art 17.
25. Ibid., art. 22.
26. Uruguay Round Agreements Act, §102(c)(1), codified at 19 U.S.C. §3512.
27. Members of the WTO include national governments, trading groups like the European Community, and separate customs areas like Hong Kong.
28. Use of the Carterfone Device in Message Toll Telephone Service, Docket No. 16942, *Decision,* 13 FCC 2d 420 (1968), *reconsideration denied by Memorandum Opinion and Order,* 14 FCC 2d 571 (1968).
29. 47 U.S.C. §§151(a) and 271.
30. 47 U.S.C. §273.
31. 47 U.S.C. §251.
32. "RBOC" and "BOC" are often used as interchangeable terms. The enterprises are actually structured as regional holding companies, each owning several Bell operating companies.
33. FCC Office of General Counsel, 251/252 *District Court Litigation,* November 25, 1997 (continually updated list released at author's request).
34. 47 U.S.C. §251(c)(2)–(4).
35. 47 U.S.C. §271(c)(2)(B)(iv)-(vi).
36. 47 U.S.C. §271(c)(2)(B)(x).
37. 47 U.S.C. §251(c)(6).
38. 47 U.S.C. §251(c)(4)(A).
39. 47 U.S.C. §251(c)(4)(B).
40. 47 U.S.C. §251(c)(5).
41. 47 U.S.C. §251(f).
42. 47 U.S.C. §271(c)(2)(B)(xi).
43. 47 U.S.C. §271(c)(2)(B)(viii).
44. 47 U.S.C. §271(c)(2)(B)(iii).
45. 47 U.S.C. §271(c)(2)(B)(xiii).
46. 47 U.S.C. §251(a).
47. 47 U.S.C. §153(r)(49).
48. See United States of America Schedule of Specific Commitments, *supra* note 10.
49. See *Iowa Utilities Board v. F.C.C.,* No. 96-3321 (8th Cir. July 18, 1997).
50. In the Matter of Federal–State Joint Board on Universal Service, CC Docket No. 96-45, *Report and Order* (FCC 97-157, released May 8, 1997).
51. Ibid., at ¶853.
52. In the Matter of Access Charge Reform, Price Cap Performance Review for Local Exchange Carriers, Transport Rate Structure and Pricing, and End User Common Line Charges, *First Report and Order* (FCC 97-158, released May 16, 1997) ("Access Reform Order").
53. See note 52, *supra.*
54. See *Computer II,* 77 FCC 2d at 477–78 (1980); *CCIA v. FCC,* 693 F.2d 198, 213 (D.C. Cir. 1982) ("CCIA").
55. Bell Operating Companies' Joint Petition for Waiver of Computer II Rules, *Memorandum Opinion and Order,* 10 FCC Rcd 1724 (1995), ¶5.
56. 47 U.S.C. §§153(41), (48), (49), and (51).
57. 47 U.S.C. §153(20).
58. 47 U.S.C. §153(43).
59. Implementation of the Non-Accounting Safeguards of Sections 271 and 272 of the Communications Act of 1934, as Amended, CC Docket No. 96-149, *First Report and Order and Further Notice of Proposed Rulemaking* (FCC 96-489, released December 24, 1996) ("Non-Accounting Safeguards Order") at ¶102.
60. 47 C.F.R. §64.702(a). Enhanced services are not regulated as common carriers. Ibid.

1.2.4 Regulation in Non-WTO Countries: Overview of Telecommunications Regulation in Africa

Raymond U. Akwule

1.2.4.1 Introduction

The last few years have brought remarkable changes in the African telecommunications regulatory horizon. The changes are part of a paradigm shift in the wider socioeconomic development approach on the continent. This section will describe this paradigm shift and then will review the context and scope of change in the African telecommunications regulatory environment. First it should be noted that Africa is not homogeneous, meaning that countries in the region, even when they are immediate neighbors, often differ in so many dimensions that oversimplified generalization would only mislead. But even as there is divergence, there are commonalties in experiences, and often in approaches to reform, which warrant some degree of generalization. Thus these similarities in trends merit as much attention from scholars and policy makers as do the divergences.

1.2.4.2 The Old Paradigm

Approximately four decades after the end of colonial rule in most of Africa, the continent can boast some modest successes, but in general, progress in socioeconomic development has fallen far short of the expectations of the Africans themselves. The hopefulness of the early 1960s, immediately following independence of many of the nations, slowly turned into despair in the 1970s and 1980s. Nation after nation experienced political strife, poor economic performance — induced partly by world economic trends but also by inappropriate national development policies — and by the poor management of national resources.[*] Today, with a total population of more than 700 million people, a land area which is three times the size of the U.S., and lots of mineral and other resources, the African continent possesses all the resources required for modern development. However, the continent lags behind other world regions in the major socioeconomic development indicators.

Some scholars have blamed this on the dominant development paradigm in most of the continent from the 1960s to the 1980s. During this period the African countries, without exception, had centralized economies and heavy government domination of the economies through fiscal and regulatory controls. In addition, there was a major reliance on various forms of foreign aid, which in turn was influenced substantially by the global politics of the Cold War. This, coupled with several missteps along to way to development, spelled dissatisfaction for many of the continent's citizens.

Meanwhile, foreign enthusiasm about Africa and especially foreign investment in the region has suffered due to decades of media headlines about famine, AIDs, ethnic killings, and corrupt dictatorships in the region. All of the foregoing has affected the development of the telecommunications sector as will be evident in the following sections.

1.2.4.3 The Telecommunications Regulatory Environment

1.2.4.3.1 Before 1990: The Dominant Paradigm

The telecommunications sector across Africa experienced very little growth between 1960 and 1990. As they emerged from the colonial era, the countries inherited telecommunications systems — mostly analog — which reflected the needs of the colonial governments and did very little to help African development in a post-colonial period. But more significantly, the governments continued an inherited pattern of government monopoly of the telecommunications sector on the grounds of its income generation and national security significance. As would be expected, the telecommunications sector policies and practices during this period reflected the wider regional macroeconomic development philosophy.

---

[*]See *21st Century Africa: Towards a New Vision of Self-Sustainable Development*, A. Seidman and F. Anang, Eds., Africa World Press, Trenton, NJ, 1992.

The existing policies and practices did little to help Africa meet the challenge of rectifying the severe inadequacy of telecommunications services in the region. As a result the continent still lags behind other continents in availability of telecommunications systems and services. Most African countries still average less than one telephone line per 100 inhabitants and several countries have only one telephone line per 1000 inhabitants. Service is typically concentrated in a few cities; for example, about 80% of Kenya's population live in places that have no telephones.

Waiting periods of up to 10 years and more to obtain a new telephone line is commonplace in many African countries. Service quality is typically poor, and in some countries it is virtually impossible to complete a telephone call during peak business hours. In addition, there is often in large parts of the continent an absence of the more advanced services, such as data transmission, electronic mail, or even facsimile, etc.

The inadequacies in the telecommunications sector have been caused by several interrelated factors, some of which are still present in many African countries. First, despite the high financial returns from investment (typically 20 to 30%, for World Bank–supported programs in the 1980s, for example), telecommunications suffers from a general shortage of state funds for investment. There are many competing demands (e.g., for more schools and hospital beds), and governments have tended to appropriate telecommunications operating surpluses for uses other than reinvestment in the sector. And given its high requirements for imported equipment, telecommunications is critically dependent on foreign currency funds.

Second, telecommunications entities are organized and managed more as parts of government administrations, than as customer-oriented, market-driven, technology-intensive, and rapidly changing businesses. The entities are thus weighed down by lack of autonomy and incentives to perform. In addition, the telecommunications services were provided as "social goods," with low and distorted tariffs. Third, given the conditions of the sector in most countries there was lack of access to capital markets to support expansion programs. Finally, and perhaps most importantly, no new entry was allowed, despite the incumbent's inability to meet demand.

1.2.4.3.2 Beyond 1990: A Paradigm Shift

Since 1990, as part of a paradigm shift in the approach to broad socioeconomic development, a majority of the African countries have taken at least tentative steps away from command economies and toward free markets. The result is that average gross domestic product (GDP) growth for sub-Saharan Africa increased steadily between 1990 and 1996. Average GDP growth was about 5% in 1996, with more than half of the continent's countries growing at rates higher than their average population growth of 3%. Foreign investment is increasing, and economic performance is improving in a wide range of areas, including telecommunications. Private capital is increasingly flowing to Africa, with healthy returns. Private investment has increased more than tenfold since 1990 to $11.7 billion in 1996. From 1990 to 1994, rates of return on foreign direct investment in Africa averaged between 24 and 30% compared with 16 to 18% for all developing countries. Indeed, a new spirit of social and economic progress has energized much of the region.

Challenges persist for the continent, however. Africa's social and economic progress is still fragile as 45% of the population still lives on less than $1.00 a day. A legacy of social unrest and ethnic rivalry continues to slow development. According to the World Bank, effective development in the region will require a hefty 8 to 10% annual GDP growth. And given the wide gap in telecommunications development between African countries and the rest of the world, it is clear that huge amounts of investment in the sector are necessary.

Realizing the opportunities provided by emerging new technologies and the need to overcome persistent past shortfalls, a growing number of African countries are reforming their telecommunications sector policies and structures along three main lines: commercialization of operations, encouraging new entrants and competition, and increasing private participation. Increasingly, the role of government is shifting from ownership and management of operations to the creation and maintenance of an enabling regulatory and policy environment. Transformation is taking place at both the regional and national levels as will be discussed in the following sections.

1.2.4.4 Regional-Level Reform Activities

It has become increasingly evident that telecommunications can help African countries to achieve development objectives and to participate more effectively in a growing global information society. African leaders have acknowledged this in three relatively recent developments, which will be described below.

1.2.4.4.1 *The African Information Society Initiative (AISI)*

In 1996, the United Nations Economic Commission for Africa (UNECA) unveiled its African Information Society Initiative (AISI), with the main goal of building an African Information Infrastructure (AII).* The AII is Africa's portion of the much talked about Global Information Infrastructure — an emerging global network of connectivity tools, infrastructure, and processes which ultimately will facilitate communication from anywhere to anywhere and anyone to anyone, and serve as the backbone of the emerging information society. It is noteworthy that African experts wrote the AISI document, which emphasized the need for major telecommunications regulatory and policy reform in the member countries.** Even more significant is the fact that it was quickly endorsed and adopted by governments across the region thus making it an official African agenda.

1.2.4.4.2 *The African Green Paper*

Also in 1996, African leaders endorsed an International Telecommunications Union–sponsored African Green Paper, which like the AISI document called for major reform of the telecommunications sector in Africa. The Green Paper, also prepared by African experts, was appropriately more specific than the AISI document about the nature of the additional required reform in the telecommunications sector.*** The following list includes some of the key recommendations of the Green Paper:

- Separation of regulatory and operational functions
- Separation of the postal and telecommunication functions
- Creation of a separate national body charged with regulating telecommunications
- Provision for financial and managerial autonomy for the telecommunications operators
- Opening to regulated competition of those market segments in which demand remains unsatisfied
- Investment in the development and management of human resources
- Creation of a consultative mechanism that allows the involvement of users and other parties concerned with improvement of the sector

1.2.4.4.3 *The Regional African Satellite Communication Organization*

In an unusual move for an African regional telecommunications organization of its size, the Regional African Satellite Communications Organization (RASCOM), whose major goal is to launch a dedicated satellite for Africa, has issued an international tender to build, operate, and transfer the satellite system to African ownership.**** RASCOM's decision to use a build, operate, and transfer strategy for achieving its goal was one of a series of decisions designed to make the organization more business and commercial oriented and denotes a changing attitude on the continent. It is significant that 24 internationally

---

*See African Information Society Initiative (AISI): An Action Framework to Build Africa's Information and Communication Infrastructure. Economic Commission for Africa, 1996.

**One notable outcome of the ECA work is the creation of an African Technical Advisory Committee, comprising six Africans of varying professional backgrounds and charged with guiding the ECA in the implementation of AISI.

***See The African Green Paper: Telecommunications Policies for Africa, International Telecommunications Union, Telecommunications development Bureau (BDT), 1996.

****RASCOM membership includes almost all the African countries, most of which are represented by their government-owned national telecommunications operators.

renowned companies responded immediately to the tender, an indication of the perceived credibility of the RASCOM project, and further evidence of renewed foreign interest in investing in the region.

The three organizations discussed below — the United Nations Economic Commission for Africa (UNECA), the International Telecommunications Union (ITU), and the Regional African Communications Organization (RASCOM) — are important because as umbrella organizations with wide African membership their activities and initiatives have wide regional impact. But beyond these regional activities, there have been specific national reform activities which clearly indicate that African leaders now recognize telecommunications as a vital part of Africa's future and which will be discussed in the next section.

1.2.4.5   National-Level Reform Activities

At the national level, words have been backed by action as just about every African country has entered into some stage of the evolution toward a market-based telecommunications sector. The two major goals of reform at the national level are universal service or access and efficiency in the provision and delivery of telecommunication services. Traditionally, there has been reluctance on the part of many African governments to open their monopoly markets to competition, based on fears about what it means for universal service, employment, and the viability of the incumbent national carrier.

1.2.4.5.1   The Direction of Reform

Since 1990 there has been a wave of telecommunications regulatory reform in Africa. The pace of reform escalated in 1995 to include the following activities: the liberalization of customer premises equipment trade; the separation of Posts and Telecommunications operations; the creation of an independent regulatory entity; the creation of new sector laws; the allowance of increased private investment in the major telecommunications sectors (Table 1.1). The reform has produced results with regard to network development and productivity as well as with finance, including the following:

More than 15 countries have raised teledensity to over one per 100 population since 1990.

Productivity, as measured by the number of main telephone lines per telecommunication employee, increased by 37% from 1990 to 1994 for the top 15 sub-Saharan African countries.

18 African nations have introduced new telecom laws that have modernized and liberalized the sector.

75% of telephone lines in sub-Saharan Africa are "private" today (25% if RSA is excluded).

Ten countries have privatized national operators — 15 more intend to do so.

Privatization proceeds have accounted for $1.7 billion since 1995, further evidence of new foreign interest in investing in the region.

Privatization has resulted in roll-out obligations of 3.8 m lines (amounting to $4 to 6 billion; Figure 1.49).

There has been significant liberalization, first in value-added services, then mobile, and finally basic services.

Cellular now accounts for 20% of the total market, and is growing fast, based mainly on private investment.

The Internet is growing fast, but remains limited in terms of access.

A frequently cited goal for Africa is to raise regional teledensity to one line per 100 inhabitants by the year 2000. The ITU estimates that approximately $28 billion will be needed to accomplish this between 1995 and 2000.* An important question is where will all this capital come from? Between 1986 and 1994, approximately 60% of telecommunications investments in African countries were financed from funds generated by the operating companies themselves. Multilateral assistance (European Investment Bank, World Bank, African Development Bank, etc.) accounted for a further 20% of telecommunications investment while the remainder came from bilateral and other sources (grants, reserve drawing,

---

*African Telecommunications Indicators, International Telecommunications Union, 1996.

TABLE 1.1   Telecommunications Liberalization in Sub-Saharan Africa*

	Liberal CPE Trade	Separate Posts and Telecoms	Independent Regulator	New Sector Law	Private Cellular	Telco Privatized	Competition in Basic Service
Angola	*	*		n.a.			
Benin		*					
Botswana	*	*		*	1997	1998	
Burkina Faso		*		1998			
Burundi	*	*		1997	*		
Cameroon	*			*		1997	
Cape Verde	*	*	*			*	
CAR		*			*		
Chad	*			1997			
Comoros		*	*				
Congo	*	*	1997	*	*	1997	
Côte d'Ivoire	*	*	*	*	*	*	2004
Djibouti	*			n.a.			
Equ. Guinea	n.a.			n.a.			
Eritrea	*			n.a.			
Ethiopia	*	*		*			
Gabon	*		1998	1997	1998	1998	
Gambia	*	*	*				
Ghana	*	*	*	*	*	*	*
Guinea	*	*	*		*	*	
Guinea-Bissau	*	*	*			*	
Kenya	*	1997	1997	1997	1997	1998	
Lesotho	*	*			*		
Liberia	n.a.	*		n.a.			
Madagascar	*	*		*	*		
Malawi	*				*		
Mali	*	*	*	n.a.			
Mauritania	*						
Mauritius	*	*	*	*	*		
Mayotte							
Mozambique	*	*	*				
Namibia	*	*	*		*		
Niger	n.a.	n.a.	n.a.	n.a.	n.a.	n.a.	n.a.
Nigeria	*	*	*				
Rwanda	*	*					
S. T. & Principe	*	*		n.a.		*	
Senegal	*	*	*	*	1998	1997	
Seychelles	*	*	*	n.a.	*	*	*
Sierra Leone	*	*		n.a.			
Somalia	n.a.	n.a.	n.a.	n.a.	n.a.	n.a.	n.a.
South Africa		*	*	*	*	*	2002
Sudan	*	*	*	*		*	
Swaziland	*						
Tanzania	*	*	*	*	*	1998	Partial
Togo	*	*	1997	1997	1998	1999	1998
Uganda	*	1997	1997	1997	*	1997	1997
Zaire	*				*		
Zambia	*	*	*		*		
Zimbabwe	*						

*Based on information available as of late 1997.

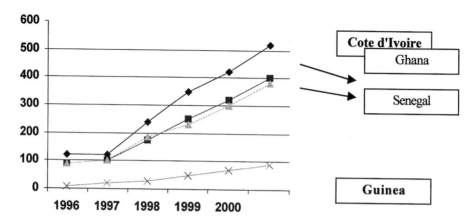

FIGURE 1.49 Roll-out obligations following privatization (thousands of new lines). (*Source:* World Bank.)

TABLE 1.2 African Telecom Sector Privatization since December 1995

Country	Operator	Strategic Partner	Stake	Proceeds ($ millions)	New Lines ('000)
Cape Verde	C.V. Tel.	Tel. Portugal	40%	20	N/A
Cote d'Ivoire	CITELCOM	France Telecom	51%	210	±400
Eritrea	Erit. Tel.	Daewoo	45%?	40?	N/A
Ghana	Ghana Tel.	Tel. Malaysia	30%	38	225
	2nd License	W. W'less/AGG	100%	10	50
Guinea	Sotelgui	Tel. Malaysia	60%	40	±50
Senegal	Sonatel	France Telecom	30%	110	±200
South Africa	Telkom	SBC/Tel. Malay.	30%	1261	2820
Uganda	2nd license	Telia/MTN	100%	6	6
Total				1735	3805

Source: World Bank.

commercial lending). Bilateral lending peaked at $431 million in 1989 but had fallen to just $97 million by 1994, and it has continued on the decline.

If the expected major increase in total funding for the telecommunications sector is to materialize, about one half or much more of the total funding requirements for the 1990s and beyond would have to come from private sources. This requires creating conditions to attract direct private investment and accessing domestic and foreign capital markets. It also stresses the importance of making operations efficient and profitable and the need to use official funds mostly to catalyze private investment and improve performance.

African leaders have acknowledged the need for infusion of private capital as reflected in the number of countries that have employed one or several privatization mechanisms (Table 1.2).

There has been a variation in the approaches to reform taken by African countries as will be evident in the following discussion of reform in three West African countries: Nigeria, Gambia, and Ghana.

1.2.4.5.2 Nigeria[*]

Next to South Africa, Nigeria is the largest telecommunications market in sub-Saharan Africa, accounting for over one fifth of telecommunications revenues in the region. The country's main telecommunications

---

[*]*Sources: Africa Communications Magazine* (Various Issues); R. Akwule, United Nations Development Program — Internet Initiative for Africa: Project for the Federal Republic of Nigeria, UNDP, November 1997.

provider, Nigeria Telecommunications PLC (Nitel) was created in 1985 from the merger of the telecommunications division of the former Posts and Telecommunications and Nigerian External Telecommunications Ltd.

The Nigeria telecommunications sector was deregulated through Decree No. 75 of 1992. The Decree converted Nitel into a public limited company — though still state-owned — and established the Nigerian Communications Commission (NCC) as an independent regulatory body. The broad objectives of the Commission include creating a regulatory environment for the supply of telecommunications services and facilities and promoting fair competition and efficient market conduct, among others. Thus three major organizations dominate telecommunications sector activities in Nigeria. They include (1) the Federal Ministry of Communications (FMOC), which sets broad policies and administers the nation's radio frequency; (2) NITEL, PLC, the semi-autonomous, government-owned national monopoly telecommunications provider; and (3) the Nigerian Communications Commission (NCC), the national telecommunications regulator. In addition, there are private commercial operators — offering telecommunications services such as private network links, fixed and mobile cellular services, voice mail, and paging services—as well as some telecom equipment vendors and manufacturers.

Telecommunications has boomed since Nitel's change of status. Nitel's revenues have risen from $275 million in 1990 to $459 million in 1994 and now account for 1.1% of Nigeria's gross domestic product. The company is now profitable once again and is no longer a drain on government finances. In 1993 it paid $25.5 million in income taxes, 19% of its profit. Revenues from international traffic contributed significantly to its turnaround. International telephony accounts for 48% of revenues and received a big boost with the installation of 5000 new international circuits in 1993.

The sector is evolving through a policy of guided deregulation and presents tremendous potentials for investment. Indeed Nitel has been successful in attracting foreign financing. It has obtained a $225 million World Bank loan covering 47% of a $484 million telecommunications project.

Since 1990, Nitel has increased the capacity of the telephone network by 80% to an estimated 850,000 by the end of 1995. The number of connected lines per 100 people has risen from 0.30 to an estimated 0.48 by 1995. Teledensity could even be higher considering that there is significant idle capacity. Indeed, the existing capacity could absorb the present waiting list of around 200,000.

Indeed there has been much progress in the Nigerian telecommunications sector in recent years though it is still plagued with problems which afflict other countries in the region, including incompatible equipment, inadequate number of skilled personnel, poor maintenance practices, erratic power supply, etc.

By mid-1997, approximately 61 private telecommunications companies had been granted licenses to provide telecom services in Nigeria. Most of the companies are based in Lagos and each of the companies is licensed to provide one or more of the following services: Internet, Value-Added Service (VAS), Paging, Pay phone, Repairs and Maintenance, Voicemail, Cabling, and Community Telephony Services. In addition, some are licensed to provide private network links and public mobile communications services.

1.2.4.5.3 *Highlights of Nigeria's Telecommunications Policy*

Government officials have proclaimed the following short- and long-range goals as part of the telecommunications component of the Vision 2010 (a blueprint for the country's development):

- Achieving an increase in teledensity from one telephone line per 200 people to one per 50
- Attaining the goal of universal coverage, "access to anywhere, at any time," global connectivity, with a communications network connected to the international information superhighway
- Intensifying current deregulation efforts in the sector, with privatization of NITEL by 1998
- Introducing a second national carrier for local and international traffic, thus introducing a competitive environment in the telecommunications sector. The decision to appoint a second national carrier to compete with NITEL is informed by the national target of 3 million telephone lines by 2005. The carrier, when approved, is expected to install a minimum of 1 million lines by 2005.

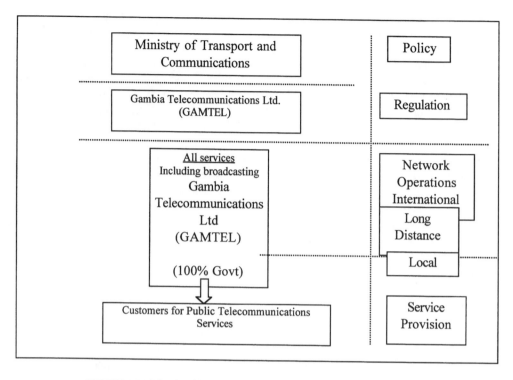

FIGURE 1.50 The Gambia — telecommunications sector structure in 1997.

1.2.4.5.4   The Gambia*

The West African nation of the Gambia is one of Africa's smallest countries in terms of population and geographical size. It also is one of many African countries that still rely almost exclusively on a government-owned national operator for the provision of all forms of telecommunications services. It is noteworthy however that the Gambia has one of the most efficient basic telecommunications services in the region.

The organization in charge of the nation's telecommunications infrastructure and services is the Gambia Telecommunications Company Limited (Gamtel) (Figure 1.50). Gamtel was formed as a state-owned limited liability company in March 1984. It immediately assumed the responsibility for the national telecommunications services (formerly under the Telecommunications Department of the Ministry of Works and Communications) and the international telecommunications services (formerly under the U.K.-based Cable and Wireless). Today, Gamtel is a limited liability company owned 99% by the Gambia Government (Ministry of Finance and Economic Affairs) and 1% by the Gambia National Insurance Company Limited. It manages both the national and international telecommunications networks, including a national cellular service, and is in charge of the national television service. Gamtel thus dominates the telecommunications sector, allowing private sector involvement only in the provisioning of telecommunications services to the public through ownership of private telephone booths (payphone/call centers). The private sector service providers resell telephone services purchased from Gamtel at discounted prices. In addition Gamtel has plans to license some private sector Internet Service Providers (ISPs) as soon as the ongoing installation of a national Internet gateway node is completed.

---

*Information on the Gambia was compiled from R. Akwule, Final Mission Report — UNDP Internet Initiative for Africa: Project for the Gambia, UNDP, September, 1997.

The network infrastructure, which was inherited by Gamtel in 1984 was obsolete and the available 2400 direct exchange lines (DELs) were in poor state. Gamtel initiated Phase I (1984–1986) and Phase II (1988–1990) telecommunications rehabilitation projects that replaced all parts of the existing telecommunications network with digital exchanges, digital microwave, and line transmission systems. Phase I was funded by the Caisse Centrale de Cooperation Economique (CCCE) of France and cost about Fr 117 million. Phase II was founded by Gamtel (28%) and by the CCCE (72%). The modernization effort resulted in an increase in national telephone penetration from 0.85 to 1.57 per 100 habitants in the rural areas and from 2.7 to 5.4 per 100 habitants in the urban area. Gamtel has a total of 21,298 DELs, 14,072 in the main centers, and 7226 in the countryside.

The government of the Gambia and the Gambia Telecommunications Co. Ltd. (Gamtel) has made provisioning of readily available, reasonably priced, and well-maintained communications a top priority. Gamtel has signed a performance contract with the Gambia government to meet profitability targets, and it has a reputation for being well managed. It is noteworthy that the Gambia has a well-developed domestic and international telecommunications network, which is exemplary in the region. At the same time it has maintained one of the lowest domestic and international telecommunications tariffs in the region. In addition, training for Gamtel staff is highly emphasized. Training is accomplished mainly through the Gambia Telecommunications and Multimedia Institute (GTMI), which is capable of providing training up to the senior technician level.

1.2.4.5.5   Vision 2020: A Program for National Development

The government has recently embarked on a "Vision 2020" program, a strategy for socioeconomic development that aims at raising the standard of living of the Gambia population by transforming the Gambia into a dynamic middle-income country by the year 2020. A government-produced "Vision 2020" document acknowledges that free flow of information is a prerequisite for the attainment of the vision. According to the document:

> The long term objectives for telecommunications are to consolidate the Gambia's achievements in the area of telecommunication by integrating the country into the Global Information Infrastructure (GII) via the global information highway, to make the Gambia a major center for data processing and training and to make telecommunication services accessible to every household and business in the country.

Gambia's national information strategy defines the following roles for Gamtel within the "Vision 2020" program:

- Provision of premium services for business and other institutions
- Provision of basic telephone service for all
- Introduction of Global Mobile Communications
- Provision of Internet services

While the telecommunications service offered in the Gambia is above average, the country has potential to become a major regional telecommunications hub for Africa or, as Gamtel literature proclaims, to become "a gateway to Africa." However, it is clear that further reform is necessary to accommodate the magnitude of expansion necessary to attain that status. The current global technological and policy environment provides opportunity for such transformation.

1.2.4.6   Ghana[*]

The West African nation of Ghana has arguably been involved in the most aggressive telecommunications reform program in Africa to date. With a 16 million population, comfortably located in the middle of

---

[*]Information on Ghana was compiled from the following sources: *Africa Communications Magazine* (various issues), Ghana GOV Web site, etc.

the 16 nations of the Economic Community of West African States, developments there have potential to be far-reaching in the region.

In 1984, Ghana commenced implementation of an Economic Recovery Program, which has given the country easier access to International Monetary Fund (IMF) and World Bank loans and boosted foreign private investment.

Up to the early 1970s, the government's Post and Telecommunications Department administered telecommunications in Ghana. In 1974 the Post and Telecommunications Department was transformed into a public corporation by National Redemption Council Decree No. 311, and placed under the authority of the Ministry of Transport and Communication. Today, the Ministry is still responsible for broad policy formulation and control of Ghana's telecommunications sector. In 1992 the Ghanaian government approved the formation of a National Communications Commission.

Meanwhile, the fast growth of Ghana's population and economy was not matched by a corresponding increase in the economy's telecommunications infrastructure. This was illustrated by the fact that Ghana's GDP grew by an average of 4.7% per annum between 1985 and 1995, while the direct telephone exchange line (DEL) penetration rate of approximately 0.3 remained static during the same period.

In 1995, Ghana Telecom was created as a successor of GPTC and thus became the major provider of both domestic and international telecommunication and related services in Ghana. It is also one of Ghana's largest companies with 1995 sales and assets of $571 million and 222 million, respectively.

To address the stagnant growth in the telecommunications sector, the government of Ghana embarked on a reform program designed to establish the market and regulatory mechanisms necessary to promote and stimulate rapid private sector–led growth and improvement of Ghana's telecommunications infrastructure and services. The government had the following objectives for its reform program:

- To increase the penetration of Ghana's telecommunications network
- To expand significantly the range of services offered to subscribers
- To improve dramatically the efficiency of telecommunications services
- To attract the requisite technical expertise and capital from within and outside of Ghana

The reform program has had several major components, including the following.

1.2.4.6.1 Privatization

A strategic equity stake of 30% and management control of Ghana Telecommunications Company Limited (GT) was sold to Malaysia Telekom as a result of an international tender. In addition, an operating license was granted to U.S.-based Western Wireless to serve as a Second National Operator (SNO) for telecommunications services throughout Ghana.

1.2.4.6.2 Telecom Law Reform

Ghana has taken great pains to create a regulatory framework, which will ensure transparency in the telecom sector. A National Communications Authority has been created whose responsibility is to ensure a transparent mechanism for the regulation of the telecom sector. This Act is designed to promote a stable operating environment for all participants, while promoting fair competition and efficiency.

Figure 1.51 shows the result of the ongoing telecom sector reform in Ghana. GT, 30% owned by Malaysia Telekom, has a 20-year, renewal license. The second national operator also has a 20-year, renewable license. During the first 5 years of the GT and SNO licenses, no other operator will be permitted to offer fixed telecommunications services in Ghana.

In addition to the sale of strategic stakes in Ghana Telecom, the government intends to sell an additional stake in Ghana Telecom to institutional and other investors at a later date to bolster private sector support of the telecommunications sector and further reduce state ownership of the sector.

The government believes that 100,000 new DELs need to be installed to meet current market demand. To reach the medium-term sector objective of 2.5 DELs per 100 inhabitants, an additional estimated 450,000 DELs is needed.

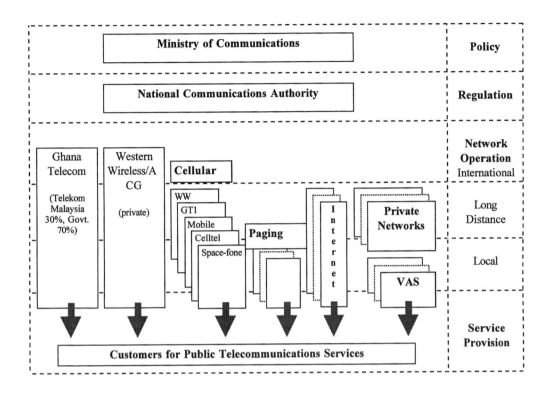

FIGURE 1.51 Ghana — telecommunications sector structure in 1997.

It is still too early for a profound assessment of the results of the recent privatization efforts in Ghana. The government has imposed substantial roll-out obligations on the new privatized operators (see Figure 1.49). In addition, Ghana's innovative Trade and Investment Gateway Program, which will incorporate export processing zones, industrial parks, and free ports to stimulate additional investment, offers investors additional expansion opportunities, particularly in the development of teleports.

Ghana offers a stable economic and political environment, with an impressive growth record and excellent future growth prospects. It also has a transparent legal framework, which guarantees the protection of investments and the ability to operate in accordance with normal business practices. The investor-oriented investment climate has not gone unnoticed as direct foreign investment has grown substantially.

1.2.4.7 Conclusion

Africa is currently experiencing a change phenomenon, which is having tremendous impact at broad macroeconomic levels as well as in the specific sectors such as telecommunications. In recent years, as part of this phenomenon, the pace of telecommunications reform in Africa has accelerated based on the realization by African leaders of the importance of telecom in Africa's future. There is a perceived need among the leadership to create a much more enabling environment for telecom growth than is generally available in many of the countries. While in some countries change has been drastic, in other countries the progress is still minimal. Nonetheless there is evidence of a new spirit of change at both the regional and national levels across the continent.

The review of three West African case studies indicates the diversity of approach to telecommunications development in Africa. Each of the three nations has a policy to enhance its National Information Infrastructure in preparation for the challenges of the global information society and the 21st century.

But each has adopted a different approach so far for coping with those challenges. A review of the remaining African countries will reveal even further diversity in approaches to reform across the continent. There is evidence everywhere on the continent of accelerated reform activities, including liberalization of customer premises equipment; separation of posts and telecommunications departments; creation of independent regulatory authorities; creation of new sector laws; liberalization of basic services; and privatization. All of these activities have resulted in significant improvements in the sector performance across the continent, which, in turn, is likely to foster international collaboration for development of much needed connectivity in Africa.

1.2.5   Satellite Technology and Regulation

Rob Frieden

1.2.5.1   Satellites as a "Bent Pipe"

Communications satellites receive and retransmit signals much like very tall radio towers. In 1945, science writer Arthur C. Clark predicted that three strategically located space stations could provide service to most of the world.[1] Mr. Clark speculated that there existed a particular orbital location where satellites would appear stationary, relative to the earth, thereby presenting a fixed, "geostationary" location for sending and receiving signals. If the satellite hovered over a particular point on earth, then signals could travel up to the known location occupied by a stationary receiver/transmitter and then downward to earth. The satellite could operate as a bent pipe: receiving signals and bending them back to earth much like what portions of the ionosphere do to propagate "skywave" radio signals over long distances.

Mr. Clark correctly predicted that objects sent into a particular orbital location could operate in synchronicity with the earth's orbit; i.e., the satellite would rotate about the earth at the same rate that the earth rotates on its axis. In other words, both communication satellites and the earth rotate at a velocity of one revolution per 24 hours. At 22,300 miles (35,800 km) satellites appear to hover in a stable location, even though they travel at 6879 miles/h in a circular orbit with a circumference of 165,000 miles![2] Satellites operate in geostationary orbits when they are positioned above the equator and thereby have no tilt (also known as 0 degree inclination) relative to a straight line. Put another way, a satellite with no inclination operates in a perfectly circular orbit relative to earth, with no deviations in its orbital plane. Other satellites, operating in inclined orbits, lack perfect circularity and thereby have orbits with a high point (apogee) farthest from earth and a low point (perigee) closest to earth.

Satellites have two fundamental characteristics that promote their usefulness for telecommunication applications:

1. With an orbital location so far from earth, satellites transmit a weak, but usable signal over a broad "footprint."
2. Large geographical coverage makes it possible to serve thousands, if not millions, of different locations from a single satellite.

One can visualize the concept of a satellite footprint and point-to-multipoint service by using a flashlight on a globe, which then shows a particular area of the globe being lit. A flashlight quite close to the globe illuminates a small area, but provides a strong, bright signal. Consider the area illuminated as the footprint. As you pull the flashlight farther from the globe the coverage area (footprint) increases, but the signal strength (light intensity) decreases. Point-to-multipoint service means that a single flashlight beam can provide signals to any location it illuminates. A single signal transmitted upward (uplinked) to the satellite can be received (downlinked) by users anywhere within the footprint. Hence, in the video program distribution marketplace, a single satellite can, for example, receive an uplinked movie originating at a Home Box Office's operations facility on Long Island, N.Y. and downlink it to cable television headends equipped with satellite receiving dishes throughout the continental U.S. Likewise, Ted Turner, the enterprising owner of a billboard services company, could launch the first cable television "super station" by first acquiring a local UHF television station in Atlanta and uplinking its programming to a satellite for nationwide reception via cable television.

Point-to-multipoint transmission capabilities change the calculus and economics of video program distribution. In lieu of a single cable system distributing a movie over an unused channel, a company can package a series of movies and other sorts of premium entertainment for distribution over the same unused channel. The programmer can aggregate audiences throughout the nation, while the cable operator can participate in a new profit center without the labor and logistical effort involved in physically receiving tape or celluloid. Once the cable television operator invests in satellite receiving technology, a number of satellite-delivered program options become available.

Because satellite footprints typically cover a wide geographical area, carriers can provide service throughout a region with the additional minor investment necessary to add an another point of communication. Such widespread coverage also means that the cost of satellite service can be spread over a number of different routes of different length and traffic density. This insensitivity to distance and traffic density means that, with proper coordination by governments, carriers, and users, satellites have the potential to achieve two important outcomes affecting video program distribution:

- Satellites can provide point-to-multipoint service, e.g., widespread distribution of a video program to a number of broadcast and cable television outlets, at roughly the same cost as a single point-to-point transmission.
- Satellites can achieve networking economies of scale by aggregating audiences who singularly would not generate the demand and revenue sufficient to support a programming venture, but who collectively do.

1.2.5.2 Achieving Interference-Free Operation

Interference-free operation of satellites requires coordination on the use of radio frequencies and the geostationary orbital arc. Nations must agree on which frequencies the different types of satellites will operate and how they will register orbital slot usage. Likewise, the satellite and associated receiving earth stations must operate with sufficient signal strength to override potential interference resulting from other terrestrial transmission like microwave relays.

Again physics plays a primary role. The receiving earth station must have an unobstructed "look angle," i.e., a direct link to the satellite unblocked by trees, terrain, and buildings and at an angle sufficiently above the plane of the earth so that the link extends above the horizon. As look angles increase to 90 degrees, directly overhead, the path to the satellite becomes more direct, meaning that the signals sent or received traverse more directly through the earth's atmosphere. Lower look angles require longer transit through the earth's atmosphere leading to greater signal attenuation.

The physics of satellite telecommunication also affect the vulnerability of earth stations to interference. As earth stations locate farther from the centerpoint of a transmission where signal strength is highest, signal quality from the intended source attenuates even as other signals from adjacent satellite increase in strength, particularly if they operate on the same frequency. Signal strength typically degrades in concentric circles or contours. The farther from the satellite's "boresight" the weaker the signal becomes with signal rolloff (degradation) accelerating as the distance from the boresight increases.

While national governments and satellite operators cannot undo the laws of physics, they can establish rules and regulations respecting such laws with an eye toward reducing signal interference and conflict. Such rule-making occurs at the International Telecommunication Union (ITU), a specialized agency of the United Nations and the world's oldest multinational forum.[3] The ITU coordinates international rules of the road on the operation of satellites. It strives for consensus decision making on such diverse issues as what frequencies satellites should use, who can operate in which satellite orbital location, and how do satellite carriers coordinate among themselves and with terrestrial operators to avoid harmful interference. Because the ITU has no enforcement power, it must rely on the shared recognition among nations that uniform rules of the road promote efficiency.

1.2.5.3   Staking Claims to the Orbital Arc

Such enlightened self-interest does not always prevail, because of the incentive to acquire a bigger piece of a shared resource like the geostationary orbital arc. Satellite carriers have incentives to use the ITU orbital slot registration process to "warehouse" slots so that other carriers cannot secure space, or must operate at less attractive orbital locations. The ITU satellite registration process generally favors first registrants of any particular orbital slot. This process favors developed countries that can amass the finances and user demand earlier than developing nations.

Developing nations have objected to a first-filed, first-protected process. They have advocated an allotment process that would reserve slots for future use, even at the expense of having space fallow that could have productive application for use by operators from another country. The registration process recently has shown signs of strain as the number of proposed registrations increases from both incumbent and newcomer operators. The process cannot easily disqualify fraudulent, unneeded, or speculative claims that have dramatically increased. Incumbent operators do not want to face a shortage of orbital slots that an increased constellation of operational satellites might require. Prospective operators start the registration process well before it becomes apparent that they have the financial resources to operate. Other registrants, like the Kingdom of Tonga, have filed for a large number of slots perhaps with an eye toward creating a private market for auctioning slots. Tonga, an island nation of about 150,000 residents, first attempted to register 31 satellites in 26 orbital slots and still pursues 6 slots. Entrepreneurial advisors to the government of this nation have articulated a business plan where orbital slot registrations can translate into cash or satellite transponder capacity.

1.2.5.4   Limits on Licenses, Frequencies, and Orbital Slots

Orbital slot claimstaking and the amount of spectrum allocated for satellite services create limitations that licensing administration must manage. This allocational scarcity can translate into financial burdens on operators who must share spectrum with other operators, agree to become part of a consortium, operate in closer proximity to other satellites or in a less desirable orbital location, and meet operational deadlines or lose the privilege to operate. In the U.S., each of the above limitations has affected the marketplace viability and upside revenue potential of operators.

1.2.5.5   Allocating Satellite Frequencies

The ITU also allocates spectrum with an eye toward establishing a global consensus that will serve as the basis for national spectrum allocations. Spectrum management involves two primary activities:

1. Identifying particular types of spectrum uses;
2. Allocating segments of spectrum propagationally appropriate for the services identified.

The ITU has identified three major satellite service types: (1) Fixed Satellite Service (FSS) where earth stations remain in one unmovable location; (2) Mobile Satellite Service (MSS) where earth stations are transportable in aeronautical, land, or maritime locations; and (3) Broadcast Satellite Service (BSS) intended for direct to home reception of video or audio programming. While striving for single, uniform spectrum allocations, the ITU divides the world into three major regions and its rules recognize the right of individual nations and nations as a block in each region to allocate spectrum inconsistent with the consensus.

The ITU has allocated various slivers of spectrum for satellite services. There are three major frequency bands used for video program distribution to cable headends and individuals:

1. The C-band 4–6 gigahertz (GHz)
2. The Ku-band 11–14 GHz
3. The Ka-band 20–30 GHz

Over time, satellite operators have found it necessary or advantageous to migrate upward in frequency from C to Ku to Ka-bands. Necessity forces the migration when terrestrial uses, like microwave radio, generate such congestion in urban locales that adequate reception at C-band proves impossible. Orbital congestion also necessitates a move upward in frequency. Satellites operating on different frequencies can occupy the same orbital slot. Commercial considerations also support the migration. Moving upward in frequency makes it possible to use smaller earth stations, because the newer generations of satellite operate with greater power and the higher frequencies means that smaller dishes can accrue the gain that results when the antenna is at a multiple or fraction of the frequency wavelength.

Each satellite frequency band has certain advantages and limitations. As the first satellite frequency allocation the C-band was carved from preexisting terrestrial allocations. This means that C-band satellite users must operate at lower powers to avoid causing interference to incumbent terrestrial users. The receiving equipment must be large to achieve the gain needed to receive an adequate signal, and most installations must be situated outside urban locales with a high volume of terrestrial microwave operations. However, C-band product lines have reached maturity, meaning that the cost is low and a secondary market for used equipment exists. The C-band does not suffer from signal attenuation caused by rain and snow. The Ku-band supports operations in urban locales, but the equipment costs more. Satellite operators have only recently started to use Ka-band frequencies for direct transmission to users. While they have needed to coordinate such usage with operators of terrestrial microwave and Local Multipoint Distribution Services, ample spectrum remains available. In addition to the "service links" providing connection from the satellite to user earth stations, satellite networks need "feeder links" that supply the satellite with information needed for tracking, telemetry and network control.

1.2.5.5.1 Spectrum Sharing

Since the onset of commercial satellite service, the U.S. Federal Communications Commission (FCC) has advocated a qualified "open skies" licensing policy. Within the limitations of spectrum and orbital slot coordination with nearby countries, the Commission will grant licenses to all financially, legally, and technically qualified applicants. The FCC wants to avoid having to make comparative judgments as to who among a pool of applicants has the best qualifications to serve the "public interest, convenience, and necessity." To accommodate the number of applicants the Commission has reduced the orbital spacing between domestic satellites, thereby requiring more expensive and selective earth stations. The Commission also imposes construction and ready-for-service deadlines that, if not met, will result in revocation of the license to operate. Such deadlines recently squelched a plan by the TCI Primestar direct broadcast satellite venture to acquire the license and orbital slot registrations of an applicant that had failed to meet the FCC construction timetable.

In the satellite arena, the FCC prefers not to hold comparative hearings, or to grant licenses by auction or lottery. Rather than deny licenses to possibly qualified applicants, the FCC will require:

- Additional proof of financial qualifications that, once licensed, the operator can afford to construct, launch, and operate the satellite even in the absence of incoming revenues for at least 1 year.
- All applicants to agree to form a single consortium as occurred in the Commission's consideration of multiple applicants to provide mobile services from geostationary satellites; the American Mobile Satellite Corporation (AMSC) arose as a single licensee, because the FCC determined the market and orbital slot coordination with other nations required its timely consideration and licensing.
- Frequency band segmentation among licensed operators as occurred when the FCC granted licenses to three low and middle earth orbiting mobile satellite operators (Iridium, Globalstar, and Odyssey); the Commission divided the available spectrum into one band that will be shared by operators using one type of transmission technology with Iridium allocated a different portion of the band, because it will operate using a different and incompatible transmission technology.

1.2.5.5.2 Signal Compression Technologies

Innovations that digitize and compress signals make it possible to derive more channels for the same amount of radio bandwidth. Technological innovations make it possible to "squeeze" as many as six digital video signals into the same amount of spectrum that used to carry only one channel. The squeezing process involves the use of integrated circuits that make it possible to examine each video frame, which change at the rate of 30 per second, and allocate bandwidth for processing only the changes that have occurred from the previous frame. A voluntary compression standard established by the Motion Pictures Expert Group (MPEG) seeks to support interoperability among the set top converters of different manufacturers.

1.2.5.5.3   Comparative Advantages of Satellites vs. Terrestrial Options

Given the emphasis on fiber-optic cables as the preferred medium for deploying broadband information superhighways, satellites seem to have been relegated to subordinate status. Many consider wireless options as inferior to wireline services in view of the broader bandwidth and interference-free operations. Yet wireless options will play a role in the National Information Infrastructure development, and they possess a comparative advantage in some respects.

1.2.5.5.4   Uniform Quality

Satellites deliver large bandwidth with virtually the same signal quality throughout a broad geographical footprint. Terrestrial facilities, particularly the cascading tree and branch cable television infrastructure, lack uniform quality in view of the multiple connections and amplifications that occur in the route from headend to individual television sets. Ironically, DBS marketers herald the superior signal quality of a digital signal that has traversed 44,600 miles relative to the cable television signal that may have run fewer than 10 miles. Until such time as fiber-optic technology provides a complete link to the consumer terminal, analog transmissions, multiple connections, and sequential amplification inject noise and degrade signal quality.

1.2.5.5.5   Distance Insensitivity

Satellite transmissions throughout a large geographical footprint means that operators incur no additional incremental costs to serve an additional point regardless of its distance from the program source. In terrestrial point-to-point networks, operators can attribute direct costs in extending a network to an additional point. Accordingly, satellites possess a comparative advantage relative to wireline networks for applications that require a large number of distribution points in diverse geographical points.

1.2.5.5.6   Efficient Point-to-Multipoint Distribution

The distance insensitivity in satellite transmission means that operators incur little if any additional cost in serving an additional point of communication within the footprint. Terrestrial networks typically use a separate line to deliver programming to each destination. Accordingly, satellites possess a comparative advantage in terms of both cost and logistics for distributing programming to geographically diverse cable headends.

1.2.5.5.7   Mobility and Ease in Reconfiguration

Electronic component miniaturization on integrated circuits and microprocessors, higher powered satellites, more sensitive receiving terminals, and orbiting satellites closer to earth make it possible to support diverse mobile, "wireless" applications. NII service providers cannot use fiber-optic cables access to users in vehicles, ships, aircraft, and sparsely populated locales. Likewise, they cannot easily add and subtract service areas and users by reconfiguring their networks. The wide satellite footprint makes it possible to add another service point simply by installing a transceiver, a process that typically takes only a few minutes.

1.2.5.5.8   One-Stop Shopping

Satellite carriers recognize the potential for footprints to traverse national boundaries and the service territories of different cable television operators and broadcasters. They can package programming in such a manner that the retailer simply plucks programming from the satellite for distribution to end users.

1.2.5.5.9   New Satellite Designs Embrace the Digital Revolution

Recently launched satellites have digital transmission capabilities that match that available from fiber-optic terrestrial networks. The satellites operate with higher power thereby reducing the size of receiving dishes to that of a pizza. They generate digital signals that can be compressed, coded, packetized, and transmitted to multiple users. More selective earth stations promote higher performance and the ability to derive more channels.

New satellite networks will convert all traffic into a digital bit stream and next generation satellites will likely have on-board processing capabilities to switch traffic to the appropriate downlink beam or link to another satellite in an operating constellation. Satellites operating as part of the information superhighway look less like simple, unintelligent bent pipes that simply relay signals and more like complex airborne analogs to terrestrial fiber-optic wide-area networks. Traffic on such networks is digital, divided into packets for efficient processing and transmission, switched and routed by the satellite down to earth stations and/or onward to another satellite via Intersatellite Links, Ka-band microwave transmissions.

1.2.5.5.10   Other Design Improvements

Satellite designers have pursued improvements in the materials that make up the satellite bus, payload, and antennas, as well as the station-keeping propulsion, primarily to reduce weight and to extend the satellite's usable life. New satellites combine aluminum with lighter metals like lithium and beryllium along with graphite, the material used to make tennis rackets lighter and more durable. The fuel used to keep satellites in the proper orbit and orientation to earth now combine fuel with electricity to generate more thrust with less fuel. Ion thrusters use an electrified metal grid to ionize xenon gas and generate a highly efficient gas plasma that results in a thrust that is ten times more efficient than previously used chemical thrusters. New satellite antenna designs make it possible to used larger, but lighter, one-piece dishes that flexibly spring into shape rather than require a mechanically activated process.

New satellite generations also take advantage of advances in lasers, component design, and amplification. Intersatellite links, which are just being implementing in satellite networks as microwave transmissions, will use higher frequency laser beams in the future. Because of the narrowness of laser beams, satellites will have to maintain a more stable location in orbit. New designs for solar cells, amplifiers, and integrated circuits will result in more efficient circuitry that will last longer, use less power, generate less heat, and weigh less.

1.2.5.6   Satellite Trends

In the near term, satellites will support development of a broadband telecommunications and information processing infrastructure by extending the reach of terrestrial networks. Satellites complement wireline technologies like fiber-optic cables, considered by most carriers and information service providers as the more basic and essential element in the development of broadband services. Satellites also can perform a "gap-filler" function by providing access to broadband telecommunication capacity in areas where fiber-optic cables do not exist. In other remote areas, satellites may constitute the only available telecommunication resource, because even the twisted copper wire pair associated with conventional narrowband telephone service has not reached into many hinterland locales.

The satellite marketplace has matured and diversified in its 25 years of commercial service. The market continues to evolve from one dominated by government-owned carriers and cooperatives with an exclusive or oligopolistic share of national, regional, or international markets, to one where private enterprise and competition predominate. Technological, business, and regulatory trends favor a more segmented, versatile, and competitive environment. The cutting edge trends in satellite development emphasize the commercial potential of a technology initially deployed for military, space exploration, and intelligence-gathering applications.

The Old World Satellite Order favored government ownership and extensive regulation. It had the view that the market can support only a few large operators, typically organized as cooperatives through agreements executed by governments. The New World Satellite Order combines deregulation, privatization, and entrepreneurialism to support competition and the view that private operators can operate efficiently and profitably.

Commercial ventures have begun to dominate the satellite marketplace. Governments have privatized the domestic or regional cooperative, and the Intelsat global cooperative anticipates creating a private commercial subsidiary to compete effectively with increasing numbers of private market entrants like PanAmSat, Orion, and Columbia Communications. These commercial ventures seek to maximize revenues rather than average rates to promote widespread access to affordable satellite technology. They are quick to respond to individual user requirements with customized applications rather than a simple "one size fits all" inventory.

Private satellite ventures have developed in such diverse markets as:

- Direct-to-home television broadcasting
- Transoceanic and transcontinental voice telecommunications and video program delivery
- Enterprise networking, which integrates both voice and data requirements with access achieved via on-site, very small aperture satellite earth station terminals (VSAT)
- Mobile telecommunications accessible anytime and anywhere

These markets niches reflect the increasing versatility of satellites. As the overall market matures and as more operators enter the market, individual carriers have begun to specialize.

1.2.5.6.1 A Changing Business Environment

The proliferation of satellites types and ownership patterns creates the opportunity for a changing and more competitive environment. Instead of a few operators with "cradle-to-grave" possession of satellites, a diverse and expanding set of operators will acquire title for a portion of a satellite's usable life. This creates the potential for the evolution of market segments to run the gamut from premium, non-interruptible service via state-of-the-art satellites in the best orbital slots, to discounted service via aging satellites perhaps also under new ownership.

Users already have the option of securing lower-cost service that lacks backup capacity and may be preempted for use by customers taking a higher grade of service from the same carrier. In the future, different carriers will target different sectors of the market on the basis of such variables as price, backup, availability of in-orbit service restoration, reliability, age of satellite, vintage of satellite, and orbital location.

The proliferation of carriers and market segments will create pressure on incumbent operators to respond to change. The current ownership structure for the major existing players restricts flexibility. Government tends to bring a simple pricing system based on averaged costs irrespective the degree of competition and preferential access to orbital slots and customers. Early on, governments helped sustain the industry by executing treaty-level documents to establish a cooperative model for the deployment of international satellite capacity through the International Telecommunications Satellite Organization (Intelsat) and its maritime counterpart the International Mobile Satellite Organization (Inmarsat). When satellite technology was making its initial cross-over to commercial applications, the cooperative model helped spread risks, achieved scale economies, and ensured that lesser developed nations could access cutting edge technology with limited investment and without financial handicaps in view of their low demand for service.

Now that the commercial satellite industry has matured, the managers at Intelsat and Inmarsat seek to replace this cooperative status through "corporatization" or outright privatization. Even as their mission remains the provision of ubiquitous access to satellite service on a globally average cost basis, they must increasingly respond to competition and marketplace imperatives. The quasi-diplomatic status of these cooperatives has accorded them privileges and immunities that have translated into financial and competitive advantages. This has allowed the cooperatives to secure commitments from participating nations to avoid causing "significant" technical or economic harm when authorizing competing satellite systems.

Countries consider it increasingly possible to satisfy their commitment to global cooperatives while still authorizing some degree of competition. This can be seen in the following alliances: PanAmSat, Orion and Constellation in the U.S.; AsiaSat and Asia Pacific Telecommunications in Hong Kong; an increasing variety of regional satellites like Indosat, Measat, Palapa, and Thaisat serving the ASEAN nations, Astra, British Sky Broadcasting, Eutelsat, Hispasat and Telecom serving Europe and beyond; and systems in such diverse nations as Canada, Brazil, Mexico, Argentina, Saudi Arabia, Korea, Japan, Australia, Russia, Turkey, Israel, Iran, India, China, and Taiwan. The managers of Intelsat and Inmarsat recognized the need to adapt to such changed circumstances. They have offered to relinquish treaty-level privileges and immunities in exchange for greater latitude to operate as commercial enterprises that can respond to competitive necessity with selective rate reductions.

1.2.5.6.2 *Satellite Service without Frontiers*

Most geostationary orbiting satellite footprints traverse national boundaries. This technological feature creates financial opportunities to aggregate traffic, but it also creates difficult political, intellectual property, and cultural challenges. Proliferating satellites may exacerbate concerns about "cultural imperialism," because more video program options may lead to audience migration from national programming to foreign programming. Despite the potential for satellites to operate without frontiers, national governments may attempt to impose border limitations by denying "landing rights" for particular satellites and by restricting the amount of foreign programming available to national cable television, broadcast, or Direct Broadcast Satellite (DBS) operators.

Satellites present both a blessing and a curse in the broadband telecommunications and information processing environment. On one hand, they can provide global access to news, information, entertainment, education, "edutainment," and "infotainment." The Global Information Infrastructure incorporates satellites to provide access in the vast regions of the world that do not qualify for broadband wireline deployment, and in most instances, have not even secured narrowband access to "Plain Old Telephone Service." On the other hand, many national governments experience great ambivalence when they recognize that satellites make boundaries porous and the citizenry more vulnerable to outside influences.

The transborder nature of a satellite footprint all but eliminates mutually exclusive domestic and international markets. National governments have found it impossible to prohibit or regulate the use of satellite terminals and the extent of citizen access to the rest of the world. Miniaturization of satellite terminals and the increasingly integrated global economy make it both technologically not feasible and commercially imprudent, even to attempt to restrict foreign investment and involvement in telecommunications ventures.

1.2.5.6.3 *Service to a Mobile, Wireless, and Networked Society*

Satellites will perform an increasingly significant role in building a terrestrial wireless infrastructure to work in conjunction with wireline options. The satellite component will fill the extensive gaps where terrestrial networks operate poorly or do not exist, but where people require access to the rest of the world in the form of another person, corporate network, company database, or transaction system.

The unprecedented marketplace success of cellular radio and other mobile technologies confirm the demand for reliable, tetherless access to the rest of the world while on the move. Mobile telecommunication networks can enhance business productivity and efficiency. For the time being, however, only "islands" of local, cellular, and special mobile radio services exist. Even when nationwide cellular roaming becomes possible, a variety of different operating standards will limit the prospect for using the same transceiver when traveling abroad.

Low and middle earth orbiting satellite projects make personal communications global in scope. These ventures include a constellation of non-geostationary orbiting satellites providing an interoperating array of beams that illuminate the entire globe. Individually and collectively these systems can provide ubiquitous, wireless, digital coverage to pocket-sized telephones.

Telecommunication planners have coined the phrases Universal Personal Telecommunications (UPT) and Global Personal Communication Services (Global PCS) to identify communication options free of cords, using available terrestrial radio options augmented by satellites. Numerous logistical and regulatory problems must be resolved to make this vision a reality, but the demand exists. The U.S. government has raised billions of dollars by auctioning off spectrum for use by personal communications service (PCS) networks that will expand the capacity and availability of terrestrial mobile radio options by reducing the size of transmission cells from several miles, as is the case in cellular radio, to several hundred yards.

The considerable optimism for terrestrial and satellite delivered telecommunications stems, in large part, from the unprecedented increase in cellular radio service demand. In the 10 years after the mid-1980s, usage has risen from near 0 to almost 30 million subscribers in the U.S. Yet even with such a steep and profitable rate of usage, cellular radio has achieved a market penetration of no greater than 10% — well short of the 30% is needed to constitute a mass market. The mobile service entrepreneurs expect terrestrial options, like PCS to achieve mass market penetration, because the infrastructure will provide ample capacity, and handset and usage costs should drop significantly below cellular radio levels.

Mobile satellite service operators can ride the coattails of terrestrial mobile service market success by providing service to "dual mode" transceivers that cut over to the satellite option when terrestrial service becomes unavailable. If terrestrial systems can achieve profitability with a market penetration of less than 10%, then it follows that global or regional satellite systems need only acquire a small portion of the total wireless market to achieve success as well. Despite the relatively small number of subscribers needed, mobile satellite ventures present substantial risk because of their cost (approximately $9 billion for Teledesic and $5 billion for Iridium) and the use of unproven technologies. LEO systems require extensive management information systems and network coordination to link as many as 228 fast-moving satellites. The Iridium satellite constellation will communicate not only with ground stations but between satellites. Consumers may balk at a $3.00/min charge, but conditions already exist where access to the rest of the world comes at a price of $10.00/min or more (from hotel rooms, the high seas, and business communication centers in countries with unreliable conventional networks). Most LEO systems will provide voice and slow-speed data transmission without capacity to meet broadband applications.

Several new satellite proposals provide broadband satellite options. Hughes has proposed the Spaceway system, a constellation of geostationary orbiting satellites that will operate in the extremely high frequencies at 20 to 30 GHz. By becoming a first-time operator in this frequency band, Hughes will have ample spectrum available so that it can provide more transponder capacity than currently available at lower frequencies. Teledesic proposes the commercial rollout of "star wars" technology with 860 refrigerator-sized satellites. These satellites will operate as global web capable of providing the same kind of broadband functionality currently available only from terrestrial options.

1.2.5.6.4 *Regulatory and Governmental Issues*

Even if quasi-diplomatic cooperatives like Intelsat and Inmarsat did not exist, the transborder operation of satellites would still generate substantial governmental involvement. Governments work together at the international level through a United Nations specialized agency known as the International Telecommunication Union (ITU), and they intervene at the domestic level through the ownership, operation, or regulation of national carriers.

1.2.5.6.5 *Satellites in a Global Information Infrastructure*

An increasing number of governments have decided to revamp the telecommunications industrial structure and regulatory process. Some nations have privatized the incumbent carrier. A larger number have retained government ownership but have sought to stimulate a more businesslike approach to the provision of telecommunication equipment and services. This is done primarily by authorizing competition in some markets, reducing regulatory oversight, and freeing the incumbent carrier to respond to competition.

When incumbents lose market share and profits in domestic markets, they typically look beyond their borders for new opportunities. In many cases, they can enhance their global market development opportunities by teaming in a strategic alliance with other carriers. A variety of new ventures like the AT&T-led Worldpartners, the British Telecom–MCI Concert system and the Sprint–France Telecom–Deutsche Telekom Global One venture aim to provide integrated, global services to multinational enterprises. Such "systems integration" requires a blending of satellite, submarine cable, and terrestrial facilities to provide "one-stop-shopping" solutions to complex and diverse service requirements.

Deregulatory, procompetitive, and efficiency-enhancing initiatives result when nations recognize that revamping the telecommunications sector can stimulate the national economy. A strong correlation exists between having a state-of-the-art telecommunication infrastructure and the ability to participate in information-age markets like financial services and data processing. The Old World Order in satellite telecommunications blended elements of diplomacy and international relations with joint business ventures. This high-minded and leisurely approach juxtaposes sharply with the current view that tele-communications simply transports bit streams. While carriers have developed reputations based on efficiency and responsiveness, the services they render increasingly become a building block upon which customization takes place either by the carrier or the customer. This view characterizes telecommunications as a commodity, subject to fierce competition with narrowing profit margins.

The Old World Order sought to ensure network reliability by mandating carrier use of both satellite and submarine cable media. Regulatory agencies required "balanced loading" of facilities to ensure that carriers would activate circuits in the currently more expensive medium, with an eye toward supporting redundancy, security, and service reliability. Policy makers opted to blunt intermodal competition and market-based resource allocation presumably because circuit activation based on the comparative merits of various transmission media might discourage investment of promising, but currently more expensive, media. Such intervention in circuit activation stemmed from the belief that regulators had to prop up a new, more costly technology even if it lacked marketplace support.

The New World Order shows that sovereign consumers demand low prices, high service reliability, and quick response to user requirements. Satellites can satisfy these requirements, particularly for mobile, point-to-multipoint, and hinterland applications. Terrestrial wireline facilities constitute the core trans-mission medium that operators will use to provide the broadband telecommunication and information processing services contemplated by architects of an expanded national and international information infrastructure. Fiber-optic cable provides the most efficient and cost-effective method for transmitting large volumes of digital bit streams. However, this preferred distribution method requires relatively high-traffic density, and a business case cannot yet be made for extending fiber-optic cable into homes, even where such facilities will reach nearby distribution points — at the curb or a neighborhood pedestal.

Wireless options will constitute an integral part in upgrading the telecommunication infrastructure to provide ubiquitous broadband services. Typically a terrestrial wireless option like cellular radio and omnidirectional microwave systems (commonly referred to as Multipoint Distribution Systems) will serve areas with moderate traffic densities. The wireless satellite option will fill in the gaps where a business case cannot support either terrestrial wireline or wireless options.

The widespread roll-out of commercial DBS services (Primestar, DirecTV, and USSB) demonstrates a viable market for broadband, albeit one-way, services via satellite. Such a distribution technology can provide many of the two-way services contemplated by the information infrastructure initiatives through integration with plain old telephone service. Most of the upstream flow of traffic from home and office to data source represents narrowband commands, e.g., requests to download data, pay-per-view movie selections, transaction authorizations, etc. A personal computer or television set top converter, which accesses a telephone line, can secure broadband delivery of content via satellite.

Already a vast array of very small aperture terminals provide businesses with satellite delivery of such diverse applications as teleconferencing, training, inventory control, and real-time transaction processing like credit card verification. Most of these applications involve closed, intracorporate networks. As information infrastructure developments reach rural locales and residences, corporations may consider migrating to quasi-public media like the Internet to make services more accessible to end users. As long as users experience no perceptible difference in service quality, which was the case with satellite-delivered voice communications, information service providers will deploy a satellite distribution option where conditions warrant an alternative to the terrestrial and wireline mainstream.

References

1. A. C. Clark, The future of world communications, *Wireless World*, p. 305, October 1945.
2. D. M. Jansky and M. C. Jeruchim, *Communications Satellites in the Geostationary Orbit*, Artech House, Boston, 1987.
3. For a more extensive outline of the ITU, see R. Frieden, *International Telecommunication Primer*, Artech House, Boston, 1996.

1.2.6 Regulation of Wireless Telecommunications in the U.S.

Charles M. Oliver

The Basic Telecom Agreement provides that "any procedures for the allocation and use of scarce resources, including frequencies ... will be carried out in an objective, timely, transparent and non-discriminatory manner. The current state of allocated frequency bands will be made publicly available, but detailed identification of frequencies allocated for specific government uses is not required."* In the U.S., the FCC regulates domestic uses of the radio spectrum through a two-step process. First, it "allocates" spectrum; i.e., it designates bands of spectrum for specified categories of use. Second, it "assigns" radio licenses to specific users — companies, private individuals, state and local governments, etc. The National Telecommunications and Information Administration (NTIA), which manages the federal government's use of radio spectrum, operates in much the same manner. Since many bands are shared between federal and nonfederal users, the two agencies must coordinate with each other.

Both the FCC and NTIA must operate within the constraints of international agreements on spectrum regulation. For most services, this means that the U.S. must have close and continuing conversations with Canada and Mexico for radio operations in border regions. Having completed such discussions, the U.S. has broad discretion to deviate from allocation tables laid down by the International Telecommunications Union (ITU), provided that it does not create interference with any other countries that follow the ITU tables.

1.2.6.1 Assigning Licenses

Traditionally, the FCC had one means of resolving mutually exclusive applications for radio licenses — comparative hearings. In 1981, Congress authorized the Commission to use lotteries. In 1993, Congress authorized the use of auctions.

1.2.6.1.1 Comparative Hearings

In principle, comparative hearings were supposed to select the best qualified and best-intentioned applicants. Sometimes they did. On other occasions, they assumed the rhetorical tone of parole board hearings. Even when participants avoided unctuous and unprovable claims of good character and beneficent purpose, outcomes were often decided on the basis of distinctions that bordered on irrelevance. One cellular telephone applicant bested a competitor by promising to deploy an additional mini-cell — in a place with no roads and no inhabitants.

Trees were often the biggest losers in comparative hearings. When the FCC announced its intention to convene comparative hearings for the first round of cellular telephone applications, the agency received so much paper that it had to call in a structural engineer to evaluate the ability of the Commission's headquarters building to support the legal verbiage and consultants' reports that accompanied the applications. The engineer advised the agency to distribute the applications away from the corner of the building where they had initially been shelved, lest the structure fall down.

Given the opportunities for corruption afforded by comparative hearings, it is remarkable that they have been accompanied by few public scandals. During the Eisenhower administration, an FCC commissioner was accused of accepting a bribe for his vote on a VHF television license, but escaped punishment by arguing that he was too stupid to know that he was being bribed. The public — and future potential clients — believed him. He was later found dead in a cheap Miami flophouse, with all the money to his name stacked in neat piles of coins on a bedside table. This embarrassment probably served as a warning to his successors. (For those who may be incredulous at the level of physical detail in this account, it is fully documented in an official report by the U.S. Senate Commerce Committee.)

---

*See United States of America Schedule of Specific Commitments, WTO Doc. GATS/SC/90/Supp. 2 (April 11, 1997) p. 6.

1.2.6.1.2 Lotteries

Congress eventually concluded that comparative hearings are not always the best means to choose among competing applicants, and authorized the use of lotteries for non-broadcast-license proceedings. The random selection process was faster than comparative hearings and reduced appearances of favoritism, but it produced anomalous results. Lucky lottery winners were rarely qualified to build and operate complex businesses like cellular telephone franchises. After holding their licenses for whatever minimum period was required by the commission, they would usually "flip" the licenses to someone else, typically a big phone company with enough money to outbid any other interested parties. A group of partners who won a cellular license for Cape Cod later benefited from a bidding war between the cellular licensees for Boston and Providence, and walked away with $40 million — after building nothing.

The Cape Cod episode was an unusual case, but the total value of spectrum given away by lotteries amounted to real money. In 1992, NTIA estimated that the market value of licenses given away for cellular telephone licenses alone, most of them in lotteries, amounted to between $40 billion and $80 billion, depending upon what appraisal method one used. Either figure, it noted, was more than the U.S. government spent to defeat Iraq in Operation Desert Storm.

The initial response among Congressmen with oversight responsibility was to urge the FCC to strengthen its so-called antitrafficking rules, to prohibit new licensees from selling out too soon. It quickly became apparent that this remedy would create an even bigger problem: unqualified and underfinanced lottery winners sitting on their licenses, engaging in the minimum amount of construction required until the waiting period expired.

1.2.6.1.3 Auctions

Despite the proliferating absurdities, Congress resisted auction proposals for 12 years after authorizing lotteries, on the ground that auctions would grant licenses to "deep pockets." This is one area where the U.S. did not take the lead — New Zealand successfully implemented spectrum auctions years before the U.S. did. Competitive bidding for radio licenses was finally authorized in the U.S. in 1993 as part of a Budget Reconciliation Act, when an incoming administration needed money and, for the first time in memory, was of the same party as the Congressional leadership. Since then, the FCC has accepted bids totaling more than $20 billion at spectrum auctions.[*]

Besides raising money for the public treasury, auctions are faster than lotteries because they can put radio licenses immediately into the hands of the companies that are most likely to develop them and bring service to consumers. The FCC has optimized the process within the constraints of existing law by arranging simultaneous auctions for multiple, potentially interdependent licenses.

In other respects, the FCC auction procedures have fallen far short of their potential for improving consumer welfare. The essence of the problem is that, while the Commission's auctions have facilitated license assignments, they have done little or nothing to improve the efficiency of spectrum allocations.

1.2.6.2 Allocating Spectrum Bands

Many of the people involved in spectrum regulation have mounted on their office walls colorful charts showing the hundreds of bands allocated to various narrowly defined service categories. This system would be relatively harmless if the amount of spectrum allocated to each service bore any rational relationship to the amount of spectrum needed. The problem is that it does not and cannot bear such a relationship.

1.2.6.2.1 The Existing Allocation Process

To appreciate the level of detail practiced by the FCC in its regulation of spectrum allocations, imagine that a federal agency had been put in charge of real estate zoning throughout the U.S., and that it had referred to headings in the Yellow Pages of telephone directories for guidance on the establishment of appropriate zoning categories. Certain parcels of land would be earmarked for dry cleaning establish-

---

[*]Some bidders have defaulted on their payment obligations or sought to reschedule payments.

ments, as one example. If space in a commercial building were vacated by a florist shop and you wanted to use it for a bookstore, you would have to ask the agency to change its allocation table or issue a waiver.

This kind of procedure applied to spectrum regulation has engendered an understandable degree of frustration among policy analysts, to the extent that some of them have called for the complete elimination of spectrum allocations and total reliance on market processes. Others have called for the abolition of the FCC. Either of those remedies would be overkill, but the situation does warrant consideration of significant changes.

As senior policy adviser to the NTIA administrator, I once asked the agency's spectrum engineers how many allocation categories the FCC would need if its only policy objective were to prevent electrical interference. The answer was, "About six." From an engineering standpoint, the FCC could establish six broadly defined spectrum categories. Within those allocations, it could conduct auctions and tell winning licensees how much electrical energy they could discharge in specified frequency bands and geographic areas. Subject to those constraints, licensees could do whatever they wanted, within broadly defined service definitions. Incumbent licensees could be grandfathered but would be liberated either to sell out or to provide different services to themselves or others. The uses to which spectrum was put would depend primarily upon market demand, changing technology, and licensees' individual needs and circumstances.

The FCC implicitly recognized the advantages of such an approach when it established ground rules for the so-called personal communications service (PCS). Frequencies with the PCS moniker can be used for almost anything the licensee wants, other than free broadcasting.

The FCC rarely provided such service flexibility in earlier allocation decisions. Part of the explanation is technological. Before computers became inexpensive and easy to use, it was simpler for spectrum engineers to coordinate like users than to coordinate disparate kinds of uses within a given frequency band. Radios built before the 1990s had limited means to screen out unwanted signals — unlike the "smart" transceivers being developed for the PCS market, some of which incorporate technologies developed by defense contractors to combat purposeful enemy jamming. The FCC also had political reasons to avoid service flexibility: many interest groups believed, and still believe, that they can obtain more spectrum by investing in lobbyists than by bidding for spectrum in a competitive market.

Some of the people who oppose simplification of the allocation table are motivated by genuine public interest concerns. If television stations were forced to bid against mobile radio operators for spectrum, most of them would go off the air. Many people believe that free over-the-air broadcasting produces public benefits that exceed the money that broadcasters can collect from advertisers. This view is especially strong among broadcasters themselves. Amateur radio hobbyists are similarly convinced that they provide significant public benefits, even though they collect no money at all for their services. Ambulance drivers and policemen also raise arguments for special treatment. Some of these arguments make sense; in other cases, the needs would be better met by subsidies in cash instead of implicit spectrum subsidies.

In analyzing the political dynamics, it is important to recognize that the people who lobby spectrum allocation matters in the U.S. do not necessarily approach the FCC on bended knee; in many cases they have persuaded the agency to delegate "coordination" within designated bands to industry advisory groups. The Commission rubber-stamps many of the decisions made by such groups and is grateful to have off-loaded a significant amount of work. The designated representatives of the advisory committees in turn take satisfaction in their contribution to the public welfare, and in some cases have amassed a significant amount of power. Many of these individuals would oppose a restructuring of the allocation process that took them out of the loop.

1.2.6.3 Proposals to Improve the Allocation Process

Given the Commission's sensitivity to spectrum coordinators and other sources of political pressure, the agency is unlikely to adopt sweeping proposals for reform in the short run. For the time being, the best we can realistically hope for is a gradual reduction in the number of allocation categories and a concomitant broadening of the degrees of freedom permitted within those allocations. Fortunately, progress on spectrum coordination issues is not an all-or-nothing proposition. Significant improvements can occur through incremental changes, even while the Commission and its critics continue to debate arguments for fundamental reform.

Although spectrum regulators operate in a high-technology context, they can use many of the same political techniques that successful reformers have used for centuries. Avoid the temptation to preach to the choir; the people who need to be won over are the unconverted. Instead of trumpeting the extent of change, emphasize continuity with past precedents. Provide concrete assurances to stakeholders who fear that incremental changes will lead to more dangerous departures down the road. Buy off politically influential opponents with targeted concessions, but give away as little as possible. Engage in well-timed saber rattling to intimidate opponents, but avoid stabbing anybody unnecessarily.

Congress, the FCC, and NTIA can apply those techniques as follows:

- *Stealth Deregulation.* Before it was merged into the newly created Wireless Bureau, the FCC Private Radio Bureau pursued an unstated policy of retaining an alphabet soup of narrowly defined allocation labels but allowed "sharing" among licensees in different categories. The Bureau culti-vated a retrograde demeanor while pursuing a progressive deregulatory policy, and gradually transformed the Specialized Mobile Radio (SMR) service into the functional equivalent of cellular telephone service. The cellular telephone industry was enraged when it discovered what was happening, and persuaded Congress to impose cellular-style regulation on SMR operators — but Congress did not roll back the *de facto* reallocation that had already been accomplished.

- *Property Rights.* Today, many incumbent radio licensees get nervous when somebody proposes to offer them more freedom, because they fear that freedom will be a preamble to expropriation. The licensees would face the prospect of change more confidently if Congress allowed the Com-mission to grant property rights in spectrum, so that radio licensees would receive the same constitutional protections against uncompensated expropriation available to other forms of prop-erty. In that context, broader allocation categories would provide incumbent licensees with an opportunity to profit from change — by attuning their services more closely to changing demands of technology and markets, or selling out to others seeking to do so.

- *Bribery.* Today, whenever the FCC conducts a spectrum auction, the Commission must return the proceeds to the U.S. Treasury. Congress should authorize the agency to keep some of the proceeds and use it to buy incumbents out — clearing bands congested with obsolete uses and making them available for refarming.

- *Warehousing.* Today, whenever markets or technology change, innovators seeking to provide new kinds of radio service must buy out incumbent licensees or persuade the Commission to do so. There is rarely an adequate supply of unused spectrum available in inventory, as there is for almost any other commodity. One of the reasons is that Congress has required the FCC to be vigilantly watchful against spectrum "warehousing" and take licenses away from anybody who is not making immediate use of them. Congress should instead allow private entrepreneurs to buy radio spectrum and use their own judgment when to put it to use, or when to hold it in reserve for possible future needs.

- *Technological Fixes.* One legitimate reason the FCC relies on spectrum allocations is that it is easier to prevent interference among comparable uses than among disparate, commingled uses of spec-trum, even though doing so inevitably results in underutilization of a scarce asset. Either the Commission or private spectrum brokers should be encouraged to apply more computer hardware, software, and engineering analysis to spectrum coordination, using proceeds from the auction process.

- *Accounting for Government Use of Spectrum.* Some of the biggest spectrum hogs are federal agencies. All such agencies should be required to put their licenses into a spectrum bank and lease them back — for cash — at prevailing market prices. Those payments should be treated like any other expenditure, appearing on the agencies' budgets and subjected to regular Congressional review in the authorization and appropriations process. Congress could fund the entire process by selling spectrum to private brokers — in effect, privatizing a resource that was nationalized in the 1920s.

- *More Bribery.* Some of the most politically sensitive reallocations involve public safety users. Congress should authorize the FCC to pay hush money to some of these groups. Public safety organizations need to communicate with police cars and ambulances, but they might find ways to do so more efficiently if they were offered cash instead of spectrum. They could use some of the cash to acquire spectrum at market clearing prices, with the rest going for other things they might need.

- *Spectrum Parks.* Increased reliance on market processes need not preclude some continuing set-asides of spectrum for public purposes, like public broadcasting. There is a viable market in real estate, even though government has set aside some lands for public parks and applies zoning regulations to property in private hands.

- *Talk about Abolishing the FCC But Don't Do It.* Eliminating the FCC, as some have proposed, would not really solve existing problems with the spectrum regulation process, but talking about eliminating it might facilitate the decision-making process. As Samuel Johnson said, "The prospect of being hanged in a fortnight has a wonderful ability to concentrate one's powers of attention."

The underlying premise of all these proposals is that markets are usually more efficient than government agencies in deciding how to use scarce resources, but that market-based transactions between private actors do not necessarily take into account all the side effects that private actions can have on society as a whole. Some government intervention in the market can therefore be justified on purely theoretical grounds.

These suggestions also reflect a philosophy of governance: that efficient solutions can sometimes be approached, but rarely achieved, and that good-hearted guile often achieves more than hard-headed sanctimony. This advice comes not from Machiavelli, but from an earlier and even wiser commentator who said, "Lo, we are as sheep among wolves. Be ye therefore wise as serpents, and gentle as doves."

1.2.6.4   Lessons in the U.S. Experience for Other Countries

Based on this brief review, it should be obvious that the U.S. cannot serve as a perfect role model for countries seeking to liberalize their telecommunications markets and derive maximum benefits from competitive entry. The American experience is worth studying both because it has involved some notable successes and because others can hope to avoid repeating its mistakes.

The most glaring problem with the U.S. regulatory regime is the fragmentation of regulatory authority. The necessity of obtaining cooperation and support from 50 state regulatory authorities is a daunting prospect for any carrier and is doubly so for foreign carriers, especially when the procedures for court review are equally fragmented. Universal service subsidies could appropriately be funded out of general tax revenues; instead, Congress has delegated untrammeled authority to the FCC and state regulatory commissions to impose quasi-taxes on telecommunications providers.

The Computer II and Computer III rules are premised on the assumption that local telephone service is a monopoly. As local facilities-based competition spreads, the U.S. will probably begin to relax some of those requirements and limit their scope. Other countries may wish to consider how much of an investment they want to make in constructing a Computer II/III-type regulatory infrastructure if they are vigorously encouraging facilities-based competition at the local level.

There is an artificial scarcity of radio spectrum in the U.S., caused in part by heavy use of the radio band by government agencies, but more significantly by arbitrary and shortsighted allocations to narrowly defined services. The FCC has recognized that problem and has begun establishing much broader definitions of permissible services when opening up new bands. That represents progress on a going-forward basis, but it still leaves the airways constricted by past allocational decisions and restrictive service operation rules.

Americans have adopted less than perfect regulatory regimes, not because the regulators involved are stupid, but because they have to navigate treacherous political currents in a complex governmental system. The U.S. was formed not by military conquest but by negotiation among sovereign governments. The compromises that were necessary to form that union still play a part in our national life, most notably in the structure of the U.S. Senate, which gives disproportionate influence to residents of rural areas. Powerful industrial lobbies also exert a significant influence. Together, these converging considerations have generated perceived entitlements that are difficult to gainsay in regulatory and legislative deliberations.

Winston Churchill once observed that the English-speaking peoples chose democracy, not because it is the best form of government, but because the alternatives that came before were even worse. Other countries can derive inspiration from the U.S. system of telecommunications regulation in the same spirit. Those who follow can seek to emulate its successes without replicating its mistakes, while adapting the principles involved to their own unique situations.

1.2.7 International Wireless Telecommunications Regulation

Pamela Riley and Chuck Cosson

This section examines common issues addressed by regulators in non-U.S. markets, primarily with regard to terrestrial mobile services including cellular, PCS, and paging services. Mobile satellite services raise similar concerns, but also raise a substantial number of additional issues not dealt with here. The topics addressed include the following:

- Questions related to the structure of the mobile market and role of regulation in that structure;
- The process of identifying spectrum frequencies, allocating them to particular uses, and licensing them to particular operators;
- The process of interconnecting mobile networks to the public switched telephone network (PSTN);
- Future developments and regulatory trends including third-generation mobile services and convergence between fixed and mobile service markets.

1.2.7.1 Introduction: The Basis for Regulation

Wireless regulation is premised on the fact that radio frequencies must be rationed among competing interests because they are a uniquely finite natural resource. Since two radio signals on similar frequencies potentially will interfere with one another, nullifying the commercial value of both, it is necessary to establish rules to prevent interference and establish a process for enforcement. The nature of mobile service requires exclusive use of spectrum in a given geographic area.

Historically, the principle of physical scarcity led to government spectrum ownership and licensing. But physical scarcity of a resource is, by itself, an insufficient basis for governmental regulation to allocate it. All resources are finite. The question unanswered by the reference to "scarcity" is whether a government role is needed to develop the rules of trading and assign rights to resources because a free market system of private property rights, enforced through judicial proceedings, would perform that function poorly. A number of scholars believe that a system of private property rights would in fact be a better means of performing these tasks with regard to wireless spectrum.[1]

On the other hand, it can be argued that the transactional costs involved in a private system would create inefficiencies. Most governments have created regulatory systems to govern the spectrum allocation, licensing, and market competition of wireless operators.[2] These government systems have varying degrees of regulatory vs. market control — some systems now use auctions to allocate licenses. But because they are premised on a system of government distribution and not private property rights in spectrum, there are a number of issues in common. It is to these systems of regulation — and those issues — that we now turn our attention.

1.2.7.2 Wireless Competition and Market Liberalization

In most, if not all, non-U.S. markets, telecommunications services were until recently provided by a single entity, established as a government department.[3] In many cases, the telecommunications network was associated with the mail delivery systems. These entities, sometimes known as the PTT for Posts, Telephone and Telegraph, were the only entities legally authorized to provide telecommunications services, including mobile wireless services. In many cases, therefore, licensing decisions were simple: the license to develop new wireless technologies such as paging and cellular was simply assigned to the PTT.

Economic and political forces worldwide have brought state ownership and monopoly regulation into disfavor.[4] Technological advances have lowered the cost of telecommunications networks, thereby eroding the case for "natural monopolies."[5] At the same time, telecommunications and information systems have increased in economic importance. Competition is viewed as superior to monopoly at creating efficient operations, due to incentives to gain market share through lower costs and higher profits.

Additionally, economic pressure has been driving a trend toward privatization. Private companies tend to be more disciplined, efficient, and customer focused than state-owned firms. Government monopolies can be slow to respond to market demand, and face difficulty competing for capital with other public programs funded through taxation, thereby hindering national economic development. Finally, privatization can be seen as a necessary condition of competition. State ownership is incompatible with competitive markets, since the government possesses incentives to distort competition in favor of the state-owned enterprise.

The wireless telecommunications sector was often the first sector to be liberalized. Market competition in the wireless sector was facilitated by the economics of wireless systems — which could be built to cover an entire country in under 2 years. Liberalization was also appealing in its own right — regulators were beginning to find traditional monopoly regulation unwieldy and were struggling with the inability of rate regulation to produce sufficient incentives for greater efficiency. Concluding that competition was a preferable mechanism for ensuring fair prices for consumers, countries later began to issue additional licenses for wireless businesses, and removed (or did not impose) the burdens of traditional rate regulation.[6]

New laws have been adopted worldwide to permit mobile services to be provided by entities other than the PTT. In Europe, market liberalization structures were a goal in the development of a European Common Market, which required European member states to break down the dominance of national PTTs in order to permit cross-border provision of services,[7] including mobile communications.[8] In particular, a 1996 Mobile Directive required, among other things, that Member States:

- Eliminate any remaining special and exclusive rights to mobile licenses;
- Liberalize infrastructures, so that mobile operators could build their own fixed or microwave transmission networks (subject to availability of frequencies) or use network owned by third companies;
- Ensure the rights of mobile operators to interconnect directly with the public network and with mobile operators in other Member States by the same date;
- Consider license requests to operate digital mobile services using DCS 1800 technologies from 1 January 1998.[9]

The process of liberalization has also begun in many less-developed countries, to acquire much-needed capital that governments cannot obtain from the domestic tax base, and to permit new investments in telecommunications. Developing countries, too, are coming to believe that state-owned businesses are not the best way to reap the benefits of the telecommunications revolution.[10]

Among the regulatory challenges faced by those who sought to invest in new mobile telephone services were restrictive foreign ownership rules. Most countries limit the percentage of equity in a telecommunications provider that can be owned by nonnational entities. Recently, national authorities have begun to relax some of these limits for wireless businesses, due to the desire for market competition and rapid network construction. For example, Article 12 of the Mexican Telecommunications Law limits foreign investment in public common-carrier telecommunications networks to 49%, but provides an exception for cellular services. In Mexico, foreign investment in cellular services can exceed 49% with the prior authorization of the National Foreign Investment Committee.[11]

The World Trade Organization agreement to liberalize international trade in basic telecommunications services, signed in 1997, will also likely lead to further relaxation of foreign equity limitations in some countries. The agreement will come into effect on February 5, 1998. The 72 WTO member governments, which have agreed to open their domestic markets to foreign companies, account for nearly 93% of the total domestic and international revenue of $600 billion generated in this sector annually.[12] Examples of the services covered by this agreement include voice telephony, data transmission, telex, telegraph, facsimile, private leased circuit services (i.e., the sale or lease of transmission capacity), fixed and mobile satellite systems and services, cellular telephony, mobile data services, paging and personal communications systems.[13] Appendix 1.2.7A lists a number of countries that have introduced schedules that will liberalize these services.

Another obstacle to robust development of competition in overseas wireless markets was the absence of an independent regulator to oversee the PTTs. Non-U.S. countries are also gradually creating independent entities to regulate the telecommunications industry. Countries in Latin America, Eastern Europe, and Asia have all looked to the U.S. FCC to some degree as a model for a sufficient regulatory institution. For example, Argentina has established the Comisión Nacional de Telecomunicaciones (CNT). Much like the U.S. FCC, the CNT is composed of six directors appointed by the executive branch, and oversees the operations of the wireline networks as well as administering radio spectrum for telecommunications services. Its decisions must be made on a public record and are subject to judicial review.[14]

In other cases, the national regulator looks quite different. In Japan, for example, the process of privatizing Nippon Telegraph and Telephone (NTT) led to expansive new powers being assumed by the Ministry of Posts and Telecommunications (MPT). For example, the MPT took the unusual step of assembling the consortia assigned licenses to compete with NTT for paging and mobile telephone services. The MPT continues to exercise a much more activist role in regulating the mobile market than has ever been undertaken by state or federal regulators in the U.S or in Europe.[15] Some countries, i.e., Portugal, do not yet have a regulatory authority that is fully independent from the PTT.

A number of regulatory issues related to a liberalized market were raised by incompatible standards. Absent a single air interface standard, it was difficult for new operators to achieve economies of scale in equipment purchases, and to establish arrangements for cross-border roaming. The Global System for Mobile (GSM) standard addressed these issues in Europe and contributed significantly to the success of international mobile operators. The issue of standards is the subject of a separate section. Here, we only mention that in many countries, regulators required (and continue to require) that applicants for mobile telephony licenses utilize a specific technology.

1.2.7.3 Licensing of Spectrum for Services

Licensing of spectrum for mobile services involves a number of questions to be addressed by regulators:

- Spectrum availability and allocation
- Number of licenses and size of serving area
- License conditions, including term of license and renewal rights
- Technology choices
- Assignment tools: auctions, comparative evaluations ("beauty contests")

1.2.7.3.1 Spectrum Allocation

A first step is to determine which sections of the radio spectrum will be allocated to particular services. International service functionality, as well as economies of scale in manufacturing, are enhanced where different countries mutually agree to use particular spectrum bands for similar, noninterfering uses. Thus, the allocation of spectrum is one of the few areas of mobile telephony in which truly international regulations play a role.

To facilitate international spectrum allocation, the International Telecommunication Union (ITU) has divided the world into a number of regions and areas, with the intent of achieving harmony in spectrum allocations to particular services within those areas. For example, within Region 1 (generally covering Africa, the Middle East, Europe, and Russia) a given segment of the radio spectrum will be allocated to one or more radio uses, the decision being based on a number of technical, political, and economic factors.[16] Within the ITU allocations, individual countries are not constrained in determining what spectrum will be allocated for specific telecommunications services.

The European Union has a central role in allocating spectrum for licenses issued by European Union member countries. Between 1987 and 1991, three directives were adopted to reserve specific frequency bands for the provision of GSM and DECT (Digital European Cordless Telephone) services, and channels for ERMES (European Radio Messaging System) paging licenses. The Council of Ministers of the European Union also made specific "recommendations" establishing the deadlines for coordinated introduction of services based on these technologies.[17]

1.2.7.3.2   *The Number of Licenses per Market*

The amount of spectrum allocated for mobile telephony uses and, in some cases, the cost of relocating existing users, is a major factor in determining the number and timing of new licenses. There is more to it than simply the amount of spectrum available. The number of licenses determines the number of facilities-based competitors in the market, thus limiting market entry. At earlier stages of market development, governments should introduce new competitors at a measured pace in order to maximize investment and growth by those seeking to establish the market.

In non-U.S. markets, most countries awarded an analog cellular license to the wireline incumbent operator; competitive licenses were not assigned until digital PCS technology was introduced in the 1990s. The first digital PCS service began operations in the U.S. in 1995. PCS encompasses a wide variety of mobile, portable, and ancillary communications services to individuals and businesses. The FCC divided the PCS spectrum into three broad categories: broadband, narrowband, and unlicensed. The majority (120 MHz of the spectrum) is for broadband PCS, which will offer primarily mobile telephone service. Although PCS was initially thought to serve a different market from cellular, PCS has become the umbrella term for digital mobile services including cellular. In fact, the only difference between cellular and PCS is the higher band in which the frequency allocation resides. Narrowband PCS is used for advanced messaging and paging while unlicensed PCS will be used for short-range communications such as local-area networks in offices. Unlicensed systems will operate with very low power and will have a limit on the duration of transmissions.

Most countries have issued licenses for cellular telephony on a national basis, but this is not universally the case. India, for example, has issued regional licenses: two operators in each of four major metropolitan areas or "metros," with two regional licenses in each of India's 20 states or "circles" (excluding the metro area within the circle).[18] When Japan introduced competition for the mobile sector in April 1994, the government allocated spectrum to four carriers in each regional block. This was followed by the similar introduction of PHS carriers in July 1995. The chart attached as Appendix 1.2.7B shows a data sample of broadband cellular telephony markets for various countries as they stood in early 1998.

Paging services and mobile data services have also generally been licensed in a similar manner: analog service licensed initially to the state-owned operator, with new entrants providing digital service granted licenses within the last 3 to 5 years. For example, in France, analog paging services are operated solely by France Telecom. Digital service was licensed by Infomobile in October 1994, and by SFR in April 1995. France has also licensed two mobile data service providers, France Telecom and TDR, who provide a service known as "Mobitex."[19]

There are no major regulatory issues concerning the provision of basic mobile data services by cellular carriers. Most cellular networks possess the ability to provide basic data services such as messaging, fax services, and Internet access, albeit at very slow speeds (usually around 9600 baud). In general, regulatory issues for mobile data revolve around third-generation issues and the ability to acquire more bandwidth to provide high-speed data services.

1.2.7.3.3 Operating Conditions

Cellular licenses included a variety of operating conditions as part of the government's offer of a license. A typical European license required a range of services to be offered (e.g., mobile voice service, call rerouting, call blocking, charge indication, closed user group, conference calls, call waiting, call hold, call transfer, caller ID and restraint of caller ID, data, etc.).

License conditions often include build-out requirements — an agreement to provide a particular degree of network coverage by a specific date, service quality standards, or other operational conditions. Build-out requirements can be either a requirement to serve a particular percentage of the license area population, or to serve a particular percentage of the geographic area. Additional public service responsibilities such as access to emergency services and wiretap capability are imposed.

Requiring an outlet of distribution of mobile services for non-facilities-based service providers (e.g., resellers of airtime) has been used in the U.K. to promote competition at the retail level. However, most other countries in Europe permit operators the freedom to establish distribution channels without specific requirements. The value that resellers bring to a market is in the form of customer choice. For example, resellers can target a market and offer specially tailored plans for that segment. However, given the competitive nature of the wireless industry, it is normally seen as unwise to force operators to enter into business arrangements with service providers. For example, regulatory requirements to "unbundle" mobile networks would create unique economic and technical problems. The uncertainty and added risk to mobile operators would in fact deter investment in more extensive network build-out, encourage companies to piggyback on the investment of others, and reduce today's rapid growth in infrastructure.

1.2.7.3.4 Licensee Selection

Governments rely on two basic mechanisms to award competitive licenses: comparative evaluations, or "beauty contests," and auctions. In a comparative hearing, the government invites firms to submit applications demonstrating how their proposal best meets established criteria. In an auction, firms are required to meet threshold financial and technical eligibility criteria, but the license is simply awarded to the party who submits the highest up-front bid for the spectrum. Often, there is a mix of the two approaches: the government will weigh both technical and operational experience as well as the amount of up-front payment, proposed retail price, or combination of such monetary offers.

A third option to assign frequencies is to use lotteries; these were used extensively in the U.S. in the 1980s to award cellular licenses, with mixed success. Lotteries have some of the same advantages as auctions in that they are objective and can be conducted quickly. Unlike auctions, however, a lottery includes no technical, commercial, or price mechanism to determine whether the licensee values the license most or will put it to best use. Outside the U.S., few countries other than Hungary have shown any interest in using lotteries.[20] The trend in licensee selection is toward a mix of comparative evaluations and auctions. Below, each process is described in detail, followed by an analysis of the advantages and disadvantages claimed for each.

1.2.7.3.5 Comparative Evaluations

An exemplary case of a comparative hearing is a 1998 tender process in Switzerland that awarded two licenses.[21] The Swiss Communications Commission awarded two new nationwide mobile radio licenses for the GSM 900 and DCS 1800 bands (a corresponding license was awarded to the PTT without a tender procedure). The tender submissions, in relevant part, requested parties to demonstrate the following items:

- Technical knowledge with respect to construction and operation of a mobile network, as well as the marketing of corresponding services.

- The applicant's "productive power," i.e., ability to implement technical and commercial planning in order to meet the conditions of the license, e.g., financial resources, project management and human resources, and technology resources such as antenna locations, etc.

- Information on the applicant's technical plans for ensuring that the full range of services required are provided, for network interconnection, number portability, call quality, emergency services, and other operational aspects. The tender also requests data on network security and reliability.

- The applicant's commercial plan, including an outline of the present net cash value of the licensed business and an estimate of a break-even point, and the assumptions and estimates upon which this calculation is based. The business plan should include a market analysis and planned balance sheets, profit and loss statements, and expected financing requirements.

Each item is weighted; for example, the last two items (technical and commercial planning) are weighted more heavily than the first. The license is to be awarded to the applicant that "appears best suited to satisfy customer demand for mobile radio services." The Swiss approach represents a classic "beauty contest" approach to the award of mobile radio licenses — the winning bidder is selected on the basis of technical and business acumen, not price.

Other forms of comparative evaluations are actually "comparative bidding." They are essentially comparative proceedings that include various types of "beauty contest" criteria, but which emphasize price factors such as commitments to provide high quality, low prices, and/or to pay a larger license fee to the government. These systems are a form of market mechanism for determining service prices, but they are not the same as an auction, which uses a market mechanism to determine the value of the spectrum itself. An example of this type of comparative proceeding is Brazil, where the Ministry of Communications issued a tender for mobile cellular service "with the combination of the lowest tariff amount for the service to be rendered and the highest offered amount to be paid for the right to exploit the service and associated radio frequencies as judgment criterion."[22]

Auctions have gained in popularity as a license assignment tool. New Zealand pioneered some of the early spectrum auctions in 1989 based in part on frustrations with the administrative burdens of other license assignment methods.[23] Other countries have since used auctions of various types to award licenses for mobile radio spectrum.

Types of auctions include:

- The "English auction," often used for works of art and agricultural commodities. The auctioneer increases the price until a single bidder is left.

- A "Dutch auction," where the auctioneer announces a high price, and then reduces it until the item is claimed by a bidder.

- "Sealed-bid auctions" of two types. In the "first-price" variation, the highest sealed bid submitted wins and pays the price bid. In the "second-price" variation, the highest bid pays the second highest amount bid.

- "Simultaneous multiple round auctions," as used in the U.S. A number of licenses are offered simultaneously, and the highest bid on each license is identified. Then another round of bids is accepted, and the process continues until no new bids are submitted on any licenses. Bidders can thereby change their strategy as the auction progresses. This process is theoretically thought to lead to a better approximation of the market value of the license than a sealed-bid auction, but practice has shown that it can result in bidders overpaying, perhaps by changing their strategy in response to the emotionally charged atmosphere of competitive bidding instead of the rational business criteria employed in a sealed-bid auction.

Although auctions are thought to avoid the long delays and deliberative processes associated with comparative evaluations, some auctions have nevertheless created litigation that has protracted the licensing process. In Europe, the introduction of auctions for second licenses raised a number of objections to be resolved by the European Commission, as the new entrants were being asked to pay for a license that the PTT mobile operators had received for free.[24] These issues arose in Ireland, Italy, Belgium, Spain, and Austria. In each case, the EC concluded that the use of auctions led to unfair conditions and threatened to thwart competition in the GSM market.

The Commission first instituted legal proceedings against Italy in December 1994. This process led to a decision in October 1995 concluding that the Italian action constituted an infringement of Article 90(1) of the Treaty of Rome, in conjunction with Article 86.[25] The issue was eventually resolved by requiring the incumbent PTT wireless operator (Telecom Italia Mobile or TIM) to make payments totaling 60 billion lire to the new entrant (Omnitel) to compensate for the license fee of 750 billion lire paid by Omnitel in 1994.[26]

Litigation has also resulted from disputes regarding the qualifications of certain license applicants or their applications. In Brazil, for example, the winning consortium for the Sao Paulo "regional" license was first disqualified by regulatory authorities, and then reinstated by the Brazilian Supreme Court in late 1997. The process delayed award of other regional licenses for about a year. In Taiwan, Jet-Tone was disqualified by the "Review Committee" from bidding on the central region because it failed to write on its bond deposit check the words "endorsement and transfer prohibited." In mid-February 1997 the Control Yuan, which supervises operations of the Taiwan authorities, reversed the Review Committee's decision. Jet-Tone was awarded the license, stripping the license originally awarded to other parties.[27]

Auctions can be seen as part of the process of regulators wrestling with how to change from older systems of regulation based on spectrum scarcity and a regulatory contract between government and licensee. In a press statement issued just before he stepped down as head of the U.K. Office of Telecommunications (Oftel), Director General Don Cruikshank noted that "the present regulatory system works by granting a privilege — the spectrum or license to operate — and being able to impose a detailed set of rules in return. This system breaks down as capacity constraints disappear and the asset becomes less valuable. And this is happening now."[28]

1.2.7.3.6 Costs and Benefits of Comparative Evaluations vs. Auctions

The observation from Oftel raises an additional point about auctions and comparative evaluations. Because auctions respond to market forces, auctions may be an efficient way of allocating licenses but they will not necessarily produce the results governments desire. Governments may view auctions simply as a new form of taxation and a source of additional revenue. But auctions, like markets, are not always "bullish." Auction revenues may fall flat, for example: where investors find that spectrum is no longer "scarce" to the same degree, where market demand for the planned services has not yet developed, or where equipment cannot be purchased with the same bulk discounts because services have not yet been defined.[29] Auctions, like markets, can produce imprudent investments, overzealous business estimations, and bankruptcies — situations less likely to occur where governments carefully plan spectrum allocation, license offerings, and select the licensees based on more complex assessments than immediate ability to pay.[30]

The debate between supporters of auctions and supporters of comparative evaluations does not always appear intellectually coherent. Policy makers who strongly favor auctions argue both that markets can most efficiently assign property rights but believe governments should be collecting and distributing wealth recovered from the sale of a public resource. Supporters of comparative evaluations argue that governments should select the business most likely to construct a high-quality, low-cost wireless network, but that competition should regulate aspects other than market entry. The chart identifies the main advantages and disadvantages of auctions and comparative evaluations.

Advantages	Disadvantages
Auctions	
Rational and transparent criteria for award	Criteria other than wealth of applicant overlooked; favors incumbents or others with internal capital — particularly a problem when one bidder is a state-owned enterprise
Administrative simplicity	High level of up-front, lump sum payments inhibit ability of winning applicants to invest in network build-out; pressure to return profits faster may distort business strategy
Recovers spectrum value for the public or the government, rather than allowing private parties to earn economic rents from scarce public resources	High level of up-front payments may raise financing costs and increase costs to consumers; this, in turn, has the potential to reduce other economic benefits that accrue to the public over the long term, e.g., jobs, tax revenue, local investment
	May also encourage financial speculation, rather than investment in service; auctions may also incense irrational bidding, disconnected from the economic value of the license
Creates incentives for efficient spectrum use	Creates incentives for government to identify for auction more spectrum for wireless services than the market will support, creating a "spectrum glut"
Comparative Evaluations	
Measures a variety of relevant criteria; relative value of different criteria can be established as a matter of public policy; where up-front payments are involved, less likely to result in poor judgment and overpayment by the license applicant	Can be less transparent and objective than auctions, although comparison based on tariff price and network coverage is equally objective
Places less financial pressure on the licensee, and may thereby allow for faster build-out and lower prices for consumers	Administratively complex, and can create delays in the introduction of service
Government can evaluate proposed technology and service plan for efficiency and establish build-out and coverage requirements	Absent safeguards, may lead to situation in which licensee does not have full incentives to use spectrum efficiently
Economic policy regarding mobile services, not political incentives for additional revenue, drive the government's decision to issue licenses.	May not fully recover value of spectrum for the public

1.2.7.4   Interconnection

Interconnection to the wireline network is a critical element of the wireless business. The value of the wireless service is largely in the ability of subscribers to call and be called from any point on the fixed network. For the mobile operator, interconnection represents the highest percentage of operational costs, between 20 and 30%. Typically a mobile operator is also dependent upon lines leased from a monopoly national PTT fixed operator to interlink elements of the mobile network and carry calls to and from the fixed network. Interconnection agreements are secured through commercial negotiation among the parties concerned, with regulatory intervention as needed. Cost-based pricing of interconnection offered by fixed operators is a critical but complex issue to resolve.

In Europe, the European Commission has adopted a number of directives that govern interconnection between networks. The European approach is entitled the "open network provision" or ONP approach, and seeks to ensure open access to public telecommunications networks and services, according to harmonized conditions. Network interfaces, usage conditions, and tariff principles are to be objective, transparent, and nondiscriminatory.[31] In 1995, the Commission prepared a Green Paper recommending a specific Directive on Interconnection, which was adopted in June 1997.[32] This Interconnection Directive applies the ONP principles to interconnection agreements between competing networks, and establishes rules to address the obligations of entities with significant market power including an obligation to provide cost-oriented interconnection tariffs, supported by transparent cost accounting systems. For mobile operators in Europe, the right to interconnect directly with the public network was established in the 1996 Mobile Directive.[33]

The key elements of an interconnection arrangement for mobile-to-fixed calling include ensuring fair rates, terms, and conditions for access, determining points of interconnection, and appropriate long-distance charges. Issues raised by interconnection agreements that require regulatory intervention also include where the fixed network owner prices interconnection to a mobile operator at a substantially higher price than it offers to an affiliated mobile carrier, or where a network operator does not "unbundle" network elements, i.e., unreasonably requires that a carrier purchase certain other services as a condition of interconnection.

Interconnection issues are often intertwined with competition law issues, since one of the central problems of liberalization in general is the problem of ensuring fair access to essential facilities of the former monopolies. In Europe, a set of specific rules was created, as well as a draft of a notice informing the public how competition rules would apply to access agreements in the telecommunications sector.[34] The specific interconnection rules include a list of the items that should be covered by a interconnection agreement, rights and obligations regarding interconnection, and cost accounting requirements. These rules, of course, must also be transposed by European Union member states into domestic law and applied by a national telecommunications regulator.[35] The Commission has also adopted a recommendation to use "long run average incremental costs" as the basis for developing cost-oriented interconnection tariffs.[36]

As far as interconnection charges go, a 1996 study prepared at the request of British Telecom compared the interconnection charges of carriers in six countries for carrying a local call from another carrier's network to a fixed network customer over a variety of distances, and at peak and off/peak hours.[37] According to this study, the average interconnection price was between 1.08 and 4.61 pence per minute. Trends observed in the study show that interconnect prices have fallen slightly, somewhat more so in Japan and the U.K. The study is also important because it demonstrates the importance of these interconnection costs to the mobile operator.

More recent data confirm that interconnect prices have fallen, albeit only slightly. In Europe, the January 8, 1998 Recommendation (referenced above) suggests that interconnection charges be aligned within the following ranges:

Local Transit: 0.6 to 1 ECU/100 per minute
Single Transit: 0.9 to 1.8 ECU/100 per minute
Double Transit: 1.5 to 2.6 ECU/100 per minute

The terms "local transit," "single transit," and "double transit" refer to the distance between the points of interconnection and the number of connections that must be made within the fixed network. "Local" means termination at a nearby local exchange, "single transit" is within a metropolitan area, while "double transit" involves termination of calls between cities.[38] Interconnection rates, for a variety of reasons, are generally not yet within this range.[39]

1.2.7.4.1 Self-Provisioning of Infrastructure

One of the vestiges of the national telecoms monopoly is that mobile operators have been forced to rely exclusively on the PTT or national operator for the leased lines and other facilities mobile operators use to connect their own switches to radio base stations and to each other. Operation of these facilities by carriers other than the PTT, including the mobile operator, represented a significant bypass opportunity, reducing PTT revenues. As part of the liberalization process, however, mobile carriers are gradually gaining the ability to self-provision some of these facilities (as well as purchase them from carriers other than the PTT). This has the potential to result in considerable cost savings for the mobile operator.

As the European Mobile Directive noted, a requirement to use the PTT or national operator exclusively gives that entity a considerable influence on the business operations of the mobile operator and constrains the growth of mobile services. The 1995 Telecommunications Law of Mexico also provides that regulations should permit "ample development" of newly authorized public telecommunications network providers.[40]

1.2.7.4.2 Interconnection and Retail Pricing

Interconnection regulations are often intertwined with retail pricing decisions. For example, in Japan, the fixed network operator (NTT) bills a retail customer for calls to a mobile subscriber. The fixed subscriber is charged a tariffed rate determined by the mobile operator (and approved by the Ministry). NTT simply keeps its interconnect fee and passes the remainder on to the mobile operator. This "calling-party-pays" system is also used in much of Europe. For calls from a mobile subscriber to a customer on the fixed network, the mobile operator also sets the tariffed rate subject to approval by a regulator, bills its mobile subscriber, and remits the terminating interconnection fee to the fixed operator.[41]

In early 1998, the European Communities Directorate-General IV, an European Union authority responsible for application of European competition law, began an investigation of these practices by submitting a data request to European mobile operators. In a press release, DG IV stated its belief that "mobile network operators have joint control amongst themselves over the termination of calls on their networks," and went on to note that the price of a fixed-to-mobile call is sometimes substantially more expensive than a mobile-to-mobile call.[42] Some European regulators have already concluded that mobile operators should be required to set interconnection rates on a cost-oriented basis.[43]

Requiring mobile operators to set interconnection rates on a cost-oriented basis raises another set of regulatory issues, including how costs are to be calculated, what costs can be included, e.g., what percentage of overhead should be allowed, and what costs allow for an adequate return. While these issues have been commonplace in the regulation of fixed operators, they have also proved to be burdensome and unwieldy. For example, in Japan, a Council of the Ministry of Posts and Telecommunications conducted a December 1996 review of interconnection practices and found that the current framework "does not function effectively," citing among other things, prolonged disputes over which costs the fixed operator should be allowed to include in its interconnect charges.[44]

Extending this system of regulation to mobile operators holds the possibility of creating further complexities in interconnection negotiations. One regulatory option is simply to benchmark the mobile operator's interconnection charges against those of other carriers, or some cost model, but this is not certain to be a simple process either. Cost-orientation of fixed-mobile interconnection and/or retail pricing also raises the possibility that regulation could inhibit the pricing flexibility mobile operators have used to respond to competitive market forces.

1.2.7.4.3 International and National Roaming

In wireless telephony, the ability of a subscriber of one network to utilize their handset on the network and/or in the service territory of another carrier is referred to as "roaming." International roaming, therefore, is the ability of subscribers of an operator licensed and operating in one country to use their handset when traveling elsewhere. Arranging for these abilities requires carriers to enter into international agreements between the network operator providing the service and the network operator with whom the customer has a network subscription.

National roaming, in contrast, is the ability of subscribers to one operator to use their handset on the network of another operator, within the same country. Where competing facilities-based operators are licensed on a national basis, however, this type of arrangement is not procompetitive. National roaming encourages an operator with less network coverage, or poorer coverage, simply to use the network of a competitor rather than building or upgrading its own facilities. For this reason, national roaming obligations are generally not part of interconnection regulations.

1.2.7.5 Future Issues and Trends

Regulatory issues worth monitoring include issues related to "third-generation" wireless services. The "third generation" refers to developments involving the convergence of mobile telephony and multimedia/Internet services. (Analog cellular is considered the first, while digital cellular is considered the second generation of mobile telephony.) The concept goes by a number of names, among them Universal Mobile Telephone Service (UMTS). The primary regulatory issues here involve addressing the proper role of current operators and the availability of new spectrum to make these services a reality.

From some perspectives, regulators believe that third-generation developments create an opportunity to open the market up for additional competitors; some may even be considering restrictions on the ability of current operators to obtain licenses for these spectrum bands. On the other hand, regulators also must consider the economies of scale and scope offered by current operators, and the fact that third-generation services could be more rapidly deployed by current operators. Additionally, all of the customary issues regarding licensing come in to play once again. Although most regulators appear to favor auctions for UMTS licenses, the issue is not resolved.

Third-generation mobile services (3G)[45] are the next iteration in the evolution of wireless systems. Rather than continue to use the divergent collection of digital wireless standards that are employed in different regions around the world today, the promise of 3G is the creation of a unified, global wireless telecommunications system. Ideally, 3G will have the capability to provide high-quality, high-speed wireless voice and multimedia services to a converged network of fixed, mobile, and satellite components regardless of the location, network, or terminal used. The intent of 3G is to provide seamless mobile services anytime, anywhere. Key features of a 3G system include (1) a common worldwide design; (2) compatible services within and among 3G and fixed networks; (2) high quality; and (4) use of a single, compact user terminal with worldwide roaming capability.

The momentum behind the development of 3G is driven by different factors in different regions, although no factor is unique to a single region. In Japan, rapidly approaching capacity constraints anticipated for existing second-generation systems have placed pressure on regulatory bodies to license spectrum for 3G as soon as possible. As a result, Japan expects 3G operations to commence in 2000/2001 time frame — in advance of the 2002 time frame envisioned by the ITU.

In Europe, the development of 3G has been spurred by the anticipated convergence between mobile communications and the Internet as well as other multimedia services. In addition, the European Commission (EC) has undertaken a proactive role in the promotion of 3G. In 1994, the EC published a comprehensive Green Paper on the subject of 3G to initiate a dialogue among the public, industry, and regulatory bodies. Subsequently, the EC formed a UMTS task force to examine further the implications for 3G in Europe. In many countries, 3G licensees may be viewed as a means of providing increased competition with existing mobile service operators.

In 1992, the ITU identified 230 MHz of spectrum in the 2-GHz band for 3G use on a worldwide basis. Since that time, a tremendous amount of work has been done to further the development of 3G worldwide. A few examples of such work are highlighted here:

- In June 1997, the UMTS Forum released the first UMTS Forum Report entitled, "A Regulatory Framework for UMTS." The UMTS Forum is an influential group of international manufacturers, operators, and regulators assembled to provide advice and recommendations to the EC and other regulatory authorities regarding UMTS. The intent of the report was to encourage prompt action from policy makers to secure and allocate UMTS frequency spectrum and to adopt a regulatory framework for the introduction of UMTS. After analyzing the evolution of the mobile market, the business and technical aspects of UMTS, and the spectrum and regulatory issues, the report made a number of key recommendations to promote the development of UMTS, including (1) a harmonized UMTS licensing framework in Europe; (2) the inclusion of second-generation operators in the UMTS licensing process; (3) the integration of different technologies and services; and (4) the removal of barriers to the transborder use of mobile terminals.

- In July 1997, the Department of Trade and Industry in the U.K. released its own proposed framework for licensing the U.K. 3G operations in a consultation document entitled, "Multimedia Communications on the Move." The U.K. has taken an active role in the development of 3G and as a result likely will be viewed as a model for Europe, if not the rest of the world.

- In February 1998, the EC released a proposal regarding the coordinated introduction of UMTS in the EC. The proposal reflects the EC efforts to create the requisite level of legal and regulatory certainty to promote the substantial investments that will be required for the successful launch of UMTS. Among the proposal requirements are (1) EC-member states must adopt a harmonized

UMTS licensing framework by January 1, 2000, to enable service commencement on January 1, 2002; (2) licensing frameworks should enable roaming to promote the development of pan-European services; (3) licensing frameworks must take into consideration technical standards adopted by European standards-setting bodies; and (4) licensing frameworks should enable end-to-end pan-European interoperability.

• Regional standards-setting bodies have also made substantial progress in the adoption of standards for 3G systems. The European Telecommunications Standards Institute (ETSI), the Japanese standards-setting body, Association of Radio Industries and Businesses (ARIB), and the U.S. Telecommunications Industry Association (TIA) will each present technical standards proposals to the ITU in an effort to facilitate a harmonized global standard for 3G services. The creation of a harmonized standard will maximize economies of scale worldwide to facilitate the introduction of the next generation of wireless services.

As national regulatory authorities turn to the task of allocating and assigning ITU-identified spectrum for 3G, they will face a myriad of crucial regulatory issues that will shape the introduction of 3G services. First, is there sufficient spectrum to accommodate expected demand for 3G services? For example, in the U.S., the FCC already has auctioned large portions of the 3G spectrum for PCS services.[46] In addition to the U.S. situation, however, the spectrum currently earmarked by the ITU for 3G in other regions may be insufficient given the phenomenal growth of wireless services combined with the expected future growth of multimedia wireless services. Second, who should be afforded access to 3G spectrum? Some regulators may choose to assign 3G licenses to existing operators under the theory that existing operators have the requisite economies of scale, established revenue sources, and capital structures to initiate this new service offering adequately. Other regulators may attempt to increase wireless competition by only offering 3G licenses to new operators.

Finally, how will licenses be assigned? Because demand for 3G licenses likely will outstrip supply, regulators will need to develop an equitable method of assigning licenses that ensures that the most competent operators are awarded licenses. Accordingly, some regulators may choose to assign licenses by auction, while other may choose a comparative licensing scheme. Still others may award 3G licenses using a method that combines both a comparative and a bidding element.

The advent of 3G brings mobile telephony and paging services closer to the issues relating to mobile data and Internet access closer together. 3G raises the possibility of a conflict between the regulatory regime traditionally applied to mobile telephony and the traditionally unregulated Internet. For example, regulatory disparity may result if 3G wireless Internet access is subject to the customary coverage, interconnection, and even rate regulation applicable to mobile telephony while wired Internet access may be no more regulated than the sale of other commercial services. On the other hand, if 3G licenses are primarily used to compete with digital GSM services (or supplement the spectrum of existing GSM operators), then equivalent regulation seems more appropriate.

3G licensees may also be viewed as an opportunity to stimulate the growth of the market for mobile data services at present, mobile data services are regulated more lightly than mobile telephony. Although many of the same requirements with regard to spectrum allocation and licensing are present (to the extent a license is required — some data services can be provided without a license), mobile data services are not always configured to connect to the public switched telephone network and therefore are not subject to the interconnection regulations described above. Remote meter-reading, internal data networks, and some types of alarm systems, for example, may only involve communications between fixed transmitters and mobile terminals. This also raises the possibility of a different regulatory regime for 3G.

1.2.7.6   Conclusion

Because of the myriad of different regimes represented around the world, drawing general conclusions about wireless regulation in non-U.S. markets is impossible. But this regulatory overview, particularly in light of the advent of 3G, presents an opportunity to assess the relative importance of each aspect of wireless regulation. There are some general lessons of which it can be said that "genius round the world stands hand in hand, and the shock of recognition runs the whole circle 'round."

The first might be that, while government licensing of spectrum is more widely adopted than a private market in spectrum licenses, this system of licensing inevitably involves government in market structure, market entry, and competition regulation. As a consequence, governments must constantly be assessing whether they have licensed too many or too few competitors to produce optimum results for consumers — a task which even savvy and well-staffed regulatory authorities are unlikely to perform as well as a market. Another might be that this system of licensing presents governments with the need to balance carefully the advantages and disadvantages between comparative hearings and auctions. Neither system consistently yields optimal results for government, consumers, and industry.

Another general observation would be that government involvement is not uniformly disruptive to all markets. In most cases, it is a fact of life that mobile telephony, paging, and some mobile data providers must depend on a monopoly fixed network operator for interconnection. In the absence of a regulatory authority independent from this fixed operator, the ability to secure such interconnection at reasonable prices, with high service quality and without unreasonable delays, is not as great as where a neutral governmental authority can establish and enforce cost-oriented interconnection principles applicable to the fixed operator. While the first assumption should be to allow markets to work, government regulation is neither inherently good nor inherently bad. It is simply a question of identifying those unusual instances where regulation is appropriate because untended market forces would not produce optimal results.

Notes

1. See, e.g., A. S. Devany et al., A property system for market allocation of the electromagnetic spectrum: a legal-economic-engineering study, 21 *Stan. L. Rev.* 1499 (1969); J. R. Minasian, Property rights in radiation: an alternative approach to radio frequency allocation, 18 *J. L. Econ.* 221 (1975); R. W. Stevens, Anarchy in the skip zone: a proposal for market allocation of high frequency spectrum, 41 *Fed. Comm. L.J.* 43 (1988). A landmark reference is R. H. Coase, "The Federal Communications Commission," 2 *J. L. Econ.* 1 (1959).

2. In some cases, national regulators may have simply concluded that private enforcement of spectrum property rights would be too expensive. In other cases, political interests in maintaining government control over natural resources have played a much larger role; this political interest is present to some degree in nearly every country's system of wireless regulation.

3. Although the U.S. is perhaps the only country in the world that has never had state ownership of its primary wireline telephone network, the AT&T Bell System was also essentially a government-controlled monopoly.

4. The American comic Lenny Bruce quipped 35 years ago, that the Soviet Union was run exactly like "one giant phone company." The comparison proved apt: the privatization of telecommunications businesses and associated market liberalization stem from some of the same causes that led to the dissolution of the Soviet Union.

5. A "natural monopoly," strictly speaking, is a condition in which costs continue to fall across all levels of production; i.e., economies of scale are present no matter how much the firm produces. In these circumstances, it is more efficient for a single firm to produce the quantity demanded by the market than to divide demand among multiple firms.

6. There are a few notable exceptions: China did not license a competing mobile operator (China Unicom) until 1993; that operator (Unicom) has since made only minimal market-share gains. See Goldman Sachs Global Research, Global Wireless Communications Industry, February 1998, p. 297.

7. See, e.g., Towards a dynamic European economy: Green Paper on the development of the common market for telecommunications services and equipment, *COM* 290 final, July 30, 1987; Communication to the Council and the European Parliament on the consultation on the review of the

situation in the telecommunications services sector, *COM* 159, April 28, 1993. These studies by the European Commission led to various resolutions by the Council of the European Union which represent the regulatory decisions to gradually liberalize the telecommunications market. See, e.g., Council Resolution of June 30, 1988 on the development of the common market for telecommunications services and equipment up to 1992 (88/C 257/01); Council Resolution of June 29, 1995 on the further development of the mobile and personal communications sector in the European Union (95/C 188/02, OJ C 188/3, July 22, 1995); Council Resolution of September 18, 1995 on the implementation of the future regulatory framework for telecommunications (95/C 258/01, OJ C 258/1, October 3, 1995).

8. See Towards a personal communications environment: Green Paper on a common approach to mobile and personal communications in the European Union, *COM* 145 final, April 27, 1994 ("Mobile Green Paper").

9. Commission Directive of January 16, 1996 amending Directive 90/388/EEC with regard to mobile and personal communications (96/2/EC; OJ L 20/59, January 26, 1996)("Mobile Directive").

10. See, e.g., Developing the regulatory footprint for newly privatized telecommunications providers in Latin America, W. M. Berenson, *Fed. Bar News J.* 400, 1991 ("Berenson").

11. Telecommunications Law, *Diario Oficial de la Federacion,* June 7, 1995, pp. 34–47.

12. WTO Press release, PRESS/87; January 26, 1998.

13. See, e.g., http://www.wto.org/archives/4prot-e.htm (4th protocol to the WTO agreement on trade in services).

14. See Berenson, *supra* note 6, p. 403.

15. See S. K. Vogel, *Freer Markets, More Rules,* Cornell University Press, Ithaca, NY, 1996, p. 163–166.

16. International Table of Frequency Allocations, ITU, Geneva; see 47 C.F.R. §2.104 (U.S. FCC regulations).

17. Council Directive 87/372/EEC; OJ L 196/85 (July 17, 1987) (GSM); Council Directive 90/544/EEC; OJ L 310/28 (November 9, 1990) (ERMES paging); Council Directive 91/287/EEC; OJ L 144/45 (June 8, 1991) (DECT).

18. Only 18 of the 20 circles received bids; the other areas were seen as uneconomic. India also created 21 basic service licenses for other areas, only 10 of which received initial bids.

19. France licensing information gathered from R. Cranston, Liberalising Telecoms in Western Europe, Financial Times Management Report 123–124, 1997.

20. See European Radiocommunications Committee Report, Doc. RR (97) 109 (October 29, 1997) at Annex I, p. 18.

21. See Invitation to Tender, available at http://www.admin.ch/bakom/tc/mob_aus/mob_aus1/ausschr_e.htm.

22. Federative Republic of Brazil, Ministry of Communications, Draft Request for Proposal, No. 001/96.

23. See, e.g., M. Mueller, Reform of Spectrum Management: Lessons from New Zealand, *Policy Insight* (Reason Foundation, November 1991); M. Shafi, et al., Experience with spectrum tendering in New Zealand, in *Proceedings of the 41st IEEE Vehicular Technology Conference,* St. Louis, MO (May 1991). The U.S. has also been a pioneer in a number of techniques for awarding mobile licenses by auction.

24. The EC emphasized some of the drawbacks of auction procedures in the Mobile Green Paper; see also, As GSM Mobile Communications Market Is Opened to Competition, the Commission Screens the Licensing Procedures, Press Release IP/95/959 (Brussels, September 13, 1995).

25. Commission Decision Concerning the Conditions Imposed on the Second Operator of GSM Radiotelephony Services in Italy, 95/489/EC, No. L. 280/49. Article 90(1) of the Treaty of Rome provides that Member States shall not enact or maintain in force any measure contrary to the rules of the Treaty, including Article 86 of the Treaty, which prohibits "abuse of a dominant position" within the common market.

26. Ibid., pp. 13–14. In 1998, TIM suspended payment in attempt to force the government to license it to operate in new 1800 MHz frequencies. See "Mobile Communications," No. 236 (March 5, 1998), p. 4. In other cases, compensation has been arranged through interconnection agreements.

27. See Bank of America Report: Taiwan, available at http://www.tradeport.org/ts/countries/taiwan/mrr/mark0086.shtml.

28. Press Release (March 19, 1998), available at http://www.open.gov.uk/radiocom/rahome.htm.

29. Whether this outcome occurs is, of course, not simply a function of auctions but of spectrum management. One effect of auctions is that the decision to identify spectrum and allocate it for mobile telephony purposes is influenced by the government's expectation of auction revenues to be used for short-term governmental or political purposes, rather than an independent assessment of market needs for spectrum.

30. This has certainly been the U.S. experience — although a 1996 PCS auction yielded bids of over $2 billion, the government has yet to collect much of it and many of the licensees have opted for bankruptcy protection. An additional spectrum auction for local loop, video, and multimedia services yielded revenues significantly below those anticipated, and a third has been canceled indefinitely.

31. See, e.g., Council Directive of June 28, 1990 on the establishment of the internal market for telecommunications services through the implementation of open network provision, 90/387/EEC; OJ L 192/1, July 24, 1990.

32. Green Paper on the liberalization of telecommunications infrastructure, COM(94) 682, January 25, 1995; Directive 97/33/EC on interconnection in telecommunications, OJ L 199, July 26, 1997.

33. To be precise, the Mobile Directive established that a refusal to interconnect or a discriminatory interconnection arrangement would constitute an abuse of dominant position by the fixed operator, and would therefore be inconsistent with the principles of the Treaty of Rome and a violation of Article 90 of that Treaty. See Mobile Directive, ¶17. The Mobile Directive also established that license conditions restricting direct interconnection between mobile systems or between mobile systems and fixed networks located in other states infringed Article 90 and must be eliminated. Ibid.

34. Draft Communication, Application of the Competition Rules to Access Agreements in the Telecommunications Sector, OJ C 76/25, March 11, 1997.

35. Directive 97/33/EC, June 30, 1997, Interconnection in telecommunications with regard to ensuring universal service and interoperability through application of the principles of Open Network Provision (ONP).

36. 98/195/EC, Commission recommendation of January 8, 1998 on interconnection in a liberalized telecommunications market (Part 1 — Interconnection pricing), OJ L 073, 12/03/1998, pp. 0042–0050.

37. D. Lewin and S. Young, International Comparison of Interconnect Prices, January 1996. The study examines interconnect prices charged by local carriers in the U.S., Sweden, Australia, Japan, and New Zealand.

38. 98/195/EC, Commission recommendation of January 8, 1998 on interconnection in a liberalized telecommunications market (Part 1 — Interconnection pricing); see Directive 97/33/EC, June 30, 1997, Interconnection in telecommunications with regard to ensuring universal service and interoperability through application of the principles of Open Network Provision (ONP).

39. For example, Telecom Italia has attempted to justify its failure to comply with the "best practice" interconnection rates on the basis that Italy's tax rate is higher than the European average, and that it must also pay a substantial annual fee for its telecommunications concessions. In September 1998, Italian officials came under pressure from the European Union but were able to convince the European Commission that an European Union proceeding was unnecessary, and promised to take steps to adjust Telecom Italia's interconnection rates.

40. See Telecommunications Law, *Diario Oficial de la Federacion,* June 7, 1995, at Article 41. Article 43 provides that all authorized network providers are required to interconnect, under terms that foresee that either party may furnish the equipment required for interconnection.

41. Additional data and a more-detailed description of the Japanese interconnection practices, as well as those of other countries, can be found in Mobile Interconnection: Strategies for Achieving World Best Practice, August 14, 1997, a study prepared by OVUM Ltd. telecommunications consultants.

42. IP98/141, Commission launches inquiry into mobile and fixed telephony prices in the European Union, Feb. 9, 1998.

43. Prices of calls to mobile phones — Statement, March 1998, available at http://www.oftel.gov.uk/pricing/ctm0398.htm.

44. Final Report of the Telecommunications Council, December 19, 1996; see August 14, 1997 OVUM Report, p. 60.

45. The International Telecommunications Union's term for 3G is International Mobile Telecommunications-2000 (IMT-2000), while the European Union refers to 3G as Universal Mobile Telecommunications Service (UMTS).

46. It should be noted that the U.S. PCS (and cellular) licensees can use that spectrum for 3G services.

Appendix 1.2.7A — Overview of the WTO Member Country Schedule of Commitments

Item: **Voice Telephone Services**

Description:	Competitive supply, public voice services	Local service	Domestic long distance	International service
Commitments:	47 schedules	41 schedules	37 schedules	42 schedules

Item: **Other Services**

Description:	Data transmission	Mobile telephone markets	Leased circuit services	PCS, mobile data, paging
Commitments:	49 schedules	46 schedules	41 schedules	45 schedules

Item: **Satellite**

Description:	Mobile satellite services/ transport capacity	Fixed satellite services/ transport capacity
Commitments:	37 schedules	36 schedules

Item: **Value-Added Telecommications Services**

Description:	E-mail, on-line, database retrieval
Commitments:	8 schedules

Note: Countries that have submitted schedules of commitments: Antigua and Barbuda, Argentina, Australia, Bangladesh, Belize, Bolivia, Brazil, Brunei Darussalam, Bulgaria, Canada, Chile, Colombia, Côte d'Ivoire, Czech Republic, Dominica, Dominican Republic, Ecuador, El Salvador, EC and its member states, Ghana, Grenada, Guatemala, Hong Kong (China), Hungary, Iceland, India, Indonesia, Israel, Jamaica, Japan, Korea, Malaysia, Mauritius, Mexico, Morocco, New Zealand, Norway, Pakistan, Papua New Guinea, Peru, Philippines, Poland, Romania, Senegal, Singapore, Sri Lanka, Switzerland, Slovak Republic, South Africa, Thailand, Trinidad and Tobago, Tunisia, Turkey, U.S., and Venezuela.

Data obtained from WTO press release, PRESS/87; January 26, 1998.

Appendix 1.2.7B — Sample Table of Worldwide Mobile Operators

Country	Operators	In-Service Date	Network Type	Method of Awarding License
Austria	PTV (*PTT*)	November 1984	Analog	PTT
	PTV	July 1990	TACS -900	PTT
	PTV	December 1993	Digital (GSM)	PTT
	Maxmobil	December 1996	Digital (GSM)	Bid
Denmark	TeleDanmark Mobile	1982	Analog	PTT
	TeleDanmark Mobile	1986	Analog	PTT
	TeleDanmark Mobile	1992	Digital (GSM)	PTT
	Sonofon	1992	Digital (GSM)	PTT

Country	Operators	In-Service Date	Network Type	Method of Awarding License
France	France Telecom	1985	Analog	PTT
	France Telecom	1992	Digital (GSM)	PTT
	France Telecom	1996	Digital (GSM)	PTT
	SFR	1989	Analog	PTT
	SFR	1992	Digital (GSM)	Given
	Bouygues Telecom	1996	Digital (GSM)	Given
Germany	T-Mobil	1985	Analog	PTT
	T-Mobil	1992	Digital (GSM)	PTT
	Mannesmann	1992	Digital (GSM)	Beauty contest
	E-Plus	1994	Digital (GSM)	Auction
Sweden	Telia Mobitel *(PTT)*	October 1981	Analog (NMT-450)	PTT
	Telia Mobitel	December 1986	Analog (NMT-900)	PTT
	Telia Mobitel	November 1992	Digital	PTT
	Comvik	September 1992	Digital	PTT
	NordicTel	September 1992	Digital	PTT
Portugal	TMN *(PTT)*	January 1989	Analog	PTT
	TMN	October 1992	Digital (GSM)	PTT
	Telecel	October 1992	Digital (GSM)	Beauty contest
South Korea	KMT *(PTT)*	1984	Analog	PTT
	Shinsegi Telecom	1996	Digital (CDMA)	Beauty contest/ Mega-Consortium
India	BPL	1995	Digital (GSM)	Bid
	Max	1995	Digital (GSM)	Bid
	Airtel	1995	Digital (GSM)	Bid
	Essar	1995	Digital (GSM)	Bid
	RPG	1995	Digital (GSM)	Bid
	Skycell	1995	Digital (GSM)	Bid
	Modi Telstra	1995	Digital (GSM)	Bid
	Usha Martin	1995	Digital (GSM)	Bid
	JT Mobile	1996	Digital (GSM)	Bid
	Tata	1996	Digital (GSM)	Bid
	Airtel	1996	Digital (GSM)	Bid
	Escotel	1996	Digital (GSM)	Bid
Romania	Telefonica Romania	—	Analog	PTT
	MobilRom	1997	Digital (GSM)	Beauty contest
	Mobifon	1997	Digital (GSM)	Beauty contest
Greece	Panofon	1993	Digital (GSM)	Bid
	TeleSTET	1993	Digital (GSM)	PTT

Note: PTT indicates where an operator is affiliated with the state-owned telecommunications network provider. Data obtained from "Global Mobile," and "Latincom" publications of Baskerville Communications Corporation, "International Regulatory Update" publication of the Financial Times, and AirTouch internal sources.

1.2.8 Universal Service Regulations in the U.S. and the European Union

Al Hammond

1.2.8.1 Introduction

For more than a decade, the U.S. and the nations of the European Union (EU) have faced increasing pressures for competition in their telecommunications industries. In response, the federal and state governments in the U.S. moved to dismantle the monopolies of the local exchange carriers remaining after the earlier dissolution of AT&T. The EU and its member states moved to open their national telecommunications markets to competition and to establish an interconnected, uniformly provisioned and priced infrastructure throughout the Union. Each has recently passed major legislation articulating the manner in which competition is to be facilitated. Each is in the throes of major litigation over the manner and the timeliness of compliance by some of their respective nations or states and firms.

A critical component of both the U.S. and EU legislation and regulatory policies is the requirement to achieve universal service. It is increasingly clear, that access to telephone service is no longer considered a luxury in the U.S. or in the EU Universal service is the assurance that every citizen of the U.S. and the EU, regardless of geographic location, economic status, or disability has access to a fully functional telecommunications network at reasonably affordable rates. Both the U.S. and the EU recognize universal service is essential to the economic and social fabric of their respective societies. Each is pursuing a number of strategies to achieve universal service and, in the process, each faces significant difficulties in achieving the goal of electronic equity.

1.2.8.2   The Telecommunications Act of 1996 and Universal Service

1.2.8.2.1   Universal Service Prior to 1996

The Communication Act of 1934, Title I, Section 1, set forth the goal of American communications policy "to make available, so far as possible, to all people of the United States a rapid, efficient, nation-wide, and world-wide wire and radio communications service with adequate facilities at reasonable charges." In telephony, this policy evolved into the requirement that monopoly telephone companies provide service to as many as possible. The companies were allowed to subsidize the cost of serving poor, rural, or other less-profitable customers with higher margin clients such as downtown businesses.

However, increasing competition in the local telephone service area generated pressure to abolish, restructure, or spread the cost of subsidies which underwrote the provision of universal phone service to poor and rural residents. For instance, long-distance telephone companies seek a reduction in access charges which comprise approximately 40 to 45% of the cost of long-distance charges. Efforts to reduce costs include bypass arrangements and efforts to enter the local market as competitors. Meanwhile, long-distance companies are joined by Competitive Access Providers (CAPs) who compete for the LEC high-end users who comprise a small number of actual customers but generate the vast majority of revenues. Finally, consumers want the benefit of any reductions in cost of service due to increased efficiencies in the network but do not want to subsidize local telephone company entry into competitive cable and information services from which their competitors are emerging. Thus revenues for competitive entry or introduction of competitive services must come from LEC profits garnered via the provision of competitive services and from shareholders. The net result is pressure on whatever subsidies exist to fund universal service.

The subsidies take two basic forms. There are direct subsidies which help telephone companies with higher-than-average costs. These companies operate in rural areas where telephone lines must be strung for many miles to reach few subscribers. There are also cross subsidies in which one group of users pays higher prices to underwrite lower prices for other groups. These include business-to-residential customers; urban-to-rural subscribers; and long distance-to-local callers.

1.2.8.2.2   The Telecommunications Act of 1996

The chief policy and legal instrument for assuring access to telephone and other network services in the U.S. is the Telecommunications Act of 1996. One of Congress's primary goals in passing the new legislation was to ensure that all Americans have affordable, nondiscriminatory access to communications services. To accomplish its goal, Congress instituted a number of regulatory changes via the Telecommunications Act of 1996. For the most part, these changes reflect a preference for marketplace- and market demand-driven methods for accomplishing the legislative goal of nondiscriminatory, affordable access.

However, while the legislation evidences a distinct and deep-seated congressional preference for reliance on the market, there are some provisions which seek to address the market's inability to assure access to all. These provisions include a revision to the nondiscrimination provision of the Communications Act of 1934, a revision to the nation's universal service policy, and the retention of Lifeline and Link-Up subsidies for the poor.

1.2.8.2.3   Reliance on the Marketplace

Among the market-reliant regulatory changes enacted in the legislation are the introduction of greater competition in the provision of local and long-distance telephone services, the deregulation of cable television rates, and the creation of opportunities for competition between telephone and cable firms in providing video programming services. Many in government and industry anticipate that increased competition in the local and long-distance telephone markets, as well as in the video market, will result in reductions in the prices consumers pay for access to existing telecommunications and video services. In addition, it is argued that increased competition will result in the more rapid deployment of new, innovative, and responsive communications services.

1.2.8.2.4 Antidiscrimination

Section 151 of the Communications Act of 1934 now states in pertinent part:

> ... to make available, so far as possible, to all people of the United States *without discrimination on the basis of race, color, religion, national origin, or sex,* a rapid, efficient, nation-wide and world-wide wire and communications service with adequate facilities at reasonable charges ...

The Act thus amends the section 151 language, which many have interpreted as the first and sole statutory basis for universal service, by prohibiting discrimination in the provision of service on the basis of race, color, religion, national origin, or sex.

1.2.8.2.5 A New Definition of Universal Service

Section 254 establishes the procedure the FCC must use to develop the new and evolving definition of universal service, including which services will be deemed "basic" or "advanced," what subsidies will be created, and who will be eligible for them. The FCC has since established the definitions for basic and advanced telecommunications services. Basic service includes voice-grade access to the public switched network, touch-tone, single party service, access to emergency services (including 911 where available), operator services, interexchange services, and directory assistance. Subsection (b)(3) of section 254 requires that the FCC must develop the nation's universal service policy such that:

> consumers in all regions of the Nation, including low-income consumers ... should have access to telecommunications and information services, including interexchange services and advanced tele-communications and information services, that are reasonably comparable to those services provided in urban areas and that are available at rates that are reasonably comparable to rates charged for similar services in urban areas.

At minimum, the subsection's guiding principle that access be provided in "all regions of the nation" is a directive assuring that rural, low-income, urban and suburban consumers are afforded access to reasonably comparable basic and advanced telecommunications and information services at comparable rates.

1.2.8.2.6 Access to Basic and Advanced Technology

Access to Basic Technology

Approximately 94% of Americans have access to basic telephone service. As a consequence, aside from establishing policies to address issues of geographic price and service disparity, the major strategy envisioned by Congress and the FCC to increase the percentage of Americans having access to a phone is to maintain and expand the Lifeline and Link-Up programs. Link-Up provides telephone installation at reduced prices to people eligible for welfare. Lifeline makes basic phone service available at a discounted rate. In this way, it is hoped that the percentage of Americans having access to a phone will be further increased. This strategy enjoys the support of a majority of the telecommunications industry. However, even with these programs, approximately 19% of rural and urban residents with incomes under $10,000 do not own a phone.

Access to Advanced Telecommunications Services

Financing Universal Service

One of the critical requirements for implementing the national policy of equitable access is that telecommunications service providers must make equitable and nondiscriminatory contributions to the preservation and advancement of universal service. These contributions are to be made pursuant to federal and state mechanisms for funding and implementing universal service which are to be "specific, predictable, and sufficient to preserve and advance universal service."

Targeted Subsidies: Schools and Libraries
Aside from the legislative intent embodied in section 151 as amended and section 254(3)(b) of the 1996 Act, Congress sought to assure that those Americans deemed poor by established criteria are protected from a loss of service as well via the continued provision of Lifeline services. In addition, the FCC, upon the advice of the Joint Board, has promulgated and seeks to implement regulatory policies which assure that low income communities are provided access to advanced telecommunications and information services via their schools and libraries. The services are to be provided to schools and libraries at a discount ranging from 20 to 90% based on established need criteria.

To date, the establishment of a universal service discount for schools and libraries has not been accomplished. Numerous legal challenges to the FCC methods for funding the Universal Service discount, and Congressional threats to the funding of the discount lead to a reasonable conclusion that the implementation of the universal service discount will most certainly be delayed in the short term.

Notes

(1) A. L. Shapiro, total access; universal telecommunications services, *The Nation,* January 6, 1997, p. 5; C. A. Gang, Reduced fees available to needy, *Commercial Appeal (Memphis),* August 2, 1991, p. C3; Proposal for a European Parliament and Council Directive Com(96)419, September 11, 1996, Summary Section 3.

(2) Communication Act of 1934, 47 U.S.C. Title I, Section 1(a).

(3) This goal has come to be known as universal service, and NTCA and its members have supported it since the beginning. J. Weikle, Ready for prime time — universal service meets universal competition, *Rural Telecommun.,* March/April 1995, pp. 50–53.

(4) In a noncompetitive telephone service environment, the universal service pricing model served to subsidize rural and low-income residential prices through higher urban business prices. R. Taylor, Leveling the field in telecommunications, *St. Louis Post-Dispatch,* Feb. 21, 1995, p. 11B. Also see C. R. Conte, Reaching for the phone, *Governing Mag.,* July 1995, p. 32; Universal Service: FCC Wants More Data, Bells Demand Change, FCC Report, Dec. 1, 1994; M. Mills, Increases in local phone rates proposed, *Record,* May 7, 1996, p. A01; H. Bray, Bill limits or protects, depending on biz size, *Crain's Detroit Business,* Oct. 16, 1995, p. 32; Bloomberg Business News, FCC may allow Baby Bells to cut access rates, *Dallas Morning News,* Sept. 15, 1995, p. 10D.

(5) "Local competition makes the inevitable adjustment in the nature of universal service more complicated, because there will no longer be one provider to subsidize whatever is deemed the proper level of universal service. Rather, the LEC will be under competitive and regulatory pressure to charge cost-based prices that eliminate the basis for internal subsidies." Competition at the Local Loop: Policies and Implications; Forum Report of the 7th Annual Aspen Conference on Telecom Policy Part 2 of 3 Parts, *Edge,* Feb. 15, 1993. Also see Universal service concept to get information-age update, *Washington Telecom News,* Feb. 7, 1994; C. Arnst, Phone frenzy, *Business Week,* Feb. 20, 1995, p. 92; A National Information Network; Changing Our Lives in the 21st Century, *Edge,* Dec. 18, 1992; Predictable camps; telephone industry debates universal service in NTIA filings, *Commun. Daily,* Dec. 20, 1994, p. 3; Testimony Feb. 9, 1994, Ivan G. Seidenberg Vice-Chairman of NYNEX Corporation, U.S. Telephone Association, House Energy/Telecommunications and Finance National Communications Competition and Information Infrastructure Act of 1993.

(6) Estimates range from 42 to 45%. See AT&T vice president testifies before Pennsylvania Senate Committee, *PR Newswire,* April 25, 1996; to 45%, see H. Bray, *supra,* note 4; Bloomberg Business News, *supra,* note 4.

(7) "… [T]he local exchange business offers real opportunities for new entrants. High rates for business telephone service, traditional pricing approaches, and the cost structure of the local service business will allow challengers to pick off the LECs profitable customers." M. Arellano, Exploiting the LECs Achilles' heel, *Telecom Strategy Letter,* July, 1995, p. 81.

(8) "Dividing subscribers into three groups of 50 million access lines each, average monthly revenue is $36.00 for the lower segment, $45.00 for the middle segment, and $72.00 for the upper segment; the overall average is $51.00 per month. There are about 100 million residential lines and 50 million business lines. Most business lines fall into the upper segment (because of higher basic charges for business lines and higher usage), while the majority of residential lines fall into the lower two segments. High revenue lines are much more profitable than low revenue lines; the upper segment brings in more access and intraLATA toll revenues, which are high-margin businesses for the LECs. Northern Business Information estimates that, holding all other factors constant, average monthly profit is $11.00 in the upper segment and $2.00 in the middle segment; in the lower segment, each access line leads to a loss of $4.00 per month, on average. Potential entrants understand these relationships, and will go after the LECs' high-margin subscribers: business customers and high-usage residential subscribers."

(9) Also see generally B. Stuck, The local loop adapts for new roles, *Business Commun. Rev.,* October, 1995, p. 55; Network Diversity, Major Cost Savings Satisfy Cap Customers, Local Competition Report, July 10, 1995; J. S. Kraemer, Local competition; W. P. Barr, Regulatory reform: recognizing market realities; FCC should move faster towards telecommunications industry reforms, *Telecommunications,* Jan., 1995, p. 30; Changing ground rules for network access, *Business Commun. Rev.,* Sept., 1994, p. S4.

(10) It has been argued that consumers of traditional telephone services should not have to finance the new or innovative services and infrastructure to be offered unless they actually use them. Competition at the local loop: policies and implications; forum report of the 7th Annual Aspen Conference on Telecom Policy, Part 2 of 3 Parts, *Edge,* Feb. 15, 1993. With this goal in mind, regulators have sought to protect consumers from footing the bill for new competitive entry. See Sparing Consumers Harm Core Competition Issue for Regulators, *State Telephone Regulation Report,* August 10, 1995. There is growing skepticism that competition in the local loop will result in lower prices for consumers. M. Mills, Increases in local phone rates proposed, *The Record,* May 7, 1996, p. A01.

(11) Testimony Feb. 9, 1994 of Ivan G. Seidenberg, *supra,* note 5.

(12) New study says target, don't expand, universal service subsidies, *State Telephone Regulation Report,* 13(2), Jan. 26, 1995.

(13) Section 254 (j) provides that "Nothing in this section shall affect the collection, distribution, or administration of the Lifeline Assistance Program provided for by the Commission under regulations set forth in section 69.117 of title 47, Code of Federal Regulations, and other related sections of such title. C. A. Gang, *supra,* note 1.

(14) The limited, evolutionary focus of the universal service policy stands in stark contrast to the congressional decision to liberalize the broadcast multiple ownership rules. The Act increases the number of broadcast stations any one person or company can own to such a great extent that small broadcasters are being driven out of the industry by the new economics of multiple owner competition. The virtual repeal of the broadcast multiple ownership rules means very few Americans can afford to own a radio or television station. Meanwhile, government auction of the spectrum has resulted in the sale of vast public resources to private interests — a sale in which most Americans have played virtually no part. As a result, Americans are left even farther behind as the courts uphold the private speech and editorial rights of media owners often to the exclusion of the public.

(15) To date, the price reduction has not occurred. Many consumers are finding that their telephone and cable subscription rates are higher, not lower. Teleco act is year old: many are willing to wait for results, *Commun. Daily,* Feb. 5, 1997. Indeed, some consumers allege that the implementation of the Act's local market competition and universal service requirements is resulting in an increased cost to consumers. Consumer organization fires shot in David and Goliath battle, *Bus. Wire,* Jan. 21, 1997.

(16) Telecommunications Act of 1996, Title I, Subtitle A — Telecommunications Services, Sec. 104. Non-Discrimination Principle.

(17) This is clearly an important revision but it fails to address the concern of minority and poor communities that they are being electronically redlined because they are perceived as being less economically desirable, not necessarily because they are composed of religious, ethnic, or racial Americans, or women.

(18) Common Carrier Action Commission Implements Telecom Act's Universal Service Provisions; Adopts Plan to Ensure Access to Affordable Telecommunications Services for All Americans (CC Docket No. 96-45), Report No. CC 97-24; CC Docket No. 96-45, May 7, 1997.

(19) The subsection still contemplates payment for a bifurcated panoply of services available at rates comparable to those available in the more desirable urban regions. These are likely to be rates that many will be unable to afford.

(20) To make telephone service available to all Americans, the federal government has mandated reduced fees for those who meet certain economic criteria. The program, Link-Up America, offers telephone installation at reduced prices. Customers who want basic telephone service but who do not use their telephones a lot may save money by using one of several measured service options: Lifeline costs $4.85 per month. Customers are charged 4 cents for the first minute of a call and 2 cents for each additional minute. The first $1.00 worth of calls is included in the basic fee. Additional discounts are given for calls made at night; local measured service is $8.50, with the first $7.50 in calling time included in the fee. For additional calling time, 4 cents is charged for the first minute and 2 cents for each additional minute; message rate service costs $6.10 per month with a 30-call allowance. Each additional call over 30 costs 10 cents. C. A. Gang, *supra,* note 1.

(21) Competition in the local loop, *Telecom Perspectives,* July 1996, p. 73; also see Statement of Jonathan B. Sallet, MCI Communications Corporation, Before the Senate Commerce, Science, and Transportation Committee Subcommittee on Communications Subject — Universal Service and Local Competition, *Federal News Service,* June 3, 1997.

(22) At least two reasons are offered for the current shortfall. First, the current telephone access deficit is created by a lack of knowledge of the discount on the part of those Americans who are eligible to receive the Lifeline discount. Second, those who may have at one time acquired access to phone service but who are currently disconnected for failure to pay may not be able to afford the reconnection fees. D. Silverman, No universal agreement on 'Net, *Houston Chronicle,* July 5, 1996, p. 1.

(23) Under section 254 (k), the Commission, with respect to interstate services, and the states, with respect to intrastate services, are required if necessary to establish cost allocation rules, accounting safeguards, and guidelines to ensure that services included in the definition of universal service bear no more than a reasonable share of the joint and common costs of facilities used to provide those services.

(24) Section 254 (b)(6), Section 254 (c)(3), and Section 254 (h)(1)(B). Also see Common Carrier Action Commission Implements Telecom Act's Universal Service Provisions; Adopts Plan to Ensure Access to Affordable Telecommunications Services for All Americans (CC Docket No. 96-45), Report No. CC 97-24; CC Docket No. 96-45, May 7, 1997.

(25) Common Carrier Action Commission, *supra,* note 24.

(26) See *Commun. Daily,* June 27, 1997 (Court of Appeals panel to hold a lottery to decide which circuit will hear the appeals of the FCC's universal service order); *Commun. Daily,* June 26, 1997; (GTE filed an appeal of the FCC universal service order because it allegedly fails to ensure that quality services will be provided at affordable prices to customers in rural areas); *Commun. Daily,* June 20, 1997 (SBC filed suit alleging that the school and library discounts place an enormous $2.25 billion burden on telephone customers to finance programs, training, equipment and services for schools that were not contemplated by Congress or authorized by the Telecommunications Act of 1996); *Commun. Daily,* June 19, 1997, (Celpage files suit asking that paging companies not be required to contribute to the universal service fund and alleging that such a requirement constitutes an illegal and discriminatory tax on the paging industry).

(27) The fiscal year 1998 budget proposed by the U.S. Senate in S-1022 includes the use of the 1998 universal service funds as a source of revenues to balance the national budget by fiscal year 2002. Meanwhile, a separate bill, S-2015, proposes that universal service payments be delayed from 2002 to 2003 for some telephone companies. Some senators are concerned that if adopted, the proposals would severely undermine the deployment of advanced telecommunications services to schools and libraries, and hence, many American communities. *Commun. Daily,* July 28, 1997.

1.2.9 Number Portability

Wouter Franx

1.2.9.1 Introduction

The history of our directory number is nearly as old as the invention of the telephone itself. Starting from the introduction of electromechanical switching systems, directory numbers have literally become the *key* to get access to our telephony services, no matter that we use them now for Plain Old Telephone Service (POTS), mobile, or enhanced 800/900 services.

Since our reachability is "connected" to directory numbers, customers attach a high value to their number. They do not like to change their number when they move from one side of the city to the other, or change from one service provider to another. Mobile networks already provide more flexibility in this matter. However, most fixed PSTN/ISDN networks have a very rigid coupling between the directory number and the physical socket through which the customers accesses their services.

Number portability will provide a more flexible coupling between the directory number and the access interface. It allows customers to change location, service and/or service provider, while retaining their number. The ability to change service provider without needing to change directory number has gained significant interest with regulatory authorities worldwide. Today, number portability is regarded as a prerequisite to stimulate competition in the telecommunications market. This section provides an overview of the challenges of number portability, including possible introduction and migration strategies for international markets. Number portability, as used in this section, focuses exclusively on the "local loop," the last "mile" of cable between the telecommunications operator and the end user, and is therefore called "number portability in the local loop."

1.2.9.1.1 Scope of Number Portability in the Local Loop

Number portability applies to both so-called geographic numbers (e.g., normal number with area code and local number) and nongeographic numbers (e.g., 800/900 numbers, mobile numbers). As far as the liberalization of the local loop market is concerned, the focus is on the portability of geographic numbers, also referred to as geographic number portability (GNP, in Europe) or local number portability (LNP, in the U.S.). Nongeographic numbers are outside the scope of this section.

Although definitions and terminology for number portability may vary from country to country, the following types of GNP are defined [ETSI, 1997]:

- *Service Provider Portability:* A service that enables a customer to resign their subscription with a service provider and to contract another subscription with another service provider without changing their geographic number, without changing location, and without changing the nature of the service offered.
- *Location Portability:* A service that allows customer to retain their directory number when changing their premises in a certain area. Four variants of location portability exist: within an exchange area, within an entire numbering area, within a charging area, and anywhere.*
- *Service Portability:* A service that allows customers to retain their directory number when they are offered a new service, e.g., from telephone service (fixed) to mobile telephone service.

---

*The boundaries of location portability may vary based on regulatory and technical constraints.

1.2.9.1.2 *Regulatory and Market Drivers*

The liberalization of the telecommunications market is the main driver for the introduction of number portability. According to the Green Paper from the Commission of the European Communities [CEC, 1996]:

> With the full liberalization of the telecommunications sector, alternative methods for local access (e.g., through cable TV networks and through new radio infrastructures such as those based on DECT and the GSM/DSC-1800 standards) are being introduced rapidly often by new entrants to the market. Number portability in the local loop is considered crucial by these new entrants to give them a fair chance to compete with the incumbent and to establish a position.

> Without number portability between local operators, new market entrants face a significant barrier to entering the market. Lack of portability may even become a disincentive to invest in alternative local loop networks with detrimental effects on the development of effective competition in Europe.

Similar regulatory drivers are also applicable in the U.S., Asia/Pacific, and other regions in the world.

Various cost/benefit studies have indicated that the economic benefits of number portability far outweigh the costs. Notwithstanding this, it is generally recognized that the initial system setup costs and the operational costs of number portability are considerable. As far as the incumbent operators are concerned, this is largely due to the fact that many legacy systems in their networks need to be upgraded. In addition to these costs, the incumbent operator will be faced with a decreasing market share and lower margins because of the increased competition.

On the other hand, there will also be some benefits for the incumbent operator. Location portability and service portability provide especially interesting business opportunities to offer new services to the installed base customers. Some examples:

- *One-number-for-life* (extending location portability to nationwide level)
- Using geographic numbers to access mobile services directly
- Upgrading from PSTN to ISDN, also if the customer needs to be connected to another (ISDN-capable) switch

In contrast to the incumbent operators, new operators consider number portability a critical success factor to enter the market, to stimulate the competition, and to grow their business more quickly.

12.9.1.3 *Status of Number Portability Worldwide*

Europe

In the U.K. the first investigations to introduce number portability were conducted by the regulator (Oftel) in the early 1990s. The initial switch-based solution* was introduced in the fourth quarter of 1996 and is considered an interim solution toward a more-advanced solution. Finland has also already introduced number portability. An IN-based solution was deployed in 1997.

The European Commission demands** that its member states make number portability available for the local loop by January 1, 2000. However, some European countries (e.g., Germany and the Netherlands) planned to introduce number portability in 1998 or 1999.

North America

In the U.S. the Federal Communications Commission (FCC) mandated a phased introduction of Local Number Portability (LNP). First deployment started in the fourth quarter of 1997, while complete deployment is scheduled by the end of 1998. The location routing number solution was adopted as the long-term solution for LNP as the result of cooperation among the FCC, state regulatory agencies, service providers, and vendors.

---

*Switch-based call routing by onward routing solution (see Section 1.2.9.4.1).

**Referring to the decision of Telecommunications Commission on Feb. 12, 1997, Brussels. Some countries are exempted and get extra time.

Asia/Pacific

In the Asia/Pacific region, Hong Kong has supported an IN-based solution of number portability since the first quarter of 1997. New Zealand and Australia are also expected to introduce number portability before the year 2000.

1.2.9.1.4   Status of Standardization

ITU-T

In ITU-T, number portability standardization is carried out in Study Group 2 (SG2) and Study Group 11 (SG11), covering general aspects, numbering, routing, network architecture, and protocols. In September 1997, a baseline document was produced on the signaling requirements for number portability. This baseline document focused on a single target solution rather than a set of different solutions. It was expected that, in the first half of 1998, the main requirements would be finalized and that the first protocol specifications would become available by the end of 1998.

ANSI

As of June 1997, Committee T1 of ANSI completed a standard for number portability called "Call Completion to a Portable Number." This standard defines the network capability for number portability. In addition, this standard provides the ISUP protocol definitions for transporting the dialed number, the routing number, and an indication that the number has already been translated (or that the database has already been dipped). This approved ANSI standard was contributed to the ITU-T meeting in September 1997 for information.

In addition, Committee T1 has formed a new working group named T1S1.6 to address number portability–related issues. This new working group consists of four subworking groups consisting of Switch and Number Portability Database, Operator, Billing, and Program Management. The goal of the new working group is to write requirements that are detailed enough for manufacturers to develop.

ETSI

At the end of 1996 ETSI created the Number Portability Task Force (NPTF) under the responsibility of the technical committee, Network Architecture (NA). A close cooperation was established with the SPS (Signaling Protocol and Switching) and SMG (Special Mobile Group) committees. The mandate of the NPTF was to study solutions in the fixed network to support service provider portability for both geographic and nongeographic numbers.

Due to the fact that number portability solutions are deployed or are under development in a number of European countries, NPTF believes that it is important to define flexible and modular solutions not based on specific technologies. NPTF does not intend to standardize a single European solution to provide number portability, but, rather, a set of modular solutions able to coexist and migrate.

At the end of 1997 three technical reports were finalized, which describe a set of varying solutions [ETSI, 1997]. ETSI planned to have available a first set of standards for the support of number portability by the end of 1998.

1.2.9.2   Challenges for Implementation

When planning the introduction and implementation of number portability a number of sometimes conflicting requirements need to be taken into account. Major challenges are related to the following:

Challenge 1: Satisfying the short time frames as set by the regulator
Challenge 2: Needing to upgrade almost the entire existing infrastructure (including many legacy systems).
Challenge 3: Meeting performance and quality of service criteria.

Challenge 3 deserves special attention with respect to selecting a long-term architecture solution which is suitable to meet the performance and quality of service criteria as set by the regulator. These criteria also play a role in the regulatory aim to safeguard fair competition between operators. For example, in the U.S. the FCC defined a clear set of performance criteria for LNP which should "guarantee" the interests of end users, operators, and fair competition. Such performance aspects are having a major influence on the targeted solution and can directly help to determine the advantages and disadvantages of any planned solution. An example selection of such performance criteria follows:

- The originating end user should not notice any difference between making a call to a ported vs. nonported destination.
- Extra call setup delays across the *entire* network should remain within certain limits.
- There should be no degradation in reliability.

In practice, Challenges 1 and 2 often necessitate a phased introduction strategy, as a compromise should be sought between the short time frames and a long-term and advanced solution. Since a number of different solutions have already been identified for number portability [ETSI, 1997], it is interesting to evaluate these solutions in terms of the ability to meet overall performance and quality of service criteria. This can be found in Section 1.2.9.4. However, to do so it is first necessary to define a generic number portability interconnection model.

1.2.9.3 Number Portability Basic Capabilities

1.2.9.3.1 Configuration Model

The ITU Study Group 11 has proposed three configuration models to reflect the typical number portability network scenarios [ITU, 1997]. These ITU configuration models can be generalized to a single generic configuration model, which is shown in Figure 1.52.

Originating exchange: The local exchange serving a calling end user.
Intermediate exchange: Any exchange(s) between originating and recipient/donor exchange.
Donor exchange: The local exchange to which a call would be routed for termination in absence of number portability.
Recipient exchange: The local exchange to which the end user is ported.

In this configuration model an individual exchange (originating, intermediate, donor, or recipient) may also reflect an entire network (e.g., originating network), each of which may belong to different operator networks. However, the different type of switches may also belong to the same operator. So, many combinations are possible given that the competitive situation will likely show a combination of local operators and long-distance (carrier) operators.

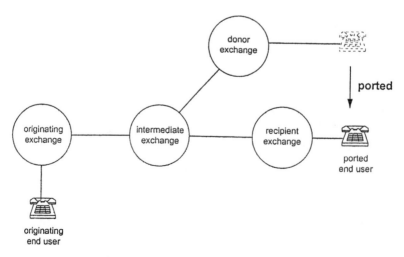

FIGURE 1.52 Generic NP configuration model.

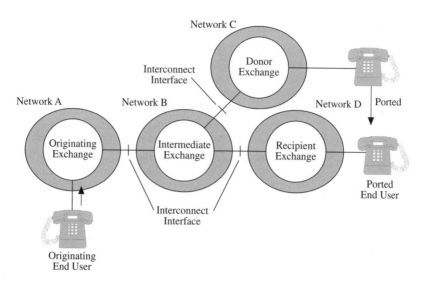

FIGURE 1.53 Possible interconnect interfaces with number portability.

1.2.9.3.2 General Number Portability Network Capabilities

Geographic numbers are part of a numbering and routing plan in which large number blocks are allocated to a specific local exchange. In absence of number portability, a call directed to a number in part of such a number block will be routed to this exchange (donor exchange). With number portability, the following general network capabilities are required to deliver the call correctly when an end user is ported from the donor exchange to a recipient exchange:

Step 1: Detect somewhere during the routing of the call (to the donor exchange) that the dialed number is ported.

Step 2: Retrieve routing information to identify the recipient exchange. This information is also referred to as network routing number (NRN).[*]

Step 3: Reroute call to recipient exchange using the obtained information from Step 2.

With regard to the targeted solution, the following aspects can be dealt with in different ways:

1. In which exchanges/networks is Step 1 performed (originating, intermediate, or donor) and what rules, if any, are applied?
2. Is the NRN routing information in Step 2 obtained from an intelligent network database or switch-based from the donor exchange?
3. When the call is rerouted in Step 3, is the NRN routing information recognizable and usable by all traversed exchanges? Also, are these exchanges part of *different* operator networks?

Assuming a multioperator environment, the above aspects (especially the first and the last) are very much dependent on the technical agreements and corresponding number portability capabilities at the interconnect interfaces between the various operators.[**] This is explained in the next section.

1.2.9.3.3 Specific Number Portability Capabilities at the Interconnect Interface

Figure 1.53 shows the possible interconnect interfaces between the different networks. Two strategies can be followed:

---

[*]Terminology proposed in ITU WP2/11 Report [1997].

[**]The national regulator can play an important role to harmonize the various technical and operational agreements between all operators in a particular country.

1. The interface remains fully standardized according to pre-number portability ISUP signaling standards (e.g., ITU-T Q.767).
2. The interface is optionally enhanced with specific number portability capabilities to allow more efficient call handling at both sides of the interface.

With respect to the latter strategy the following enhancements are proposed in international standardization [ITU, 1997]:

Interconnect Capability 1	Ability to transfer NRN information across the interconnect interface
Rationale and advantage	No need for multiple retrieval of NRN information in subsequent networks. Instead, only a single NRN retrieval at any point in one of the networks will suffice. This will reduce call setup delays and improve the quality of service. The NRNs should be unique for all operators in the country (administration on a nationwide basis).
Interconnect Capability 2	Ability to transfer number portability status information to indicate that a call has (already) been determined to be ported or nonported
Rationale and advantage	This information allows the receiving network to control how to handle the call. In case the status information indicates a nonported call (after the query), there is no need anymore to perform a database query by the receiving network.
Interconnect Capability 3	A mechanism should be applied to prevent potential looping of calls to ported numbers
Rationale and advantage	If this is not properly done, significant network problems may occur in case calls loop as result of missynchronization of porting data between different networks.

1.2.9.4 Introduction Strategies

In the absence of any standardized solution for number portability, many countries are choosing (or have chosen already) an introduction strategy which best suits their immediate needs and technological abilities. The initial solution is often a compromise between the conflicting challenges/requirements as described in Section 1.2.9.2. Looking at current implementations (U.K., U.S., Hong Kong, Finland) and soon to be expected implementations (The Netherlands, Germany, France), an evolution of subsequent technology solutions can be recognized (Figure 1.54).

The category of temporary solutions is primarily driven by very short time frames to deploy number portability in the existing network. The target solutions are more driven by optimal network performance and quality of service. The intermediate solutions bridge the gap by moving toward the target solution. Note that individual countries may opt directly for an intermediate solution (e.g., The Netherlands) or even the target solution (e.g., U.S., Hong Kong) as initial implementation. The following characterization can be given to each of these solutions[*]:

	Technology	Time Frame	Interconnect Interface	Network Efficiency and Performance
Temporary solutions	Switch-based	1996–2000	Standard	Low
Intermediate "Step-Up" solutions	IN-based (Query on Release)	1998–2002	Standard	Medium
Target solutions	IN-based (All Query)	1998+	NP enhanced	High

In the subsections below, the technologies, which are regarded to be representative for the temporary, intermediate "step-up," and target solutions, are explained. These are, respectively, call rerouting by onward routing, query on release, and all query technologies.

---

[*]Some variants may also use some IN functions (e.g., call forwarding of ported number to IN), but this does not influence the characterization as such.

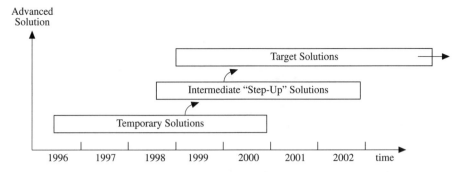

FIGURE 1.54 Evolution of technology solutions.

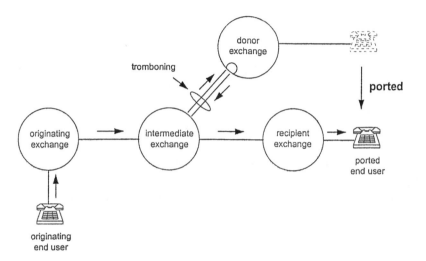

FIGURE 1.55 Call rerouting by onward routing solution.

1.2.9.4.1 Temporary Solutions

Call Rerouting by Onward Routing Method

The call is first routed to the donor exchange (Figure 1.55.). This exchange detects that the dialed number is ported. The new routing number is retrieved from the switch and the call is rerouted accordingly.* This rerouting capability may be based on existing call-forwarding capabilities. Since the donor exchange remains within the call path, this solution normally results in "tromboning" in the trunk network.

Advantages

- Based on existing rerouting capabilities
- No or limited impacts on signaling at interconnect interface
- Can be quickly introduced

Disadvantages

- High operational costs because of inefficient use of resources (tromboning)
- High provisioning and maintenance costs due to decentralized routing information and subscriber data management actions

---

*Usually the new routing number is prefixed to the called party number, but new information elements may also be added in the signaling.

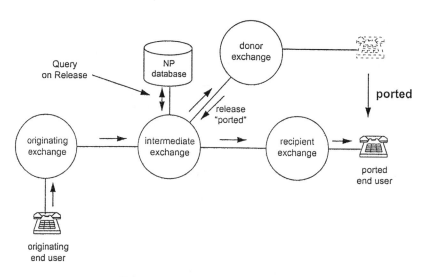

FIGURE 1.56 Query on release solution.

- High call setup delays
- Feature transparency problems, e.g., some features (e.g., UUS) may not work anymore for ported calls
- When the donor exchange is not ISDN-capable, service portability from PSTN to ISDN is not possible
- Very strong dependency on capacity and operation of the donor exchange

1.2.9.4.2 Intermediate "Step-Up" Solutions

Query on Release Method

In the basic scenario (not showing interconnect interfaces) the call is first routed to the donor exchange (Figure 1.56). This exchange detects that the dialed number is ported and releases the call to the previous exchange with a special cause value "ported." The intermediate exchange triggers to the IN when receiving this cause value. The IN database supplies the NRN routing number and instructs the exchange to reroute the call on the basis of this number.

Note that, if the donor exchange is not ISDN-capable, the query on release solution allows feature transparency for the ISDN service in case a PSTN customer is moved to ISDN on another exchange. This is a significant advantage compared with the call routing by onward routing solution.

Query on Release with Call Screening

Since all ported numbers will be stored in an IN database, this solution allows for call screening of incoming and outgoing calls at the interconnect interface. Call screening is best performed at the incoming and/or outgoing gateway. All calls or particular calls with a called number belonging to certain number blocks will trigger the IN to verify whether the call is ported. This allows for a more efficient delivery of ported calls, because in some scenarios a ported call no longer needs to be routed to the donor exchange (Figure 1.57). If the IN query indicates that the call is not ported, the call should proceed as normal and no IN translation performed.

It is expected that the initial implementations of query on release do not (operationally) employ the specific number portability capabilities at the interconnect interface as discussed in Section 1.2.9.3.3. Lack of this functionality may lead to some drawbacks related to the call screening at the interconnect interface. In this respect, call screening may lead to multiple subsequent database queries on a single call, on both ported and nonported calls. For particular call scenarios, this has a negative impact on performance and call setup delays.

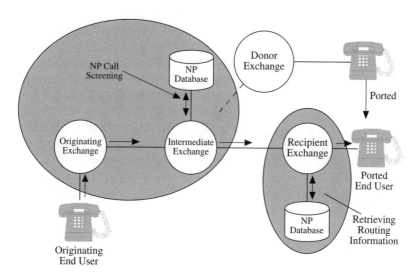

FIGURE 1.57 Query on release solution with number portability call screening.

Advantages

- Feature transparency
- Based on flexibility of IN architecture
- Centralized management of routing information in an IN database
- Increased call routing efficiency from call screening
- No or limited impact on signaling at interconnect interface
- Limited impact on local exchanges

Disadvantages

- Call setup delays may increase because of multiple subsequent database queries
- Nonoptimal use of resources because of call attempts to donor exchange
- Continued (but less) dependency on capacity and operation of the donor exchange

1.2.9.4.3   Target Solutions

All Query Method
This architecture (Figure 1.58) assumes that specific network arrangements are implemented and agreed upon by all involved operators:

1. A centralized number administration center is the master database for assignment of the network routing number. This eliminates the risk of internetwork looping due to conflicting data in an independent database by providing all operators the same routing information for ported subscribers. It also helps coordinate the porting process between operators with appropriate checks to prevent unauthorized switching of subscribers (known as "slamming" in the U.S.). Operators may choose to deploy their own private databases for call routing but must synchronize with the number administration center.
2. Specific number portability capabilities at the interconnect interface (as defined in Section 1.2.9.3.3) are agreed upon and used by *all* operator networks.

Having the above arrangements and agreements, the architecture makes it possible that a single database query is always performed at the next to last network ("N-1" network) in the call. This database query is performed without knowledge of whether or not the call is ported. Effectively, this means that all (interswitch) calls need to be queried (all query). Two outcomes of the database query are possible:

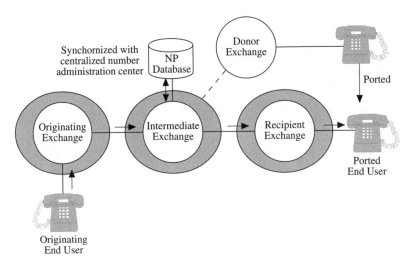

FIGURE 1.58 All query solution.

1. Ported call: The call is further routed on basis of NRN information to the recipient exchange. Since the NRN is defined and transferred on nationwide level, the recipient network does not need to perform an IN query to deliver the call to the in-ported customer.
2. Nonported call: The call is further routed on the basis of the dialed directory number. The number portability status information will inform the recipient network that an IN query has already been performed. Therefore, the recipient network can immediately deliver the call.

Unquestionably, there will be flexibility in the final implementation of the all query method for a particular country or network. However, the concept of a centralized number administration center and the concept that every call becomes an IN call characterize this architecture.

Advantages
- Feature transparency
- Based on flexibility of IN architecture
- Centralized database storage of porting information
- Optimal overall network efficiency and performance
- No dependency on donor exchange

Disadvantages
- Requires a single nationwide IN database system to be accessed and used by *all* operators
- Requires *all* operators to support specific number portability capabilities at the interconnect interface
- Requires extensive usage of and experience with IN

1.2.9.5 Management and Customer Care

The introduction of number portability is having a significant impact on the management and customer care systems of operators. Not only are existing mechanisms influenced, but new processes and procedures are also required to support the daily operation of the number portability service for end users and other operators.

The areas mainly affected are

- Service ordering and provisioning
- Network traffic management

- Billing and accounting
- Network maintenance

Brief descriptions of these areas are provided in the following subsections.

1.2.9.5.1 *Service Ordering and Provisioning*

The number portability service starts with a porting request from a customer. The general approach is that a porting request is made by the customer of the current operator (often the donor) at a single point of contact — the new operator (recipient). New processes, information flows, and cross-checking are required between the involved operators to support this service ordering and provisioning process.

Aspects covered in the provisioning flow include:

- Receipt of customer porting request by recipient
- Notifying donor and requesting needed data
- Checking feasibility
- Agreeing on porting date/time
- Preparing routing databases (IN solutions)
- Preparing internal processes and data settings
- Checking/testing preparations
- Cutting over
- Checking result with the customer

Usually a large number of operational support systems and customer care systems must be enhanced for this to occur.

1.2.9.5.2 *Network Traffic Management*

Network traffic management is affected in a number of ways. First, a new routing mechanism will be introduced based on the NRN routing number. This may impact existing (pre-number-portability) traffic management systems which currently manage on the basis of the dialed number only. Second, number portability will change the traffic patterns in the network. Also, new categories of traffic will be introduced such as ported (translated) calls to a recipient network. As a result, the overall requirements of the traffic management functionality are likely to change as well.

1.2.9.5.3 *Billing and Accounting*

Although billing and charging of the calling end user may not be directly impacted by number portability functionality, special consideration is indeed required for the call accounting mechanisms between the various operator networks. Call detail records should be generated for ported calls with all necessary information to support the accounting process.

1.2.9.5.4 *Network Maintenance*

Specific network maintenance activities should be done to ensure proper operation of the number portability service over time. Various "resources" need to managed, such as allocation and administration of number blocks in the network, administration of NRN routing information, management of digit analysis and routing tables, arming of IN triggers, etc.

1.2.9.6 Conclusion

Regulators all over the world are emphasizing introduction of number portability. By the year 2000, many incumbent and new operators were to have initial support of number portability available in their network.

But, number portability cannot be introduced overnight and most implementations were not finished in early 2000. Number portability is having a significant impact, both technically and operationally, on "traditional" network infrastructures. Different starting points and priorities in various countries may lead to different solutions. It is in the interest of the telecommunications industry as a whole to migrate and converge to a target solution that can be commonly deployed and that meets the overall requirements of number portability in an efficient and structural way.

The regulators, operators, and suppliers will play an important role in the process. Cooperation among all parties is required to ensure that all views are taken into account and that parties buy into the proposed solutions, both from a requirements and an implementation point of view.

All query architecture is widely recognized as a target solution that meets the various needs of number portability, including overall network performance and quality of service. Efforts are currently being undertaken in ITU-T to standardize this solution. The query on release solution allows a relatively smooth migration to this target architecture.

Abbreviations

ANSI	American National Standards Institute
DECT	Digital European Cordless Telecommunications
ETSI	European Telecommunications Standards Institute
FCC	Federal Communications Commission
GNP	Geographic Number Portability
GSM	Global System for Mobile communications
IN	Intelligent Network
ISDN	Integrated Services Digital Network
ITU-T	International Telecommunications Union — Telecommunications
LNP	Local Number Portability
NPTF	Number Portability Task Force
NRN	Network Routing Number
POTS	Plain Old Telephone Service
PSTN	Public Switched Telephone Network
SG	Study Group
UUS	User-to-User Signaling

References

CEC. 1996. Green Paper on a Numbering Policy for Telecommunications Services in Europe: Brussels, Nov. 20, 1996 — COM(96)590.

ETSI. 1997. NPTF Deliverables ETSI SPS1 meeting in Edinburgh Nov. 10–14, 1997: Number Portability Task Force deliverables WP0, WP1, and WP2.

ITU. 1997. WP2/11 Report, ITU Study Group 11 meeting in Geneva, Sept. 1–17, 1997: Discussion Minutes of the Q.21 (including number portability).

1.3   Standardization

1.3.1   Telecommunications and Information Systems Standards

Maarten Looijen

1.3.1.1   Scope

This section focuses on *standardization* at the level of operational products: a way of defining and accepting certain equal types and dimensions to get great uniformity of products. This general definition tells us that telecommunications and information systems standards describe the components of these products unambiguously, so that each component can be understood in the same way regardless of the place it is installed. Because telecommunications and information systems consist of many different components it is very important to know if these components are based on standards or not. If they are based on standards, it is much easier to maintain them than it would be if there were no standards at all and each component had to be seen as a stand-alone product.

As we all know, telecommunications and information systems consist of hardware and software, which belong to a constantly changing technological world. In that world, design, implementation, management, control, and maintenance play important roles. A poor approach to one or more of these aspects influences operational telecommunications and information systems in an unacceptable way. That is the reason we need standardization on the level of design, implementation, management, control, and maintenance. Within the context of this section we focus on the last three topics and introduce standardized models to get a good understanding of management, control, and maintenance tasks, how to organize these tasks, and how to fulfill them.

First, we give a sound specification of the concepts of normalization and standardization. Second, we give a description of telecommunications and information systems. Third, we mention the management, control, and maintenance aspects in relation to the necessity for standards. Fourth, we present a few standardized models to support the management, control, and maintenance of telecommunications and information systems. We conclude with presentation of a practical situation, which reflects the necessity for standards and how to implement them.

1.3.1.2 Normalization and Standardization

The terms *normalization* and *standardization* (norm and standards) are often, both nationally and internationally, used to mean the same thing. Nevertheless, they are two different concepts. Normalization indicates the determination by an official and independent body of technical specifications of products, methods, and the like, which are being considered for application. The standards applicable to these specifications are produced according to agreed-upon procedures. At the national level, in the Netherlands standards are produced by the Dutch Normalization Institute (NNI), in England by the British Standards Institute (BSI), in Germany by the Deutsches Institut für Normung (DIN). At the international level, standards are produced, for example, by the International Standards Organization (ISO).

Standardization means that design regulations are laid down by producing organizations or organizations applying them. Computer networks and communication networks, for example, must meet these design regulations when they are implemented. The way and the extent to which standardization arises define the level of a standard. Although actual practice often suggests otherwise, it appears that standardization does not, in general, have much legal validity. The standards may, but need not, agree with the official norms (Figure 1.59).

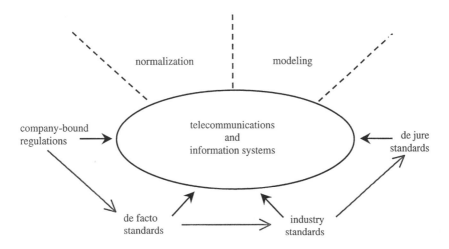

FIGURE 1.59 A helicopter view concerning standardization.

At the lowest level of standardization are the *standards specific to a company.* These are company-bound regulations which, when implemented, allow only products (hardware and software) of the company in question and those of companies that are in conformity with the standard to be used. Examples are Systems Network Architecture (SNA) of IBM and Digital Network Architecture (DNA) of Digital.

On the level above this are the *de facto* standards. These are the regulations that a product, such as an operating system, complies with so that the product can perform with products from various other companies. Examples are the operating systems Windows NT and UNIX, which can be implemented on several hardware platforms, coming from various suppliers.

On the next level are the *industry standards.* These are regulations accepted by various, more or less official groups of companies. Examples of such groups are X/Open group with regard to the operating system UNIX and the Institute of Electrical and Electronics Engineers (IEEE) with regard to electronics.

On the highest level are the *de jure standards.* These are regulations determined by official bodies of government and drawn up and managed by international standardization commissions such as the ISO, the American National Standards Institute (ANSI), the Comité Consultatif International pour Télégraphique et Téléphonique (CCITT), and the International Electrotechnical Commission (IEC). A product will often take a long time to become a de jure standard. For example, the programming language COBOL was first a de facto standard developed at a university; then it became an industry standard, and finally was elevated to a de jure standard by the U.S. Department of Defense (DoD). Figure 1.59 shows a "helicopter" view of telecommunications and information systems related to normalization, modeling, and standards (Looijen, 1998).

1.3.1.3 Telecommunications and Information Systems

Telecommunications (TC) can be defined as the use of communication equipment to transport signals from nodes to nodes. *Equipment* include:

- Front-end processors
- Modems
- Multiplexes
- Private automated branch exchanges
- Bridges
- Routers
- Repeaters
- Transceivers
- Hubs and gateways
- Communication software

Connections include:

- Coaxial cables
- Twisted pair
- Shielded twisted pair
- Unshielded twisted pair
- Fiber optics
- Radio transmissions
- Satellite circuits

The main function is the transportation of data, voice, and pictures.

The utilization of TC involves a number of *requirements*:

- Availability
- Maintainability

- Performance
- Reliability
- Robustness
- Security
- Others

and a number of *preconditions*:

- Finance
- Dispersion
- Standardization
- Others

All of these requirements and preconditions are the result of extreme dependency on high-quality telecommunications.

To meet these requirements and preconditions it is necessary to have a good understanding of the *characteristics of telecommunications*. Examples of communication equipment characteristics are

- The level of normalization and standardization
- The interface with other facilities

Examples of connection characteristics are

- Capacity
- Speed
- Reliability

Examples of communication software characteristics are

- The way in which, and the frequency with which, new versions are delivered
- The interface with other software

All these examples indicate that it is necessary to have a diversity of knowledge regarding the characteristics of the different components.

Information systems (IS) should be defined as the use of *hardware,* such as

- Computers
- Storage devices
- Input equipment
- Output equipment

and *software,* such as

- Basic software
- Application software
- Database management software
- Programming tools, data sets, procedures, and people involved in the utilization

Just as in the case with TC, IS encompasses a number of requirements in relation with all above-mentioned aspects. To meet these requirements and preconditions it is necessary to have a good understanding of the *characteristics of information systems*. Examples of computer characteristics are

- The size and the number of buffers
- The data transfer capacity
- The extent of the main memory

- The computer architecture
- Heat dissipation

Examples of storage device characteristics are

- Storage capacity
- The speed of accessibility to the data
- The read–write possibilities vs. the read possibility alone

An example of input equipment characteristics is

- Sensitivity

Examples of output equipment are

- Technology
- Printing speed
- Print quality
- The interface with other facilities

Examples of basic software characteristics are

- The way in which, and the frequency with which, new versions and new releases are delivered
- The way in which diagnosis of errors is carried out

Examples of application software characteristics are

- The installation method
- The programming method
- The age
- The maintainability method

Examples of database management software are

- The way in which recovery activities in databases are carried out
- The backup facility
- The logging or the registration
- Reporting of processing

Examples of programming tool characteristics are

- Generation
- The operating method
- The way in which diagnosis of errors is carried out

Examples of data set characteristics are

- The extent
- The data structure
- The data definition
- The accessibility

Examples of procedure characteristics are

- The number
- The extent

- The user-friendliness
- The language in which procedures are formulated

Examples of user characteristics are

- The number
- The skills
- The experience
- The stability/instability of the personnel turnover

All these examples indicate that it is necessary to have a diversity of knowledge regarding the characteristics of the different components.

This scope of characteristics is one of the starting points illustrating the need for a standardized approach to the management, control, and maintenance of TC and IS to avoid chaos and all manner of individual inventions. Another starting point for such standardized approaches is the requirements and preconditions formulated by the user organization. Fulfilling these demands a structured approach in the form of well-defined models.

1.3.1.4 Management, Control, and Maintenance of TC and IS

Management, control, and maintenance of TC and IS encounter a number of factors. All these factors typify TC and All Routes Explorer (ARE). The ultimate profile defines to a great extent precisely what kind of TC and IS has to be managed, controlled, and maintained. The factors refer to the individual components of TC and IS as well as to their relationships. The factors are *quantity, diversity, distribution, dynamics, ownership,* and *utilization.* The clarification of these factors is as follows: TC and IS may include large numbers (quantity) of hardware, software, data sets, and procedures, characterized by all kinds of types (diversity). These products can be decentralized to a high degree (distribution) and regularly be subject to changes (dynamics). The products can belong to various owners (ownership); and the users can make very divergent demands and preconditions (utilization).

All these factors, together with the earlier mentioned requirements; preconditions, and TC and IS characteristics, are the pillars for the content and the organization of management, control, and maintenance of TC and IS. Implementation requires a structured, standardized approach.

Within the context of a management, control, and maintenance of TC and IS are all tasks that must be fulfilled for operationalization, supporting and influencing the goals of the organization in a positive way. This must be done in such a way that it corresponds with the requirements and the preconditions imposed by the utilization and the characteristics of TC and IS.

To exclude all kinds of *ad hoc* solutions and fragmentary approaches the next sections introduce several models. They all focus on a way to deal with a structured approach of defining and organizing management, control, and maintenance of TC and ARE.

1.3.1.5 Models (Standards) Supporting the Management, Control, and Maintenance of TC as Part of IS

The title of this section emphasizes that TC is not an isolated combination of hardware, software, and connections to transport data, voice, and pictures. TC is part of one or more information systems. And as we know, the functionality of an IS is to receive data, to transfer data, to collect data, to transport data, and to produce data. All these functions define in a nutshell an automated IS. But it is more comprehensible, as we have seen before, to focus on TC and to focus on IS in a separate way; and that is what we will do in this section and the next.

In this section we look at some models that focus mainly on TC, are available and have proved to be applicable in practice (van Hemmen, 1997).

OSI Network Management is designed for the management, control, and maintenance of OSI stocks in open systems, systems that can communicate with other systems in accordance with the principles of

Open Systems Interconnection (OSI) defined by the ISO. The OSI Network Management reference model describes five functional areas for network management (Terplan, 1992):

1. Configuration management: an ongoing process providing at anytime a complete survey of the network, the location of its components, and its status.
2. Fault management: focusing on fault detection, fault diagnosis, and fault repair.
3. Performance management: the ongoing evaluation of the network to verify the balance between service levels and actual performance.
4. Security management: ensuring an ongoing protection of all components of the network and the data transported through the network.
5. Accounting management: costing and changing activities in relation with investments, obligations, and utilization of the network.

Telecommunications Management Networks (TMNs) aim at the coordination of activities on standardization in the area of telecommunications management to meet growing management needs, such as multivendor network management, interoperability of management systems, and extension of management functions. To fulfill these needs, CCITT is developing a model based on an architecture that meets such requirements as minimizing management reaction times to events, providing isolation mechanisms to locate network faults, and improving service assistance and interaction with customers.

The TMN includes three architectures:

1. TMN functional architecture: achieving standardized exchange of management information.
2. TMN physical architecture: derived from the above-mentioned architecture and defining five connected systems, each system comprising one or more function blocks.
3. TMN information architecture: defining the principles for exchanging management information between function blocks such as the syntax and semantics of their messages that have to be exchanged

In addition to this, the TMN architecture consists of four layers:

1. Network element management: offering functions for the management, control, and maintenance of a set of network elements that belong to one supplier or vendor of network components. The functions are made into a uniform standard format.
2. Network management: presenting a total overview of the network; in other words, the topology of the network is visible.
3. Service management: offering functions for the management, control, and maintenance of the services related to the network. There are functions supporting operational customer processes and functions supporting service managers.
4. Business management: supporting the business managers by emphasizing market, economy, trends, and cost-effectiveness.

Open Management Interoperability Point (OMNIPoint) is a development and adjustment of the Network Management Forum (1990). It is a set of standards, implementation specifications, testing methods, and tools making possible the development of interoperable management systems and applications. It defines the elements that must be implemented to achieve effective exchange of management information:

- Common Information Management Services: defining the protocols necessary to transfer management information.
- External Communication Services: providing specifications for an OSI transport network.
- Managed Objects: specifying the method by which the details of the resources being managed are recorded so that the information can be communicated between systems.

Dunet Management Model, developed by the computing center of Delft University of Technology, the Netherlands, orders the network services and defines the service interfaces. Because networks are mostly

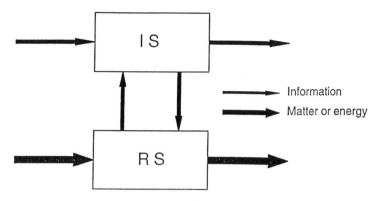

FIGURE 1.60 The Information paradigm.

very large and have various parties (many service providers and customers are involved), it is worthwhile to define the responsibilities of the different parties. This model concerns this. It can be used to determine all relationships between the network and management, control, and maintenance. Therefore, it offers six layers to perform all network services in the same way as in the OSI model:

1. Services: presenting the network services to the users.
2. Applications: consisting of software specific to the network services.
3. Application architecture: containing basic software.
4. Network architecture: providing transport services over the network.
5. Connections: providing a faultless transmission of information.
6. Cables: consisting of cables which are able to transmit signals.

The models described contain differences and similarities. All models try to present a structured (standardized) approach to management, control, and maintenance of TC. On the conceptual level, we recognize the most similarities. As soon as the models become more detailed, we discover the differences. This is a quite normal situation. The more management, control, and maintenance of TC approach the state of implementation in a real-world contingency, the more factors initiate and influence the defined levels of standardization. This fact is an important reason to put telecommunications into a broader context, namely, the context of information systems. As stated earlier we can see TS as a part, a very important part, of IS.

1.3.1.6 Models (Standards)

This section, elaborate in comparison with the others, emphasizes the modeling of organizing and fulfilling the management, control, and maintenance of IS/IT. This way of standardization is meant to support the many activities that have to be carried out in relation to the management, control, and maintenance of IS/IT. In practice, this happens all too often in an unstructured way.

Organizations are viewed as dynamic systems. Any such system can be modeled in terms of a *real system* (RS) and an *information system* (IS), where the RS determines the behavior of the IS. Information systems, like the organization of which they are a part, are complex, multileveled, dynamic entities. At any level of abstraction an RS/IS combination has an environment comprising everything the information system does not control. Input to the information system from the environment includes, for example, managerial directives, and output toward the environment is in the form of messages. Application of this so-called information paradigm (Figure 1.60) forces analysts to structure and position RS and IS components that seem to be, or are, strongly intertwined. On this basis, information systems can be developed systematically by defining all components, object types, and processes within both RS and IS. In more detail, each RS and IS can be seen by another RS/IS combination. This means that there are more information paradigms in themselves; this is called the recursion principle: an IS and/or an RS is specified by one or more RS/IS combinations.

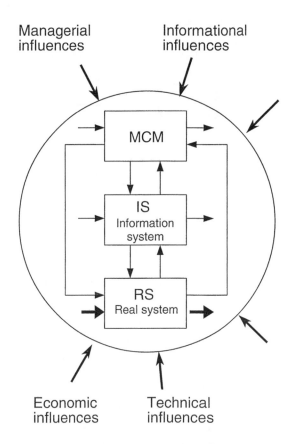

FIGURE 1.61 The MCM paradigm.

To position the management, control, and maintenance (MCM) of information systems the MCM paradigm is introduced (Figure 1.61). In this paradigm, the RS/IS combination forms the real system and MCM is the information system needed to control this combination. This paradigm underlines that technical and economic influences from the environment do not directly affect the operations of existing information systems, but are first evaluated by MCM processes. This approach can be explained by an arbitrarily chosen example.

An educational institute owns a student information system to control its student administration. There are also several information systems for the personnel and salary/payroll administration. Furthermore the complete inventory of the institute has been computerized, and several office automation environments have been installed. As a result of several mergers with other institutes in the past, the organization has become geographically widespread. Communication facilities were set up to connect all the local units of the institute. Students and employees have become increasingly dependent on the information systems. These systems do not solely support secondary administrative processes but also primary educational activities. Instruction rooms have been provided with computer-aided instruction applications running on personal computers connected via local-area networks. The strong dependence of educational processes (RS) on automated data processing (IS) means that the use of information systems is necessary to satisfy all kinds of requirements. These requirements address a high level of availability, security of databases, hardware, and software, and high performance in terms of efficient throughput and short response times. When there are errors or failures of hardware or software, corrective maintenance should be performed as quickly as possible. Serious deterioration when a disaster occurs should be tackled by emergency facilities to ensure the continuity of primary processes. These requirements make demands on the capacity of human and financial resources. For instance, 100% availability and security, if feasible, requires extensive financial resources and human efforts. To provide some balance between realistic requirements and adequate control tasks, it is essential to analyze the major operational risks and investigate the consequences if some requirements are not met.

At the management level it is therefore necessary to formulate constraints and preconditions. Here constraints may address the way in which maintenance will take place, or the extensiveness of business resumption facilities. These requirements and constraints have to be met during the entire life cycle of the information system, especially after the system is delivered and designers and constructors have left.

All system components have their own technical characteristics. These characteristics arise during several stages of the development and the implementation process. Each computer has its own specific internal memory and processing speed; data storage devices have their access time and storage capacity; databases are reflected by their data structure; procedures can be either easy or difficult to access; and people differ in their skills and expertise. After implementation, these characteristics and others should be specified in a clear and understandable way. After information systems become operational and/or are integrated with other systems, there is a new dimension of characteristics to be considered. This dimension refers to the performance characteristics of an information system, where it should be anticipated that actual service levels may deviate from the requested standards. For instance, a computer may be used in such a way that its processing capacity is not optimally utilized, and the maximum storage capacity of magnetic disks is not used completely most of the time due to performance constraints. So, there can be all kinds of environmental conditions or contingencies, which result in the optimal potential performance not being met.

Finally, the function of information systems is another important field of interest. Functions of an IS address data input, data manipulation, graphical representation (user interface), report generation, and so on. The above is what really counts for information system users. Before these users can exploit these functions, they need training and advice. After some period of operation it often becomes clear which functions are used frequently and which functions are ignored. In fact, the latter necessitates the investigation of the underlying causes, for instance, problems users experience in getting access to certain functions, or simply the uselessness of certain functions. It is also possible that some functions are missing and need to be added. Changes in personnel require permanent efforts directed toward training and user support. Utilization management should track permanently to what extent application software and packages are used and which applications appear to have become obsolete.

From the observations above, it can be derived that developments in both RS and IS initiate continuous control tasks which should guarantee that information systems operate in accordance with the preconditions and requirements imposed on the utilization and the technical characteristics of the system components. All these tasks are covered by MCM and can be modeled by the MCM paradigm. As mentioned before, this paradigm demarcates the environmental forces that influence MCM. External influences differ in nature as follows:

- Managerial: As a result of mergers, alliances, joint ventures, or disposal of certain business units the organizational structure of the real system may be changed.
- Informational: New governmental regulations/laws may cause redesign of privacy control mechanisms.
- Economic: Changes in the economic situation/conjuncture at national or international level may cause additional cost reduction measures or the provisioning of additional financial resources.
- Technical: New technological innovations and improved price/performance ratios may lead to replacement of existing hardware and software.

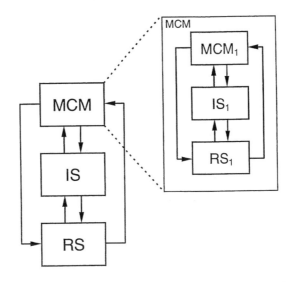

FIGURE 1.62 The MCM paradigm and the recursion principle.

It may be clear that the MCM paradigm includes the same recursion principles as described for the information paradigm. To illustrate the MCM recursion principle, the following example is used. Information system control includes incident handling and solving problems arising during operations. These incidents and problems can be associated with both software or hardware components. This incident handling process can be considered the RS. To support this process in a more or less computerized way, IS can be developed in such a way that it supports specific units which are responsible for this process, and this information system itself has also to be managed, controlled, and maintained (Figure 1.62).

Elaboration of MCM is depicted by two approaches. The first approach is a short description of a frame of reference of the tasks that have to be performed when managing, controlling, and maintaining an information system. This includes a structured ordering of the tasks in relation to the organizational levels and the kind of object. It also implies a sound look at the life cycle of an IS in the form of an extended state model. The combination of all this is indicated as MCM of IS.

The second approach is a short description of the IT Infrastructure Library (ITIL) (CCTA, 1990). This is a set of modules that describes MCM by a number of processes.

First, *MCM of IS* covers in accordance with the definition stated before, all tasks that relate to the separate information system components as well as to the relationships among them, and also to the data-processing procedures and information management processes. The many tasks involved require first of all an arrangement (a standard approach) whereby clusters of tasks are formed. Each cluster includes tasks on main lines relating to their nature and coherence. Such a cluster is called a task area. Then, for each task area one or more subclusters are identified; we call such subclusters task fields. This involves a further refinement by nature and coherence. We distinguish the following task areas and mention for each task area the accompanying task fields; this will be done by very short descriptions.

The *management* (M) task area determines policy, draws up plans, and coordinates the organization and maintenance of the MCM of information systems. The task fields identified are

- Strategic management
- Tactical management
- Operational management

The *personnel management* (PM) task area covers the management of the personnel within an organization. The task field identified is:

- Personnel management

The *technical support* (TSu) task area aims at examining, evaluating, and optimizing the availability of hardware and software. The task fields identified are

- Hardware and basic software support
- Communication support
- Database management systems support
- Management of PC application packages
- Applied research

The *general business support* (GBS) task area consists of general management support tasks. The task fields identified are

- Administrative management
- Quality control
- Capacity planning
- Order control
- Budgeting
- Charging back
- Acquisition of IT resources

The *operational control* (OC) task area covers the continuous control and the operational management of hardware, software, and data processing procedures. The task fields identified are

- Acceptance
- Operating
- Hardware management
- Technical software management
- Physical data management
- Utilization analysis
- Performance management
- Tuning

The *maintenance of the technical infrastructure and operational support* (MTI-OS) task area is responsible for making changes/modifications and removing problems related to the technical infrastructure. Furthermore, it supports availability, disaster recovery, and security. The task fields identified are

- Changing the technical infrastructure
- Problem management
- Availability control
- Disaster recovery
- Security

The *technical services* (TSe) task area consists of all services provided to the users of information systems and the users of hardware and software for development purposes on the basis of the preceding task areas. The task fields identified are

- Management of the range and cost of services
- Data processing
- Advice and participation
- Information center

The *utilization management* (UM) task area consists of the tasks that directly support the users of information systems. The task fields identified are

- User support
- Functional system management
- Management business data

The *functional maintenance* (FM) task area consists of the tasks related to the maintenance of procedures, specifications, and definitions for which the user is responsible. The task fields identified are

- Maintenance of manual procedures
- Functional maintenance of information systems
- Data definition control

The *application maintenance* (AM) task area consists of the tasks related to maintenance of application software and data sets of information systems once accepted and adopted. The task fields identified are

- Maintenance of application software
- Database management

Second, MCM of IS relates these task areas and task fields to the life cycle of an information system. Therefore, we introduce the *state model* followed by the *extended state model*. The *state model* distinguishes the following states:

- *State IPP* (information policy and information planning): In this state, information policy and information planning are determined, which leads to the information system (or telecommunications) being developed.
- *State D* (development): In this state, the information system (or telecommunications) is designed and constructed; then it is either accepted or not in the following state.
- *State AI* (acceptance and implementation): In this state, the information system (or telecommunications) is either accepted or not. If not accepted, it goes back to the previous state. If accepted, implementation takes place, whereupon it will be put into use as well as exploited.
- *State U* (utilization): In this state, the functions of the information system (or telecommunications) are being used.
- *State E* (exploitation): In this state, the information system (or telecommunications) is kept operational or exploited for utilization.
- State M (maintenance): In this state, the information system (or telecommunications) or part of it is changed or modified as a result of maintenance, initiated from the states U and E.

These changes or modifications will not only alter the information system or telecommunications, but utilization and control as well. This is the reason to classify the changes or modifications into two categories. The first category consists of changes or modifications that, after implementation, result in hardly any or no changes or modifications at all in states U and E. The second category consists of changes or modifications that, after implementation, result in a change or modification of at least one of the states U and E. In connection with this, state M has two substates, i.e., M1 (minor or no impact) and M2 (major impact) (Figure 1.63). In substate M1 the changes or modifications found, in terms of maintenance, belong to

- Corrective maintenance
- Preventive maintenance
- Improvement maintenance (dependent on nature and extent)

The changes or modifications found in substate M2 are quite different from those belonging to M1. They affect more than what is normally understood by maintenance. In fact they may concern changes

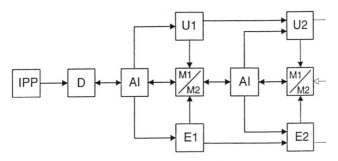

FIGURE 1.63 Extended state model.

or modifications at the level of software development. They may also imply that information policy and information planning have to be adjusted. In substate M2 the changes or modifications found, in terms of maintenance, belong to:

- Improvement maintenance (dependent on nature and extent)
- Adaptive maintenance
- Additive maintenance

So far, the MCM has been described as one set of management tasks classified into task areas and task fields. All these tasks can be related to the extended state model. Each state includes a number of relevant tasks that correspond with the scope of the state.

Third, we derive from the extended state model three kinds of MCM:

- *Functional management* (FM) is responsible for the maintenance and control of the functionality of the information system (or telecommunications). We meet the corresponding tasks in the states U and M1/M2.
- *Application management* (AM) is responsible for the maintenance and control of the application software and the databases. We meet the corresponding tasks in the state M1/M2.
- *Technical management* (TM) is responsible for the maintenance and control of the operational-ization of the information system (or telecommunications) consisting of hardware, software, and data sets. We meet the corresponding tasks in the states E and M1/M2.

These three kinds of MCM are the ultimate base for the organization of MCM in a practical organization. They are similar to the organizational units and are related to each other. Depending on the real situation each organizational unit can appear more times (Figure 1.64). Part of this organizational approach is the introduction of three managerial levels for each kind of MCM:

- The strategic level
- The tactical level
- The operational level

Each level has a specific responsibility for a number of tasks of MCM.

The *IT Infrastructure Library* (ITIL) is the second approach to realize MCM in such a way that it corresponds with user requirements and preconditions with regard to IT. ITIL consists of approximately 40 booklets, each discussing a subject (module) related to MCM of IT. The library was developed by the Central Computer & Telecommunications Agency, U.K. The main purpose of ITIL is described as: "to facilitate improvements in efficiency and effectiveness in the provision of quality IT services, and the management of the IT infrastructure within any organization."

To realize this objective, ITIL describes a range of subjects, which try to cover the whole field of management of IT. The line of approach of ITIL is process-based and has a strong relation with computing centers. The whole library is intended to provide logical coherence to the many aspects of MCM. This coherence, however, is not always apparent in the text.

This means that at the application level, where this coherence has to be present, a few things have still to be done. Thus, implementation of ITIL is more than simply carrying out the actions prescribed.

Similar subjects are clustered in nine sets. Six of these sets concentrate on IT services and the management of the IT infrastructure. Three sets include advice on the organization of technical facilities which are inherent to an IT infrastructure. This approach should be regarded as a first structuring of the many management subjects, with one focus on "service provision and IT-infrastructure management" and one focus on "environmental."

The following six sets belong to IT Service Provision and IT Infrastructure Management:

Manager's Set: This set comprises six subjects which cover the management of internal as well as external IT services. The subjects are assigned to senior managers responsible for one or more fields of management. It concerns subjects that involve coordination and deal with the whole of the management:

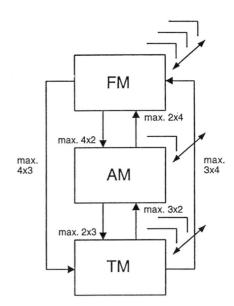

FIGURE 1.64 Multiple relationships among functional management, application management, and technical management.

- Planning and control for IT services
- IT services organization
- Quality management for IT services
- Managing facilities management
- Managing supplier relationships
- Customer liaison

Service Support Set: This set comprises five subjects which are clearly interrelated and deal with achieving a stable and flexible service provision:

- Configuration management
- Problem management
- Change management
- Help desk
- Software control and distribution

Service Delivery Set: This set comprises five subjects which refer to a qualitative and cost-effective service provision:

- Service level management
- Capacity management
- Contingency planning
- Availability management
- Cost management for IT services

Software Support Set: This set comprises two subjects describing the relationship with the developers of software:

- Software life cycle support
- Testing an IT service for operational use

Networks Set: This set comprises two subjects which refer to the management of an IT infrastructure distributed over various locations:

- Network services management
- Management of local processors and terminals

Computer Operations Set: This set comprises four subjects which are of direct importance to all those who are responsible for the operationalization of large computer systems:

- Computer operations management
- Unattended operating
- Third-party and single-source maintenance
- Computer installation and acceptance

The following three sets belong to the environmental category:

Environmental Strategy Set: This set comprises two subjects which are intended for those who are responsible for the strategic management (with respect to place and nature) of the technical facilities as a specific part of the entire IT (Daalen, 1993):

- Cable infrastructure strategy
- Environmental services policy

Environmental Management Set: This set comprises eight subjects which refer to the maintenance of the technical facilities supporting IT:

- Management of acoustic noise
- Secure power supplies
- Fire precautions in IT installations
- Management of electrical interference
- Accommodation specification
- Environmental standards for equipment accommodation
- Specification and management of a cable infrastructure
- Maintaining a quality environment for IT auditing and cleaning

Office Environment Set: This set comprises four subjects which provide advice and guidance on the workplace environment where IT is used:

- The office working environment and IT
- Office design and planning
- Human factors in the office environment
- Managing a quality working environment for IT users

1.3.1.7 A Practical Approach

A managerial step-by-step plan (MSP), is based on the preceding subjects and provides a number of steps by means of which, in a systematic way, the organization of the MCM of information systems (and telecommunications) can be realized. In addition, these steps express the relationship existing between the various subjects. MSP assumes that the building blocks for the realization of MCM are brought forward and therefore it will be possible to erect a complete building. Moreover, MSP aims at critically analyzing the existing MCM situations. On the basis of the results obtained it will then be possible to decide whether or not to come to adaptation of MCM or parts of it. This means that MSP aims at the realization of a totally new MCM organization as well as at a possible adjustment of existing MCM organizations and everything involved. Each step includes a description of the activities to be done.

Step 1: Describe the Object of MCM

Describe the object to be managed, controlled, and maintained, namely, one or more information systems and the business processes connected. The object is expressed in hardware, software, data sets, procedures, and the people making use of it. It concerns the elaboration of all the characteristics going with it. Moreover, it is important to know in what way the object presents itself physically; this often happens in terms as distribution, diversity, and quantity. Also, the technical facilities to be responsible for electricity, cooling, and the like have to be known. As to the business processes, of which the information systems are derivatives, type and size have to be mapped.

Step 2: Describe the User Organization

Describe the demands on hardware, software, data sets, procedures, and the relevant data processing procedures being made in relation to practical applications. Also, the technical facilities should be involved here. In addition, the preconditions that are set upon the financial, material, and human resources being related to the demands formulated in this step have to be known. An aspect that is also important in this step relates to the situational factors (or contingency factors). They describe to a considerable degree the company to which the object of MCM belongs. Typical of the company are nature, size, location, and age. Together with the requirements and the preconditions, they form the starting points for the following step, which anticipates the details of it more fully.

Step 3: Select the MCM Tasks

Select the task fields by means of which the demands made by the user organization in Step 2 can be met. The prevailing preconditions as well as the situational factors of the company have to be taken into account here. This may imply that the entire task field need not be selected, but just one or more MCM tasks belonging to the task field in question. When selecting, the relationship between task fields has to be immediately identified as soon as it appears that an independent task field cannot meet a certain demand. For, in practice, there is a great coherence between task fields.

Step 4: Translate into MCM Processes

Translate the task fields of Step 3 into MCM processes. This implies a further elaboration of the tasks fields in order to present them coherently as processes. ITIL, can be used here. In practice it almost always concerns two out of the nine ITIL sets. It concerns the service support set (configuration management, problem management, change management, help desk, and software control and distribution) and the service delivery set (service level management, capacity management, contingency planning, availability management, and cost management for IT services).

Step 5: Realize Generic and Specific Templates

All processes of Step 5 are worked out by using the so-called basic process model which functions as a generic template. The template, in its most elementary form, consists of a transformation. This is preceded by input which is transformed into output. Input as well as output pass a filter which filters input and output respectively. What is passed as input for transformation is buffered. What is not passed to be transformed goes elsewhere (another transformation), whether or not to return, to be filtered as output. The result of the filtering may be that there is a feedback to the input filter in order again to be considered for transformation. If there is no feedback, buffering takes place and the process is then brought to an end or continued elsewhere (Figure 1.65). Proceeding from the generic template, a specific template is constructed for each MCM process. On the basis of this, each MCM process can be worked out unambiguously. Relationships between the processes are established wherever necessary. These relationships need not be alike for each management situation. However, for a number of processes this is the case, because certain processes are clearly interrelated, for instance, problem management and change management. For other processes this need not be the case.

Step 6: Describe the Three MCM Forms

Describe the three MCM forms functional management, application management, and technical management on the basis of the task fields selected in Step 3. Distinguish here the three levels, i.e., the strategic

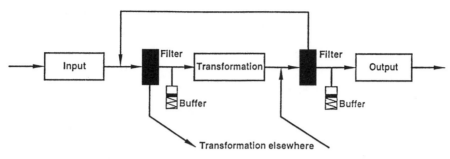

FIGURE 1.65 The generic template.

level, the tactical level, and the operational level. The approach is two dimensional, i.e., each management form occurs only once. The result is rather theoretical and must serve as model for a more practical elaboration.

Step 7: Realize the Organization of MCM
Realize the organic MCM units that are responsible for functional management, application management, and technical management. This may lead to several occurrences of the three MCM forms distinguished in Step 6. To be taken into consideration are the problems referring to the centralization, decentralization, and outsourcing of MCM. The result is an overall picture of how MCM is distributed over organic MCM units.

Step 8: Design the Integration Model
Put the result of Step 5, namely, the MCM processes, into relation with the organization of MCM. The result is the embedding of the MCM processes in the organic MCM units of Step 7. Here, the responsibilities of the MCM units are expressed and what MCM tasks in the MCM processes have to be performed.

Step 9: Describe Functions and Employees
Establish the functions to be performed, and also the required knowledge, insight, and skills, in order to be able to carry out the MCM tasks within the MCM processes. Assign the functions to employees who will be responsible for the performance of the MCM tasks. The ultimate result is an MCM organization which may be considered capable of interpreting MCM as expressed in the preceding steps. This concerns an MCM being attuned to the requirements and preconditions made in relation to the user organization and where the identified situational factors are taken into account. At the same time, the MCM organization has to be considered capable of contributing to the realization of organizational objectives considering the fact that information systems increasingly form an integral part of the business processes. To sharpen this and give it a professional complexion, the following step is necessary.

Step 10: Formulate Service Management
Draw up the service-level agreements between utilization and MCM of information systems on the basis of the requirements and preconditions formulated in Step 2. Equip the MCM organization with methods and techniques to measure and, if necessary, to adjust the services agreed upon. Wherever possible, MCM has to be proactive. This implies observation and analysis of tendencies in the progress of the quality level of the services. Along with quality and control, economic and juridical aspects play a part. A service-level agreement is a contract where the rights and duties of utilization and management are described. Furthermore, measuring and comparing play an important part.

Step 11: Perform Evaluation and Simulation
Evaluate regularly the functioning of MCM and take account of changes inside as well as outside the user organization. For MCM, that is managing, maintaining, and controlling information systems in accordance with requirements and preconditions agreed upon in relation to practical applications and the characteristics of the information system components, as well as contributing to the realization of business

objectives. An important aid here is the application of MCM simulation. With the help of this the management can, in its totality, be investigated at the animation level.

The steps are a reflection of a standardized approach to analyze and to realize, totally or partly, management, control, and maintenance of TC and IS.

1.3.1.8 Summary and Conclusions

Telecommunications and information systems standards are not only related to hardware and software, such as *standards specific to a company, de facto standards, industry standards,* and *de jure standards,* but are also related to management, control, and maintenance. It is very important to understand that the organization and the fulfillment of all tasks, based on requirements and preconditions and related to a variety of hardware and software characteristics, need a well-structured approach. Such an approach must avoid ineffective and inefficient effort of personnel, material, and finance. In practice, the management, control, and maintenance of TC/IS still occurs in an unstructured way, far from conceptualization and modeling.

References

CCTA, The IT Infrastructure Library: An Introduction, 1990.
Daalen, P.J.J. et al., Strategisch NetwerkPlan; de visie op netwerkdiensten, deel 1, 1993 [in Dutch].
Looijen, M., *Information Systems: Management, Control and Maintenance,* Ten Hagen Stam, the Netherlands, 1998.
Network Management Forum, *Discovering OMNI Point: A Common Approach to the Integrated Management of Networked Information Systems,* Prentice-Hall, Englewood Cliffs, NJ, 1990.
Terplan, K., *Communication Networks Management,* 2nd ed., Prentice-Hall, Englewood Cliffs, NJ, 1992.
Van Hemmen, L.J.G.T., Modeling Change Management of Evolving Heterogeneous Networks, Dissertation, Delft University of Technology, the Netherlands, 1997.

1.3.2   In the Trenches of the Browser Wars: Standards in the Real World

David Allen

The browser wars, particularly the conflict over Internet Explorer and Windows, left us in the midst of a dilemma. Policy was suspended between a standard such as Windows, which represents an implicit monopoly, and dictates for competition policy, which eschews monopoly. On this note, we turn to actual practice in the attempt to find some new light, and perhaps policy guidance as well.

1.3.2.1   Information Product — A Choice

If a group of people want to set up communications, they must agree on a common dimension. Language is an example of a dimension at a most fundamental level. An "information product," is the result of reaching an agreement on a dimension.

To explore the dimensions of an information product, consider two contrasting examples — Java and Windows.* Java is a layered information product, while Windows is integrated vertically on purpose. Java, for example, intermediates between the functionality, which an applet may provide, and the particularities of the operating system (OS) on which the applet will run. Windows, by contrast, is to be taken as a (vertical) whole. The struggle between the U.S. Department of Justice and Microsoft over Internet

---

*How are these part of communications? Computer protocols serve well as examples since computing is an intermediate case, relative to network protocols. Agreement on a common approach is a necessity for a network, if it is to operate; in computing, however, the effect of interoperability requirements is felt quite strongly but there is still some leeway for variation. As a result, computing is a useful intermediate example which permits seeing a range of effects. Of course, computing also increasingly *converges* with telecommunications.

FIGURE 1.66 Access for variety (innovation).

Explorer amply demonstrates this. The conflict centers on whether Microsoft can be allowed to *integrate* whatever it chooses into the Windows OS. We portray that contrast graphically in Figure 1.66.

1.3.2.2 Access for Variety (Innovation)

An essential distinction emerges (Figure 1.66). A layered information product offers more points of entry, for later innovation. A vertically integrated solution, on the other side, offers fewer such opportunities; but vertical integration may bring better performance.[*] Consider one of our examples, Java. Simply because it is an intermediary — layered — solution, it invites more input from the developer community. But Java is notoriously a "low(er) performer," particularly since by design it is interposed as intermediate, between software commands and calls to the chip.

Object-oriented programming, particularly its structural relationships, offers a useful metaphor. Objects are each a part in a functional whole. An essential feature is how tight or loose are the linkages among objects. The interface between objects may afford more or less flexibility to the internal functioning for each object. In loose linkage among objects, the functioning of a given object is afforded wider latitude, with a greater range of function possible internally for that object (which requires, perhaps ironically, more specification, and therefore rigidity, at the interface — the point of *interconnection* — among objects). Conversely, tighter linkage entails less flexibility for a particular object (with simpler specification for the interface, or interconnection). The extreme is no flexibility and so is analog to the case of *vertical integration* (among ideas) above.

Object-oriented structure is particularly useful as a metaphor to show that, whether layered or vertical, an information product is constituted by parts in a whole (or you may prefer to speak of "elements in context" or "foreground in gestalt"). A central feature is the tightness/looseness of structural, or "architectural," relations among the parts and whole. Now we can return to our group of people who are choosing a common method to make communications possible. At least implicitly, this group must choose between layered and vertical (of course, the choice is actually some point along a dimension between the two extremes). Layered opens new points to innovation,[**] but vertical may perform better.

We can notice now that balancing and choosing between the two seems to fly in the face of calls for "openness" in standards. The layered case represents what often is referred to as open. Where does the paradox lie? We traffic in a new notion such as information product in the hope of giving ourselves such new levers on understanding.

1.3.2.3 Standards in the Real World

To answer that question, we turn to ask how choices of information product are made — how, in the real world, does a group arrive at standards?[***] We inspect two not-unrelated cases, taken from networking and computers respectively (they come together in the browser wars). In part, these are defining instances

---

[*]Since vertical integration usually applies in the same breath with thoughts about human organization, we have to be careful to notice that this vertical integration applies to the structure of ideas in an information product.

[**]In the object oriented metaphor, more objects — which is one form of a "looser" linkage (perhaps tighter interface!) — mean more opportunities for application to novel uses.

[***]We have already been clear that communications necessitates some agreement (with "information product" the more general form of views held jointly and "standard" one set of them).

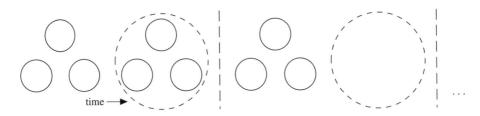

FIGURE 1.67 Basic cycle with repeated progression from atomized actors to coalescence.

in the recent practice of standardization; in part, by the contrast between them, they frame the central issues. One is the Internet Engineering Task Force (IETF); the other is Microsoft.

1.3.2.4   The IETF

The IETF has overseen one of the more startling runs in the annals of innovation. The rate of successful innovation on the Internet and Web is hardly paralleled, particularly as it has been sustained across years. In rough terms, the software engineers — the hackers — who comprise the Internet community first advance some new idea. Under the aegis of the IETF, that new notion is implemented in a test form and tried out, perhaps spurring other related novelties. After gathering some experience to test the, perhaps now competing, proposition(s), deliberations in the group proceed, culminating in a new standard. This may involve melding more than one of the mooted good ideas into a composite better approach. Then the process starts over, with the next new idea. The pace is torrid, and the rate of change breathtaking by all accounts, with this scenario is repeated over and over.

Boiled down to its essentials, the process has a simple, building-block cycle at its core. The cycle begins with innovation; it completes with a standardization. The implications in terms of industry behavior are what cause surprise. The innovation phase requires competition among the participants, over the new ideas put forward. The standardization phase, by contrast, requires the opposite behavior, namely, consensus. The industry structure must, perforce, take on a dynamic character. During innovation, the actors in the industry are atomized and act individually on behalf of the new idea that they separately have put forward. During standardization, these same actors now assume a new cloak and coalesce, accepting roles within the loose hierarchy formed by the IETF. In this mode they consider jointly what will serve a "better" Internet and establish the new standard — they arrive at a new information product.*

Through the IETF process, this transition, between behavioral opposites, is effected seamlessly numerous times, in a short span of time. The flip-flop between individual innovator and IETF member repeats frequently, seeming effortlessly. Industry structure alternates between full competition, with no market power disturbing tests on the merits, and full vertical integration,** in effect monopoly control (though by the whole group). This alternation between competition and monopoly occurs repeatedly, again seemingly effortlessly.

We can depict the basic cycle, with a repeated progression from atomized actors (in Figure 1.67 there are three industry members) to an IETF-style coalescence to deliberate some new standard.

How is it possible that individuals in a group can assume opposite roles, first serving each individual's separate interests, then switching to concern for betterment of the group as a whole; first competing, then finding consensus and alternating in this way repeatedly? Clearly this is what happens. Just as clearly,

---

*The development of ethernet is another example, from about 25 years earlier, of the basic cycle at work. After being invented privately, ethernet — after some years — came to be publicly held. The counterpart organization is the 802.3 standards committee of the Institute of Electrical and Electronics Engineers (IEEE). Though the cycle was not immediately repeated, the impact was profound with over 100 million users worldwide by the late 1990s. Subsequently, the cycle even repeated, to produce higher capacity versions of the LAN protocol. There are other examples in the U.S. as well, and certainly outside the U.S. I am indebted to John Markoff for his article bringing the ethernet case to attention, *The New York Times,* May 18, 1998, p. D1.

**This vertical organization does regard human organization (rather than the structure of ideas).

the answer appears to be in the ground rules for behavior shared across the group. Although such social protocols are by-and-large informal, they are if anything even more binding, as proudly held as the glue that keeps the community together. This suggests an enlightened self-interest, where both individual aggrandizement and preservation/betterment of the group serve the self. Then strongly held protocols serve to implement relatively complex behavior necessary to meet this enlightened objective. But does it make sense to speak of a dynamic industry structure, with individuals and companies alternately separate, then together, repeatedly? Consider the social architecture of everyday life. We are each a member of several groups: family, work or school, play, perhaps religion, and so forth. For each group, there is an expanding ring of concentric, increasingly wider membership. The felt linkage may lessen as the circle widens; the connection extends nonetheless from the center outward. For work, as an example, the expanding circles may be the work group, a division in the company, the company itself, even the industry. Moment-to-moment an individual mentally switches among each of those levels, or memberships, to deal with different problems. Dynamic borders are endemic to our experience of daily life, in other words, despite our hesitance to visualize such variability.

This social architecture also serves as the template for the loose hierarchy into which the IETF process periodically congeals. So now we can notice that both the human organization and its information product are characterized as being both part and whole — there is sound reason vertical integration (and layering) apply as basic ideas to both organization and to information product. Though these are two quite different phenomena — one regards social structure, the other the structure of shared ideas — the same notions characterize both. (We can also notice that variety* not subsumed by a standard continues to be maintained at "lower" levels in both the social and the idea structures.)

1.3.2.5 Microsoft

The contrasting case is Microsoft. Microsoft has succeeded in becoming the epitome of the competitive ideal. Surely the success of its chairman, Bill Gates, is the fulfillment of the American dream, as conventionally defined. When it comes to standards, Microsoft maintains control, at all costs. It has now even been revealed that Gates was willing to "put a bullet through the head"** of his own online service, Microsoft Network, to crush Netscape and maintain the supremacy of Windows as the standard. This is in contrast with the IETF process, where all are invited to contribute with new ideas. Microsoft's obsession with control means the company takes all available steps to limit discretion to those inside the Microsoft circle.

Microsoft control over the standards process has predictable effects. The input of new ideas onto turf controlled by Microsoft has progressively chilled. Reports indicate that venture capital simply is not available for innovations that would tread on territory where Microsoft holds sway.*** The operating system is of course at that heart, with spreading borders progressively proscribed from innovative input by industry participants outside Microsoft. Whole swaths of new ideas are not even stillborn. The standards we get serve the economic interests of only the one party, Microsoft. Among the very many examples, a prominent one is the fate of OpenDoc. Though widely acknowledged to be far better as an object technology than Microsoft's Object Linking and Embedding (OLE) and progeny, the Microsoft juggernaut helped to seal the fate of OpenDoc, removing the (much) better technology as an option for any of us. Because vertical integration — in ideas — may serve to reinforce control of, for instance, the operating system, the Microsoft penchant for control leads to over-choosing vertical integration for the technology it permits the rest of us

---

*In other words, the trace left by an earlier innovation.

**"[I]n order to induce America Online to promote Internet Explorer instead of Netscape's Navigator [and so protect Windows], Microsoft agreed to promote AOL in Windows at the expense of Microsoft's own online service … Gates purportedly [described this tactic in the terms quoted in the text]," Smoking gun in Microsoft memos? D. Goodin and J. Pelline, CNET NEWS.COM, May 18, 1998, 5:40 p.m. PT.

***"Because of Microsoft dominance, in recent years venture capital has simply not been available for software start-ups focused on desktop applications, programming tools or, in your dreams, operating systems." *The New York Times,* May 24, 1998, Money & Business Section, First page.

to enjoy. As an ironic (if not canonical) case, the difficulties presented by the open Intel-based hardware platform impel Microsoft even further to integrate its Windows software OS vertically.

1.3.2.6 Policy Lessons from the Real World

There is a fundamental opposition — some would say a war — between these two contrasting approaches to innovation and its Janus face, standardization. The two approaches see, in effect, radically different social organization to be appropriate for the inflow of innovation into a society. One insists absolutely on the control by one company of the process. The other, a substantially more complex social dance, emphasizes both inclusiveness of ideas from all, as well as rapid progress through joint choice to serve the entire group. If we would consider the latter for policy, we need not only competition, but also a dynamic alternation with consensus.

Can we decipher our original paradox, where the choice might potentially select what is ordinarily considered not open? With the information product separated from the industry organization which produces it, we can now treat the question of openness more accurately. "Open," in this expanded understanding, refers to inclusiveness in organization. Each industry participant may have a useful good idea to contribute. In the fairly complex scenario, with industry structure alternating between atomization and coalescence, the cardinal requirement for openness is the inclusion of ideas from all quarters. For the information product, by contrast, either vertical or layered may be chosen — in fact, balancing vertical for performance against layering for more access points to later innovation is typically a challenging task. By separating the information product analytically, and specifying the dynamics of industry organization, we try to see the process clearly.

Perhaps Robert Metcalfe, one of the inventors of ethernet, put "open" in the most accurate real-world terms: "Standards are genuinely open … only when they are publicly documented, nonproprietary, and when there is a public forum for updating the standard [– something, in other words, which] the public owns."*

---

*The Markoff article referenced in footnote** on page 161. The author's "Microsoft vs. Netscape: Policy for Dynamic Models" treats the theory of the two opposed models (chapter in *The Limits of Government: On Policy, Competence and Economic Growth*, G. Eliasson and N. Karlson, Eds., City University Press, Stockholm, 1998).

2

Basic Communication Principles

Rolf Stadler, Editor
Columbia University

Jean-Pierre Hubaux
EPFL-ICA

Simon Znaty
ENST-Bretagne

Steven Minzer
Lucent Technologies

Hisaya Hadama
NTT Telecommunications Network Laboratory Group

Michele Zorzi
University of California, San Diego

2.1 Telecommunications Services Engineering: Definitions, Architectures, and Tools

Jean-Pierre Hubaux and Simon Znaty

2.1.1 Introduction to Telecommunications Services Engineering

The demand for advanced telecommunications services has increased enormously over the last few years. This has led to situations where network operators must deploy new services at a rapid pace when satisfying customer needs. The telecommunications monopolies have disappeared, and the fight for market shares has become fiercer than ever before. Furthermore, the demand for ever more specialized end-user services keeps growing, along with the demand for having the new services deployed within shorter and shorter time frames. The structure and function of networks must change, in order to cope with these new challenges. The telecommunications industry is witnessing a changeover from being interconnection driven to being service driven. A new discipline called *telecommunications services engineering* is emerging. It encompasses the set of principles, architectures, and tools required to tackle activities ranging from service specification to service implementation, service deployment, and exploitation.

In this section, we survey the different facets of this new area. In the second subsection, we define the term *service*. The third subsection introduces *service engineering*. In the fourth section, we emphasize the network architectures used to ease the introduction of telecommunications and management services. We particularly focus on the IN, TMN, and TINA architectures and compare their features. The fifth subsection positions IN and TINA with regard to the Internet. The sixth subsection deals with the methods, techniques, and tools used to build telecommunication-services such as the object-oriented approach, open distributed processing, and the agent technology.

2.1.2   Service Definition

The word *service* has become magic in the telecommunications world these last years. This word is somewhat fuzzy and ambiguous since there are so many aspects of services, and therefore, there is a need for explanation of some of the terms most often used. Telecommunications services is a common name for all services offered by, or over, a telecommunications network.

The word service is used in several different contexts with somewhat different meanings. In the ISDN (Integrated Services Digital Network) world, three types of network services can be distinguished[1]:

Support services define the transmission capabilities between two access points, including routing and switching across the network. They correspond to *bearer services*.

Teleservices include capabilities for communication between applications. Teleservices are supplied by communication software in the terminals. Telephony, audio, fax, videotex, video telephony are examples of teleservices.

Supplementary services, also called features, complement support services or teleservices. Most well-known supplementary services are related to the telephony teleservice (call forwarding, three party conference, etc.), but they could, of course, be generalized to other teleservices.

Value-added services are services that require storage/transformation of data in the network and are often marketed as stand-alone products. Examples of such services are freephone, premium rate, virtual private network, and telebanking. Many value-added services can be offered by special service providers connected to the network.

2.1.3   Service Engineering Definition

Since the early 1980s, the major trend in the area of service provision has been toward dissociating service control from the underlying network equipment. As a result, services have been seen as sets of interactions among software pieces running on top of the network infrastructure. Consequently, the concepts, principles, and rules of service engineering were borrowed to a large extent from the software engineering area.

The telecommunications community was faced with a new challenge, which was to bring the telecommunications specific requirements together with software engineering. Interests grew to integrate results from other disciplines such as security, verification and validation, database management systems, communication management, etc.

Service engineering can be defined as the set of methods, techniques, and tools to specify, design, implement, verify, and validate services that meet user needs and deploy and exploit these services over the current or future networks. Service engineering is a young discipline, but is a discipline in itself, as is protocol engineering.

Three important components are considered within the framework of service engineering (Figure 2.1):

- *Service creation environment:* A software engineering platform specialized for the development of telecommunications services.
- *Telecommunications network:* Contains the transmission and switching equipment. Each of these pieces of equipment may be seen as one black box that offers an application programming interface (API); this may be a signaling or management interface.
- *Network architecture:* Responsible for controlling the network in such a way that a service's specific requirements get satisfied.

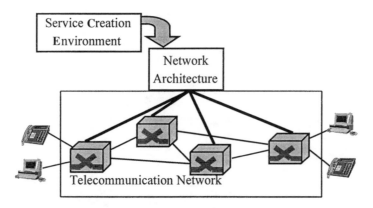

FIGURE 2.1 Components of telecommunications services engineering.

Service engineering covers three important domains:

- *Service creation:* The service is considered as a distributed application running on the multiple nodes of a telecommunication network.
- *Service management:* The way a service is operated throughout its life cycle.
- *Network management:* The management of network resources used to provide telecommunication services.

Therefore, two kinds of services are involved, telecommunications services and management services.

2.1.4   Network Architectures

2.1.4.1   The Intelligent Network (IN)

The term *Intelligent Network* (IN) was first introduced by Bellcore in the 1980s following the deployment of the green number service in the U.S. The IN is an architectural concept allowing a rapid, smooth, and easy introduction of new telecommunications services in the network.[2] These services may be customized.

The architecture chosen is based on a centralized control. Service control is completely separated from call control. It is based on the existence of a signaling network linking all the switches. In the modern digital telephone network this signaling network does exist — it is called Common Channel Signaling No. 7 (CCS7) network.

The ITU-T has defined the *IN conceptual model* (INCM)[3] which is not exactly an architecture but rather a methodology to describe and design an IN architecture. The INCM consists in four planes (Figure 2.2).

The *Service Plane* (SP) describes services from a user's perspective, and therefore is of primary interest to service users and providers. It consists of one or more service features. A service feature is a service component; it may correspond to a complete service or part of a service. This composition principle enables the customization of services, i.e., the creation of services by the subscriber and not necessarily by the telecom operator.

The *Global Functional Plane* (GFP) deals with service creation and models the network as a unique and global virtual machine. This plane is of primary interest to service designers. It contains the Service Independent Building Blocks (SIBs) that are used as standard reusable capabilities to build features and services. There exists a particular SIB called Basic Call Process (BCP) from which a service is launched. In this plane, a service consists in a chain of SIBs which can be viewed as a script.

The *Distributed Functional Plane* (DFP) models a distributed view of an IN and is of interest mainly to network designers and providers. It describes the functional architecture of the IN which is composed of a set of functional entities (FEs) executing actions. A functional entity is a network functionality. The

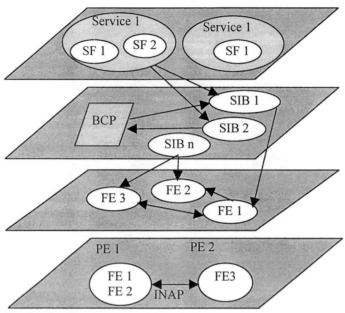

SF: Service Feature
SIB: Service-Independent Building Block
BCP: Basic Call Process
FE: Functional Entity
PE: Physical Entity
INAP: Intelligent Network Application Protocol

FIGURE 2.2 IN conceptual model.

two main functions are the Service Control Function (SCF) which contains the IN service logic and controls the overall execution of the service and the Service Switching Function (SSF) which provides a standard-ized interface between the SCF and the switch for the control of this switch. The other functional entities are the Specialized Resource Function (SRF) which performs user interaction functions (e.g., playing announcements or prompting and collecting user information) via established connections; the Service Data Function (SDF) which performs related data processing functions such as retrieving or updating user information; the Service Management Function (SMF) which handles the activities of service deployment, service provisioning, service control, service billing and service monitoring, and finally, the Service Creation Environment Function (SCEF), which allows a service to be defined, developed, and tested on an IN structured network. Each SIB of the GFP is decomposed in the DFP into a set of client/server relationships between one or more functional entity.

The *Physical Plane* (PP) corresponds to the physical architecture of the IN which consists in a set of physical entities and interfaces among them. This plane is of primary interest to network operators and equipment providers. The functional entities in the DFP are implemented into the physical entities. For example, the SCF becomes the SCP (Service Control Point) and the SSF is translated into an SSP (Service Switching Point). The interface between SSP and SCP for IN service execution is called INAP (Intelligent Network Application Protocol). INAP messages are encapsulated into SS7 messages that are exchanged between SS7 signaling points over 56 or 64 kbit/s bidirectional channels called signaling links. Signaling takes place out-of-band on dedicated channels rather than in-band on voice channels.

Figure 2.3 shows a simplified IN architecture. Such an architecture is well adapted to services needing a centralized database like the green number (sometimes called freephone) service.

The IN will play a major role in the provision of mobile telecommunications services. Indeed, they raise some specific problems to the tracking of mobile users and terminals. A key enabler for providing

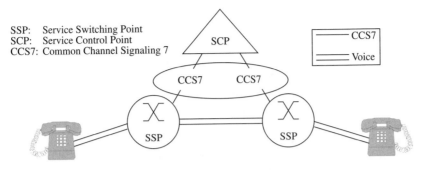

FIGURE 2.3 Simplified IN architecture.

these kinds of services is the IN capability. Mobile services need IN capabilities for mobility management and service control and, as these services expand, they will put new demands on IN.

Each phase of development in the definition of the IN architecture is intended to produce a particular set of IN capabilities, known as a Capability Set (CS). Each CS is compatible with the previous CS and is enhanced to ensure that it is one stage closer to the final IN target. CS-1 was the first development phase. CS-2, which was completed at the end of the first quarter of 1997, is the second standardized stage of the intelligent network evolution; it addresses the limitations of CS-1. CS-2 enables interworking between IN architectures to provide international services, allows the management of both IN services and IN equipment through the TMN, and supports enhanced IN services such as mobility services. CS-3 addresses issues such as full IN/TMN integration, full IN/B-ISDN integration, and full support for mobile/personal communications systems.

2.1.4.2 The Telecommunications Management Network (TMN)

Parallel to the IN standardization, the ITU-T has defined the *Telecommunications Management Network* (TMN). TMN enables federation of the equipment that constitute the telecommunications network, produced generally by different telecommunications vendors, to enable their control in a uniform, global, and efficient way.[4]

Management of telecommunications networks may be defined as the set of activities of monitoring, analysis, control, and planning of the operation of telecommunications network resources to provide services to customers with a certain level of quality and cost.

- *Monitoring* is defined as the process of dynamic collection, interpretation, and presentation of information concerning objects under scrutiny.[5] It is used for general management activities which have a permanent continuous nature such as the systems management functional areas (i.e., performance, configuration, fault, accounting, security).
- *Analysis* is applied to monitoring information to determine average or mean variance values of particular status variables. Analysis is application specific. It can range from very simple gathering of statistics to very sophisticated model-based analysis.[6]
- *Control* is the process by which changes in the managed network are effected.
- *Planning* is defining the network topology and sizing every network element in order for the user to obtain any given service in optimal conditions with regard to quality and price.

The set of capabilities necessary for network management relies on a reference structure which identifies the main TMN components and interfaces. The TMN can be considered according to three views: information architecture, functional architecture, and physical architecture.

The *TMN information architecture* provides a data representation of the network resources for the purpose of monitoring, control, and management. The approach adopted for the specification of the information model is object oriented. The TMN information architecture also defines management layers which correspond to levels where decisions are made and management information resides (business management layer, service management layer, network management layer, element management layer). The ITU-T proposed a generic network information model.[7] Genericity enables the model to be applicable to different network technologies, e.g., Asynchronous Transfer Mode (ATM), Synchronous Digital Hierarchy (SDH), Plesiochronous Digital Hierarchy (PDH). The model is currently applicable to both network element and network management layers.

The *TMN functional architecture* describes the realization of a TMN in terms of different categories of function blocks and different classes of interconnection among these function blocks, called reference points.

The *TMN physical architecture* corresponds to the physical realization of the functional architecture. Each function block becomes a physical block or a set of physical blocks or Operation System (OS) and reference points are transformed into interfaces. Among these interfaces, we can find the Q3 interface between an OS and the managed resource or between two OSs of a given management domain and the X interface between two OSs belonging to different TMN domains. The TMN is seen as a set of connected physical blocks, each of them executing a set of TMN functions. To ensure interoperability, the specification of an interface requires the use of compatible communication protocols and compatible data representation. The exchanges of information between two management systems are performed by means of management operations and notifications through the Common Management Information Service (CMIS) service and Common Management Information Protocol (CMIP) protocol.

Although the TMN was defined with network management in mind, it can be used to provide a multitude of services. One of the most sophisticated of these services is the Virtual Private Network (VPN). VPN is a telecommunication service that provides corporate networking among geographically dispersed customer premises, based on a shared public switched network infrastructure.

Figure 2.4 shows the physical architecture of a VPN configuration management system.[8] The configuration management architecture consists of a set of OSs, the CPN OS that manages the CPN resources, the PN OS that manages the public network resources, the PN-service OS which is responsible for the management of the services offered over the public network (e.g., a virtual path service in an ATM network), the CPN-service OS whose role is to administer the services provided over the CPN, and finally the VPN-service OS for the management of the VPN service. The X interface enables interactions among the VPN service actors, i.e., the customer, the service provider, and the network provider. The Q3 interface takes place between OSs of a given management domain.

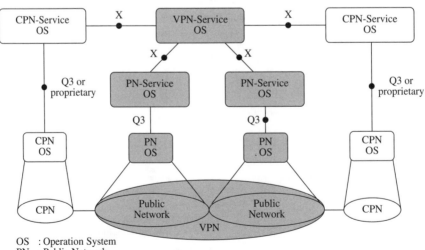

FIGURE 2.4 VPN physical management architecture.

Obviously, the IN and TMN architectures overlap.[9] For instance, one TMN application such as billing, and one IN application such as Freephone, must be tightly related because VPN billing should be handled in a consistent way with TMN billing. This shows that unless both IN and TMN architectures are made more consistent, the interworking of IN and TMN applications would be very difficult. Moreover, it will be difficult to support two independent architectures while applications of both architectures must interoperate. The TINA architecture encompasses an integrated IN/TMN architecture.

2.1.4.3 Telecommunications Information Networking Architecture (TINA)

The evolution of the IN calls for new facilities such as flexible control of emerging multimedia, multi-session, multipoint, broadband network resources, and services interoperability across diverse network domains. To meet these requirements, the TINA consortium defined a global architecture that enables the creation, deployment, exploitation, and management of services worldwide.[10]

The goal is to build a reference model for open telecommunications architectures that incorporate telecommunications services and management services, and integrate the IN and TMN domains. TINA makes use of the latest advances in distributed computing Open Distributed Processing (ODP)[11] and Object Management Group (OMG),[12] and in object orientation to ensure interoperability, software reuse, flexible distribution of software, and homogeneity in the design of services and their management.

The layers of the TINA architecture divide application objects into different domains (Figure 2.5): the service layer, where service components provide value-added services with their integrated management, and the resource layer, where resource management components provide an abstraction of the network resources used to supply the service (e.g., components that enable services to establish, maintain, and release connections). Service and resource management components run over a Distributed Processing Environment (DPE).[13] At the lowest layer of the architecture, we can find the physical resources such as transmission links, switches, and terminals.

TINA is composed of three architectures:

The Computing Architecture defines the concepts, principles, and rules for telecommunication software reusability and interoperability by relying on ODP. These concepts are applied for the design of both telecommunications and management services. The computing architecture also provides a prototype of a DPE for TINA services. It describes the function of this DPE, its main components, and its programming interface. The TINA DPE may be regarded as an abstraction of distributed systems such as CORBA.

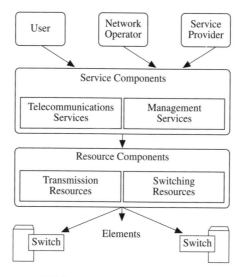

FIGURE 2.5 The TINA architecture.

TABLE 2.1   Comparison of the Three Telecommunication Architectures

Criteria	IN	TMN	TINA
Types of services	Telephony-based services (UPT, freephone service)	Management services (VPN configuration management service)	Multimedia services, management services
Support networks	All types	All types	Broadband networks
Method for service creation	Functional approach	Object-oriented approach	Distributed object-oriented approach
Components for service creation	SIBs	SMFs	USCM
Main elements of the architecture	SCP, SSP	OS, NE	SSM, CSM
Communication	SS7 (INAP)	X, Q3 (CMIS/CMIP)	CORBA (IDL)
Standardization body	ITU-T	ITU-T	TINA-C

Note: UPT: Universal Personal Telecommunications; SIB: Service Independent Building Block; SMF: Systems Management Function; USCM: Universal Service Component Model; SCP: Service Control Point; SSP: Service Switching Point; OS: Operation System; NE: Network Element; SSM: Service Session Manager; CSM: Communication Session Manager; SS7: Signaling System 7; INAP: Intelligent Network Application Protocol; CMIS/CMIP: Common Management Information Service/Common Management Information Protocol; CORBA: Common Object Request Broker Architecture; IDL: Interface Definition Language; ITU-T: International Telecommunications Union — Telecommunication sector; TINA-C: TINA Consortium.

The Service Architecture provides a set of concepts to build and deploy telecommunication services. In a TINA system, a service consists of a set of components that interact with each other and are deployed over a DPE. The service architecture defines the required objects for the realization of a service, their composition, and interactions. Moreover, a universal service component model (USCM) has been proposed to promote reusability during service development. The three important concepts in this architecture are:

- The concept of session which refers to service activity[14]
- The concept of access which relates to the associations between the user and the service
- The concept of management for service management

The Network Resource Architecture defines a set of generic concepts for the realization of network resource management applications. Among these generic concepts, we can find a new area added to the TMN management functional areas, namely, connection management. Another important concept is the "computational object" to model both managers and agents. Finally, the network resource architecture proposes a generic network resource information model (NRIM) based on Reference 7.

2.1.4.4 Network Architectures Comparison

Table 2.1 summarizes the different characteristics of the three network architectures presented above.

The IN architecture, in its present state of development, is confined to basic telephony call-control capabilities. This architecture is mostly deployed over the public switched telephone network, ISDN, and mobile networks. TMN provides management services for the management of telecommunications networks and services. It could be deployed over any network. TINA is an architecture that embraces IN and TMN within a framework based on ODP. This architecture enables the deployment of complex services along with their management (e.g., multimedia services).

The way services are built with the IN is functional. A service is seen as a chain of elementary instructions called SIBs (e.g., translate, algorithm, and compare). The TMN applies the object-oriented approach since resources that are managed are represented as objects. To build a management application, the designer may make use of basic software units called systems management functions which provide some capabilities (e.g., state management, log control, and alarm reporting). Within TINA, a service consists of a set of service components that will run over CORBA. Every service component specification derives from a universal service component model (USCM) which may be seen as a skeleton to which every service component should conform.

The IN architecture consists mainly of SCPs that control the execution of a service and SSPs that correspond to switches with a standardized interface for the dialogue with SCPs. The TMN architecture is composed of OSs that contain the management applications and NEs that are the resources to be managed. The TINA architecture makes use of two important objects, the service session manager (SSM), which is responsible for the control of the execution of a service of a given type, and the communication session manager (CSM), which supplies an end-to-end connectivity to the SSM.

The control network in the case of IN, TMN, and TINA is different for each architecture. Signaling System 7 is considered in the case of IN for exchanges of INAP messages between SCPs and SSPs; a Data communication network may be used for CMIS/CMIP interactions between the OSs and NEs. In the case of TINA, CORBA messages are exchanged between the SSMs and CSM.

2.1.5   IN, TMN, TINA, and the Internet

The architectures described so far have been defined with telecommunications services in mind, without really taking into account the development of the Internet. However, it is now clear that the growing popularity of the Internet is dramatically changing the landscape of the communications marketplace. The two separate worlds of the Internet and telecommunications need to converge and should be integrated to fulfill the promise of the information superhighways. This section evaluates the current status of this convergence.

2.1.5.1   IN and the Internet

Today, several architectures have already been proposed, promoting the integration of the IN and the Internet and in particular the World Wide Web. Among them, we can find the reference model proposed by the IETF PINT group[15] and the WebIN architecture designed by HP.[16]

Faynberg et al.[15] propose connection at Internet-based devices such as WEB servers and SNMP-based management systems to IN devices. Four services are considered as case studies. These are Click-to-Dial, Click-to-Fax, Click-to-Fax-Back, and Voice-Access-to-Content. The Click-to-Dial service enables a user to initiate a PSTN outgoing as well as incoming call by hitting a button during a Web session, the Click-to-Fax service enables the user to click on a button to send a fax; the Click-to-Fax-Back allows the user to receive a fax by clicking on a button; and finally, the Voice-Access-to-Content service enables the user to access content-based services such as Web pages.

Low[16] proposes a WebIN architecture resembling in different aspects the classical IN architecture. The major differences are (1) the Service Data Point (SDP) and the Service Management Point (SMP) are connected to the Internet, (2) SCPs are connected to SDPs through a distributed processing environment (DPE) that might be the Web (HTTP/CGI) or CORBA, and (3) the role of the SCP is confined to finding out the correct reference of the SDF that corresponds to the service actually invoked.

In addition, the SCE relies on the Web service creation environment (HTML, Java). After writing the service logic using these technologies, the service creator simply uploads the service logic into the Web service dedicated to the creator.

The goal of the Web IN architecture is to show how World Wide Web (www) technology and IN technology complement each other. It proposes a hybrid architecture where service logic and subscriber information (SDF) are implemented on an Internet-based signaling and distribution network and interfaced to the SCF for bearer channel and resource control.

2.1.5.2   TMN and the Internet

The complexity of telecommunications networks, i.e., enterprise and carrier networks, has grown over the last two decades. The management of these networks has followed different approaches because carrier networks apply TMN, while enterprise networks consider a Simple Network Management Protocol (SNMP)-based management approach.

The borders between public (carrier) and private (enterprise) networks are becoming increasingly transparent, the distinction between both types of networks may soon be irrelevant from a network management point of view.[17] The challenge that the progressing convergence of networks presents is to manage the concerned network management worlds in a consistent way, while still preserving the vast investments in existing networks and network management solutions.

The new management framework should consider:

- The use of www technology for representing management tasks. This requirement addresses the fact that the user interfaces provided by www browsers have received wide acceptance and are the common interface to server-based information services.
- The introduction of cooperative sessions for network management. This requirement acknowledges the fact that management in modern networks is shared among many parties.
- The interworking with SNMP-based, CMIP-based, and proprietary management systems. This requirement is a consequence of the need to interoperate with existing management systems, especially at the element management layer.

2.1.5.3 TINA and the Internet

From its very beginning, TINA has been a project funded and carried out by the traditional telecommunications companies, network operators, and telecommunications vendors. The main assumptions on which TINA were based are:

- The services will continue to be primarily provided by the network or by network-based servers rather than by the terminals.
- The B-ISDN will rely on end-to-end-ATM.
- The pace of the telecommunication evolution will remain under the control of the main telecommunications stakeholders.

The first assumption is inherited from the IN. For TINA, it led to a high level of complexity in the service architecture. Meanwhile, Internet services became extremely popular; the generalization of browsers, the deployment of Internet telephony, and the multiparty session over the MBone showed that it was possible to implement a high number of services over a network of limited capabilities. Indeed, the role of the terminal became completely different with its increasing processing power which enables the implementation of sophisticated functions in software such as audio and video decoders, and with the advent of new software paradigms thanks to Java (e.g., applets) which permits downloading executable code during a given service session. This shows that network operators should concentrate on the provision of services that cannot be supported by the terminals, such as UPT and routing-based services, in order not to confine themselves to being mere connectivity providers.

The second assumption explains why so much effort has been put on the specification of the connection management architecture. However, it becomes obvious that the B-ISDN will in fact be based on the Internet/intranet technologies, provided that the appropriate resource reservation, billing, and security mechanisms are implemented. Therefore, the TINA architecture should also encompass connectionless networks such as the Internet.

The third assumption is clearly based on the monopolistic tradition in this domain. However, in the meantime, the development of the Internet protocols has proved that it is possible to "do first and standardize later." Incremental development is now the usual practice.

This analysis shows that most of the assumptions that gave birth to TINA are no longer valid because of the complete change of landscape brought on by the Internet. However, the Internet has not fully met the challenge of the provision of multiparty, multimedia services. In particular, the guarantee of a given quality of service is still a problem that needs to be solved. That means that, although TINA will not be implemented as was initially intended, some good concepts could be inherited from it. Recently, several TINA proponents have been studying the applicability of TINA to the Internet.[18-20]

2.1.6   Techniques

The advanced information processing techniques are playing a major role in the realization of telecommunications services and the underlying network architectures. Among these techniques, we find object-oriented methods, open distributed processing, and the agent technology.

2.1.6.1. Open Distributed Processing

A telecommunications service is a distributed application that runs over the multiple nodes of a telecommunications network. The ODP reference model jointly defined by ISO and ITU-T provides a framework for the design of distributed systems with the introduction of viewpoints. Each viewpoint represents a different abstraction of the original system. Informally, a viewpoint leads to a representation of the system with emphasis on a specific concern. Five viewpoints were identified: enterprise, information, computation, engineering, and technology (Figure 2.6).

The *enterprise* viewpoint is concerned with the overall environment within which an ODP system is to operate. The *information* viewpoint focuses on the information requirements of the system, and deals with information object types, together with their states and permitted state changes. The *computational* viewpoint shows processing functions and data types, abstracting away from the underlying hardware structures via transparency functions. The *engineering* viewpoint establishes transparency services utilizing concepts from operating systems and communications. The *technology* viewpoint is concerned with the realization of an ODP system in terms of specific hardware and software components. ODP has been extensively used for the definition of TINA.[21]

2.1.6.2   Mobile Agents

An agent is a program, which, with a certain degree of autonomy, performs tasks on behalf of a user or an application. An agent may move between network sites and cooperate with other agents to achieve its goals.[22]

Agent development finds its roots in two research domains: intelligent agents stemming from artificial intelligence, which studies the capabilities of learning and decision making of cooperative autonomous entities; and mobile code technology that enables programs to migrate from a machine to another, while preserving their execution environment. This latter domain is evolving at a fast pace because of the emergence of languages such as Tcl[23] and Java,[24] and of their portable execution environment.

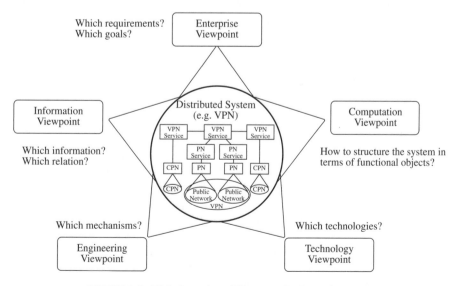

FIGURE 2.6 ODP viewpoints: different projections of a system.

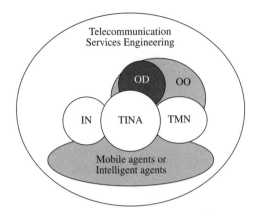

FIGURE 2.7   Interrelations between network architectures and impact of techniques.

The use of agent technology for telecommunications services engineering is a very hot topic in particular in the area of service and network management.[25,26] It lies within the boundaries of areas such as telecommunications, artificial intelligence, and software engineering. This can be seen as an advantage because it promotes the convergence of research results from different communities.

2.1.6.3   Other Techniques

Among the other techniques not detailed in this survey, we mention formal methods for the verification, validation, and testing of telecommunications services before deploying them. The goal of formal methods is to improve the reliability of these services.[27] Indeed, the rapid growth of the number of services makes the problem of proving that the services conform to their specification more acute. In fact, reacting rapidly to customer or market needs requires introducing new services only a few months or even a few weeks after the first specification; such a short interval makes it quite impossible to go through the tedious and long (several months) tests usually performed for new services. This problem is getting worse since there are more services continually added to the networks, contributing to the overall complexity. Services must all work correctly without hindering the function of other services; this last problem is often referred to as the "feature interactions problem." These obstacles on the road to rapid service introduction call for new approaches to increase confidence in the service.

Figure 2.7 summarizes the interrelations among the different network architectures of telecommunication services engineering and the impact of techniques on these architectures. TMN, TINA, and ODP follow the object-oriented approach. TINA applies the ODP concepts, principles, and viewpoints and integrates the IN and TMN architectures. Finally, mobile or intelligent agents may be perceived as an emerging technology for the next generation of telecommunications.[28]

2.1.7   Conclusion

As we have seen, telecommunications services engineering is composed of two major parts.

One part is related to the network architecture which is in charge of executing the service in the network. The IN evolution will notably encompass its integration with the International Mobile Telecommunication 2000 architecture[29] for rapid introduction of services and efficient service control. In addition, there is an increasing interest in bringing telephone services provided by PSTNs to Internet users through the IN.

The TMN evolution is integrating CORBA-based management particularly for the new task of service management. Interoperability will play a very important role. Indeed, if a service extends over multiple networks, network operators of these networks should be able to negotiate service provision and contract establishment with each other. Although the effort spent toward the provision of an integrated IN/TMN architecture called TINA has been important, TINA in its current status is not deployed. Indeed, TINA is not evolutionary but rather revolutionary with regard to IN and TMN. This would lead a network operator to reconsider the investment in IN and TMN architectures, which is not acceptable. Moreover, CORBA is immature for the deployment and execution of telecommunications services which require real-time constraints. CORBA is currently foreseen for management aspects only. Finally, TINA did not take seriously into consideration its possible deployment over the Internet. The IN CS-3, which is the IN long-term architecture, should reuse the TINA computing architecture and part of the TINA service architecture.

The other part is related to methods, techniques and tools for the analysis, design, implementation, verification, and validation of telecommunication services. These methods, techniques, and tools rely heavily on the object-oriented approach and open distributed processing. Unfortunately, there is little theoretical background behind the topics we have discussed; therefore, experimentation and prototyping are the only means available today to show the feasibility of the proposed concepts and scenarios. More research is needed to establish solid foundations.

Telecommunications services engineering is an important research domain at the boundaries of software engineering and telecommunications systems. It draws much attention from network operators, service providers, and telecommunications vendors since it is an important source of income. Specifically, how the Internet will develop within or in parallel with these architectures is still an open question.

Acknowledgments

The authors are indebted to Holly Cogliati and Constant Gbaguidi for their valuable comments on early drafts of this section.

References

1. ITU-T Recommendation I.210, Principles of Telecommunications Services supported by an ISDN and the means to describe them, 1988.
2. ITU-T Recommendations Q120x, Q121x, Q122x, The Intelligent Network, 1993.
3. ITU-T Recommendation Q1201, Principles of Intelligent Network Architecture, 1992.
4. ITU-T Recommendation M.3010, Principles for a Telecommunications Management Network, 1993.
5. Joyce, J. et al., Monitoring distributed systems, *ACM Trans. Comput. Syst.*, 5 (2), May 1987.
6. Mansouri-Samani, M. and Sloman, M., Monitoring Distributed Systems (A Survey), Report, Imperial College, Deptartment of Computing, 1993.
7. ITU-T Recommendation M.3100, Generic Network Information Model, 1992.
8. Gaspoz, J. P., Hubaux, J. P., and Znaty, S., A generic architecture for VPN configuration management, *J. Network Sys. Manage.*, 4(4), 375, 1996.
9. Pavlou J. and Griffin, D., An evolutionary approach towards the future of IN and TMN, *Baltzer ICN*, Special issue on telecommunications services, January 1998.
10. Barr, W. J., Boyd, T., and Inoue, Y., The TINA initiative, *IEEE Commun. Mag.*, 70, March 1993.
11. ITU-T Recommendation X901, Basic Reference Model of Open Distributed Processing — Part 1: Overview and Guide to Use, 1994.
12. OMG, The Common Object Request Broker: Architecture and Specification. OMG Document 91.12.1, Rev. 1.1, December 1991.
13. TINA-C, TINA Distributed Processing Environment (TINA-DPE), TINA-C Deliverable TB-PL001-1.0-95, July 1995.
14. Koerner, E. and Danthine, A., Towards a TINA-based framework for collaborative work. *Baltzer ICN*, Special issue on telecommunications services, January 1998.
15. Faynberg, I., Krishnaswamy, M., and Lu., H., A Proposal for Internet and Public Switched Telephone Networks (PSTN) Internetworking, Internet Draft, 1997.
16. Low, C., The Internet telephony red herring, in *Proc. of Conf. IEEE GLOBECOM*, IEEE Press, New York, 1996.

17. Bogler, G., Internet Technology for Integration of Carrier Network Management (TMN) and Enterprise Network Management, Internet Draft, 1998.
18. Smith, C., Applying TINA-C service architecture to the Internet and intranets, *Proc. Conf. TINA'97*, Santiago, 1997.
19. Zen, G. et al., Value added Internet: a pragmatic TINA-based path to the Internet and PSTN integration, *Proc. Conf. TINA'97*, Santiago, 1997.
20. Licciardi, C. A., Minerva, R., Moiso, C., Spinelli, G., and Spinolo, M., TINA and Internet: an evaluation of some scenarios, *Proc. Conf. TINA'97*, Santiago, 1997.
21. TINA-C, TINA Computational Modeling Concepts, TINA-C Deliverable TB-A2.HC.012-1.2-94, February 1995.
22. Wooldridge, M., Agent based computing, *Baltzer ICN*, Special issue on telecommunications services, January 1998.
23. Ousterhout, J. K., TCL: an embeddable command language, *Conf. Winter USENIX'90*, 1990, 133-146.
24. Sun Microsystems, The Java Language Environment: A White Paper, 1995 available at http://javasoft.com/whitePaper/java-whitepaper_1.html.
25. Goldszmidt, G., Yemini, Y., Meyer, K., Erlinger, M., Betser, J., and Sunshine, C., Decentralizing control and intelligence in network management, in *Proc. Int. Symp. Integrated Network Management*, Chapman & Hall, New York, 1995.
26. Mountzia, M. A., A distributed management approach based on flexible agents, *Baltzer ICN*, Special issue on telecommunications services, January 1998.
27. Etique, P. A., Hubaux, J. P., and Logean, X., Service specification and validation for the intelligent network, *Baltzer ICN*, Special issue on telecommunications services, October 1997.
28. Magedanz, T., Rothermel, K., and Krause, S., Intelligent agents: an emerging technology for next generation telecommunications, *Proc. Conf. IEEE INFOCOM 96*, San Francisco, 1996.
29. Panda, R., Grillo, D., Lycksell, E., Mieybégué, P., Okinaka, H., and Yabusaki, M., IMT-2000 standards: network aspects, *IEEE Personal Commun. Mag.*, August 1997.

2.2 Signaling in Telecommunications Systems

Steven Minzer

2.2.1   Introduction

Signaling has evolved with advances in telecommunications and computing technologies, and the new service offerings made possible by these advances. These advances, in fact, have changed the definition of signaling over time. A traditional definition limits signaling to the exchange of information specifically concerned with the establishment and control of a call. This is no longer adequate. A modern definition must explicitly differentiate connections from call. It is no longer assumed that a call implies a single connection or just two parties. Call control (CC) coordinates and manages the negotiation of telecommunications services among a set of end users. Connection control, also referred to as bearer control (BC), deals with physical resources on a link-by-link basis. The traditional definition also excludes the signaling for a variety of other telecommunications functions related to the support of supplementary services (e.g., 800 number translation, authentication, and calling line identification services) and terminal/personal mobility.

This section analyzes trends characterizing this development. It discusses how functional signaling and remote operations protocols have supplanted stimulus signaling systems. These trends include shifts from:

- Hardware-intensive systems requiring manual operator intervention to signaling application software systems;
- Exclusive implementation of network control in switches to distributed functionality over a variety of network elements;

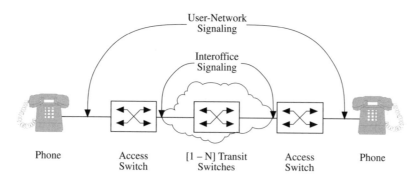

FIGURE 2.8 Stimulus signaling network architecture.

- Provision of plain old telephony to voice, video, and data enhanced with supplementary service, mobility management, and personal communications capabilities;
- Single-function, monolithic systems to modular systems that reuse generic components.

Demand for customized services in conjunction with advances in distributed object computing technology appear to be major drivers behind further evolution. Object-orientation technology, while not fully matured, has begun to influence cutting-edge telecommunications systems. Its full potential has yet to be realized.

Our discussion is limited to the exchange of information pertaining to control for providing network-based telecommunications services. This includes signaling between users and the network over a User–Network Interface (UNI) and signaling between network control elements over a Network Node Interface (NNI). It excludes exchange of information among network management entities.

2.2.2 Stimulus Signaling

Classical signaling systems are based on stimulus signaling. It is called stimulus signaling because of the close association between the signaling language and the electrical stimuli that are its basis. A network architecture for a telecommunications network relying exclusively on stimulus signaling is depicted in Figure 2.8. It is a monolithic. That is, there is virtually no functional differentiation among network control elements. It is characterized by a single-level model with two kinds of entities: customer phones and switches. The latter can be broken down into access switches and transit switches; however, both types primarily engage in the basic call control functions.

There are three distinct kinds of signaling that are apparent in stimulus systems: supervisory, addressing, and information. Off-hook and on-hook, which, respectively, open and close the access loop circuit over a subscriber line, are called supervisory signals.

In the earliest signaling systems, "ring down" systems, supervisory signals were the only nonverbal component of signaling. A human user turned a crank on a phone, which generated a high voltage signal on the access line. The electrical current produced a visual sign, changing the position of a drop indicator on a shutter or lighting a lamp at a switchboard, getting the attention of an operator, who manually connected a voice path to the caller that was used by the operator to obtain addressing information from the caller. Then the operator might need to contact an operator at another switch, establishing a route to the called party. Interoffice signaling required negotiations among operators along the route using the "order wire system," a special network of lines for setting up voice paths. Cords were then manually plugged into a series of switchboards to allocate resources for the end-to-end connection and the operator at the called party's side generated an electrical signal on the called line to ring the phone.

Nonverbal encoding of the called party address was the next major innovation. The step-by-step switch, a fully automated device driven by electrical impulses, made it possible for calls to be set up without operator intervention. Upon going off-hook, using a switchhook instead of a crank, the generated current caused a line finder in the switch to connect the access line to an idle first selector in the step-by-step switch, which waited to receive the first digit. Dial pulses, momentary breaks in the current, generated by the movement of a rotary dial, conveyed digits encoded as a series of precisely timed pulses, the number of pulses in a series defining the digit. Each digit was separated by longer gaps. The number of pulses between each gap governed the rotation of each of a series of selectors in the step-by-step switch, one selector per digit in the address space.

The next innovation came in the 1950s with common control systems having shared directors and markers used to position crossbars of the crossbar switches. In addition to accepting dial pulses, for backward compatibility, the crossbar switches could use multifrequency (MF) alternating current (AC) signals to carry addressing. These signals reduced signaling latency. MF signaling was first used for interoffice signaling and later for user-network signaling with the introduction of push button tone dialing.

Information signals, the third category of stimulus signals, are the audible tones and announcements, such as dial tone and busy signals, that were produced by the switch.

On-hook, off-hook, altering, dial pulses, and MF digits are the vocabulary of stimulus signaling. Human users have become universally adept in this language. Its longevity is remarkable, considering its drawbacks: relatively high latency and human factors anomalies.

Stimulus signals are semantically ambiguous. For example, off-hook is used to originate an outgoing call and to answer an incoming call. A user can unintentionally answer an incoming call when meaning to place an outgoing call. Similar signals convey different kinds of control information, using context and timing to determine their semantics. The evolution of stimulus signaling to support supplementary services exacerbated human factors problem. For example, three-way calling requires "flashing" the switchhook, putting the active call on hold and acquiring a dial tone. Next, the user inputs digits indicating the address of the third party, followed by another flash to join the held party after the third party answers. Mishaps in setting up three-way calls, forwarding calls, etc. are commonly encountered human factors problems. These occur because opening and closing a loop circuit can signify a dial pulse, a flash,[*] or disconnect signal followed by an origination, depending on the timing and context of the signals.

A fundamental technological assumption underlying stimulus signaling is that a user terminal has relatively little intelligence. Phones require no memory and retain no call state. All intelligence resides in human users and network equipment. The human directly deals with the elements of stimulus signaling. The network hardware interprets signals generated by the human user and controls the information signals at the human interface. The terminal is not concerned with the semantics of these signals.

2.2.3 Functional Signaling

Advances in digital computing created the technological basis for stored-program controlled switches with call processing logic implemented in software. Digital technology also served as a foundation for functional signaling, realized as call control entities exchanging digitally encoded signaling messages. Digital messages can carry more information and can be transmitted much faster than stimulus signals, reducing signaling latency. Functional signaling protocols are generally defined as layered protocol stacks, each layer providing a well-defined service to the layer above via a cleanly defined interface. The procedures at a given layer are frequently modeled using finite-state machines.

2.2.3.1 Layered Protocols

Functional signaling protocols are generally a suite of layered protocol based more or less on the Open Systems Interconnection (OSI) protocol reference model[1-3] for packet data networking. Figure 2.9 shows this protocol reference model. Protocol specifications concentrate on the interaction between *peer* entities at the same level in the model, represented by dotted lines in Figure 2.9. A layer offers a service to a higher layer through an interface, hiding implementation details below the interface. This property is

---

[*]Some newer phones have a special button called a tap button to generate a properly timed flash.

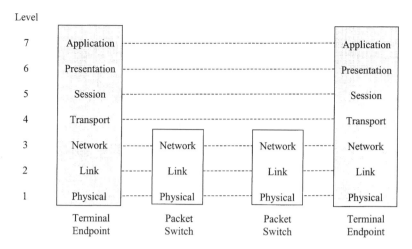

FIGURE 2.9 OSI protocol reference model.

called *encapsulation*. Layering decouples applications, such as call control, from the characteristics of the physical signals and makes it possible for a variety of higher layer protocols to reuse the same layers. It is also possible to adapt higher layers to a variety of lower layer protocols.

Both terminal end points and packet switches implement the three lowest layers. Starting from the bottom:

Level 1   The physical layer defines the physical encoding of data on a physical link
Level 2   The link layer provides for insured or uninsured delivery of frames of data
Level 3   The network layer transports a packet between end points, performing switching functions at intermediate nodes when required by the network topology

Network end points also implement layers 4 to 7:

Level 4   The transport layer provides reliable transparent data transfer between endpoints, performing error recovery and flow control
Level 5   The session layer manages communications sessions; establishing, managing, and terminating connections
Level 6   The presentation layer provides common operations on the format of exchanged data, hiding syntax conversion due to discrepancies in how different machines may order bytes of application data types. ISO has defined a general abstract syntax notation number called ASN.1[3,4] for defining implementation-independent data types and structures for specifying application level protocols.
Level 7   The application layer facilitates communication between applications. An OSI application layer entity can consist of multiple Application Service Elements (ASE).[3] An ASE provides a reusable communications mechanism. For example, a Remote Operations Service Element (ROSE) is a tool for managing request/response transactions.

2.2.3.2   Finite-State Machines and Call Models

Finite-state machines can be used to define procedures for protocol entities at most layers above the physical layer in a protocol stack. Standards frequently use the Specification Definition Language (SDL),[5,6] based on finite-state machine concepts, to specify procedures formally. This sections concentrates on the procedures for call control, embodied in a state machine referred to as a *call model*.

A finite-state machine defines a set of states that can be assumed by a state variable, a set of input signals, and a set of transition rules. The transition rules specify actions to be performed by the entity

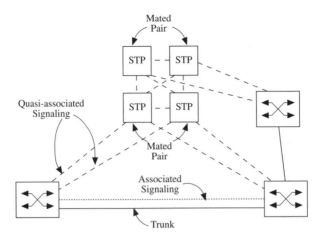

FIGURE 2.10 SS7 architecture for call control.

and the next state it should enter, given a current state and an input signal. Call processing extends the finite-state machine by allowing condition testing of additional extrinsic variables, such as customer profile, to enter into the transition and affect its outcome.

A call control signaling message is treated as an input signal to a call state machine. Each message contains a data element defining a function corresponding to an input signal. The sender invokes the function (e.g., call setup, disconnect, etc.), which is performed by a peer entity receiving it. A message generally contains an identifier to specify a state machine instance (e.g., a call identifier allowing multiple simultaneous calls), and additional parameters required by the function (e.g., service-related attributed, such as bandwidth requirements, service type).

2.2.3.3 Functional Call/Connection Control Protocols

In theory, functional messages afford a limitless, unambiguous signaling vocabulary that can be used to define new features and services. However, supporting a large number of inputs signals generally increases the number of states and complicates the state machine.

The call model underlying processing in commercial switches accommodates both stimulus and function signaling protocols. Functional messages combine many discrete stimulus signals; e.g., the entire called party address is contained as a field in a single message, not requiring separate stimulus signals for each digit.

Functional signaling is *out-of-band*, that is, the signaling links are separate from channels that transport user information. This separation is sometimes physical, using different physical transport facilities, or logical, based on multiplexing, e.g., time division, logical channel label, code division multiplexing. Out-of-band signaling does not interfere with active communication, eliminating "talkoff," the misinterpretation of user information as a signal. It simplifies terminating user information channels since they do not require monitoring for signaling. Functional signaling is also frequently known as *common channel signaling* since the signaling end points use a signaling link as a shared resource for controlling multiple service instances.

Functional signaling first displaced stimulus at the NNI for interoffice call control. Almost all interoffice signaling now uses Signaling System 7 (SS7), an internal packet-switching network supporting a variety of telecommunications applications. Figure 2.10 depicts the SS7 architecture for supporting basic call control. The SS7 packet switching fabric is based on a network of mated pairs of Signaling Transfer Points (STPs) that are accessible to the switches. Mating provides redundant links for reliability.

SS7 supports two signaling modes:

1. Associated mode: The signaling and trunks carrying user information take the same route, using the same physical facilities between adjacent switches.
2. Quasi-associated mode: The user information takes a direct route between adjacent switches, but the signaling is via one of more STPs.

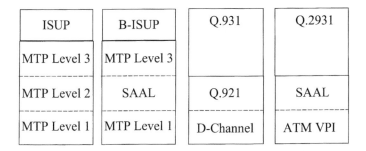

FIGURE 2.11 Call control protocol stacks.

TABLE 2.2 ISDN User Part Messages

ISUP Messages	Direction	Function
Initial Address Message (IAM)	From caller	Sets up call and reserves outgoing trunk
Address Complete Message (ACM)	Toward caller	Indicates called party received call offering
Answer Message (ANM)	Toward caller	Call is answered
Release (REL)	Either	Circuit is being released
Release Complete (RLC)	Opposite REL	Circuit is ready for reallocation

TABLE 2.3 ISDN/B-ISDN Functional Messages

ISDN/B-ISDN Message	ISUP Message
SETUP	IAM
ALERTING	ACM
CONNECT	ANM
RELEASE	REL
RELEASE COMPLETE	RLC

Figure 2.11 shows the SS7 layered protocol stack used for call control. The three lowest levels, the levels terminated by STPs, are called the Message Transfer Part (MTP).[7] MTP correspond to the three lowest levels in the OSI model. MTP Level 1, also called the data link level, defines the physical and electrical characteristics of the 56 or 64 kbs signaling links. MTP Level 2, the link control level, guarantees error-free transfer of signaling data. MTP Level 3, the network level, routes signaling packets from an SS7 source to an SS7 destination. The ISDN User Part (ISUP)[8] is a signaling application that sets up and disconnect a call/connection. ISUP uses MTP Level 3 directly, not using any separate layer 4 to 7 protocols. Table 2.2 provides a list of significant ISUP messages and their functions.

Integrated Services Digital Network (ISDN) and Broadband ISDN (B-ISDN)[9] extend digital transport capabilities for voice, video, and data services along with functional signaling capabilities to the UNI. The ISDN Q.931[10] functionally corresponds to ISUP. The B-ISDN equivalent, Q.2931,[11] is based on the same set of functions as Q.931, altering the messages to control connections based on Asynchronous Transfer Mode (ATM).[12] Table 2.3 lists the most significant Q.931/Q.2931 functions for call control and their corresponding ISUP messages. Figure 2.12 shows the end-to-end signaling message flows to set up and release an ISDN/B-ISDN call.

Q.931 runs over Q.921,[13] a link layer protocol performing functions similar to MTP-2.* Q.921 is transferred over an ISDN D-channel (16 kbs on a basic ISDN interface and 64 kbs on a primary rate).

---

*Although Q.931 is a signaling application protocol, it is referred to as a layer 3 protocol because it runs directly over a layer 2 protocol and is used to establish layer 3 network connectivity. Functionally, they are equivalent to other signaling application protocols, like ISUP, which is above level 3.

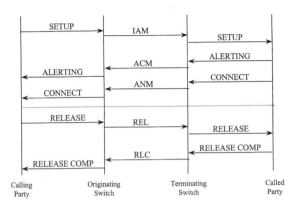

FIGURE 2.12 Q.931/Q.2931 message flows.

Q.2931 runs on the UNI Signaling ATM adaptation layer (SAAL, consisting of Q.2110[14]/Q.2130[15]), which runs on an ATM channel dedicated for signaling. B-ISUP* is the call/connection control protocol for an ATM NNI. It is similar to ISUP and also runs over MTP Level 3. The preferred Level 2 option for B-ISUP is the NNI SAAL (Q.2110/Q.2140[16]) defined for ATM links. Network administrators can optionally use MTP Levels 2 and 1 for B-ISUP. The protocol stacks for ISUP, B-ISUP, Q.931, and Q.2931 appear in Figure 2.11. Note that the two lowest layers of the signaling protocols for the UNI protocols functional correspond to the two lowest levels of SS7 and ISUP corresponds with Q.931 and Q.2931. There is no protocol layer corresponding to MTP Level 3 on the UNI.

Functional signaling terminals are more intelligent than stimulus terminals. ISDN terminals digitally encode user information (e.g., voice and/or video). They also implement logic decoupling the human user (or data application interface) from the functional protocol. While ISDN voice phones emulate the stimulus phone at a human user interface, they also include many additional features that are less directly related to Q.931 signaling. Greater intelligence is also required in the terminals because functional signaling messages carry more information. Q.931 messages contain an identifier, a call reference, which allows a user to engage in multiple simultaneous calls. A terminal needs a separate call state variable for each call.

Penetration of functional signaling over the wired access has been less successful than in the interoffice domain. Functional signaling, however, has grown rapidly for wireless access with second-generation digital radio interfaces. This will be discussed in Section 2.2.4.2.

2.2.4 Functionally Distributed Control

Basic call services have been supported using monolithic network architectures, those depicted in Figures 2.8 and 2.10, with all signaling involving network switches and user terminals. The intelligent network, wireless mobile services and the separation of call and connection introduce specialized network elements, distributing control over a control architecture with functionally differentiated elements. This section examines this functional distribution and its impact on signaling protocols. Basic call control protocols described in Section 2.2.3.3 are defined as a single layer above the network or link layer, not relying on protocol mechanisms between Levels 4 and 7. Common, reusable communication mechanisms have been defined at these levels to facilitate interaction among many of the newer network elements. These mechanisms mark the early appearance of what is now being called middleware.

2.2.4.1 Supplementary Services and Intelligent Networking

Software-controlled switches connected by the SS7 packet infrastructure significantly improved the performance of basic call processing functions. It also made it easier to incorporate software and

---

*The ATM Forum has specified an NNI for connecting ATM switches in a private network that uses the Q.2931 stack instead of B-ISUP.

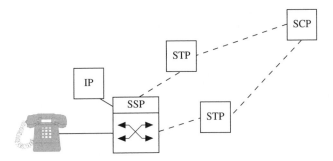

FIGURE 2.13 Intelligent network architecture.

hardware for supplementary services, such as call waiting, call forwarding, and three-way calling. Implementation of some supplementary services could be incorporated entirely within the switch. But others, such as 800 number translation and alternative billing systems, needed access to centralized data, leading to the introduction of separate network elements. Moreover, network service providers preferred IN over switch-based solutions for many supplementary services, because IN is switch vendor independent.

2.2.4.1.1 Intelligent Network Architecture

The IN[18] was conceived as a way for network operators to facilitate introduction of supplementary services and custom features. Service logic is implemented on external processors called Service Control Points (SCP), which can have access to central databases. SS7 makes it possible for IN-equipped switches, called Service Switching Points (SSPs), to invoke the services of an SCP. Figure 2.13 depicts the IN network architecture. IN also defines Intelligent Peripherals (IP), specialized resources for processing in-band user information. An IP is directly attached to a switch, generally using standard ISDN interfaces. Examples of IPs include voice recognition devices for voice dialing and IN-controlled bridges.

The ITU-T has been standardizing the IN architecture and protocols in the Q.1200 series.[19] A key objective is to make it possible for SCPs to invoke new IN services and custom features without affecting switch logic, protecting investment in existing switches. SCPs support a flexible, service creation environment.

IN services are single ended with a single point of control. This means that a service is offered on behalf of a single customer and can only involve SCP interaction with a single SSP. This excludes services where SCPs, acting as agents of different users, can interact. It constrains the possible signaling flows. Foreign-looking IN work is looking for ways to overcome this constraint.

2.2.4.1.2 The Basic Call State Model

SSP interaction with IN is governed by the Basic Call State Model (BCSM). BCSM is derived from basic call processing procedures for a two-party, single-connection call. IN defines two BCSMs, one for the originating side, the other for the terminating side. A BCSM defines a set of call states, referred to as Points In Call (PICs), and a set of discrete points between PICs, called Detection Points (DPs). DPs can be enabled on a per-customer basis to trigger queries to an SSP. Upon encountering an enabled DP, the SSP invokes SCP-based service logic and suspends processing. After executing the service logic, the SCP responds to the SSP, which normally resumes execution after the DP. Figure 2.14 modifies the call setup flow in Figure 2.12 to include queries to SCPs both by an originating and terminating SSP. In this example, the originating SSP may be executing logic for 800 number translation, the terminating side performing call screening. Note that these are independent supplementary services. Undesired features interactions are possible. This is a serious concern to service providers.

These examples are relatively simple since each service requires a single query. Other services may require multiple interactions between the SSP and SCP, with triggers at several DPs. Defining the more complex services, with requirements in terms of the predefined standard DPs, can be challenging.

Call Setup with IN Triggers

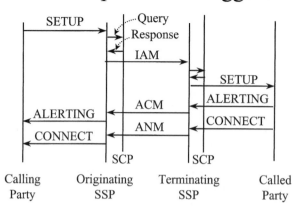

FIGURE 2.14 Call setup flow with IN triggers.

2.2.4.1.3 Remote Operations Protocols

Provision of services in the manner described in Section 2.2.4.1.2 can be simplified if the SSP can take advantage of a general, relatively simple mechanism to invoke SCP-based services. IN seeks to define such a mechanism based on concepts derived from the upper layers of the OSI model. IN signaling is based on remote operations.

Remote operations provide the basis for a flexible, open-ended approach to introducing new services. It is based on the client/server concept popular in computing, using the within-process procedure call single as a model for a remote call over a network. The function invoking the procedure call executes is a client, the invoked function runs on a server. Generally, it is assumed that the remote call procedure, like the single-process procedure call, is synchronous, which means that the client "blocks," is waiting for a server to respond before resuming execution. After responding, the server retains no information about a procedure call, though there may be side effects, such as changes to persistent data in a database. But this information is not local to any particular server. Thus, binding between the client and server only exists for the duration of a procedure call.

The signaling protocol defined for interaction between an SSP and an SCP is called Intelligent Network Application Part (INAP). INAP uses Transaction Capabilities Application Part (TCAP),[20] which is a remote operations protocol defined for telecommunications applications. TCAP runs on top of the Signaling Connection Control Part (SCCP).[*21]

TCAP supports asynchronous unidirectional messages as well query/response sequences. A TCAP "package," the term used for the TCAP data unit, contains one or more "components." Each component has a type (query, response, unidirectional, or abort), an operation, and parameters. An SSP chooses an operation based on its BCSM state and passes other parameters based on call attributes. Thus, the SSP logic is general and service independent. An SCP can change the value of the passed attributes in its response and can even instruct the SSP to resume execution at a different PIC.

TCAP applications are defined using ASN.1,[4] making the definition of INAP protocol data units independent of the physical encoding, which is defined by a presentation layer protocol, such as Basic Encoding Rules (BER).[22]

Many supplementary services require UNI signaling as well. Stimulus signaling UNIs constrains the information that a user can convey. Q.932[23] and Q.2932,[24] which supplement Q.931 and Q.2931, respectively, define a number of ways for supporting supplementary services. One is by embedding stimulus signal

---

*SCCP is a Level 4 protocol that enhances MTP addressing, supporting global title translation, used in 800 number translation and other applications. It also provides transport layer services, supporting connection-oriented and connectionless transport services and the segmentation/reassembly of packages into MTP packets.

information using the keypad protocol within a call- or non-call-related* message. The semantics for these embedded signals are not standardized and defined by the local operator. A second way is to define new functional message types. Q.932 defines hold and resume functions for placing a call on hold and reactivating it. Q.932 defines an "orthogonal state machine" for these functions to avoid complicating the Q.931 call model. And third, Q.931 and Q.2931 define a registration/facility feature, which can be used to embed remote operations in call- or non-call-related messages. It allows multiple supplementary service invocations within one message and supports services with a large number of variations without proliferating new messages. While the functional message may affect call state, embedded operation do not.

2.2.4.1.4 Call Party Handling
Thus far, the supplementary services described can be provided using typical interaction between an SSP client and an SCP server. The SSP retains high-level control of the signaling flow, as in the basic call flows. An SSP temporarily suspends the basic flow pattern at an enabled DP to invoke an SCP in a remote procedure call and continues that pattern after the response, as depicted in Figure 2.14. Other SSP-SCP relationships are possible. Call Party Handling (CPH) is an example, In CPH, the SCP assumes a larger role in the overall end-to-end signaling flow pattern. It retains a call-related state beyond a simple SSP–SCP interaction, being actively involved in managing the association of multiple connections.

CPH can be used to provide multiparty connections. The SCP controls bridging of the parties, using basic call control of the switches to provide "legs." Each leg comprises a distinct connection between the bridging point and a user end point. CPH can associate the legs, with the SCP acting as a signaling end point for each leg. An SCP can use an SSP to connect an added party to a bridge in response to an add-party request, which can be invoked by a user in the call through a non-call-associated UNI signaling message. In CPH, switches have call state for each leg and the SCP keeps the state that associates bridged legs.

This is not the only way to provide multiparty connections. Other approaches that functionally separate call control agents from connection control will be discussed in Section 2.2.4.3 and Section 2.2.5.3. This section, however, illustrates that it is possible to model complex services by extending basic IN concepts.

2.2.4.2   Cellular Mobile Communications Services
Network support of cellular mobile communications[25] is another example of a service that is handled more effectively using functionally distributed network control. Mobile services includes both terminal and/or personal mobility. Many characteristics of both are similar. This section focuses on terminal mobility, which is the predominant mobile service today.

Cellular mobile systems are fundamentally different from wireline access. The configuration of the wireline network is static. A subscriber is attached to the network at a fixed location and uses dedicated access resources. Cellular systems, on the other hand, are dynamic. A mobile user changes its network access point and shares radio resources that must be dynamically allocated at call setup and during a call as the mobile user changes location. Wireless systems need specialized network elements and special signaling capabilities to support mobility and radio resource management.

The topology of a cellular access network is dictated by the need to accommodate a large number of mobile users using scarce radio resources. The network houses its radio transceivers in base stations (BS). Each cell has a BS near its geographic center. The geographic coverage of radio signals between a BS and mobile users is relatively small, so other BSs can reuse the same radio spectrum. To reduce system costs, BS control activities are largely limited to radio resource management functions. A group of BSs are connected to a mobile switching center (MSC)** via the wired portion of the mobile access network

---

*Noncall messages can be used to change user profiles.

**In newer mobile system architectures, a Base Station Controller (BSC) assumes some control functions for a group of base stations. The interface between the BSC and MSC, called the A-interface, has been standardized in GSM[26] and IS-634,[27] allowing network providers to purchase components from different manufacturers. When Code Division Multiplexed Access (CDMA) technology is used on the radio access, frame selection is required to cope with transmission diversity from the mobile. The unit can be physically located in the MSC, the BSC, or stand-alone. Its location affects the interfaces and protocols required in the mobile access network.

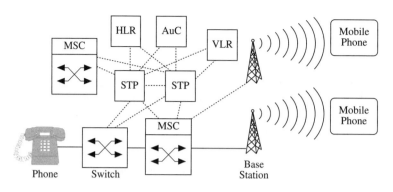

FIGURE 2.15 Mobile network architecture.

(Figure 2.15). The MSC supports higher-level mobility management and call control functions and interworks the mobile system to the wireline network.

First-generation mobile systems, based on analog technology, use in-band signaling. Second-generation digital systems are rapidly replacing the first generation because of significant improvement in voice quality and a greater number of calls per cell can be supported. The remaining discussion deals with digital systems.

Second-generation radio interfaces have shared control channels in both the forward and reverse directions* that are used for mobile paging, synchronization, signal strength measurement and traffic channel allocation. A traffic channel, which contains separate control and user data subchannels, is allocated to a mobile on a per-call basis.

When a mobile powers up, it scans a group of shared forward control channels, seeking a strong radio signal from a nearby base station. It can then signal on a corresponding shared reverse control channel, called an access channel. A mobile periodically registers with a local MSC over the access channel to receive incoming calls. The MSC authenticates a registering mobile, which may be a local subscriber or a visitor.

Information about each subscriber is stored by its mobile service provider in a Home Location Register (HLR). An HLR is informed when one of its subscribers registers so that calls can be forwarded to a roaming subscriber. Authentication information is kept in an Authentication Center (AuC). Information needed by the visited MSC to accommodate roaming mobiles, such as user profile information, is cached in a Visiting Location Register (VLR).

When a mobile station is called, the request to set up the call is first sent to an MSC in the mobile's home area. This MSC checks with its HLR to discover the location of a registered called mobile. If the mobile is roaming, it forwards the call request to an MSC in the visited area. The MSC must then locate the mobile more precisely within its area, since the mobile may have moved since last registering. The MSC distributes a paging message to BSs in its area, which broadcast it over shared forward control channels. The mobile responds over an appropriate reverse control channel. Then the network selects forward and reverse traffic channels for the call. At this point, the network can signal over the control part of the traffic channel to ring the mobile.

As the mobile station moves, the quality of the signal is monitored by both the MS and BS in order to determine if the quality of service can be improved if it were served by another BS. The procedure for changing base stations during a call is called handoff or handover. A handoff requires dynamic allocation of traffic channels. In Code Division Multiple Access (CDMA) systems, a mobile can be connected to multiple base stations at the same time. This diversity improves the quality of the service when a mobile does not have a strong signal with a single base station.

---

*The forward direction is from the base station to the mobile station. The reverse direction is the opposite.

The MSC generally assumes call control as well as mobility management functions. An MSC, however, can incorporate standard switches as components and use the switch capabilities not only for connections but also supplementary services, such as three-way calling, call forwarding, and call waiting. Partly in order to exploit the commonality of call control in both wireline and wireless domains, the predominantly European Global System for Mobile Telecommunications (GSM)[26] has defined separate Call Control (CC), Mobility Management (MM), and Radio Transmission (RT) management modules in its signaling application protocol. Each module has a unique protocol identifier and a separate set of messages. CC is based on Q.931,[10] simplifying ISDN interworking to the wireline world. Modularizing also cleanly separates message components that are processed by a BS from those that are relayed to and processed by the MSC. A mobile and MSC are peers for CC messages, but the mobile and base station exchange RT messages. Other UNI protocols, e.g., ANSI-95,[27] the radio interface for CDMA, are not currently modular. However, ways of making it modular are being considered for future evolution.

The MSCs, HLRs, VLRs, and AuCs are connected using TCAP. Different mobile service providers use different mobile application protocols on top of TCAP. North America uses predominantly ANSI-41, Europe uses predominantly GSM-MAP (Mobile Application Part)[26] and Japan uses predominantly PDC (Personal Digital Cellular)-MAP, though there are mobile providers that do not use the predominant standard in their area. Wireless Intelligent Network (WIN) capabilities are also being defined that allow mobile network elements to use SCPs. These augment features associated with the wireline system with features that may be triggered by mobility management functions, such as registration.

Differences in radio interfaces and in signaling applications are an impediment to global roaming. International standards bodies are trying to overcome this problem the standardization of third-generation systems, International Mobile Telephony for the Year 2000 (IMT-2000),[29] though there are obstacles due to a desire for backward compatibility with regional second-generation systems.

2.2.4.3   B-ISDN and the Functional Separation of Call and Connection Control

An objective of the future evolution of B-ISDN signaling is functional separation of call and connection control protocols. Separation makes call control independent of the functions controlling distribution and utilization of network resources. This facilitates third-party signaling, independent routing of multiple connection calls, route optimization of forwarded calls, and end-to-end service negotiation prior to resource allocation (lookahead). Separation is useful for multimedia services supported by multiparty/multiple connection calls.

Figure 2.16 shows a network architecture that functionally separates call and connection control. Call Control (CC) resides in the user terminals and in CC entities serving as a user's agent for the negotiation and realization of communications sessions and supplementary services at the edges of the network. Connection (or Bearer) Control (BC) manages network resources on a link-by-link basis. Note that there is no separate connection control protocol shown for the UNI in Figure 2.16. There is no substantial benefit in separating call and connection control protocols for a wireline UNI because access bandwidth is dedicated. However, separation of CC for wireless access is beneficial since connections can be modified independently of existing call associations.

It is envisioned that a separate CC can implement a structured call model that can support multiparty/multimedia calls requiring multiple connections provided by BC. Ideas discussed in Section 2.2.5.3 are one possible approach for defining a new call model based on object-oriented concepts. Opening an interface between CC and BC would make it possible to run CC on general-purpose computers that could use a BC entity that may manage different ATM switch products. Although there have been some discussions, standards bodies have not begun work to realize an open interface between call and connection control.

In anticipation of functional separation, B-ISUP[17] has been modularly structured containing separate CC and BC ASEs. Currently, B-ISUP uses both modules in a monolithic protocol, but it is possible to use the CC part in a separate protocol.

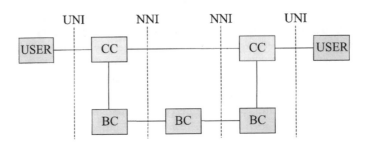

CC - Call Control
BC - Bearer Control

FIGURE 2.16 Separation of call and connection control.

The ITU-T has chosen not to define a comprehensive call model supporting multimedia services for B-ISDN Capability Set 2 (CS/2). Instead they are extending the existing call models and protocols in a step-by-step fashion. Table 2.4 lists some additional B-ISDN CS/2 signaling recommendations. As first steps, CS/2 will support a single connection unidirectional point-to-multipoint call and point-to-point multiple connection calls, but not multiple connection calls with multiple parties. The limitations of lookahead procedures defined in Q.2724[30] illustrates the limited nature of the functional separation. A network CC entity at the caller's side sends a TCAP-based lookahead query to obtain the called party's busy/idle state and terminal service capability parameters. The query does not create a call association nor affect the call state of the queried party. Lookahead does not change the busy/idle status. The state of the called party could change by the time the real call setup is sent using B-ISUP. The query does not impact call state.

TABLE 2.4   Additional B-ISDN Signaling Capabilities

Capability	UNI	NNI
Variable rate connections	Q.2961	Q.2723
Bandwidth negotiation	Q.2962	Q.2725.1
Lookahead	Q.2964	Q.2724
Connection modification	Q.2963	Q.2725.2
Point-to-multipoint	Q.2971	Q.2722.1
Multiconnection point-to-point	Q.2972	Q.2722.2

2.2.5   Object-Oriented Signaling

Packet switching technology promoted layered protocol architectures, resulting in the definition of general protocols for SS7 applications, like MTP, SCCP, ASN.1, and TCAP. Functional signaling application protocols for call/connection control, like Q.931 and ISUP, and TCAP-based applications also appeared, promoting the growth of a functionally distributed control network for IN and mobile services. This section discusses exploratory work in object-oriented technology that may accelerate the growth and significantly shape the evolution of signaling, allowing for even greater reuse and customization.

Middleware is one of the hottest topics in computer networking today. Middleware deals with Level 4 to 7 functionality. TCAP and ASN.1 are examples of middleware for distributed computing. Distributed object computing (DOC) technology is more ambitious. It aims at extending reusability by providing distributed applications with network and computing platform transparency and a reusable set of common services. DOC is supported by Object Request Broker (ORB) middleware, like the Object

Management Group's (OMG) Common Object Request Broken Architecture (CORBA.),[31] Microsoft's Distributed Component Object Model (DCOM), and Java Remote Method Invocation (RMI). And, at an even higher level, this technology may provide the foundation for object-based frameworks defining reusable components geared for specific application domains, such as telecommunications, making it easier and cheaper to define flexible, evolvable, and customizable telecommunications applications.

2.2.5.1   Distributed Objects

2.2.5.1.1   *What Is Distributed Object Computing?*

Distributed Object Computing (DOC) is the convergence of remote operations and object-oriented technology. Douglas Schmidt writes that DOC "is the discipline that uses [object-oriented] techniques to distribute reusable services and applications, efficiently, flexibly and robustly over multiple, often heterogeneous, computing and networking elements."[32] It was stated in Section 2.2.4.1.3 that in classical remote operations, the model generally used for IN services, a server can act as an agent of a client to retrieve data from a file or database, but retains no intrinsic state or data of its own. Another server can retrieve the same data because the data is separated from the server. In the classical client/server model, there is no binding or association between a client and server outside the context of a remote procedure call transaction. In contrast, a server *object* can have its own internal data, maintaining state and retaining an association with a client that persists beyond the scope of a single transaction.

Objects, like protocol layers, support encapsulation, hiding the implementation of servers from their clients. Two additional properties, inheritance and polymorphism, give objects properties enhancing that can further enhance reuse. Inheritance allows new objects to be based on previously defined objects. It allows more specific object types to be derived from generic objects or abstract classes.* Polymorphism allows the interaction between objects to be defined at an abstract level, making generalized patterns or procedures available to specific applications. The designers of a generalized framework do not need to envision application details, though they must be knowledgeable of the application domain to create a useful framework. Inheritance and polymorphism make it possible to define generic object-based design patterns and frameworks geared to specific application domains (e.g., telecommunications).

2.2.5.1.2   *Object Request Brokers and Common Facilities*

An ORB is middleware that facilitates location-independent client/server object interaction. CORBA, the OMG standard, is strongly supported by major computer industry companies, including Compaq Digital Equipment Corporation, Hewlett-Packard, SunSoft, and NCR. CORBA makes it possible for application objects to be defined independently of the underlying computing and network platform. These details are hidden because a CORBA-compliant server object defines its interface using CORBA Interface Definition Language (IDL). IDL compilers are written for specific platforms and generate code that adapts objects to a generic ORB, which isolates the application's interface over the ORB from platform considerations. Thus, IDL object definitions provide a platform-independent language for specifying the externally visible characteristics of an object. CORBA also defines common facilities, generic objects that support commonly required services, such as two-phase commit transactions, event handling, and persistence objects.

2.2.5.2   Telecommunication Information Network Architecture Consortium (TINA-C)

The Telecommunication Information Network Architecture Consortium (TINA-C)[34] is an international group of network providers and suppliers looking at using DOC for network operations and control. It is specifying an open architecture for future telecommunications applications, including broadband, multimedia, mobile, and information services capabilities.

TINA-C aims to provision customizable services rapidly based on reusable service components. Like service creation in IN, new services can be based on composition of generic and/or customized

---

*An abstract class is not an object itself, but can be used as a basis for deriving objects.

implementations of standard component objects. Subscribers can be involved in service definitions by dynamically defining services using standard components.

TINA-C has defined a set of computational objects at a level of abstraction that correspond to the functional entities in the Telecommunications Management Network (TMN), IN, and the separated B-ISDN functional architectures. These include user agents and session managers objects that are involved in service and call control and resource managers that provide connection control services. TINA-C makes it possible to implement control objects on general-purpose computing platforms, moving control logic out of the switches.

TINA-C envisions a network evolution with DOC-based systems that are adapted to standard message-based protocols, with distributed objects being initially used for management systems and supplementary service processing and, later, for call control. In theory, assuming all of the security issues are tackled, a DOC-based telecommunications system could ultimately encompass the UNI, replacing message-based protocols with IDL interfaces. TINA-C has thus far focused on principles and high-level requirements. Its goal is the definition TINA-compliant, interoperable application objects.

2.2.5.3　Examples of Exploratory Work on Object-Oriented Telecommunications

There are a number of other noteworthy directions in research and forward-looking work involving distributed objects and telecommunications applications:

- Support for real-time interaction among distributed objects;
- Distributed object frameworks and libraries to facilitate introduction of telecommunication network services;
- Telecommunications protocols based on object-oriented concepts.

Real-time applications with deadlines and stringent quality of service requirements, especially when objects are distributed over a wide-area network, cannot be supported by ORB implementations based on the current CORBA 2.0 specifications. OMG and the Distributed Object Computing group at the University of Washington have been looking at supporting real-time interaction among distributed objects. OMG has created a number of Domain Technical Committees (DTC) to investigate specific application areas, including a telecommunications committee. DTC has been working on supporting isochronous streams between objects that can meet quality of service requirements for interactive audio and video. The DOC group at Washington University has implemented TAO (The ACE ORB), a real-time platform running on top of their ADAPTIVE Communication Environment (ACE) software, which supports hard real-time applications over ATM networks. They also use their platform for telecommunications applications.

Wireless Distributed Call Processing Architecture (W-DCPA) is an exploratory project at Lucent Technologies that supports call processing and mobility management for multimedia services using DOC.[35,36] It defines a distributed object framework for MSCs, BSCs, and BSs consisting of server objects that are supported by an object class library. Server objects communicate over a CORBA ORB. Interface module servers have also been defined to adapt DOC-based servers to standard message-based signaling protocols.

Object-oriented concepts can be used to define new call models and message-based signaling protocols for multimedia.[37] Using a comprehensive object-based call control framework may prove to be a more fruitful approach for managing complex multimedia/multiparty services than the piecemeal enhancements to basic two-party, single-connection protocols discussed in Section 2.2.4.3. B-ISDN CS/2 requirements did some modeling based on this approach and a number of experimental prototypes (RACE MAGIC,[38] Object-Oriented Call Model,[39] and Bellcore's EXPANSE[40]) have been implemented based on similar call models. Complex services can be defined in terms of a set of related building block component objects (e.g., calls, parties, connections, etc.) with each object containing part of the call state instead of trying to capture all the call state in a single variable. These elementary call objects are combined and organized into composite structures that logically represent a service.

Inheritance can be used to simplify the definition of elementary objects. All elementary objects, for example, have a simple state variable. These objects can belong to subclasses differing according to their mode of control, that is, the generic procedures for their creating and deletion. For example, creating a confirmed elementary call object requires concurrence by a "called" party, but a local elementary call object is "private" and can be controlled by one user (e.g., putting a local access channel on "hold"). In this way, it is possible to derive a relatively simple set of objects* for a generic call control framework. Deriving definitions of services from a generic structure facilitates the introduction of new services and capabilities.

Call control can be based on remote operations on the elementary call objects. Remote operations can be embedded in generic atomic action transactions,** a set of distributed operations that succeeds or fails as a unit, to further promote reusability and simplify call control. TCAP and registration/facility can be extended to support atomic actions. This approach, in contrast to the functional approach, does not require an explicit definition of a signaling flow for each capability. A generic flow based on recursive atomic actions can be reused to support all capabilities. This mechanism promotes flexibility and reusability, empowering users to create services by composing a transaction containing a set of operations meeting instantaneous service requirements.

A high-performance middleware system may ultimately hide some aspects of message-based signaling just described, including generic object-based mechanism for supporting atomic actions and handling events. But it is not clear that the future will result in the total disappearance of message-based signaling protocol applications by client/server distributed object computing. DOC assumes that location independence is a desired objective; some control applications may perform better if they are location sensitive. The existing signaling architectures are related to some extent to the location of specific functionality. This distribution is a reflection of rigid performance requirements that must be met. The know-how to make DOC systems meet these requirements must first be acquired.

2.2.6 Conclusions

This section has demonstrated how software, packet switching, distributed processing, and object-oriented technologies have been shaping the evolution of signaling in public telecommunications network. It has summarized the principles that have shaped currently deployed signaling systems and introduced forward-looking work based on distributed object concepts that may significantly affect the future, accelerating the migration of logic from switching to general-purpose computing equipment. It is ironic that packet switching, the underlying technology for modern signaling, is also the basis of the Internet,[41] which is potentially a competing technology for providing networking services.

Both the public telephony network and the Internet have plans for integrated services that overlap. The public network traditionally provided connections with guaranteed quality of service, while the Internet provided best-effort connectionless service. Classical Internet protocols do not separate signaling from the transmission of packets, all packets carrying network addressing information for routing purposes. Many signaling-related functions associated with call, connection, and service control are performed by Internet terminal equipment (hosts) without network intervention or knowledge. However, users are now attempting to transport real-time voice and video data over the Internet and the Internet Engineering Task Force (IETF) is exploring ways of offering better qualities of service, incorporating traditional signaling functions into the network. The Resource ReSerVation Protocol (RSVP)[42] is an example of a newly defined Internet protocol providing a function traditionally associated only with connection-oriented signaling. RSVP guarantees network resources along a path, contradicting the textbook notion that connectionless packets can arrive out of sequence because they travel different paths.

---

*A set would most likely include objects such as a call object, a party object, an access channel object, a connection object, and an attachment object that associates a party with a connection.

**A generic transaction mechanism can be used to support functions besides call control.

While there appears to be some growing convergence on service requirements, the resolution of the competition between the public telephony and Internet is not yet clear. In some contexts, there is potential for harmonization and interworking. Signaling can be used to obtain "Level 2" connections for switching between Internet nodes and services, such as IP telephony, which can cross Internet/public phone network boundaries. Many contending ideas are now being explored in standards organizations and ongoing technical workshops, such as Columbia University's Center for Telecommunications Research OPENSIG (Open Signaling for ATM, Internet and Mobile Networks) Workshop.

References

1. International Standards Organization (ISO), *ISO Open Systems Interconnection — Basic Reference Model*, 2nd ed., ISO/TC 97/SC 16, No. ISO CD 7498-1, 1992 (also ITU-T Recommendation X.700).
2. Zimmermann, H., OSI Reference Model—The ISO model of architecture for open systems interconnection, *IEEE Trans. Commun.*, 28 (4), 425–432.
3. Rose, M., *The Open Book, A Practical Perspective on OSI*, Prentice-Hall, Englewood Cliffs, NJ, 1990.
4. International Standards Organization (ISO*)*, Specifications of Abstract Syntax Notation One (ASN.1), ISO/SEC 8824: 1988 (Also ITU-T Recommendation X.208).
5. ITU-T Recommendation, CCITT Specification and Description Language (SDL), March 1993.
6. Belina, F., Hogrefe, D., and Sarma, A., SDL with Applications from Protocol Specification, Prentice-Hall, Englewood Cliffs, NJ, 1991.
7. ITU-T Recommendation Q.701, Functional Description of the Message Transfer Part (MTP) of Signalling System No. 7.
8. ITU-T Recommendation Q.761, Functional Description of the ISDN User Part of Signalling System No. 7.
9. Stallings, W., *ISDN and Broadband ISDN with Frame Relay and ATM*. 3rd ed., Prentice-Hall, Englewood Cliffs, NJ, 1995.
10. ITU-T Recommendation Q.931, ISDN User-network Interface Layer 3 Specification for Basic Call Control.
11. ITU-T Recommendation Q.2931, Broadband Integrated Services Digital Network (B-ISDN) — Digital Subscriber Signalling System No. 2 (DSS2) — User-Network Interface (UNI) — Layer 3 Specification for Basic Call/connection Control.
12. De Prycker, Martin, *Asynchronous Transfer Mode: Solution for Broadband ISDN*, 3d ed., Prentice-Hall, Englewood Cliffs, NJ, August 1995.
13. ITU-T Recommendation Q.921, ISDN User-network Interface — Data Link Layer Specification.
14. ITU-T Recommendation Q.2110, B-ISDN ATM Adaptation Layer — Service Specific Connection Oriented Protocol (SSCOP).
15. ITU-T Recommendation Q.2130, B-ISDN ATM Adaptation Layer — Service Specific Coordination Functions for Supporting Signaling at the User-Network Interface (SSFC at UNI).
16. ITU-T Recommendation Q.2140, B-ISDN ATM Adaptation Layer — Service Specific Coordination Functions for Supporting Signaling at the Network Node Interface (SSFC at NNI).
17. ITU-T Recommendation Q.2721.1, B-ISDN User Part — Overview of the B-ISDN Network Node Interface Signalling Capability Set 2, Step 1.
18. Faynberg, I., ed., *The Intelligent Network Standards: Their Application to Services*, Series on Telecommunications, McGraw-Hill, New York, November 1996.
19. ITU-T Recommendations Q.1200 Series, Intelligent Network, Capability Set 1.
20. ITU-T Recommendations Q.771-778, Transaction Capabilities Application Part (TCAP).
21. ITU-T Recommendation Q.711, Functional Description of the Signalling Connection Control Part.
22. ITU-T Recommendation X.209, Specification of Basic Encoding Rules for Abstract Syntax Notation One (ASN.1).
23. ITU-T Recommendation Q.932, Digital Subscriber Signalling System No. 1 (DSS 1) — Generic Procedures for the Control of ISDN Supplementary Services Q.932.

24. ITU-T Recommendation Q.2932.1, Digital Subscriber Signalling System No. 2 — Generic Functional Protocol: Core Functions.

25. Rappaport, T. S., *Wireless Communications: Principles and Practice*, Prentice-Hall, Englewood Cliffs, NJ, 1996.

26. Mouli, M. and Pautet, M.-G., *The GSM System for Mobile Communications*, published by the authors, France, 1992.

27. Garg, V. K., Smolik, K. F., and Wilkes, J. E., *Applications of Cdma in Wireless/Personal Communications*, Prentice-Hall, Englewood Cliffs, NJ, 1996.

28. Gallagher, M. D. and Snyder, R. A., *Mobile Telecommunications Networking with IS-41*, McGraw-Hill, New York, 1997.

29. Pandya, R., Grillo, D., Lyckssell, L., Mieybegue, P., and Yabusaki, M., IMT-2000 standards: network aspects, *IEEE Personal Commun.*, August 20-29, 1997.

30. ITU-T Recommendation Q.2724-1, B-ISDN User Part — Look-Ahead Without State Change for the Network Node Interface (NNI).

31. Orfali, R., Harkey, D., and Edwards, J., *The Essential Distributed Object Survival Guide*, John Wiley & Sons, New York, 1996.

32. Schmidt, D. C., Distributed Object Computing, *IEEE Commun. Mag.*, 35 (2), 1997.

33. Object Management Group (OMG), Control and Management of A/V Streams Request for Proposals, OMG Document telecom/96-08-01 ed., August 1996.

34. Dupuy, F., Nilsson, G., and Inoue, Y., The TINA Consortium: towards networking telecommunications services, in *Proceedings of International Switching Symposium (ISS '95)*, 1995.

35. LaPorta, T. F., Veeraraghavan, M., Treventi, P. A., and Ramjee, R., Distributed call processing for personal communication services, *IEEE Commun. Mag.*, 33 (6), 66–95, 1995.

36. LaPorta, T. F., Sawkar, A., and Strom, W., The role of new technologies in wireless acess network evolution, in *Proceedings of International Switching Symposium (ISS '97)*, IS-03.18, 1997.

37. Minzer, S., Bussey, H., Porter, R., and Ratta G., Evolutionary trends in call control, in *Proc. Int. Switching Symp. (ISS '95)*, 2, April 1995, 300–304.

38. Hellemans, P., Reyniers, G., An object-oriented approach to controlling complex communication configurations, in *Proc. Int. Switching Symp. (ISS '95)*, 2, April 1995, 72–76.

39. Grebeno, J., Hanberger, N., and Ohlman, B., Experiences from implementing an object oriented call model, in *Proc. Int. Switching Symp. (ISS '95)*, 2, April 1995, 77–81.

40. Minzer, S., A signaling protocol for complex multimedia services, *IEEE J. Selected Areas Commun.*, 9 (9), 1383–1394, 1991.

41. Comer, D. E., *Internetworking with TCP/IP: Principles, Protocols, and Architecture*, Vol. 1, 3rd ed., Prentice-Hall, Englewood Cliffs, NJ, 1995.

42. Braden, R., Ed., Resource ReSerVation Protocol (RSVP) — *Version 1 Functional Specification*, IETF RFC 2205.

2.3 Telecommunications Services on Broadband Networks

Hisaya Hadama

2.3.1 Introduction

It is recommended that the broadband Integrated Services Digital Network (B-ISDN) be based on the Asynchronous Transfer Mode (ATM).[1] An ATM network can efficiently support various multimedia applications as well as traditional telephone services.

Some examples of such multimedia applications are electronic mail with multimedia contents (text, voice, moving pictures, programs, etc.), videoconferences, an electronic museum, electronic commerce, and multimedia data delivery. ATM allows a network to carry any kind of digital data as a stream of

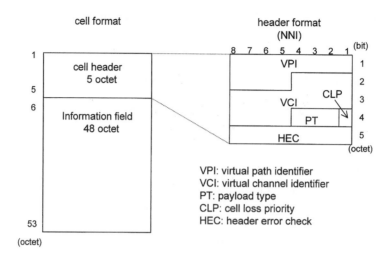

FIGURE 2.17 ATM cell format at network node interface (NNI).

fixed-sized "cells." A cell is a standardized data transmission unit in an ATM network. Figure 2.17 shows the cell format in a network node interface (NNI) recommended by ITU-T.[2]

In a traditional Synchronous Transfer Mode (STM)[3-5] network for public telephone service, it is necessary to use various types of switches to handle hierarchically multiplexed digital paths (i.e., virtual containers of VC-11 and VC-4[4]). An essential function of an ATM switch is cell-by-cell data transmission according to routing information written in each cell header. This mechanism allows an ATM switch to handle connections of any bit rate. This feature is suitable to construct a cost-effective network infrastructure for multimedia communication.[6]

Broadband service includes two categories: (1) bearer service which provides information transfer capability between end terminals and (2) teleservice which provides end-to-end service of specific broadband applications. Examples of teleservices[7] are teleconference and broadband videotex.

An ATM network is connection oriented by nature. All bearer services provided by the ATM network are based on connections, which are categorized into Virtual Channels (VCs) and Virtual Paths (VPs); thus the most fundamental broadband services of an ATM network are VC and VP services. ATM can also provide connectionless data transfer services using a connectionless server function,[8] which provides connectionless packet transfer with an ATM network.

A high-speed IP network and frame relay network can be created over an ATM infrastructure, utilizing the high-speed ATM data transmission capability.

This section first explains the fundamental data transfer mechanism of ATM. Then it explains connection-oriented services. Finally, it explains connectionless services of ATM networks.

2.3.2 Basics of ATM

2.3.2.1 Cell Assembly and Disassembly

To transmit analog (e.g., a voice signal) information between end terminals, a sending terminal must have an Analog-to-Digital (A-D) converter to digitize the analog sound wave. To transmit the digitized signal through an ATM network, the terminal divides the digital signal into fixed-sized cells and attaches such necessary overhead information as a cell header for each cell, as outlined in Figure 2.17. The receiving terminal needs to have functions to reproduce the digital signal from the received cell stream and to reproduce the original analog sound wave from it.

To transfer data packets within ATM networks, the terminal equipment must also have another adaptation function. It must divide the packet into fixed-sized cells and then assemble them. The receiver terminal needs to have a function to reproduce the original packets from the received cell stream.

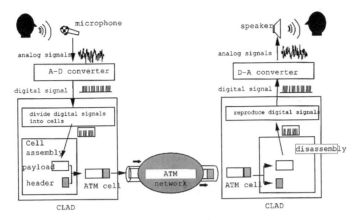

FIGURE 2.18 CLAD function for voice signal transfer. (*Source:* Aoyama, T. et al., in *ATM Basics,* The Telecommunications Association, Tokyo, 1998. With permission.)

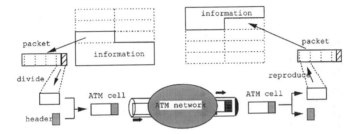

FIGURE 2.19 CLAD function for data transfer. (*Source:* Aoyama, T. et al., in *ATM Basics,* The Telecommunications Association, Tokyo, 1998. With permission.)

A function that assembles ATM cells from the original format of information and a function that reproduces the original data format are implemented in Cell Assembly and Disassembly (CLAD). Figure 2.18[9] shows an example of CLAD for voice communication.

In traditional digitized telephone networks, a voice signal is converted to a 64-kb/s digital signal, so a delay of 6 ms (= 48 bytes × 8 bits ÷ 64 kb/s) is needed to fill up the payload of a cell (48 bytes) with a 64-kb/s voice signal at a CLAD. This delay is proportional to the length of the cell payload. If the payload is much larger than the present value, this delay increases tremendously, which degrades the voice quality in telephone service.

One important reason for the current small cell size (53 bytes = 48 bytes of payload + a 5-byte header) is due to this delay value for voice communication. This problem is more serious when we use a codec having a low bit rate (e.g., 16 kb/s, 8 kb/s, etc.). The delay for cell assembly is inversely proportional to the bit rate of a voice channel. For example, we need a 24-ms delay for 16-kb/s signal, and a 48-ms delay for an 8-kb/s signal. To solve this degradation of voice communication, a new capability for transmitting a bundle of voice channels has been developed as Voice and Telephony Over ATM (VTOA) techniques with ATM Adaptation Layer Type 2 (AAL2) standard.[10]

Figure 2.19[9] shows a CLAD for data communication through an ATM network. Functions for mapping various data formats onto ATM cells are standardized as ATM Adaptation Layer (AAL) functions[11] by ITU-T. The following section explains AAL functions.

2.3.2.2 ATM Adaptation Layer

The roles of ATM Adaptation Layer (AAL) functions are (1) mapping various Protocol Data Units (PDUs) of user's data (or control signals and management signals) onto the cell stream and (2) reproducing the original PDU format after receiving cells from an ATM network.[11]

bearer service category	Broadband connection-oriented bearer service (BCOB)				Broadband connection-less data bearer service (BCLB)
bearer service sub-category	A	B	C	X	D
Connection mode	Connection	Connection	Connection	Connection	Connectionless
Bit rate	Fixed	variable	variable	defined by a user	variable
End end timing	necessary	necessary	unnecessary	defined by a user	unnecessary

FIGURE 2.20 Classification of B-ISDN bearer service. (*Source:* ITU-T, F811.)

From the viewpoint of ATM Adaptation layer function, five ATM classes of broadband bearer service are described in ITU-T Recommendations in References 8 and 12 (Figure 2.20). There is a timing relationship between sender and receiver sides in traffic class A. Class A service has a constant bit rate (CBR) and provides connection-oriented services. An example of a class A service is transmission of a 64-kb/s telephone channel over an ATM network. Class B service also has a timing relationship between sender and receiver. It also provides connection-oriented services, but they are variable bit rate (VBR) services.

An example of a class B service is voice traffic and video signals with a variable bit rate coding scheme. Class C service does not have any timing relationship between sender and receiver. It provides connection-oriented service with variable bit rates. An example of this class of service is data transmission with ATM connections. Class X is defined as a class which uses ATM adaptation layer defined by user. Class D provides connectionless type variable bit rate services[8] with no timing relationship between the two terminals. Each of those classes of information flow may be provided on point-to-point, multipoint, broadcast, and multicast connections.

ITU-T defines four AAL types (AAL1, AAL2, AAL3/4, and AAL5)[11] for broadband bearer service classes. AAL1 is recommended for the transfer of CBR traffic.[13] AAL2 has been standardized for supporting voice and telephony over ATM (VTOA) technique. While the correspondence between service and AAL is still under study except for CBR traffic, in practice, AAL5 is widely used for IP packet transmission and frame relay service.

2.3.2.3 Virtual Channel

ATM uses a very small header compared with other packet networks, e.g., the Internet, in order to make efficient use of network resources, and minimize transmission delay. To minimize the header size, ATM networks use the following cell transmission method, which is different from packet networks. In an ATM network, the VCI value attached to a cell in a certain link is not the same as the VCI value of the same cell in the other links. In other words, when an ATM switch transfers a cell from an input port to an output port, the VCI value of the cell is changed according to the switch routing table. The VCI does not indicate a destination address, while an IP address in an IP packet does. This technique leads to a connection-oriented network. It enables one to shorten the cell header size, but makes the necessary connection setup between end terminals before they start communication.

Each cell header has a routing information field, which consists of a virtual channel identifier (VCI) and a virtual path identifier (VPI).[2] An ATM switch transmits cells according to the VCI and VPI values. For simplicity, here we assume that the routing decision of an ATM switch is based on only VCI. In other words, VPIs of all cells always have a fixed value in the following explanation. The role and usage of VPI is explained in Section 2.3.2.4.

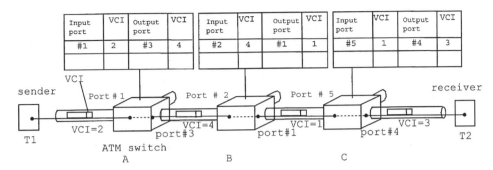

FIGURE 2.21 Classification of B-ISDN bearer service. (*Source:* Aoyama, T. et al., in *ATM Basics,* The Telecommunications Association, Tokyo, 1998. With permission.)

To create a connection in an ATM network, we need to set up routing tables of switching nodes along the connection. A connection that is based on cell routing information in routing tables of ATM switches is called a virtual channel. The setting up of routing tables of ATM switches along a VC route is called "VC setup," and the removal of all information concerning the VC from routing tables is called "VC release." Figure 2.21[9] shows the cell transmission mechanism in an ATM network. In this network model, we have three ATM switches: A, B, and C. Optical fibers (links) are installed between those switches. Each switch has some output ports, each of which is connected to an input port of another switch by an optical fiber. For example, an output port 1 of switch B is connected to port 5 of switching node C.

To transfer a cell from terminal equipment T1 to T2, you need to complete routing information in routing tables at ATM switches along a routing path. In a routing table, there is necessary information to transfer input cells to the neighboring switch through an adequate output port. In Figure 2.21, the routing table of switching node A has the following information: "If a cell came from input port #1 with VCI = 2, transfer it to output port #3 with VCI = 4." The routing table of switch B has this information: "If a cell came from input port #2 with VCI = 4, transfer it to output port #1 with VCI = 1." The routing table of switch C has different information: "If a cell came from input port #5 with VCI = 1, transfer it to output port #4 with VCI = 3." In this way, an ATM switch has a mechanism to transfer cells to an adequate output port according to the routing information in its routing table. Therefore, if terminal T1 sends a cell with VCI = 2, terminal T2 receives it with VCI = 3.

From the viewpoint of VC setup and release, VC service can be categorized into two types: switched VC (SVC) and permanent VC (PVC) services. SVC service needs a signaling system[14-16] to set up and release a VC dynamically in response to customers' demands. PVC services do not require dynamic setup and release because they exist in permanent (or semipermanent) fashion. The network operation system may be used to set up and release PVCs, but it is also possible to use a signaling system for setting up PVCs. While the requirements for VC setup and release techniques differ between those two services, their cell transmission mechanisms in an ATM switch are the same.

ATM is connection-oriented because the cell transmission function of an ATM switch is based on connections as described above. However, connectionless services have been developed by introducing a connectionless server function with cell routing capability based on the destination address.[17] This type of connectionless service is standardized in Reference 8.

2.3.2.4 Virtual Path (VP)

The VP concept[2,18-21] is necessary to construct a large-scale ATM network whose size is comparable to that of the existing telephone networks.

The number of VCs in such a network is tremendous. If the network handles each VC independently, the total processing load for control and management of all VCs will be extremely large, which will lead to slow-moving VC control and very complex operation for VCs. A VP is defined as a preestablished logical "pipe" within which VCs can be set up. Figure 2.22[22] shows VPs allocated in a physical layer (optical

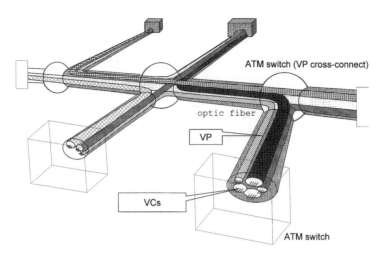

FIGURE 2.22 Virtual paths in a physical layer network. (*Source:* The Committee, *Handbook for Electronics, Information and Communication Engineers,* Institute of Electronics Information and Communication Engineers, Tokyo, 1998. With permission.)

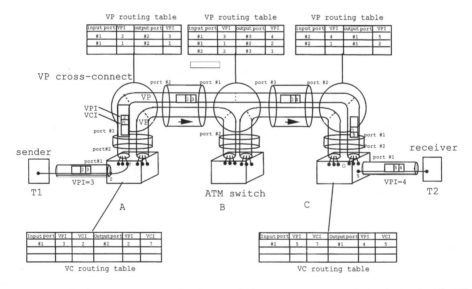

VP routing table

input port	VPI	output port	VPI
#1	2	#2	3
#1	1	#2	1

VP routing table

input port	VPI	output port	VPI
#1	3	#3	4
#1	1	#2	2
#2	2	#3	1

VP routing table

input port	VPI	output port	VPI
#2	4	#1	5
#2	1	#1	2

VC routing table

Input port	VPI	VCI	Output port	VPI	VCI
#1	3	2	#2	2	7

VC routing table

Input port	VPI	VCI	Output port	VPI	VCI
#2	5	7	#1	4	5

FIGURE 2.23 Cell transmission in a virtual path network. (*Source:* Aoyama, T. et al., in *ATM Basics,* The Telecommunications Association, Tokyo, 1998. With permission.)

fiber) network. A VP enables us to remove all processing loads concerning VC setup and release at all transit nodes of the VP.[21]

This reduces the processing load for VC setup and reduces VC setup delay. Moreover, a VP can be a broadband leased line service for a business customer who needs many channels between his or her sites because a VP can provide a bundle of ATM channels all at once.[19]

Figure 2.23[9] shows three VPs (between ATM switch A and B, B and C, and A and C), allocated in an ATM network. Those VPs exist in permanent fashion and they can accommodate many VC connections. A VP is recognized by the VPI of the cell header at ATM switches. Transit switches of a VP (VP cross-connects) do not need to read VCIs. Those switches transfer cells by utilizing the VP routing table referencing the VPI of the cell header. Cell transmission mechanism within a VP is achieved by creating entries of a VP routing table in all cross-connects along the VP route. For example, in Figure 2.23, the routing table of a cross-connect A has the following VP routing information: "If a cell came from input

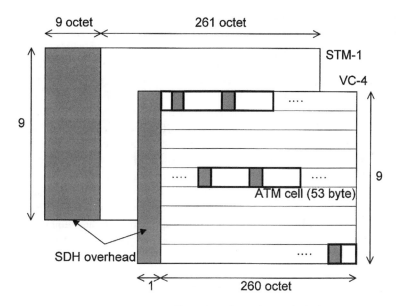

FIGURE 2.24 ATM cells on SDH frame format (VC-4).

port #1 with VPI = 2, transfer it to output port #2 with VPI = 3," and "if a cell came from input port #1 with VPI = 1, transfer it to output port #2 with VPI = 1." With those entries of the routing table, two VPs are set up through a cross-connect at node A. For the VP between node A and C, the routing table of a cross-connect B has the information: "If a cell came from input port #1 with VPI = 3, transfer it to output port #3 with VPI = 4," and the cross-connect C has the information: "If a cell came from input port #2 with VPI = 4, transfer it to output port #1 with VPI = 5." In this way, cross-connects transfer cells to an adequate output port according to the VP routing information in their routing tables. While a cell is being transferred through a VP via those cross-connects, the VCI value is not used for routing and the value is preserved. In this way, "direct information pipes" of VPs are constructed between network provider's or users' ATM switches through an ATM network.

2.3.2.5 Physical Layer

ITU-T recommended three categories of frame format for the ATM physical layer. They are based on (1) SDH (Synchronous Digital Hierarchy) format,[23,24] (2) PDH (Plesiochronous Digital Hierarchy) format,[25-27] which has been used for telephone networks, and (3) cell-based format.

SDH-based network node interfaces (NNI) of VC-3 (its capacity is 48.384 Mb/s), VC-4 (149.760 Mb/s), VC-Xc (X × 149.760 Mb/s), VC-2 (6.784 Mb/s), VC-12 (2.176 Mb/s), and VC11 (1.600 Mb/s) are standardized in recommendation G.707.[23] Figure 2.24 shows a mapping of ATM cells on VC-4 frame. User network interfaces (UNI) of 155.520, 622.080, 1.544, 2.048, 51.840, and 25.6 Mb/s (with a twisted pair cable) are also standardized in Recommendation I.432.[24]

Mapping of ATM cells into PDH frame format of 1.544, 2.048, 6.312, 8.448, 34,368, 44,736, 97.728, and 139.264 Mb/s are described in Recommendation G.804.[25] With those standardized interfaces, PHD-based traditional networks can be infrastructure for ATM transmission capability.

Although a cell-based interface is also recommended, the frame format has not been finalized.

2.3.3 Connection-Oriented Services

2.3.3.1 VC and VP Services

Two kinds of connections, VC and VP, exist in ATM networks. A VC connection provides a single ATM channel for users, while VP connection provides a bundle of VCs all together for a user. Despite this difference, the same traffic controls are used for VP and VC services.

The Quality of Service (QoS) of an ATM connection is described by three parameters: cell loss ratio, cell transmission delay, and cell delay variation.[13] To maintain adequate QoS of all connections, traffic control plays a key role. The traffic control consists of Connection Admission Control (CAC) and Usage Parameter Control (UPC).[13] When a new connection setup is required, CAC decides whether to accept or reject the connection setup according to the available network resources and required QoS. For this purpose, user and network need to exchange information about traffic and the connection QoS. CAC has not been standardized by ITU-T and it is left to the network provider's discretion. UPC is provided at the entrance of an ATM network (user network interface) to identify any violations of the traffic contract between a user and the network. Two different reactions are recommended for the UPC function when it finds excessive cell flow coming into a network, either to (1) discard excessive cells or (2) attach a violation tag to the headers of the excessive cells.[13] To make CAC and UPC functions work, a description of the traffic of a connection is indispensable. When a connection is set up, the user and the network must agree to a contract based on the connection traffic descriptors, which include peak cell rate and cell delay variation tolerance. For VBR connections, the sustainable cell rate and burst tolerance are also considered traffic descriptors. The available 16,384 values of peak cell rate are described by the following equation in I.371.[13]

$$\Lambda = 2^m \left(1 + \kappa/512\right) \quad \text{[cells/second]} \tag{2.1}$$

$$0 \leq m \leq 31$$

$$0 \leq \kappa \leq 511$$

The maximum bit rate provided by ATM connections is in practice bounded by bandwidth limitations of physical layer techniques.

In addition to the fundamental CBR and VBR class, ITU-T defined three other classes of connections — Undefined Bit Rate (UBR), Available Bit Rate (ABR), and ATM Block Transfer (ABT) — as ATM Transfer Capabilities (ATCs).[13] UBR is a class that has no traffic descriptors except peak cell rate. QoS is not guaranteed for UBR connections. An ABR connection has a special feedback congestion control mechanism to adapt the available bit rate to the network status. ABT aims to achieve rapid bandwidth reservation with resource management cells. However, the detailed specifications of ABT connections are under study.

To provide SVC services, a signaling protocol was developed to provide dynamic VC setup and release. Signaling protocols[14-16] for broadband networks have standardized by extending the existing signaling protocol for (narrow band) ISDN. SVC setup delay and connection blocking probability are important quality parameters for users. On a commercial broadband service, practical values of those parameters depend on network designing by service providers.

2.3.3.2 Virtual Path Group Services

One of the most important differences between an ATM network and an STM-based telephone network is the flexibility of connection bandwidths.[21,28] ATMs cell transmission mechanism allows network operators to change the connection bandwidth dynamically. Making the best use of the bandwidth flexibility of VPs, a virtual path group (VPG)-based enterprise networking service was proposed by Chan[29] et al. (Figure 2.25) and is being studied as an attractive broadband service for enterprise networking.

A VPG[29,30] is defined as a logical link within the public ATM network. One is permanently set up between two VP switches or between a VP switch and a customer premises network switch that acts as a customer access point for the service. It accommodates a bundle of VPs that interconnect customer access points. The service provider allocates bandwidth to a VPG, which defines the maximum total capacity for all VPs within the VPG. A VPG-based enterprise networking service consists of interconnected VPGs.

In order to guarantee QoS in the carrier's network, UPC functions are required at the entrance of each VPG. Note that there is no need for a VPG identifier in the ATM cell header, since cells are transmitted

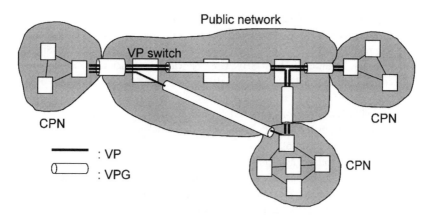

FIGURE 2.25 VPG-based enterprise networking service.

by VP switches based on their VPI. Only the network management system must know about the routes of the VPGs, their assigned bandwidth, and the VPs associated with them. VPs and VPGs are set up by the network management system of the service provider during the service configuration phase.

The VPG concept enhances the customer's ability to control VP bandwidth. A customer can change the VP bandwidth, within the limits of the VPG capacities, without affecting other customers or network providers. The VPG bandwidth can be shared by VPs with different source destination pairs, without negotiation between the customer and the service provider.

2.3.4   Connectionless Service

A B-ISDN connectionless data service by definition allows the transfer of information among service subscribers without the need for end-to-end call establishment procedures between users.[8] This service provides high-speed transfer of variable-length data units without connection establishment procedures between users. These data units can be transferred from a single source to a single destination, or from a single source to multiple destinations. Each data unit includes a source address validated by the network.

This service uses addresses based on the E.164[31] ISDN numbering plan. Multicast communication requires an addressing scheme using group addressing, which is a mechanism that allows the same data unit to be transmitted to several intended recipients.

This service is offered on B-ISDN over ATM connections between the subscriber and a connectionless service node called a "connectionless server."[7] This service will be supported on the B-ISDN user–network interface based on ATM.

References

1. ITU-T Recommendation I.121, Broadband Aspects of ISDN.
2. ITU-T Recommendation I.361, B-ISDN ATM Layer Specification.
3. ITU-T Recommendation G.707, Synchronous Digital Hierarchy Bit Rates.
4. ITU-T Recommendation 708, Network Node Interface for the Synchronous Digital Hierarchy.
5. ITU-T Recommendation 709, Synchronous Multiplexing Structure.
6. Sato, K. and Tokizawa, I., Flexible asynchronous transfer mode networks utilizing virtual paths, in *Proc. ICC'90*, 318.4, 1–8, 1990.
7. ITU-T Recommendation I.240, Definition of Teleservices.
8. ITU-T Recommendation F.812, Broadband Connectionless Data Bearer Service.
9. Aoyama, T., Tsuboi, T., Yamanaka, N., Kanayama, Y., Uematsu, H., Ohta, H., Hadama, H., Hirano, M., and Ito, T., *Basic ATM*, The Telecommunications Association, Tokyo, 1998.
10. ITU-T Recommendation I.363, B-ISDN ATM Adaptation Layer (AAL) Specification.

11. ITU-T Recommendation I.363.2, B-ISDN AAL, Type 2 Specification.
12. ITU-T Recommendation F.811, Broadband Connection Oriented Bearer Service.
13. ITU-T Recommendation I.371, Traffic Control and Congestion Control in B-ISDN
14. ITU-T Recommendation Q.2931, B-ISDN User–Network Interface Layer 3 Protocol.
15. ITU-T Recommendation Q.2761, Functional Description of the B-ISDN User Part of Signaling No 7.
16. ATM Forum, ATM User-Network Interface Specification Version 3.1, 1994.
17. ITU-T Recommendation B.191, B-ISDN Numbering and Addressing.
18. Kanada, T., Sato K., and Tsuboi, T., An ATM based transport network architecture, in *Proc. Int. Workshop of Burst/Packetized Multimedia Commun.,* Osaka, Nov. 1987.
19. Tokizawa, I., Kanada, T., and Sato, K., A new transport network architecture based on asynchronous transfer mode technique, in *Proc. ISSLS'88,* 11.2, 217–221, Sept. 1988.
20. ITU-T Recommendation F.813, Virtual Path Service for Reserved and Permanent Communications.
21. Ohta, S. and Sato, K., Dynamic bandwidth control of the virtual path in an asynchronous transfer mode network, *IEEE Trans. Commun.,* 40(7), 1239–1247, 1992.
22. The Handbook Committee, *Handbook for Electronics, Information and Communication Engineers,* IEICE, Tokyo, 1998.
23. ITU-T Recommendation G.707, Network Node Interface for the Synchronous Digital Hierarchy (SDH).
24. ITU-T Recommendation I.432, B-ISDN User-Network Interface — Physical Layer Specification: General Characteristics.
25. ITU-T Recommendation G.804, ATM Cell Mapping into Plesiochronous Digital Hierarchy (PDH).
26. ITU-T Recommendation G.704, Synchronous Frame Structures used at 1544, 6312, 2048, 8488 and 44736 kbit/s Hierarchical Levels.
27. ITU-T Recommendation G.832, Transport of SDH Elements on PDH Networks — Frame and Multiplexing Structures.
28. Wang, W., Sadawi, T. N., and Aihara, K., Bandwidth Allocation for ATM Networks, in *Proc. ICC'90,* 306B, 2.1–2.4, Atlanta, April 1990.
29. Chan, M. C., Hadama, H., and Stadler, R., An architecture for broadband virtual networks under customer control, in *Proc. NOMS'96,* Kyoto, April 1996.
30. Hadama, H., Kawamura, R., Izaki, T., and Tokizawa, I., Direct virtual path configuration in large-scale ATM networks, in *Proc. INFOCOM'94,* Toronto, 2b.3, 201–206, April 1994.
31. ITU-T Recommendation E.164, The International Public Telecommunication Numbering Plan.

2.4 Mobile and Wireless Telecommunications Networks

Michele Zorzi

2.4.1 Introduction

Even though the radio was invented more than 100 years ago, it was 10 to 15 years ago that widespread use of wireless communications began. Before then it was mainly used by police and military, but now the capability of communicating without being constrained to a fixed physical location is within every-one's reach. The booming industry, which has been growing at such an astounding rate in the past couple of decades, is a very good indicator of how much such capability meets the human desire for freedom.

Besides being a very exciting and profitable business, wireless and mobile communications has been an extremely rich field for research, due to the many difficulties that the wireless environment presents, and due to the ever-increasing users' demand for more, newer, and better services. People want to benefit from wireless services, and as the services become more and more widespread, increasing expectations challenge manufacturers and providers to search continuously for new solutions.

From the technical point of view, the past few years have seen tremendous advances in the research, development, and design of mobile radio systems, with many more are expected in the near future. The purpose of this section is to give a brief overview of this exciting field and to provide some basic information about the problems inherent in the wireless environment, the solutions on which the implementation of the systems in use today are based, and the research ideas that are being developed and will be the foundation of the telecommunications systems of tomorrow.

2.4.2 The Key Technical Challenges

The idea of being able to communicate with other people or with machines (e.g., databases or video servers) without being constrained to a fixed location is of course very appealing. Unfortunately, mobility introduces new management issues and the mobile wireless environment is very hostile when compared with classic telecommunications systems. In particular, radio channels are inherently prone to severe impairments (especially when the transmitter and/or the receiver are moving) and have limited band-width. In this section, we describe the key features which make the wireless environment so challenging.

2.4.2.1 Channel Unreliability

Probably the most notorious characteristic of wireless channels in mobile communications is their unreliability. Due to a number of physical factors, signals propagating through such channels are subject to severe impairments. The effect of propagation on the transmitted signal can be adequately described by three phenomena, namely, path loss, shadowing, and multipath fading.

Path loss accounts for the signal attenuation due to the physical distance, r, between transmitter and receiver. In free space, the received power is proportional to r^{-2}, whereas in groundwave propagation there is an additional effect due to the combination of the direct (free-space) propagated wave and its replica as reflected by the earth surface. In this case, the path loss behavior is better described by the function $r^{-\eta}$, where η depends on a number of properties of the environment, and usually takes values between 2 and 5. More accurate models account for other effects, by introducing some correction factors in the relationship between received power and distance to model, for example, effects similar to guided propagation (as happens along streets with tall buildings). Such models are usually employed in system simulations, and their parameters can be chosen by direct measurements of the environment to be studied.

Another macroscopic effect, i.e., an effect which does not change much over distances on the order of many wavelengths, is the shadowing effect (sometimes called shadow fading) which, as the name suggests, accounts for obstructions along the propagation path (e.g., buildings or trees), which may cause significant signal attenuations. As these attenuations combine with each other, shadowing is often modeled as having Gaussian distribution when path loss is expressed in decibels. This model (also called *log-normal shadowing*), has been found to match field measurements very well.

On a much finer scale, signals are subject to multipath fading. Due to various objects and reflectors, the signal does not directly propagate to the receiver, but rather is scattered in many directions so that multiple versions of it, coming from different angles and with different attenuations and delays, are received. If the delay differences among the various reflections are much smaller than the timescale at which the signal is detected (e.g., the bit duration), then the channel impulse response is very short (in time), and the effect of multipath in this case is only an additional attenuation (often modeled as a Rayleigh or Rician random variable). Such a situation corresponds to the transfer function of the channel being essentially constant over the signal bandwidth, and is called *flat fading*. If, on the other hand, the delay distribution on the different paths is spread out (e.g, it exceeds the bit duration), then the channel changes the spectral shape of the transmitted signal, which is therefore distorted. As different frequencies may undergo different attenuations, this case is called *frequency-selective fading*, and is typical of wideband signaling. For digital communications, some sort of equalization is usually needed in this case. Also, the fact that not all spectral components fade together can be exploited as a form of frequency diversity.

Since in a wireless system terminals are allowed to move, the channel impulse response changes in time in general. Even in the case in which terminals do not move, the environment around them may change (e.g., vehicles or people passing by). Path loss and shadowing are relatively constant over significant distances (tens of meters or more), whereas multipath fading can usually be considered independent at distances on the order of the wavelength (which is equal to the speed of light divided by the carrier frequency, and is $\lambda = 33$ cm at 900 MHz), and therefore changes very rapidly. This means that the channel, besides being subject to significant impairments, is also very unstable and sometimes may even suddenly disappear. This makes wireless communications very different from communications over more stable and predictable channels such as cables and fibers or even microwave links. Stabilizing the radio channel and overcoming problems posed by the propagation environment are major tasks in the design of mobile communications systems.

Typical techniques which are used in today's systems to counteract the channel behavior are power control and diversity. Power control is based on the concept of trying to boost the power if the channel is currently undergoing large attenuations and to reduce the power when the channel experiences good conditions. This ideally results in all users transmitting just as much power as they need in order to meet their quality objectives, and therefore minimizes the interference level caused to others, under quality of service constraints. Implementation of the power control mechanism requires knowledge of the channel attenuation. In open-loop power control schemes,[*] the transmitter autonomously estimates the channel attenuation (either measuring the received signal in full duplex communications, or through the use of a pilot channel) and adjusts the transmitted power to compensate for such attenuation. In closed-loop power control, based on the quality of the received signal the receiver sends feedback to the transmitter indicating whether the power should be increased or decreased. Open-loop power control is easier to implement but cannot usually compensate for multipath fading, which is not reciprocal if the transmitted and received signals occupy different frequency bands.[**] Therefore, to be able to compensate for multi-path, closed-loop power control is usually envisioned. On the other hand, since multipath fading is rapidly varying, this results in potentially inaccurate channel estimation and significant overhead due to the need for frequent updates.

A classic means of counteracting multipath fading is through diversity, which is based on the fact that multiple copies of a signal received at antennas sufficiently apart from each other (a wavelength is usually considered sufficient, at least at the mobile terminal) experience almost independent fading, so that it is very unlikely that all antennas receive weak signals at the same time. More-advanced diversity systems try to combine all received signals optimally rather than just selecting one of the signals.

Robust modulation techniques can also be used to maintain transmission quality even under severe propagation conditions. A well-known example for digital communications is spread spectrum modulation, where signal processing techniques are used at the transmitter to spread the signal power over a wide band and at the receiver to despread it. At the same time, the power spectral density of other spread spectrum users is essentially unchanged, whereas the power of narrowband interferers is spread over the whole bandwidth. Therefore, when filtering over the intended signal bandwidth after despreading, the power interference is greatly reduced. The robustness of this technique to interference comes at the expense of bandwidth expansion.

Finally, another well-established means to fix the channel unreliability in digital communications is through the use of error control techniques, whereby redundant symbols are added to the information stream to be transmitted, and are used at the receiver for enhanced detection. Error correction schemes, traditionally designed for channels with additive white Gaussian noise, are applied in this environment jointly with interleaving, whereby the bits are scrambled prior to transmission and descrambled at the receiver, thus breaking the error correlation inherently present because of the channel memory. The interleaving/deinterleaving operation introduces some delay which effectively limits the extent to which

---

[*]The term *open-loop* is used here to indicate that there is no *explicit* signaling in the return direction to control the transmitted power, even though a return channel is in fact present and used to estimate the attenuation.

[**]This approach to separating transmitted and received signals is referred to as Frequency Division Duplexing (FDD).

correlation can be removed: in the presence of fast varying channels interleaving usually works very well, whereas for slow channels it is basically ineffective (but in this case very accurate power control can be performed due to high channel predictability).

2.4.2.2 Limited Bandwidth

Another important difference between wireline and wireless networks is the amount of bandwidth available. One reason for this is that practical communications schemes today operate at UHF frequencies (typical bands are 900 MHz and 1.8 GHz) and additional frequency bands being considered for future systems are mostly in the 1 to 6 GHz range (even though 17 GHz has been designated for wireless LANs in Europe, and there are proposals for systems at frequencies up to 60 GHz). The amount of bandwidth available in these ranges is not large, since the radio spectrum itself is inherently a public resource and is already very crowded due to the presence of other services (e.g., broadcast TV, military communications, and point-to-point radio links). Therefore, wireless systems are always limited by interference, which dictates the amount of bandwidth available and imposes restrictions on how it can be used, whereas interference effects in wireline networks, although present (e.g., cross talk), are less significant in limiting the capacity. Important consequences of this fact are (1) the limited bandwidth available for wireless services, which constrains both the capacity (i.e., the number of users the system can accommodate) and the user's capabilities (i.e., the bandwidth which may be instantaneously available to a user) and (2) the need to conform to regulatory constraints. The fact that the transmission medium in radio communications is inherently public also raises concerns about its security and privacy. Means need to be provided to guarantee that telephone conversations or data transactions cannot be eavesdropped on (matching at least the security achieved in wireline networks). In addition, in this environment, privacy about the user's location and equipment security (e.g., to avoid unauthorized use of cellular telephones or of billing identification numbers) needs to be ensured.

Because of the shortage of the radio spectrum, a major research objective toward improved design of wireless networks has been to provide more-bandwidth-efficient techniques, to increase the number of users who can be accommodated for a given system bandwidth. A major breakthrough in this sense was the invention of the cellular concept, devised at Bell Laboratories in the 1970s. Before then, the number of available channels was very limited, because of the limited spectrum available. Cellular systems are based on the same concept used for many years in broadcasting, where the same channel can be used in noninterfering geographic locations. This allows a capacity increase which is in principle unlimited, even though other issues make it infeasible to reduce the cell size beyond a certain limit.

Typical cellular systems today are based on circuit switching, which has been a very successful paradigm for speech communications. The total system bandwidth is split into a number of channels, and these channels are in turn assigned to cells, according to the frequency reuse plan, if any. Therefore, each cell has a number of channels, and cells using the same channels are separated geographically by a sufficiently large distance, thereby causing small interference to each other. Cells using the same radio channels belong to the same cochannel set, and the number of distinct cochannel sets is sometimes called *cluster size* (Figure 2.26). How the system bandwidth is subdivided into channels depends on the multiple access technology of choice. The early systems used Frequency Division Multiple Access (FDMA), in which the system bandwidth is sliced in the frequency domain to achieve orthogonal (nonoverlapping in frequency) channels. With the introduction of digital transmission in wireless communications systems, other techniques became possible, most notably Time Division Multiple Access (TDMA) and Code Division Multiple Access (CDMA). In the former, the time axis has a framed structure, and different channels use the same bandwidth but at different times, thereby achieving orthogonality in the time domain. In CDMA, spread spectrum modulation is used to allow multiple users to communicate simultaneously in the same radio channel, and the separation is obtained through signal processing techniques: if all signals are synchronously received, and in the absence of multipath, it is possible to achieve perfect orthogonality in the code domain, whereas some residual interference will be present if multipath is present or synchronization is not possible (due, for example, to different propagation delays). Hybrid multiple access

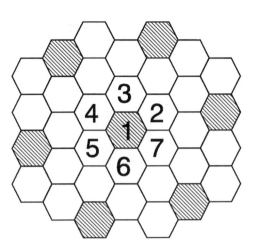

FIGURE 2.26 Example of cellular layout with seven-cell frequency reuse plan. Cells with different numbers use different radio channel sets. The shaded cells all use channel set number 1. (*Source:* Borgonovo, F. et al., *IEEE JSAC*, vol. 14, pp. 609–623, May 1996. With permission.)

schemes can also be envisioned, and are actually used in many practical system implementations. A very important advantage of CDMA with respect to TDMA and FDMA is its inherent robustness to interference, which makes it possible to use the same frequency band in all cells. This eliminates the need for frequency planning (which is a major undertaking, especially in systems with small cells) and achieves higher capacity, since the fact that the same frequency can only be used in distant cells results in a significant capacity penalty in TDMA and FDMA, where the bandwidth assigned to each cell is only a fraction of the total.

A major improvement in the spectrum utilization efficiency has been achieved with the introduction of digital communications systems. Digitized speech can be easily compressed, so that, for example, the 30 kHz necessary for a single FM analog voice channel (used in the early systems) can support three digital voice channels with data rate of 16 kbps. The original pulse code modulation technique, where voice signals were sampled and quantized, has been successively refined, based on exploiting the large amount of redundancy of the source, and schemes requiring fewer and fewer bits per second to transmit a single telephone call reliably have been developed. An even more dramatic compression rate can be achieved by vocoders which, instead of compressing the sampled version of the actual speech waveform, model the mechanics of how the sounds are produced as linear filters, and transmit their parameters which are then used at the receiver to generate the same sounds. The currently used VSELP and RPE belong to this latter class.

The increased compression rates, along with the use of bandwidth-efficient digital modulation schemes such as QPSK or Gaussian MSK, make it possible, for example, for the replacement of each 30-kHz analog channel with three TDMA channels, thus tripling the system capacity. An even larger gain may be obtained when using CDMA, because of the possibility of reusing the spectrum everywhere, so that the capacity gain offered by CDMA with respect to the early systems is estimated to be on the order of three to five in circuit switched systems.

2.4.2.3 Support for Mobility

The third important technical challenge besides channel instability and limited spectrum is the need to support user mobility. Even though mobility in wired networks (as in Mobile IP) and wireless systems with no mobility (as in some wireless LAN and wireless local loop applications) have been considered, the issue of mobility remains largely an essential feature of wireless networks. A whole new approach is needed to be developed in order to locate and track users and to provide a means to change the point of attachment to the network, even during an active connection. Implementation of these capabilities requires a considerable amount of signaling and network management software.

Two basic problems need to be solved to guarantee effective mobility management, i.e., location and handoff. The location problem arises because the system needs to know where a user is at any time, in order to be able to forward incoming calls correctly. In current cellular systems, the network is divided into service areas, each composed of a number of cells managed by a central unit, called a mobile switching center (MSC).* Each MSC has two databases, called home location register (HLR) and visitor location register (VLR). The HLR contains information regarding users who are registered as subscribers in the area, whereas the VLR contains information regarding users who are registered as subscribers somewhere else but happen to be roaming in the area. Every time a user registers as a roamer in a new area, some signaling takes place between the MSCs involved, and the databases are appropriately updated. Basically, the information contained in those registers allows the network to locate the user, i.e., to identify the set of cells where the user can be found at any given time. In fact, every time the network needs to locate a user, the HLR at the user's home MSC will be looked up. If the user is not in that area, the HLR contains information on where the user can be found. If an incoming call for that user arrives, it may be redirected to the MSC of the area in which the user can be found. The MSC will then instruct the central transmitters, called Base Stations (BS), serving all cells in that area to send a paging message for the user, who will reply to the BS of the appropriate cell. This response completes the location procedure, and the call setup can be completed at the designated BS.

To know where a terminal is at any time, the network must be able to track it while it is moving. For this purpose, signaling messages need to be frequently sent to update the network's information about the user's location. In particular, the system is divided into location areas, and every time the user crosses the boundary of one of those, a location update message needs to be sent. Large location areas result in little update signaling but large paging overhead (all cells in the area must send a page for the user), whereas the opposite is true for small location areas.

The above location procedures take place when the user is not sending information (even though the terminal is, but just for signaling purposes). However, users may move while communicating as well. The second major problem to be solved in a mobile communications system is therefore to handle ongoing calls when, because of mobility and/or changing channel conditions, the user needs to change the point of attachment to the network. A typical situation, although by no means the only possible, occurs when a user moves away from the BS at which the call was established, going toward another BS. At some point, the attenuation experienced on the path toward the first BS may become too severe, to the extent that the transmission quality is no longer adequate. A system with no support for mobility would drop the call at that point. In cellular systems, this situation would trigger a handoff procedure, whereby the mobile terminal and/or the serving BS look for an alternative BS which can guarantee adequate transmission quality. If no such alternative can be found, the call is dropped. Otherwise a handoff occurs; i.e., the connection to the old BS is released and a connection with the new BS is established.

The details of the handoff procedure depend on the system implementation. In general, however, it involves periodic measurements of the quality (typically, the signal strength or the signal-to-interference ratio) of the channel currently being used and of the paths to a number of candidate BS. Whenever the need arises, those measurements make it possible to determine which BS should be chosen as the destination of the handoff procedure. The decision about when a handoff is performed and which BS is selected can be made either by the network alone, as in classic cellular systems, or with the cooperation of the mobile terminal, as in the case of mobile assisted handoff.

2.4.3   Current Wireless Cellular Systems

The following are the main elements of all cellular systems (Figure 2.27):

- Mobile terminal, which contains a radio transceiver and a processor;
- Base station, which manages communications to and from the users belonging to its own cells, and is composed of a radio part and a base station controller;

---

*Different names are used in different systems, but the basic functionalities remain essentially the same.

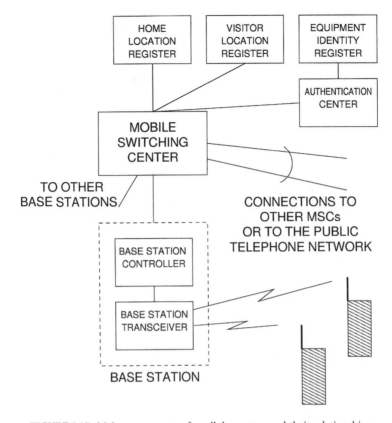

FIGURE 2.27 Main components of a cellular system and their relationships.

- Mobile switching center, which controls a number of BS (typically a few tens), provides additional functionalities such as call management and diagnostics and fault recovery, and is the interface to and from the public switched telephone network;
- Databases, such as the already mentioned HLR and VLR, plus the equipment identity register and authentication center, used for security and authentication purposes.

Also, the following control channels are present in all systems:

- Broadcast channels (base-to-mobile), where general control information is transmitted to all mobiles;
- Paging channels (base-to-mobile), used to notify a mobile user of an incoming call;
- Random access channels (mobile-to-base), used for mobile-initiated call requests.

These elements provide essential communications services and network management functionalities, which enable proper system operation. In particular, they provide location and mobility management. Therefore, although in different forms and possibly with different names, they are present in all cellular systems.

2.4.3.1 First-Generation Cellular Systems

The early cellular systems were based on analog FM. Examples of such systems are the Advanced Mobile Phone Service (AMPS) in the U.S. and the Total Access Communications System (TACS) in Europe. As an example, the AMPS system operates in the vicinity of 850 MHz. Two competing operators are allocated 25 MHz of bandwidth each, corresponding to 833 30-kHz radio channels. Since in these systems the modulation format for the voice channels is analog, FDMA must be used for multiple access. With a seven-cell frequency reuse plan (which guarantees adequately low interference), this results in about 60 full-duplex voice channels per cell.

First-generation systems provide supervisory signaling during voice transmission, to verify proper connection between base station and the mobile terminal. This is achieved by sending tones at frequencies above the audio bands. In addition, a wideband signaling capability is provided (e.g., for use during handoffs) by means of the *blank-and-burst* technique, in which bursts of data are sent for short (almost unnoticeable) durations while speech is muted.

During a call, the BS makes signal-strength measurements of its own mobiles. Also, a receiver is used to monitor signal strength of users in other cells. Based on the measurements forwarded by the BS, the MSC makes decisions about call handoffs whenever needed.

A major drawback of first-generation standards was their incompatibility, due, for example, to different channel spacing or frequency band of operation. This problem was especially severe in Europe, where many different countries had different standards, preventing users from roaming abroad. This is one of the reasons that the second-generation pan-European GSM has been so successful both in the standardization process and commercially.

2.4.3.2 Second-Generation Cellular Systems

The main difference between first- and second-generation cellular systems is that the modulation format used in the latter is digital rather than analog. This introduces considerable flexibility in handling the information signals, allowing the use of error control coding; source coding, and compression, TDMA, and CDMA. Also, digital modulation is inherently more robust to channel impairments than frequency modulation, even though channel errors may result in artifacts which are less tolerated by human perception. Also, digital transmission increases the compatibility with the fixed portion of the telephone network (which is mostly digital), allows greater switching and routing flexibility, and lends itself to implementation in VLSI. Most importantly, the digital transmission format is more suitable for data communications, which is expected to be a major player in future systems and is already supported by some second-generation systems. The coverage provided by second-generation systems is not yet universal, but roaming is fully supported at least within Europe (this was a major goal of GSM), within North America, and within Japan.

There are basically two different types of second-generation systems, depending on whether they are TDMA based or CDMA based. The first class includes standards such as IS-54/136* (the "digital AMPS") in the U.S., GSM in Europe, and PDC in Japan. The only CDMA-based standard is IS-95. All these systems use hybrid multiple-access schemes. For example, in GSM each radio channel (with a data rate of 270 kbps) accommodates eight users in a TDMA structure, but radio channels are divided through FDMA. Also, in IS-95 the basic radio channel is 1.25 MHz wide, so that ten such channels in each direction can be supported in an FDMA arrangement for a total system bandwidth of 25 Mhz.

2.4.3.2.1 TDMA-Based Systems

In GSM, which is an example of TDMA-based second-generation system, the voice signal consists of a 13-kbps stream produced by the RPE speech coder. This bit stream is subdivided into two classes. The bits that are more critical in speech reconstruction at the receiver are given protection and encoded with a rate one half convolutional code, whereas the others are transmitted unprotected. Additional protection is provided by the interleaving mechanism which breaks the channel error correlation and makes the use of the error correction code more effective. This coding scheme introduces an additional 10 kbps to the original bit stream which, with the further addition of synchronization bits, training sequences for equalization, and guard times, sums up to a total of about 33.8 kbps, which is the channel rate for each user (a 156-bit time slot every 4.6 ms).

---

*The only significant differences of IS-136 from IS-54 are higher rate control channels and a richer set of services offered.

TABLE 2.5   Summary and Comparison of Second-Generation TDMA-Based System Parameters

	Europe (GSM)	North America (IS-54/136)	Japan (PDC)
Access method	TDMA	TDMA	TDMA
Carrier spacing	200 kHz	30 kHz	25 kHz
Users per carrier	8 (16)	3 (6)	3 (tbd)
Modulation	GMSK	$\pi/4$-DQPSK	$\pi/4$-DQPSK
Voice codec	RPE 13 kbps	VSELP 7.95 kbps	VSELP 6.7 kbps
Voice frame	20 ms	20 ms	20 ms
Channel code	Convolutional	Convolutional	Convolutional
Coded bit rate	22.8 kbps	13 kbps	11.2 kbps
TDMA frame duration	4.6 ms	20 ms	20 ms
Interleaving	40 ms	27 ms	27 ms
ACCH	Extra slot	In slot	In slot
Handoff	MAHO	MAHO	MAHO

In addition to the common control channels already mentioned (broadcast, paging, and random access), GSM provides two additional channels, called Associated Control Channels (ACCH). The slow ACCH (SACCH) is assigned as part of a connection and is implemented by using dedicated frames in the GSM multiframe structure. The data rate on the SACCH sums up to a little less than 1 kbps. A typical use of the SACCH is the transmission of information about the signal strength received by the mobile from a number of surrounding BSs, to be used in the event that a handoff is needed. The fast ACCH is not continuously present, and is activated only under exceptional circumstances (e.g., while performing a handoff) when the transmission capacity normally assigned to the voice signal is preempted and reassigned to control traffic. Therefore, unlike in the analog systems, when a call is initiated a user is assigned a two-way control channel (SACCH) in addition to the two-way voice channel.

Mobile Assisted Handoff (MAHO) is implemented in GSM. Each cell has a broadcast channel that sends a pilot signal. Via its SACCH, a user learns the identities of the broadcast channels of the neighboring cells. During idle time, taking advantage of the TDMA structure where a user is active only one eighth of the time in either direction, each user listens to the broadcast channels of the closest 32 cells and reports data about the six strongest to its own BS via the SACCH. This information is periodically updated and used to determine candidate BS if the need for a handoff arises.

The various base stations are interconnected through a backbone network, which also includes mobile switching centers and their databases. To manage a complex network with a variety of features, GSM proponents have chosen to rely on some already defined procedures and interfaces, possibly with some adaptation for the mobile environment. GSM uses the CCITT signaling system number 7, originally defined for signaling in wireline ISDN.

An evolution of GSM is the DCS1800 system, whose architecture is very similar to what has been described so far. Differences with GSM are the frequency range (1800 MHz rather than 900 MHz), the amount of spectrum allocated (150 MHz rather than 50 MHz), the cell size (microcells rather than macrocells). The DCS1800 is targeted toward the so-called PCS services (which potentially include data, fax, E-mail, and various other services), whereas GSM has as its primary goal cellular telephony.

Other TDMA-based standards include IS-54 and PDC. Similar to GSM, they use speech compression and channel coding, organize transmissions in frames, provide associated control channels, and support MAHO. Specific differences among the three systems are detailed in Table 2.5.

2.4.3.2.2 IS-95
As already mentioned, IS-95 is based on CDMA, which takes advantage of spread spectrum modulation to guarantee reliable communication in the presence of interference and multipath. The particular interference rejection properties of spread spectrum (whose description is beyond the scope of this section) allow in principle the use of the same frequency band everywhere throughout the system, so that user channel separation is achieved by the appropriate use of quasi-orthogonal spreading sequences.

Because the bandwidth of the transmitted signal is much larger than the bandwidth of the information signal (a factor of 128 is specified in IS-95), CDMA can enjoy the benefit of frequency diversity, since different spectral components fade independently.* In the time domain, this corresponds to being able to discriminate among the various replicas of the signal arriving at the receiver with different delays, which can therefore be constructively added rather than randomly interfere. The particular structure that is able to combine signals in this way is called the RAKE receiver.

Another advantage of spread spectrum systems is the possibility of trading bandwidth expansion for coding redundancy. For example, if the channel bandwidth is 100 times larger than the user's signal bandwidth, it is possible to transmit such signal uncoded with a spreading factor of 100, or coded with a rate-1/2 code and a spreading factor of 50. This allows the use of coding (which is essential in certain applications) without further increase in the bandwidth of the transmitted signal.**

Due to the fact that all users share the same bandwidth and that interferers are essentially seen by the intended receiver as Gaussian noise, any techniques to reduce interference or improve detection result automatically into a capacity gain. That is, if one can find a means of reducing the interference by 50%, twice as many users can be admitted to the system. Similarly, if a detection scheme achieves the same performance with a signal-to-interference ratio 3 dB lower, then twice as many users can be admitted. Also, this characteristic provides graceful performance degradation as users join the system, so that CDMA systems have "soft capacity," in the sense that a user can always be admitted to the system if a slight performance degradation for all users is accepted. This is not possible in static resource assignment systems (such as TDMA and FDMA with fixed channel allocation), where there is a constant number of physical channels per cell, and one more cannot be added.***

Typical techniques to enhance detection are the use of antenna diversity and of error correction coding (convolutional codes with rate 1/2 on the base-to-mobile link and rate 1/3 on the mobile-to-base link are specified in IS-95). Interleaving is also used to improve the error correction coding performance. Typical techniques to reduce interference are cell sectorization (120° antennas are used, which pick up one third as much interference as omnidirectional antennas) and silence suppression (during periods when the user is silent, which amount to more than half the time in typical telephone conversations, the transmitter switches to a low-rate mode used for signaling and synchronization purposes only).

A key requirement for spread spectrum systems to work properly is that all signals reach the receiver with the same power, as power imbalances may greatly reduce the system capacity. To this purpose, power control is to be used in the mobile-to-base direction, to compensate for the channel attenuation. If perfect power control were possible, all signals coming from mobiles in the same cell would have exactly the same power at the base station. On the other hand, power control errors do occur and interference from other cells cannot be controlled, and these effects degrade the performance of the system. In any event, accurate power control is a critical requirement of CDMA-based systems for proper operation, unlike in other systems where power control may increase the performance to some extent, but is by no means critical. Open-loop power control is implemented based on the measurements of the pilot signal transmitted by the BS. Closed-loop power control is also specified in IS-95, with the BS sending one-bit power control commands every 1.25 ms, indicating whether the mobile should turn up or down the transmitted power. A power dynamic range of 80 dB is specified. In the other direction (base-to-mobile), a simpler power control scheme is used, whose main goals are improved transmission quality and minimized interference.

---

*In IS-95, the transmission bandwidth is only 1.25 MHz, and therefore the advantages of frequency diversity cannot be fully exploited, since typical values of the coherence bandwidth (which gives the distance between frequency components fading independently) of the wireless channels are in the range 1 to 5 MHz. Other wider-band CDMA systems will experience larger diversity gains.

**On the other hand, in TDMA-based systems the use of coding requires bandwidth expansion, unless Trellis Coded Modulation is used.

***However, dynamic resource assignment has the potential to offer soft capacity even in these systems.[1]

As in other systems, paging and random access channels are present. The broadcast channels are used by the base stations to transmit a pilot signal, which is used by the mobiles as a power control reference and also as a coherent reference to be used for signal detection. A synchronization channel is also provided, which is used in the acquisition phase as a timing reference.

MAHO is specified in IS-95, where a mobile monitors the pilot signals coming from a number of cells and reports the results of these measurements to the controller. A unique feature of IS-95 is the so-called soft handoff, where during the handoff procedure a mobile may be connected to more than one BS simultaneously. This allows the use of macrodiversity reception (the network can combine signals received by multiple BS) and provides a means to avoid interruptions as, unlike in other systems, the connection to the new BS is set up before the connection with the old BS is torn down ("make before break").

2.4.3.3   Cordless Systems

Another class of very popular devices is that of cordless systems. The first generation of such systems is very simple, consisting of a base which communicates with a single portable unit, and basically represents a wireless extension of the household telephone. For their convenience, such systems have been very successful, but they are subject to interference and lack of security. Also, with the emergence of the concept of Personal Communication Services (PCS), the desire for digital devices and for increased capability (including some mobility) has been growing.

Second-generation cordless systems have therefore been introduced, in which the cordless phone becomes part of a geographically distributed network and can connect to various attachment points, instead of being constrained to its own base equipment. For example, a service called Telepoint has been offered in the U.K., where a cordless phone can connect to the wireless equivalent of public telephone booths.

Probably the most popular of second-generation systems is the Digital European (Enhanced) Cordless Telecommunication (DECT) system. It uses radio channels spaced by 1.728 MHz, each with a data rate of 1.152 Mbps and capable of supporting 12 TDMA channels. All carriers are used throughout the system (no frequency reuse plan), and channels are dynamically selected based on measurements at the time of connection setup. Speech coding uses ADPCM which does not offer good compression performance (32 kbps per voice channel) but has very good robustness, especially needed in the very hostile environment (there is neither a prior separation between cochannel interferers nor spread spectrum protection). The two directions of communications share the same channel by Time Division Duplexing (TDD).*Control channels are provided by means of a dedicated field in each time slot.

Applications of DECT include residential systems (as with today's cordless phones), cordless PBXs with switching capability, telepoint public access (especially in high density areas such as airports, train stations, shopping and business centers), and radio in the local loop. The emergence of these applications will likely facilitate the start-up of DECT-based PCS facilities.

2.4.3.4   Personal Communications Services

An important concept that emerged in recent years is PCS. Even though PCS systems are expected to provide a variety of services and therefore may be considered as belonging to a third generation, they are mainly based on the current second-generation cellular technology, and are therefore somewhat between second and third generation. These systems are also referred to as "low-tier" and differ from cellular systems ("high-tier") in that they envision smaller cells and reduced power levels; they focus mainly on handheld devices with limited mobility and aim to provide a rich variety of services.

Key features of PCS will be easy-to-use high-functionality terminals and the "personal" dimension, which involves essentially two ideas. The first idea is that of "personal telephone number," where people will have their own number at which it will be possible to reach them at any time and regardless of their physical location. That is, telecommunications numbers will correspond to persons instead of

---

*A DECT frame is made up of 24 slots, the first 12 carrying base-to-mobile traffic and the latter 12 carrying the mobile-to-base traffic.

fixed installations. The second aspect of the "personal" character of these new services is that they will be "personalized" through the use of user profiles where subscribers will be allowed to build their own set of preferences so that the network will be able to deal with traffic addressed to or originating from them in an appropriately customized manner. These preferences could include services with which we are already familiar, such as call forwarding, call blocking, voice mail, paging, billing options, or other more advanced features such as QoS specifications. Much of this relates to efforts trying to achieve QoS guarantees at the application layer when dealing with channel impairments or hostile environments, where, for example, a user may be given the choice of having the call dropped or accepting a lower-quality video resolution (e.g., by decreasing the video data rate and switching to black and white or to slow motion) or only keeping the most critical media (e.g., just the audio in a teleconferencing session). As different people may make different choices in these situations, it cannot be assumed that "one-service-fits-all."

A very successful example of PCS is the Person Handiphone System, which was deployed in Japan in 1995 and has experienced a tremendous growth rate since then. Other examples include DCS1800 (which essentially retains the GSM architecture), DECT (described in the previous subsection), and PACS (mainly intended for wireless local loop applications).

2.4.3.5 Wireless Data

With the increasing demand for data services, cellular telephony is not adequate to meet all the users' needs, and therefore provision of data services is gaining importance. In this subsection, we will briefly address wide-area wireless data systems, whereas wireless local-area networks will be discussed in a later section.

In 1993, the U.S. cellular industry developed the Cellular Digital Packet Data (CDPD) standard, intended to coexist with the first- and second-generation cellular systems in the U.S., and to use the 30-kHz radio channels on a shared basis to provide packet-switched data communications. CDPD takes advantage of the fact that some radio channels in cellular systems are often left unused during normal operation,* and in this case the corresponding radio resources can be devoted to data transmission. By doing so, CDPD is able to provide packet data services without additional bandwidth. CDPD uses the existing cellular base stations and air interface, but not the MSCs, as it has its own routing capabilities.

CDPD uses the 30-kHz AMPS radio channels to provide 19.2 kbps data. The use of the cellular radio channels is only allowed when such channels are not supporting telephone calls, which have preemptive priority over the data traffic. In the standard specifications, when a channel used by CDPD is assigned to a telephone call, another empty channel is sought, so that the CDPD connections "hop" among various channels according to the channel assignment dynamics of the cellular network. In practice, however, in many cases a voice channel per cell is permanently reserved for CDPD use, as this is much simpler to manage (although less efficient).

The multiple access protocol in CDPD is DSMA/CD, which belongs to the class of carrier-sense random access schemes described in the next section. Data-link functions are provided by the Mobile Data Link Protocol, whereas the Radio Resource Management Protocol is responsible for finding and managing unused cellular radio channels. Services provided by CDPD include broadcast, E-mail, monitoring, and dispatch. Error control capability is also provided, through the use of a Reed Solomon code.

Another standard for cellular data transmission is IS-99, which uses the CDMA-based physical layer of IS-95. IS-99 supports TCP on top of a radio link protocol (RLP) which partially recovers from channel errors via selective repeat ARQ. Other wireless data systems include the General Packet Radio Service (GPRS) supported in GSM; ARDIS, by Motorola, which provides a very good in-building coverage in urban areas; and RAM, based on Ericsson's MOBITEX standard.

---

*Note, in fact, that the number of available channels must be chosen to guarantee adequate blocking performance under peak conditions.

2.4.4   Wireless Local-Area Networks

With the need for mobility and the positive experience with wireless telephony, it is natural for users to be attracted by wireless capabilities in the computer communications domain, so that the demand for wireless LANs has fueled a considerable amount of research and a number of commercial products.

Wireless LANs have many advantages over wired LANs, including quick and simple installation (no need for wiring), reconfigurability, application in environments where wiring is difficult or impractical (e.g., historical buildings), support for *ad hoc* and temporary facilities (e.g., during a convention). Moreover, they can be used in environments where mobility is desired (e.g., inventory of stocks, car rental check-in). Two types of wireless LANs can be essentially envisioned: (1) LANs with infrastructure, used, for example, to provide coverage in a building or campus environment by multiple access points attached to a common backbone, and (2) LANs without infrastructure, used, for example, to provide *ad hoc* networking in a conference room.

There are several differences between wireless LANs and cellular telephone systems. Cellular systems today are targeted toward circuit-switched low-rate voice communications, provide wide-area coverage, support mobility of terminals, and have switching capabilities. On the other hand, wireless LANs are targeted toward computer and data applications, support high transmission speeds (up to several Mbps), are based on packet-switching technology, have limited switching capability and support no or limited mobility, and are intended as local-area facilities. It is therefore not surprising that a completely separate set of technological solutions has been devised for cellular systems and wireless LANs (as happened for wireline telephone and computer networks). On the other hand, it must be said that the current trend toward integration of communications and computing will bring the two fields closer and closer together, and future wireless systems (addressed in a later section) will grow out of a combination of these technologies, as already testified by many proposals about third-generation systems which share many aspects of computer communications protocols.

The major difference between wireline and wireless LANs is in the data rates and quality they can support. Providing high transmission speeds in the wireless environment is very challenging because of the very hostile propagation environment. The IEEE 802.11 and HIPERLAN standards offer channel speeds of 1 to 2 Mbps and 25 Mbps, respectively (to be compared with 10 Mbps of ethernet and 100 Mbps of fast ethernet), and can operate over limited distances (tens of meters or less, much smaller than those allowed in wireline). Higher data rates can be achieved by means of infrared communications (as opposed to radio transmission, implicitly considered so far), which on the other hand operates on shorter distances and usually requires line-of-sight propagation.

2.4.4.1   Multiple Access Schemes

In an environment where data users are present and generate traffic in packet bursts, it is well known that circuit switching and fixed resource assignment techniques (e.g., TDMA or FDMA) may be very inefficient. For this reason, other multiple access techniques have been devised, which are more suitable for this environment. The range of possible solutions has at one extreme complete coordination among users, and at the other extreme no coordination at all. An example of complete coordination may be polling systems, where a central controller tells users when they should transmit, based on the knowledge of their communications needs. This kind of structure may be more flexible than fixed resource assignment, but requires coordination among users and controller, and in particular, some signaling is needed and some overhead is involved. On the other hand, the best example of noncoordinated access is the ALOHA protocol, in which users transmit whenever they have packets, regardless of what other users are doing. Overlapping transmissions (collisions) may lead to packet losses, which are recovered by retransmission.

The idea underlying the ALOHA protocol, and in particular its extreme simplicity, has been very successful, and has originated the whole research field of random access protocols.* Several modifications

---

*The ALOHA protocol has also great practical relevance by itself and is used in a variety of situations, including the random access channels of cellular systems.

of the original idea have been proposed, trying to solve some problems that random access presents. The first improvement was to slot the time axis, so that packets can be transmitted only at some fixed instants of time, thereby reducing the probability of overlapping transmissions. The recognition of the fact that the throughput performance greatly depends on the occurrence of collisions led to more-advanced schemes, where transmitters listen to the channel before attempting transmission, and refrain from sending packets if the channel is sensed busy. In the presence of zero signal propagation delay, this would lead to perfect sensing and would completely eliminate collisions. On the other hand, since signals take time to propagate, the sensing information may be incorrect. For example, just after a transmission is initiated a transmitter would sense the channel as idle and also start transmitting, thus causing a collision. Another improvement led to schemes where overlapping transmissions can be detected and aborted, thus minimizing the waste of communications resources caused by collisions. This last idea is implemented in the Carrier Sense Multiple Access with Collision Detection (CSMA/CD) protocol, which is the scheme used in ethernet. Other more elaborate improvements have also been proposed, addressing other more subtle issues such as protocol stability.

Given the large amount of research performed in the 1970s and 1980s on multiple access schemes for LANs, trying to extend those protocols as they are to the wireless domain seems a reasonable thing to do. However, this gives rise to problems which are not present in wireline LANs, and which effectively prevent using the well-established LAN protocols in the wireless environment. A typical issue is "hidden terminal" problem. Basically, in wireline communications, all users are attached to a cable, and therefore they can all hear each other's signals. This condition is at the basis of the carrier-sensing protocols. Unfortunately, in radio communications this is not necessarily the case. In fact, the erratic behavior of the radio channel may not guarantee connectivity. Moreover, the fact that two mobile users A and B are within range of the same BS does not necessarily mean that they are within range of each other (e.g., if they are located at the extremes of a diameter of the BS coverage region). Therefore, CSMA protocols fail to work in the wireless environment, and other provisions must be made to guarantee proper operation. One possibility is the use of a busy tone, which is activated by the BS when some transmission is active, so that all mobiles which could potentially start communications to that BS (and are therefore within its range) will be informed about the channel being busy.[*] The multiple access protocol specified in CDPD (called DSMA/CD) uses a busy/idle flag in the base-to-mobile direction to notify mobiles about the mobile-to-base channel occupancy, thus overcoming the hidden terminal problem.

In the two recent standards for wireless LANs, the IEEE 802.11 standard and the European HIPERLAN, another protocol is used, which is called CSMA with Collision Avoidance (CSMA/CA). The basic concept relies on the observation that collisions are most likely to occur at the end of a transmission, since during that transmission communications requests may have accumulated and they are likely to be colliding as soon as the channel becomes idle. To avoid this effect, additional backoff mechanisms are introduced to limit the number of terminals which will eventually attempt transmission.

In IEEE 802.11, the concept of InterFrame Space (IFS) has been introduced. IFS is a time delay that a terminal has to wait before attempting transmission. Different lengths for IFS allow the introduction of different priorities. In particular, three IFS are specified, i.e., short IFS (SIFS) for high-priority control packets such as acknowledgments, point-coordination-function IFS (PIFS) for time-sensitive traffic, and distributed-coordination-function IFS (DIFS, the longest) for asynchronous traffic. If at the end of its IFS a terminal still senses the channel as idle (i.e., there is no higher-priority traffic), it chooses a random time uniformly in a contention window (which is dynamically updated), and if after that time still no transmission is sensed, then the terminal sends its message. The hidden terminal problem can be solved by using some handshake procedure, where prior to message transmission there is a dialogue between mobile terminal and BS, which improves reliability at the expense of some overhead.

---

[*]It should be noted here that the concept of "radio range" is not as straightforward (early models based on circular areas are rather gross oversimplifications), but is rather a probabilistic concept, and operation can still suffer from radio propagation effects.

A more elaborate version of CSMA/CA is used in HIPERLAN. If upon message generation a terminal senses the channel as idle for a given amount of time (1700 bits in this case), then transmission can immediately take place. Otherwise, a three-phase collision avoidance mechanism is performed. In the prioritization phase there are up to five time slots, one for each priority class, $p = 0,1, \ldots 4$ (with 0 the highest priority), and a terminal with priority p transmits a signal burst in slot $p + 1$. This phase terminates upon the first transmission, so that only the terminals belonging to the highest priority class will survive, whereas all the others will yield. In the elimination phase, each node transmits a signal burst for a geometrically distributed number of slots, and then listens for one more slot: if any signal is detected in that slot, the terminal drops out, and only those which chose the highest number survive. Finally, during the yield phase, each of the surviving nodes defers transmission for a geometrically distributed number of slots and listens to the channel in the meantime, dropping out if any signal is detected. This whole procedure has as its objective to guarantee with high probability that only one terminal will eventually transmit, as it is very unlikely that collisions will occur after all these random choices.

2.4.5 Third Generation and Beyond

Everything described so far refers to either commercially available products or to standard documents already approved or in the process of being approved. In addition, a large amount of work has been and is being done on proposals for future systems. This body of work includes basic research on fundamental issues, as well as standardization and regulatory efforts currently being undertaken.

Standard activities on future wireless systems include Future Public Land Mobile Telecommunications Systems (FPLMTS, supported by the International Telecommunications Union and recently renamed IMT-2000) and Universal Mobile Telecommunications Services (UMTS, supported by the European Union).

Third-generation wireless networks are expected to be deployed as a further evolution of current second-generation and personal communications technologies. Even though the phrase "third-generation systems" may still mean different things to different people, there seems to be general consensus about some of the key requirements of these new systems, which include global coverage, high capacity, improved quality, and increased variety of offered services (including data, video, and multimedia). The distinction among various services in existence today (paging, cordless, cellular) will blur and all of them are expected to be offered by a single personal device. The broadband ISDN will be used to provide access to information networks and servers (e.g., the Internet) and satellites will be part of this picture as well. Applications including multimedia sessions are envisioned, with telephone conversations accounting for the minority of the total traffic. Applications that are already a reality in the wireline domain, such as Web browsing, multimedia communications, teleconferencing, intelligent networking, virtual offices, and so on, will find their way into the wireless domain, allowing people to be set free from location constraints.

A great deal of research is being done on all aspects of third-generation wireless systems. In the following subsections, we give an quick overview of the many issues involved. Details about any of them can be found in the technical literature (see Further Reading for a basic set of references).

2.4.5.1 Physical Layer and Error Control

Advanced requirements envisioned in third-generation wireless networks include data rates which, if not quantitatively comparable with those of the broadband ISDN, at least qualitatively are broadband and intended to support multimedia.[*] This often leads to specifications of channel data rates in the range of 20 Mbps or higher. Such data rates are hard to achieve on a radio channel, especially because of the need for equalization due to severe time dispersion of the transmitted pulses. Due to the short packet length (53 bytes in ATM), considerable overhead may be involved when using standard equalization techniques.

---

[*]It should be clear at this point that third-generation systems will be mostly (if not exclusively) based on digital transmission and packet switching.

Use of advanced signal processing may be needed, including antenna array adaptation, and combining which is being considered a feasible means to counteract multipath and interference.

As to the carrier frequency range to be used, current systems occupy the 900 MHz band and PCS systems are being developed at 1.8 to 1.9 GHz, whereas new allocations are being made at higher frequencies (see Reference 2 for a chart with specific information about spectrum allocations). Issues that arise when choosing the operating frequency include regulatory policies (some bands do not require licensing and are therefore attractive both for the small cost and for the immediate availability, whereas others require licensing and FCC approval, which is a long and expensive process but guarantees dedicated channel use), cost of the equipment (which is higher at higher frequencies in general), propagation characteristics (e.g., at 60 GHz electromagnetic signals do not propagate through walls, and this may be an advantage for indoor systems due to reduced interference but also a serious drawback for global coverage), amount of bandwidth available (usually, the higher the carrier frequency, the larger the available bandwidth). The use of diffuse infrared propagation does not seem to be a good alternative to radio systems due to the effect of multipath and the very limited range (line of sight is typically required), although it may be preferable when concerned with electromagnetic interference (e.g., in hospitals).

Linear modulation techniques can be used in wireless systems, e.g., BPSK, QPSK (with their differential counterparts), and QAM, even though demodulation of multilevel signaling is usually hard to implement in fading environments. Schemes such as GMSK or π/4-DQPSK achieve (or approximate) constant envelope characteristics, thus enjoying the advantage of better efficiency of the amplifiers (which need not be linear). Much attention is being given to orthogonal frequency division multiplexing (OFDM), a technique which uses many narrowband carriers instead of a single broadband channel, thus simplifying considerably the need for equalization.[3] Multicarrier modulation techniques are also attractive for multimedia traffic, as they enjoy additional flexibility as to variable data rates, and therefore can easily support bandwidth-upon-demand. The use of spread spectrum modulation is also being considered with great interest, due to its inherent ability to deal with multipath propagation and cochannel interference, especially in its wideband version.[*,4,5] The main objection to the use of CDMA for broadband services is that at very high data rates and with the current spectrum allocations there is simply no room for bandwidth spreading. Also, in the presence of small to moderate spreading factors, it has been reported that TDMA schemes can be more efficient than CDMA.[6] Further research remains to be done in order to clear up many of these issues and to make advanced techniques commercially available.

Regarding error control, two basic techniques exist, namely, Forward Error Correction (FEC) and automatic repeat request (ARQ).[7] In FEC schemes, redundant symbols are added at the transmitter in such a way that the receiver is able to decode the message correctly even in the presence of transmission errors (to a certain extent). These techniques provide deterministic transmission delay,[**] but usually involve considerable overhead. In ARQ schemes, on the other hand, only few redundant symbols are added before transmission, giving the receiver the capability to detect errors and to request retransmission of the message portions which get corrupted on the channel. A much smaller overhead is paid in this case, but no guarantees can be made about transmission delays, as a packet may need to be retransmitted many times before it is finally received correctly. Also, ARQ systems require a feedback channel which is not needed for FEC. For future services, where small radio cells will lead to very short propagation delays and will allow almost immediate feedback, ARQ techniques may gain renewed interest. Studies have been recently published showing that for packet communications in the local wireless environment ARQ schemes can successfully compete with FEC, due to bandwidth efficiency and additional features such as the possibility of selective packet dropping in delay-constrained communications, easier QoS support, enhanced flexibility, and lower complexity.[8,9] It is likely that hybrid schemes (possibly adaptive) will be

---

[*]That is, when the spread bandwidth significantly exceeds the coherence bandwidth of the channel, which is usually of the order of a few megahertz. This would typically require spread bandwidths in excess of 10 MHz.

[**]This delay is due to the encoding/decoding algorithms and to the use of interleaving for improved error correction performance.

employed in these systems, according to the trade-offs between data reliability on the one hand and complexity and delay performance on the other. More-advanced techniques, such as Trellis Coded Modulation[10] and Turbo Codes,[11] which provide much better protection against interference and noise, suffer from other problems (e.g., decoding complexity for TCM and interleaving delay for Turbo Codes), which might delay their application to broadband wireless communications.

2.4.5.2 Multiple Access

In cellular systems, two issues must be solved at the multiple access level. The *channel reuse* scheme determines how channels are allocated to cells throughout the system, whereas the *access scheme* solves the problem of resource sharing within a radio cell. Different systems make different choices on these two issues. For example, classic cellular systems use some frequency reuse plan, so that the channel reuse leads to tolerable cochannel interference. Also, in circuit-switched systems, the access scheme is usually achieved by orthogonal channels (in the time or frequency domain). These channels can be rigidly assigned to cells (as in first- and second-generation systems), or more flexible schemes can be used, ranging from borrowing schemes to fully dynamic channel allocation. Various schemes have been proposed, achieving better resource utilization efficiency at the expense of additional complexity and the need for coordination among network elements.[12] One attractive feature of dynamic resource allocation is soft capacity, which provides additional flexibility in traffic management and improved teletraffic performance.[1] In cellular CDMA, spread spectrum is used to achieve both multiple access and channel reuse. The inherent interference rejection capability of this technique makes it possible to use the same radio channel in all cells, thus avoiding the need for a frequency reuse plan.

Recent proposals for multiple access schemes make the implicit assumption that information is carried in packets. The flexibility of packet communications and the increased delay tolerance of some data traffic open up opportunities for more efficient design of multiple access protocols. Many of these protocols resemble those used successfully in LANs, with modifications to make them suitable to the wireless environment. An early proposal, Packet Reservation Multiple Access (PRMA),[13] is essentially a reservation ALOHA scheme, where slots in a frame (the "channels") can be either reserved or available. Multipacket messages are transmitted on a contention basis in available slots, and a successful transmission acts as a reservation, guaranteeing contention-free access (in reserved slots) during the rest of the message. More elaborate access techniques have been proposed,[14-17] and a detailed discussion of them is out of the scope of this description.

The basic structure of most proposed multiple access protocols is based on reservation. It is assumed that some provision is given to the users to make reservations (as soon as they become active), thereby notifying the central BS of their need for communications resources. This can be seen as a separate channel, even though it can be dynamically implemented based on the instantaneous channel load (as in the case of PRMA, where increased traffic reduces the number of available slots). Information packets, on the other hand, are carried based on reservation and polling: users who gained access to the system (by successfully placing a reservation) are polled by the BS, according to some service policy which in general depends on the QoS requirements of each connection. Flexible protocols, capable of managing different traffic types and QoS requirements, can be envisioned in this environment. Examples of these protocols can be found in References 14 through 16.

Most studies of multiple access protocols for third-generation systems are concerned only with the access scheme, neglecting the channel reuse part. This typically means that some frequency reuse plan is assumed that guarantees that cochannel cells are sufficiently far apart and interference is not an issue. This is also the reason most studies in this field rely on very similar assumptions as their LAN counterparts, e.g., that the channel is error-free and only collisions lead to packet loss. Unfortunately, besides leading to analyses which do not reflect reality (where packet errors do occur due to noise and interference, which cannot be neglected even in the presence of the classic seven-cell frequency reuse), protocols designed under these assumptions in general fail to work well in the presence of errors (for example, no provision is given for packet retransmission in PRMA). An even more serious problem is that protocols which rely on separation of cochannel radio cells result in poor capacity performance. It is apparent that systems which can avoid paying the large penalty involved in any frequency reuse plan will have a major advantage with respect to systems that still rely on classic cellular concepts. For example, it was shown in Reference 18 that even the simple Slotted ALOHA can work in a cellular system with full frequency reuse, yielding a throughput significantly larger than 14% (which is the maximum achievable in systems using the popular seven-cell frequency reuse pattern). More-advanced protocols, which are able to cope with fairly large error rates, would provide even higher efficiency.[14] Based on these concepts, a cellular architecture with full frequency reuse has been proposed, whose capacity is competitive with CDMA, and which can support various traffic classes in the presence of delay constraints.[19]

Other interesting issues regard the compatibility of multiple access schemes with advanced physical layer techniques. For example, the use of array antennas may require users to transmit pilot signals periodically or on demand, so that the BS can estimate the channel impulse response and determine the optimum weighting coefficients to be used in the diversity combining block. This requirement must be integrated in the design of the multiple access scheme.[20]

2.4.5.3 Quality of Service Support

With the advent of broadband networking and ATM, Quality of Service (QoS) provisioning has become a key concept in telecommunications.[21] Typical QoS metrics in ATM networks are the packet loss rate, delay, and delay jitter.[22] Packet networks, unlike circuit-switched networks, are based on statistical multiplexing and resource sharing, and are prone to congestion, which occurs when the instantaneous user demand collectively exceeds the network capabilities. A major concern in broadband networking is to try to avoid congestion situations as much as possible and to cope with them when they occur. Admission control techniques are used to determine whether or not incoming connection requests can be accepted.[23] A connection request can only be accepted if its QoS requirements can be met and if the QoS requirements of all connections already admitted can be maintained. This concept of admitting users based on a QoS calculation at the time the connection request is presented to the network implicitly relies on a "static" network paradigm, in which the topology of the network, the capacity of its links, and the geographic distribution of the admitted traffic are fixed. In this situation, if the admission control algorithm indicates that QoS is guaranteed when a connection is set up, then those guarantees are in general not violated in the future.

It is clear that this paradigm does not apply to wireless networks, where neither the network topology and the capacity of its links nor the geographic distribution of the admitted traffic are fixed. In fact, link capacity fluctuates due to fading and mobility, and sometimes links become unavailable. Also, mobility is a primary requirement in wireless communications, and users can roam anywhere after being admitted, possibly driving some radio cells to overload. In this environment, guaranteeing QoS is an exceedingly hard task, if it is at all possible. On the other hand, as QoS is an essential feature of modern telecommunications services, a way must be found to offer this capability in the wireless domain as well.[24]

Of course, a very strict design of the network architecture and of its protocols, where everything is overdesigned in such a way that worst-case situations are still manageable, is a legitimate solution to that problem (using essentially the same philosophy of circuit switching). However, its inefficiency makes it virtually useless for advanced wireless systems. A better solution may be to accept a different definition of QoS guarantees, where a user negotiates a *set* of QoS levels, and the network tries to provide the highest possible level at any given time (and bills the user accordingly). For example, if the bandwidth of a link drops because of degraded propagation conditions, the quality is downgraded and a lower service is provided. This may sound like contradiction of the concept of QoS guarantees, but it should be compared with the dropping of the connection altogether due to inability to maintain the required QoS level, a solution which in many cases may be less desirable.

An even more elaborate solution can be found in cooperative QoS management strategies involving both the application and the network.[25] The network in general is requested to be application aware, in the sense that it knows the application needs (as specified in the QoS profile) and tries to meet them.

The idea here is to also make the application aware of what happens in the network, giving it a chance to react and adapt to changed conditions. For example, this has the advantage that the application itself may choose how to modify the level of service, instead of letting the network (which in most cases does not have enough knowledge, e.g., of the application contents) do it. A typical example is layered encoding, as used, for example, in video streams, in which bits (packets) are divided into classes of different priorities. Highest-priority information is essential for any understanding of the message, whereas a lot of low-priority information, although desirable for high-quality playback, can be lost while still retaining sufficient understanding. Techniques like this can be used to adapt the perceived quality dynamically to the channel conditions. As long as the user/application is willing to tolerate these fluctuations, such techniques may provide a dramatic improvement in connection-dropping performance as caused by radio channel impairments.

2.4.5.4 Networking and Transport Issues

Much research has been devoted to the analysis of higher layer protocols and to the design of schemes that can deal with the unique issues arising in the wireless environment.

One of the characterizing aspects of wireless networks is mobility, and this requires a whole set of new functionalities which need to be provided. Current cellular systems keep databases and offer handoff capabilities, which allow users to roam freely throughout the system area. Performing a handoff is not a major undertaking in current systems, due to traffic uniformity and low data rates. With broadband networking capabilities and QoS support, handoff will be much more challenging for a number of reasons. First of all, a brief interruption of speech communications, as occurs in hard handoff in current cellular systems, may be barely noticeable and is certainly accepted in general. For a broadband data connection, this interruption may involve a large amount of data which, in the case of reliable connections, must be retransmitted or recovered. Error recovery, as well as duplication and reordering problems, must be addressed in this case. Furthermore, in current systems, handoff requests are usually accepted or rejected based on channel occupancy in the target radio cell, since cellular telephone networks have a very primitive concept of QoS, if any. On the other hand, when heterogeneous traffic classes are to be served, the QoS calculations to be performed by the admission controller are much more complicated and time-consuming. If these calculations are required every time a handoff is requested, the burden on the controllers may be exceedingly large, possibly causing a dramatic increase of signaling traffic and of the number of calls that are dropped because a handoff cannot be completed within reasonable time. Three types of handoff schemes exist:[26] anchor rerouting, dynamic rerouting, and connection tree routing. Unlike the first two, the third class of schemes does not require tearing down and setting up the connection, and has been actively studied for broadband wireless applications.[27,28]

Other issues at the networking and higher layers have to do with the performance of established network protocols over mobile and wireless systems. A very good example is the TCP/IP protocol suite, which supports popular Internet applications such as Web browsing, ftp, and E-mail. The Internet protocol as it is does not support mobility, and therefore cannot be successfully used in a wireless environment. On the other hand, due to the widespread penetration of the World Wide Web, it is impossible to change the protocols in use in the whole Internet, and considerable research has gone into studying how to devise schemes that can support mobility while being compatible with the protocols currently implemented throughout the world. An example is Mobile IP, a version of the Internet protocol that supports host mobility (but not handoff) through the provision of a care-of address capability used by the "home network" to forward traffic for one of its hosts which may temporarily reside in a different ("foreign") network.[29]

Another important problem to be solved before popular computer applications find their way in the wireless domain concerns the suitability of TCP as a reliable transport protocol over connections comprising wireless segments. It has been reported in several papers (e.g., see Reference 30 and the references therein) that current releases of TCP do not work properly on wireless channels because they misinterpret packet losses due to channel impairments and assume that congestion occurred. The action triggered by the protocol rules in this case simply addresses the problem in the wrong way, and catastrophic throughput losses have been observed. Possible solutions to this problem include changing the rules of TCP (with serious compatibility problems), terminating the TCP connection at the base station (which violates the semantics of TCP, which is an end-to-end protocol), and introducing some link control mechanism to enhance the quality of the radio channel (see Reference 30 for a comparative study of various schemes). In particular, the last solution is attractive since it does not involve any change to current TCP implementations and relies on ARQ techniques, which are able to recover erroneous packets before a timeout expires and avoid congestion control protocol actions.[31] Other issues related to the use of wireless channels are discussed in Reference 32.

2.4.5.5 Wireless ATM

ATM is viewed as a strong candidate for the support of high-bandwidth, high-quality multimedia applications through the provision of bandwidth and QoS upon demand. Extension of such services to the wireless environment is, of course, of great interest, and the concept of wireless ATM has gained a lot of attention in recent years. This topic is the focus of a number of research efforts throughout the world, involving universities and industrial companies. The ATM Forum (a standards body for ATM) has created a wireless ATM working group to address the many technical issues which are still open.

Major issues to be solved are the extension of the networking capabilities of ATM and the enhancement of the quality of the radio link. The first issue involves extension of the ATM architectural model to allow the implementation of concepts such as admission and call control, QoS, routing, resource allocation, and signaling for connections over mobile and wireless links. The second issue deals mainly with transmission and access technologies able to stabilize the wireless channel making it suitable for the support of broadband applications.[33]

2.4.5.6 Energy and Power Management

Battery life has always been a concern in the wireless communications industry, since improved battery performance translates in smaller and lighter devices. Unfortunately, the rate at which battery performance improves (in terms of available energy per unit size or weight) is fairly slow, despite the great interest generated by the booming wireless business. Other means to reduce power consumption in wireless devices have therefore been receiving interest in the research community. Besides the traditional search for better circuits and power amplifiers, which directly improve the power consumption efficiency of the whole device, some recent research results have taken a broader view of the problem. In particular, it has been recognized that energy conservation is a task which can be performed at multiple levels, and energy-efficient protocol design criteria can be identified and applied. This concept has resulted in studies on energy-efficient error control schemes (where bad channel conditions cause the transmitter to enter a sleep mode), multiple access protocols (by minimizing collisions and the need for retransmission), software techniques (such as various low-power modes at the operating system level), and signal processing algorithms (via more careful software circuit design). Another traditional topic for wireless communications, namely, power control, has been recently revisited in a network resource allocation perspective (as opposed to simple attenuation compensation), opening up new optimization problems and networking issues. Advances on these topics can be found in References 34 through 38.

2.4.5.7 Wireless Local Loop

Even though most wireless applications involve mobile users, there has been increasing interest in wireless fixed systems. In particular, it is clear that in the absence of a wired infrastructure and distribution network, delivering telecommunications services to residential or business users may require large investments and a considerable amount of time, due to the need to put cables or fibers into the ground. In this scenario, which is typical of countries with little or no telecommunications facilities (e.g., developing countries), a wireless solution may provide the rapid development of communications services and may turn out to be economically attractive.[39] As a matter of fact, the great interest shown by major telecommunications providers suggests that even in developed countries the wireless solution may be competitive and should be seriously considered. This is especially true if wireless broadband access will become a reality, so that the type of service that can be offered through wireless is better than what can be achieved, e.g., through twisted pair or even cable. It remains to be seen whether or not wireless local loop distribution will result in a large business segment, but at the moment much research is being done in that direction.

2.4.5.8   The Role of Satellites

In future wireless communications systems, satellites are expected to play a significant role. The main reason for this is the requirement that third-generation systems be truly universal, i.e., that they provide global coverage. It is clear that coverage of sparsely populated areas or regions of the world such as deserts and oceans by means of terrestrial cellular facilities does not make economic sense, and satellites in this case may represent an attractive (or even the only feasible) alternative. Also, satellites may provide umbrella coverage to be used as stand-by resources for traffic that overflows from cellular systems, and are also useful in disaster situations where terrestrial facilities may be destroyed.

Different options for the orbits are possible. The Geostationary Orbit (GEO) is widely used in satellite communications systems (e.g., mobile services provided by INMARSAT), but it involves large distances between transmitter and receiver, which imply significant round-trip delays and large attenuations. This may make it difficult to provide high-quality high-data-rate services to portable terminals, where highly directive antennas are difficult to implement. On the other hand, GEOs have the advantage of continuous coverage without requiring handoff capability. The problems of GEOs can be alleviated by choosing medium or low earth orbits (MEO/LEO). LEOs, in particular, with circular orbits at 1500 km or less from the earth's surface, appear suitable for personal communications services (e.g., Globalstar and Iridium are LEO systems). The main drawback of LEOs is the relative movement between satellites and earth, which induces some Doppler shift and requires a whole constellation of satellites (for example, Iridium has 66) equipped with handoff and intersatellite communications capabilities.

Until now, satellite systems have been studied for the provision of stand-alone services. However, there is increasing awareness that satellites should be integrated with the terrestrial cellular networks and be able to interwork with them. Even though it may be easier to keep the two elements (satellite and terrestrial) logically separate and connect them through the fixed network, a more integrated solution where they have the same air interface and may even share network elements (e.g., MSCs) has recently received much attention (see Reference 40 for some details). Many technical and regulatory issues still need to be resolved, and research is being done in this direction.

2.4.5.9   Toward Multimedia Radios

In the above description, adaptive techniques have been mentioned several times as a means to face the erratic behavior of the wireless channel. Also, adaptation of applications to changing network conditions may be a feasible means to overcome such difficulty without compromising the network efficiency.

Another good reason for adaptivity stems from the fact that applications may have widely different needs, and choices that are good (or even optimal) for one type of service may be suboptimal (or even catastrophic) for others. The current wireless infrastructure has been designed based on the assumption of (uniform) voice traffic, and many communications functions have been optimized under these conditions. It may be expected that with the emergence of data services this kind of network infrastructure will not be adequate. For example, packet switching needs to replace circuit switching. Error control, performed by FEC and interleaving in current voice systems, may be done more efficiently by ARQ schemes for many data applications. There is evidence that some services may tolerate correlated errors better than dispersed errors (while the opposite is true for current voice communication systems).[41] Also, it is likely that there exists no solution which is optimal for everything, but rather various solutions which best address the needs of specific classes of applications.

In current implementations, being able to reach down to the radio level and to access and reprogram the hardware is very difficult, and in many cases impossible. However, for a flexible device that can be used to support a variety of applications, ranging from Web browsing to plain telephony and from E-mail to full motion video transmission, being able to select the transmission scheme that best adapts to the specific application needs may have a great value. Besides giving the possibility of selecting at any time the best possible scheme (e.g., by being able to turn off some blocks or to select the modulation/error control scheme by simple software instructions), such a "software radio" could be upgraded easily via software downloading. The military started thinking about this a few years ago, because of the need to achieve interoperability across different frequency bands, channel modulations, and data formats. Commercial interest has been growing as well in recent years.[42]

Even protocol stacks could be programmed and their functionalities negotiated, so that the user (or the application, without need for direct human intervention) can select the protocol specifications that best suit the needs of the service being sought. The negotiation process would then extend to protocol specifications, beyond a simple set of connection parameters. The paradigm of intelligent software agents and Java, where code segments and modules can be downloaded on the fly, may apply in the context of communications systems as well.

This vision promises full liberation of radio systems from the slavery of "hard-wired" features, including data rates, channel coding, and frequency bands, and is based on the use of multiband antennas, wideband A/D and D/A conversions, and digital signal processing capabilities through general-purpose programmable processors. This new radio architectural concept (which can reach up to higher layers, as briefly described above) will lead to more efficient systems, where no hardware designer makes choices *a priori* for the user or the application. Truly universal services and devices will then be realized.

Abbreviations

ACCH	Associated Control Channels
ADPCM	Adaptive Differential Pulse Code Modulation
AMPS	Advanced Mobile Phone Service
ARDIS	Advanced Radio Data Information System
ARQ	Automatic Repeat Request
ATM	Asynchronous Transfer Mode
BPSK	Binary Phase Shift Keying
BS	Base Station
CCITT	International Telegraph and Telephone Consultative Committee
CDMA	Code Division Multiple Access
CDPD	Cellular Digital Packet Data
CSMA/CA	Carrier Sense Multiple Access with Collision Avoidance
CSMA/CD	Carrier Sense Multiple Access with Collision Detection
DCS1800	Digital Communications System — 1800
DECT	Digital European (Enhanced) Cordless Telecommunication
DIFS	Distributed Coordination Function Interframe Space
DQPSK	Differential Quadrature Phase Shift Keying
DSMA/CD	Digital Sense Multiple Access with Collision Detection
FCC	Federal Communications Commission
FDD	Frequency Division Duplexing
FDMA	Frequency Division Multiple Access
FEC	Forward Error Correction
FM	Frequency Modulation
FPLMTS	Future Public Land Mobile Telecommunications Systems
GEO	Geostationary Orbit
GMSK	Gaussian Minimum Shift Keying
GPRS	General Packet Radio Service
GSM	Global System for Mobile Communication
HIPERLAN	High-Performance Radio Local-Area Network
HLR	Home Location Register
IFS	Interframe Space
IMT-2000	International Mobile Telecommunication — 2000
IP	Internet Protocol

ISDN   Integrated Services Digital Network
LAN   Local-Area Network
LEO   Low Earth Orbit
MAHO   Mobile Assisted Handoff
MEO   Medium Earth Orbit
MSC   Mobile Switching Center
MSK   Minimum Shift Keying
OFDM   Orthogonal Frequency Division Multiplexing
PACS   Personal Access Communication System
PBX   Private Branch Exchange
PCS   Personal Communications Services
PDC   Pacific Digital Cellular
PIFS   Point Coordination Function Interframe Space
PRMA   Packet Reservation Multiple Access
QPSK   Quadrature Phase Shift Keying
QoS   Quality of Service
RPE   Regular Pulse Excited
SACCH   Slow Associated Control Channels
SIFS   Short Interframe Space
TACS   Total Access Communications System
TCM   Trellis Coded Modulation
TCP   Transport Control Protocol
TDD   Time Division Duplexing
TDMA   Time Division Multiple Access
UMTS   Universal Mobile Telecommunications Services
VLR   Visitor Location Register
VSELP   Vector Sum Excited Linear Predictor

Acknowledgments

The author gratefully acknowledges the contribution of Ender Ayanoglu, Justin Chuang, David Goodman, Larry Milstein, Silvano Pupolin, and Mischa Schwartz, who helped improve this section with their comments on an early draft. This work was supported by the Center for Wireless Communications, University of California, San Diego.

References

1. S.A. Grandhi, R.D. Yates, and D.J. Goodman, Resource allocation for cellular radio systems, *IEEE Trans. Veh. Technol.*, vol. VT-46, 581–588, August 1997.

2. E. Ayanoglu, K.Y. Eng, and M.J. Karol, Wireless ATM: limits, challenges, and proposals, *IEEE Personal Commun.*, 3, 18–34, August 1996.

3. J. Bingham, Multicarrier modulation for data transmission: an idea whose time has come, *IEEE Commun. Mag.*, 28, 5–14, May 1990.

4. A. Fukasawa et al.,Wideband CDMA system for personal radio communications, *IEEE Commun. Mag.*, 34 (10), 116–123, October 1996.

5. D.L. Schilling et al., Broadband CDMA for personal communications systems, *IEEE Commun. Mag.*, 29, 86–93, November 1991.

6. D. Falconer, A system architecture for broadband millimeter-wave access to an ATM LAN, *IEEE Personal Commun.*, 3, 36–41, August 1996.

7. S. Lin and D.J. Costello, *Error Control Coding: Fundamentals and Applications*, Prentice-Hall, Englewood Cliffs, NJ, 1983.

8. M. Khansari et al., Low bit-rate video transmission over fading channels for wireless microcellular systems, *IEEE Trans. Circuits Sys. Video Technol.*, 6, 1–11, February 1996.

9. M. Zorzi and R.R. Rao, ARQ Error Control for Delay-Constrained Communications on Short-Range Burst-Error Channels, in *Proc. VTC'97*, 1528–32, May 1997.

10. E. Biglieri et al., *Introduction to Trellis-Coded Modulation, with Applications,* Macmillan, New York, 1991.

11. C. Berrou and A. Glavieux, Near optimum error correcting coding and decoding: turbo-codes, *IEEE Trans. Commun.,* 44, 1261–71, October 1996.

12. I. Katzela and M. Naghshineh, Channel assignment schemes for cellular mobile telecommunication systems: a comprehensive survey, *IEEE Personal Commun.,* 3, 10–31, June 1996.

13. D.J. Goodman et al., Packet reservation multiple access for local wireless communications, *IEEE Trans. Commun.,* 37, 885–890, August 1989.

14. G. Bianchi, F. Borgonovo, L. Fratta, L. Musumeci, and M. Zorzi, C-PRMA: a centralized packet reservation multiple access for local wireless communications, *IEEE Trans. Veh. Technol.,* 46, 422–436, May 1997.

15. M.J. Karol, Z. Liu, and K.Y. Eng, Distributed-queueing request update multiple access (DQRUMA) for wireless packet (ATM) networks, in *Proc. ICC'95,* 1224–1231, June 1995.

16. N.M. Mitrou, T.D. Orinos, and E.N. Protonotarios, A reservation multiple access protocol for microcellular mobile-communication system, *IEEE Trans. Veh. Technol.,* VT-39, 340–351, November 1990.

17. G. Wu, K. Mukumoto, and A. Fukuda, Performance evaluation of reserved idle signal multiple access scheme for local wireless communications, *IEEE Trans. Veh. Technol.,* VT-43, 653–658, August 1994.

18. M. Zorzi and S. Pupolin, Slotted ALOHA for high capacity voice cellular communications, *IEEE Trans. Veh. Technol.,* 43, 1011–1021, November 1994.

19. F. Borgonovo, L. Fratta, M. Zorzi, and A.S. Acampora, Capture Division Packet Access — A new cellular access architecture for future PCNs, *IEEE Commun. Mag.,* 34 (9), 154–162, September 1996.

20. A.S. Acampora, S. Krishnamurthy, and M. Zorzi, Media access protocols for use with smart array antennas to enable wireless multimedia applications, in *Proc. 9th Tyrrhenian Workshop on Digital Communications,* September 1997.

21. M. Schwartz, *Broadband Integrated Networks,* Prentice-Hall, Upper Saddle River, NJ, 1996.

22. R. Onvural, *Asynchronous Transfer Mode Networks: Performance Issues,* Artech House, Boston, 1995.

23. A.S. Acampora, *An Introduction to Broadband Networks: LANs, MANs, ATM, B-ISDN, and Optical Networks for Integrated Multimedia Telecommunications,* Plenum Press, New York, 1994.

24. M. Naghshineh and A.S. Acampora, QoS provisioning in micro-cellular networks supporting multiple classes of traffic, *Wireless Networks,* 2, 195–203, 1996.

25. M. Naghshineh and M. Willebeek-LeMair, End to end QoS provisioning multimedia wireless/mobile networks using an adaptive framework, *IEEE Commun. Mag.,* special issue on Introduction to mobile and wireless ATM, November 1997.

26. T.-H. Wu, and L.Chang, Architectures for PCS mobility management on ATM transport networks, Fourth IEEE International Conference on Universal Personal Communications, 1995.

27. A.S. Acampora and M. Naghshineh, An architecture and methodology for mobile-executed handoff in cellular ATM networks, *IEEE J. Selected Areas Commun.,* 12, 1365–1375, October 1994.

28. M. Veeraraghavan, M. Karol, and K. Eng, Mobility and connection management in a wireless ATM LAN, *IEEE J. Selected Areas Commun.,* 15, 50–68, January 1997.

29. C. Perkins, *Mobile IP: Design Principles and Practices,* Addison-Wesley, Reading, MA, 1998.

30. H. Balakrishnan, V. N. Padmanabhan, S. Seshan, and R. H. Katz, A comparison of mechanisms for improving TCP performance over wireless links, *ACM/IEEE Trans. Networking,* December 1997.

31. H. Chaskar, T.V. Lakshman, and U. Madhow, On the design of interfaces for TCP/IP over wireless, in *Proc. MILCOM'96,* November 1996.

32. H. Balakrishnan et al., TCP improvements for heterogeneous networks: the Daedalus approach, in *Proc. 35th Annual Allerton Conference,* September 1997.

33. A.S. Acampora, Wireless ATM: a perspective on issues and prospects, *IEEE Personal Communi.,* 3, 8–17, August 1996.

34. *Proceedings of the IEEE*, special issue on Low-Power Electronics, April 1995.

35. *IEEE Personal Communications Magazine*, special issue on Energy Management in Personal Communications/Mobile Computing, June 1998.

36. N. Bambos and J. M. Rulnick, Mobile power management for wireless communication networks, *Wireless Networks*, 3, 3–14, 1997.

37. K.M. Sivalingam, M.B. Srivastava, and P. Agrawal, Low power link and access protocols for wireless multimedia networks, in *Proc. IEEE VTC'97*, Phoenix, AZ, May 1997, 1331–1335.

38. M. Zorzi and R.R. Rao, Error control and energy consumption in communications for nomadic computing, *IEEE Trans. Comput.* (special issue on Mobile Computing), 46, 279–289, March 1997.

39. V.K. Garg and E.L. Sneed, Digital wireless local loop system, *IEEE Commun. Mag.*, 34 (10), 112–115, October 1996.

40. J. Gardiner, The role of satellites in PCS, in *Personal Communication Systems and Technologies*, J. Gardiner and B. West, Eds., Artech House, Boston, 1995.

41. M. Zorzi and R.R. Rao, The role of error correlations in the design of protocols for packet switched services, in *Proc. 35th Annual Allerton Conference*, September 1997.

42. *IEEE Communications Magazine*, special issue on Software Radios, May 1995.

43. V.K. Garg and J.E. Wilkes, *Wireless and Personal Communications Systems*, Prentice-Hall, Upper Saddle River, NJ, 1996.

44. D.J. Goodman, *Wireless Personal Communications Systems*, Addison-Wesley, Reading, MA, 1997.

45. W.C. Jakes, Jr., *Microwave Mobile Communications*, John Wiley & Sons, New York, 1974.

46. K. Pahlavan and A.H. Levesque, *Wireless Information Networks*, Wiley, New York, 1995.

47. W.C.Y. Lee, *Mobile Communications Design Fundamentals*, 2nd ed., Wiley, New York, 1993.

48. T.S. Rappaport, *Wireless Communications: Principles and Practice*, Prentice-Hall, Upper Saddle River, NJ, 1996.

49. K.-C. Chen, Medium access control of wireless LANs for mobile computing, *IEEE Network*, 8, 50–63, 1994.

50. D. Cox, Wireless personal communications: what is it? *IEEE Personal Commun.*, 2, 20–35, April 1995.

51. K.S. Gilhousen et al., On the capacity of a cellular CDMA system, *IEEE Trans. Veh. Technol.*, VT-40, 303–312, May 1991.

52. H. Hashemi, The indoor radio propagation channel, *Proc. of the IEEE*, 81, 943–968, July 1993.

53. V.O.K. Li and X. Qiu, Personal communication systems (PCS), *Proc. IEEE*, 83, 1210–1243, September 1995.

54. R.O. LaMaire et al., Wireless LANs and mobile networking: standards and future directions, *IEEE Commun. Mag.*, 34, 86–94, August 1996.

55. K. Pahlavan and A.H. Levesque, Wireless data communcations, *Proc. IEEE*, 82, 1398–1430, September 1994.

56. R.L. Pickholtz, L.B. Milstein, and D.L. Schilling, Spread spectrum for mobile communications, *IEEE Trans. Vehic. Technol.*, 40 (2), 313–322, May 1991.

57. M. Schwartz, Network management and control issues in multimedia wireless networks, *IEEE Personal Commun.*, 2, 8–16, June 1995.

Further Reading

The purpose of this section was essentially to give a broad overview of the issues involved in the provision of telecommunications services in wireless and mobile networks, by describing the key technical challenges, current system designs, and future trends. Because of the great variety of issues and the large amount of research which has been done on this topic (fueled by the tremendous interest in wireless communications), it is clearly impossible to give any level of detail in such a limited space. The interested reader will find many in-depth books and technical articles which address many specific aspects of this extremely rich field. Books on the fundamentals of mobile communications and of cellular systems include References 43 through 48. Tutorial articles covering various aspects of this topic and reporting extensive literature citations include References 11, 33, and 49 through 57.

Articles dealing with various aspects of wireless communications systems can be found in the technical literature. Journals such as the *IEEE Transactions on Communications,* the *IEEE Journal on Selected Areas in Communications* (which publishes single-topic special issues), and the *IEEE Communications Magazine* (intended for a broader audience and more of a tutorial nature) deal with all aspects of communication theory and networking. Of particular interest are the following recent special issues. In the *IEEE Journal on Selected Areas in Communications:* Mobile Satellite Communications (Feb. 1995), Mobile and Wireless Computing Networks (June 1995), Wireless Local Communications (April/May 1996), CDMA Networks (Oct./Dec. 1996), Wireless ATM (Jan. 1997), Networking Issues for Mobile Communications (Sept. 1997), Personal Communications Services (Oct. 1997). Issues to be published in 1998 include Wireless Access Broadband Networks, Signal Processing for Wireless Communications, Software Radios, Network Multi-Media Radios. In the *IEEE Communications Magazine:* Wireless Personal Communications (Jan. 1995), High-Speed Wireless Networks (March 1995), Mobility and Intelligent Networks (Jun. 1995), European Mobile Communications (Feb. 1996), Wireless Networks (Sept. 1996, August 1997), Wireless Broadband Communications Systems (Jan. 1997), Introduction to Mobile and Wireless ATM (Nov. 1997).

Periodicals entirely devoted to research on wireless communications and networks include the *IEEE Transactions on Vehicular Technology,* the *IEEE Personal Communications* magazine, the *ACM/Baltzer Journal of Wireless Networks,* and several other new journals started in the past few years to collect the many technical contributions in the field. Many conferences on wireless systems are also held every year.

3

Communication Technologies

Biao Chen
University of Texas at Dallas

Wei Zhao
Texas A&M University

Nicholas Malcolm
Hewlett-Packard (Canada) Ltd.

Izhak Rubin
University of California, Los Angeles

John (Toby) Jessup
US West Communications

Chris B. Autry
Lucent Technologies

Henry L. Owen
Georgia Institute of Technology

Zygmunt Haas
Cornell University

3.1 FDDI/CDDI and Real-Time Communications

Biao Chen, Wei Zhao, and Nicholas Malcolm

3.1.1 Introduction

The Fiber Distributed Data Interface (FDDI) standard was produced by the ANSI X3T9.5 standards committee in mid-1980s and later approved by International Organization for Standardization (ISO) as an international standard. FDDI specifies a 100-Mbps, token-passing, dual-ring local-area network (LAN) using a fiber-optic transmission medium. High transmission speed and a bounded access time[19] make FDDI very suitable for supporting real-time applications. Furthermore, the FDDI built-in support for fault-tolerant communication makes it possible to migrate mission-critical applications from large computers to networks. Several new civil and military networks have adopted FDDI as the backbone network.

To achieve efficiency, a high-speed network such as an FDDI network requires simple (that is, low overhead) protocols. In a token ring network, the simplest protocol is the *token-passing* protocol. In this protocol, a node transmits its message whenever it receives the token. After it completes its transmission, the node passes on the token to the next neighboring node. Although it has the least overhead, the token-passing protocol cannot bound the time between two consecutive visits of the token to a node (called *token rotation time*), which makes it incapable of guaranteeing message delay requirements. The *timed token protocol*, proposed by Grow,[8] overcomes this problem. This protocol has been adopted as a standard for the FDDI networks. The idea behind the timed token protocol is to control the token rotation time. As a result, FDDI is able to support real-time applications.

In the rest of this section, we will first introduce the architecture of FDDI networks and discuss its fault management capabilities. We then address how to use FDDI to support time-critical applications.

3.1.2 FDDI/CDDI Network Architecture

3.1.2.1 FDDI Specifications

FDDI is defined by four separate specifications:

- *Media Access Control* (MAC): Defines how the medium is accessed, including frame format, token handling, addressing, algorithm for calculating a cyclic redundancy check value, and error recovery mechanisms.
- *Physical Layer Protocol* (PHY): Defines data encoding/decoding procedures, clocking requirements, framing, and other functions.
- *Physical Layer* Medium (PMD): Defines the characteristics of the transmission medium, including the fiber-optic link, power levels, bit error rates, optical components, and connectors.
- *Station Management* (SMT): Defines the FDDI station configuration, ring configuration, and ring control features, including station insertation and removal, initialization, fault isolation and recovery, scheduling, and collection of statics.

Although it operates at faster speeds, FDDI is similar in many ways to token ring. The basic architecture of an FDDI network consists of nodes connected by two counterrotating loops as illustrated in Figure 3.1a. A node in an FDDI network can be a single attachment station (SAS), a dual attachment station (DAS), a single attachment concentrator (SAC), or a dual attachment concentrator (DAC). Whereas stations constitute the sources and destinations for user frames, the concentrators provide attachments to other stations. The single attachment stations and concentrators are so called because they connect to only one of the two loops. The two fiber loops of an FDDI network are usually enclosed in a single cable. These two loops will be collectively referred to as the FDDI *trunk ring*. Physically, the rings consist of two or more point-to-point connections between adjacent stations. One of the two FDDI rings is

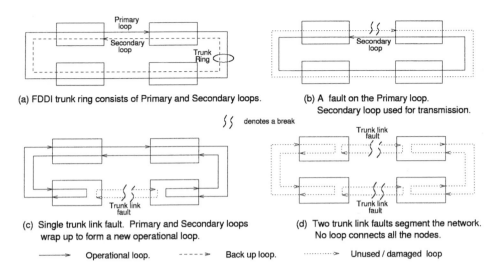

(a) FDDI trunk ring consists of Primary and Secondary loops.

(b) A fault on the Primary loop.
Secondary loop used for transmission.

ʃ ʃ denotes a break

(c) Single trunk link fault. Primary and Secondary loops wrap up to form a new operational loop.

(d) Two trunk link faults segment the network. No loop connects all the nodes.

⟶ Operational loop. - - - - ➤ Back up loop. ⋯⋯➤ Unused / damaged loop

FIGURE 3.1 FDDI/CDDI architecture.

called the *primary* ring; the other is called the *secondary* ring. The primary ring is used for data transmission, while the secondary ring is generally used as a backup.

3.1.2.2 Fault Management

The FDDI standards have been influenced by the need to provide a certain degree of built-in fault tolerance. In ring networks such as FDDT, faults are broadly classified into two categories: node faults and link faults. The FDDI standard specifies explicit mechanisms and procedures for detection of and recovery from both kinds of faults. These fault management capabilities of FDDI provide a foundation for capabilities of FDDI enable fault-tolerant operation. We present here a brief sketch of FDDIs fault management capabilities. For a comprehensive discussion, see Reference 1.

To deal with node faults, each node in FDDI is equipped with an optical bypass mechanism. Using this mechanism, a faulty node can be isolated from the ring, thus letting the network recover from a node fault. Link faults are handled by exploiting the dual-loop architecture of FDDI. In the event of a fault on the primary loop, the backup loop is used for transmission (Figure 3.1b).

Because of the proximity of the two fiber loops, a link fault on one loop is quite likely to be accompanied by a fault on the second loop as well, at approximately the same physical location. This is particularly true of faults caused by destructive forces. Such link faults, with both loops damaged, may be termed *trunk link faults.*

An FDDI network can recover from a single trunk link fault using the so-called wrap-up operation. This operation is illustrated in Figure 1.3c. The wrap-up operation consists of connecting the primary loop to the secondary loop. Once a link fault is detected, the two nodes on either side of the fault perform the wrap-up operation. This process isolates the fault and restores a single closed loop.

The fault detection and the wrap-up operation is performed by a sequence of steps defined by the FDDI link fault management. Once the network is initialized, each node continuously executes the link fault management procedure. The flowchart of the link fault management procedure is shown in Figure 3.2. Figure 3.2 only presents the gist of the fault management procedure. For details, the reader is referred to Reference 1. The major steps of this procedure are as follows:

- An FDDI node detects a fault by measuring the time elapsed since the previous token visit to that node. At network initialization time, a constant value is chosen for a protocol parameter called the target token rotation time (TTBI). The FDDI protocol ensures that under fault-free operating conditions, the duration between two consecutive token visits to any node, is not more than two

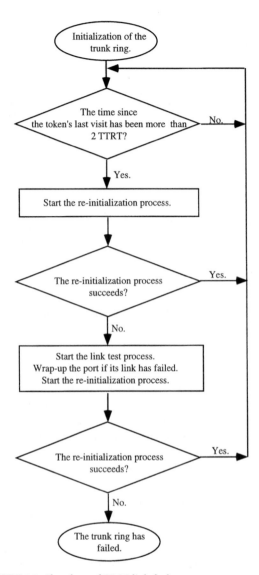

FIGURE 3.2 Flowchart of FDDI link fault management procedure.

TTBT. Hence, a node initiates the fault tracing procedure when the elapsed time since the previous token arrival at the node exceeds two TTRL

- The node first tries to reinitialize the network (using claim frames). If the fault was temporary, the network will recover. Several nodes may initiate the reinitialization procedure asynchronously. However, the procedure converges within TTRT time (which is usually on the order of a few milliseconds). The details of this procedure are beyond the scope of this section. The reader is referred to Reference 1.

- If the reinitialization procedure times out, then the node suspects a permanent fault in the trunk ring. The nodes then execute a series of further steps to locate the fault domain of the ring. The nodes within the fault domain perform link tests to check the status of the links on all their ports. If a link is faulty, the ring is wrapped up at the corresponding port. The node then tries to reinitialize the wrapped up ring. Once again, note that these operations are carried out concurrently and asynchronously by all the nodes in the trunk ring. That is, any of them may initiate the series of fault recovery steps but their actions will eventually converge to locate and recover from the fault(s).

- In the event that the last step has also failed, the network is considered unrecoverable. It is possible that operator intervention is required to solve the problem.

With the basic fault management mechanism just described, an FDDI network may not be able to tolerate two or more faults. This is illustrated in Figure 3.1d where two trunk link faults leave the network disconnected. Hence, the basic FDDI architecture needs to be augmented to meet the more stringent fault tolerance requirements of mission-critical applications. In addition, the bandwidth of a single FDDI ring may not be sufficient for some high-bandwidth applications. FDDI-based reconfigurable networks (FBRNs) provide increased fault tolerance and transmission bandwidth as compared with standard FDDI networks (see References 3, 6, and 12) for a discussion on this subject.

3.1.2.3 CDDI

The high cost of fiber-optic cable has been a major impediment to the widespread deployment of FDDI to desktop computers. At the same time, shielded twisted-pair (STP) and unshielded twisted-pair (UTP) copper wire is relatively inexpensive and has been widely deployed. The implementation of FDDI over copper wire is known as Copper Distributed Data Interface (CDDI).

Before FDDI could be implemented over copper wire, a problem had to be solved. When signals strong enough to be reliably interpreted as data are transmitted over twisted-pair wire, the wire radiates electromagnetic interference (EM). Any attempt to implement FDDI over twisted-pair wire had to ensure that the resulting energy radiation did not exceed the specifications set in the U.S. by the Federal Communications Commission (FCC) and in Europe by the European Economic Council (ECC). Three technologies have been used to reduce energy radiation: scrambling, encoding, and equalization. In June 1990, ANSI established a subgroup called the Twisted Pair-Physical Mediam Dependent (TP-PMD) working group to develop a specification for implementing FDDI protocols over twisted-pair wire. ANSI approved the TP-PMD standard in February 1994.

3.1.3 The Protocol and Its Timing Properties

FDDJ uses the timed token protocol.[8] The timed token protocol is a token-passing protocol in which each node gets a guaranteed share of the network bandwidth. Messages in the timed token protocol are segregated into two separate classes: the *synchronous* class and the *asynchronous* class.[8] Synchronous messages are given a guaranteed share of the network bandwidth, and are used for real-time communication.

The idea behind the timed token protocol is to control the token rotation time. During network initialization, a protocol parameter called the target token rotation time (TTRT) is determined which indicates the expected token rotation time. Each node is assigned a portion of the TTRT, known as its *synchronous bandwidth* (H_i), which is the maximum time a node is permitted to transmit synchronous messages *every time* it receives the token. When a node receives the token, it transmits its synchronous messages, if any, for a time no more than its allocated synchronous bandwidth. It can then transmit its asynchronous messages only if the time elapsed since the previous token departure from the same node is less than the value of TTRT, that is, only if the token arrives earlier than expected.

To use the timed token protocol for real-time communications, that is, to guarantee that the deadlines of synchronous messages are met, parameters such as the synchronous bandwidth, the target token rotation time, and the buffer size must be chosen carefully.

- The synchronous bandwidth is the most critical parameter in determining whether message deadlines will be met. If the synchronous bandwidth allocated to a node is too small, then the node may not have enough network access time to transmit messages before their deadlines. Conversely, large synchronous bandwidths can result in a long token rotation time, which can also cause message deadlines to be missed.

- Proper selection of TTRT is also important. Let τ be the token walk time around the network. The proportion of time taken due to token walking is given by τ/TTRT. The maximum network utilization available to user applications is then 1 − τ/TTRT.[24] A smaller TTRT results in less available utilization and limits the network capacity. On the other hand, if TTRT is too large, the token may not arrive at a node soon enough to meet message deadlines.
- Each node has buffers for outgoing and incoming synchronous messages. The sizes of these buffers also affect the real-time performance of the network. A buffer that is too small can result in messages being lost due to buffer overflow. A buffer that is too large is wasteful of memory.

In Section 3.1.4, we will discuss parameter selection for real-time applications. Before that, we need to define network and message models and investigate protocol properties.

3.1.3.1   The Network and the Message Models

The network contains n nodes arranged in a ring. Message transmission is controlled by the timed token protocol, and the network is assumed to operate without any faults. Outgoing messages at a node are assumed to be queued in first-in and first-out (FIFO) order.

Recall that the token walk time is denoted by τ, which includes the node-to-node delay and the token transmission time. Here τ is the portion of TTRT that is not available for message transmission. Let α be the ratio of τ to TTRT, that is, $\alpha = \tau/TTRT$. Then α represents the proportion of time that is not available for transmitting messages.

There are n streams of synchronous messages, $S_1, ..., S_n$, with stream S_i originating at node i. Each synchronous message stream S_i may be characterized as $S_i = (C_i, P_i, D_i)$:

- C_i is the maximum amount of time required to transmit a message in the stream. This includes the time to transmit both the payload data and the message headers.
- P_i is the interarrival period between messages in the synchronous message stream. Let the first message in stream S_i arrive at node i at time $t_{i,1}$. The jth message in stream S_i will arrive at node i at time $t_{i,j} = t_{i,1} + (j-1)P_i$, where $j \geq 1$.
- D_i is the *relative deadline* of messages in the stream. The relative deadline is the maximum amount of time that may elapse between a message arrival and completion of its transmission. Thus, the transmission of the jth message in stream S_i, which arrives at $t_{i,j}$, must be completed by $t_{i,j} + D_i$, which is the *absolute deadline* of the message. To simplify the discussion, the terms *relative* and *absolute* will be omitted in the remainder of the section when the meaning is clear from the context.

Throughout this section, we make no assumptions regarding the destination of synchronous message streams. Several streams may be sent to one node. Alternatively, multicasting may occur in which messages from one stream are sent to several nodes.

If the parameters (C_i, P_i, D_i) are not completely deterministic, then their worst-case values must be used. For example, if the period varies between 80 and 100 ms, a period of 80 ms must be assumed. Asynchronous messages that have time constraints can have their deadlines guaranteed by using pseudo synchronous streams. For each source of time-constrained asynchronous messages, a corresponding synchronous message stream is created. The time-constrained asynchronous messages are then promoted to the synchronous class and sent as synchronous messages in the corresponding synchronous message stream.

Each synchronous message stream places a certain load on the system. We need a measure for the load. We define the *effective utilization*, U_i, of stream S_i as follows:

$$U_i = \frac{C_i}{\min(P_i, D_i)}. \tag{3.1}$$

Because a message of length C_i arrives every P_i time, $U_i = C_i/P_i$ is usually regarded as the load presented by stream S_i. An exception occurs when $D_i < P_i$. A message with such a small deadline must be sent relatively urgently, even if the period is very large. Thus $U_i = C_i/D_i$ is used to reflect the load in this case. The total effective utilization U of a synchronous message set can now be defined as

$$U = \sum_{i=1}^{n} U_i.$$ (3.2)

The total effective utilization is a measure of the demands placed on the system by the entire synchronous message set. To meet message deadlines, it is necessary that $U \leq 1$.

3.1.3.2 Constraints

The timed token protocol requires several parameters to be set. The synchronous bandwidths, the target token rotation time, and the buffer sizes are crucial in guaranteeing the deadlines of synchronous messages. Any choice of these parameters must satisfy the following constraints.

The Protocol Constraint. This constraint states that the total bandwidth allocated to synchronous messages must be less than the available network bandwidth, that is,

$$\sum_{i=1}^{n} H_i \leq \text{TTRT} - \tau.$$ (3.3)

This constraint is necessary to ensure stable operation of the timed token protocol.

The Deadline Constraint. This constraint simply states that every synchronous message must be transmitted before its (absolute) deadline. Formally, let $s_{i,j}$ be the time that the transmission of the j-th message in stream S_i is completed. The deadline constraint requires that for $i = 1, \ldots, n$ and $j = 1, 2, \ldots,$

$$s_{i,j} \leq t_{i,j} + D_i,$$ (3.4)

where $t_{i,j}$ is the arrival time and D_i is the (relative) deadline. Note that in the above inequality, $t_{i,j}$ and D_i are given by the application, but $s_{i,j}$ depends on the synchronous bandwidth allocation and the choice of TTRT.

The Buffer Constraint. This constraint states that the size of the buffers at each node must be sufficient to hold the maximum number of outgoing or incoming synchronous messages that could be queued at the node. This constraint is necessary to ensure that messages are not lost due to buffer overflow.

3.1.3.3 Timing Properties

According to Equation 3.4 the deadline constraint of message M is satisfied if and only if the minimum synchronous bandwidth available to M within its deadline is bigger than or equal to its message size. That is, for a given TTRT, whether or not deadline constraints can be satisfied is solely determined by bandwidth allocation. To allocate appropriate synchronous bandwidth to a message stream, we need to know the worst-case available synchronous bandwidth for a stream within any time period.

Sevcik and Johnson[19] showed that the maximum amount of time that may pass between two consecutive token arrivals at a node can approach two TTRT. This bound holds regardless of the behavior of asynchronous messages in the network. To satisfy the deadline constraint, it is necessary for a node to have at least one opportunity to send each synchronous message before the message deadline expires. Therefore, in order for the deadline constraint to be satisfied, it is necessary that for $i = 1, \ldots, n,$

$$D_i \geq 2\text{TTRT}.$$ (3.5)

It is important to note that Equation 3.5 is only a necessary but not a sufficient condition for the deadline constraint to be satisfied. For all message deadlines to be met it is also crucial to choose the synchronous bandwidths H_i appropriately.

Further studies on timing properties have been performed in Reference 5 and the results can be summarized by the following theorem.

Theorem 1 Let $X_i(t, t + I, \vec{H})$ be the minimum total transmission time available for node i to transmit its synchronous message within the time interval $(t, t + I)$ under bandwidth allocation $\vec{H} = (H_1, H_2, \ldots, H_n)_{T2}$ then

$$X_i\left(t, t+I, \vec{H}\right) = \begin{cases} \left[\dfrac{I}{\text{TTRT}} - 1\right] \cdot H_i + \max\left(0, \min\left(r_i - \tau - \Sigma_{j=1,\ldots,n,\, j\neq i} H_j, H_i\right)\right) & \text{if } I \geq \text{TTRT}, \\ 0 & \text{otherwise} \end{cases}$$

where $r_i = I - \lfloor I/\text{TTRT} \rfloor \cdot \text{TTRT}$.

By Theorem 1, deadline constraint (Equation 3.4) is satisfied if and only if for $i = 1, \ldots, n$,

$$X_i\left(t, t+D_i, \vec{H}\right) \geq C_i. \tag{3.6}$$

3.1.4 Parameter Selection for Real-Time Applications

To support real-time applications on FDDI, we have to set the following three types of parameters properly: synchronous bandwidth, TTRT, and buffer size. We address this parameter selection problem in this section.

3.1.4.1 Synchronous Bandwidth Allocation

In FDDI synchronous bandwidths are assigned by a synchronous bandwidth allocation scheme. This section examines synchronous bandwidth allocation schemes, and discusses how to evaluate their effectiveness.

3.1.4.1.1 A Classification of Allocation Schemes

Synchronous bandwidth allocation schemes may be divided into two classes: local allocation schemes and global allocation schemes. These schemes differ in the type of information they may use. A *local* synchronous bandwidth allocation scheme can only use information available locally to node i in allocating H_i. Locally available information at node i includes the parameters of stream S_i (that is, C_i, P_i, and D_i). TTRT and τ are also locally available at node i, because these values are known to all nodes. On the other hand, a *global* synchronous bandwidth allocation scheme can use global information in its allocation of synchronous bandwidth to a node. Global information includes both local information and information regarding the parameters of synchronous message streams originating at other nodes.

A local scheme is preferable from a network management perspective. If the parameters of stream S_i change, then only the synchronous bandwidth H_i of node i needs to be recalculated. The synchronous bandwidths at other nodes do not need to be changed because they were calculated independently of S_i. This makes a local scheme flexible and suited for use in dynamic environments where synchronous message streams are dynamically initiated or terminated.

In a global scheme, if the parameters of S_i change, it may be necessary to recompute the synchronous bandwidths for all nodes. Therefore, a global scheme is not well suited for a dynamic environment. In addition, the extra information employed by a global scheme may cause it to handle more traffic than a local scheme. However, it is known that local schemes can perform *very closely* to the optimal synchronous bandwidth allocation scheme when message deadlines are equal to message periods. Consequently, given the previously demonstrated good performance of local schemes and their desirable network management properties, we concentrate on local synchronous bandwidth allocation schemes in this chapter.

3.1.4.1.2 A Local Allocation Scheme

Several local synchronous bandwidth allocation schemes have been proposed, for both the case of $D_i = P_i^2$ and the case of $D_i \neq P_i$.[13,26] These schemes all have similar worst-case performance. Here we will consider the scheme proposed in Reference 13. With this scheme, the synchronous bandwidth for node i is allocated according to the following formula:

$$H_i = \frac{U_i D_i}{\dfrac{D_i}{\text{TTRT}} - 1}. \tag{3.7}$$

Intuitively, this scheme follows the flow conservation principle. Between the arrival of a message and its absolute deadline, which is D_i time later, node i will have at least $[D_i/\text{TTRT} - 1]H_i$ of transmission time available for synchronous messages by Equation 3.6. This transmission time is available regardless of the number of asynchronous messages in the network. During the D_i time, $U_i D_i$ can loosely be regarded as the load on node i. Thus the synchronous bandwidth in Equation 3.7 is just sufficient to handle the load on node i between the arrival of a message and its deadline.

The scheme defined in Equation 3.7 is a simplified version of those in References 13 and 26. In the rest of this section, we assume this scheme is used because it is simple, intuitive, and well understood. Another reason for concentrating on this scheme is that it has been adopted for use with the SAFENET standard, and thus will be used in distributed real-time systems in the future.

3.1.4.1.3 Schedulability Testing

We now consider schedulability testing. A message set is schedulable if the deadlines of its synchronous messages can be satisfied. This can be determined by referring to the tests that both the protocol and deadline constraints (Equations 3.3 and 3.4 or Equation 3.6) are satisfied.

Testing if the protocol constraint is satisfied is very straightforward. But the test of deadline constraints is more complicated and requires more information. This test can be greatly simplified if the bandwidths are allocated according to Equation 3.7. It was shown that if this scheme is used, the protocol constraint condition defined in Equation 3.3 implies the deadline constraint condition (Equation 3.4). Testing the protocol constraint alone is sufficient to ensure that both constraints are satisfied. This is a big advantage of using the allocation scheme defined in Equation 3.7.

A second method of schedulability testing is to use the *worst-case achievable utilization* criteria. This criteria has been widely used in real-time systems. For a synchronous bandwidth allocation scheme, the worst-case achievable utilization U^* defines an upper bound on the effective utilization of a message set: if the effective utilization is no more than the upper bound, both the protocol and the deadline constraints are always satisfied. The worst-case achievable utilization U^* for the scheme defined in Equation 3.7 is

$$U = \frac{\left[\dfrac{D_{\min}}{\text{TTRT}}\right] - 1}{\left[\dfrac{D_{\min}}{\text{TTRT}}\right] + 1}(1 - \alpha), \tag{3.8}$$

where TTRT is the target token rotation time and $D_{\min}$ is the minimum deadline. For any message set, if its effective utilization (Equation 3.2) is less than U^*, both protocol and deadline constraints are guaranteed to be satisfied.

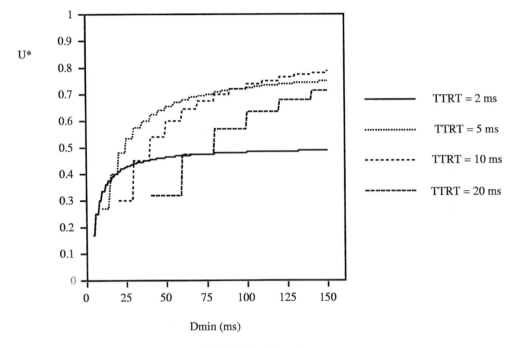

FIGURE 3.3 U^* vs. D_{min}.

We would like to emphasize that in practice, using these criteria can simplify network management considerably. The parameters of a synchronous message set can be freely modified while still maintaining schedulability, provided that the effective utilization remains less than U^*.

Let us examine the impact of TTRT and D_{min} on the worst-case achievable utilization given in Equation 3.8. Figure 3.3 shows the worst-case achievable utilization vs. D_{min} for several different values of TTRT. This figure was obtained by plotting Equation 3.8, with τ taken to be 1 ms (a typical value for an FDDI network). Several observations can be made from Figure 3.3 and Equation 3.8.

1. For a fixed value of TTRT, the worst-case achievable utilization increases as D_{min} increases. From Equation 3.8 it can be shown that when D_{min} approaches infinity, U^* approaches $(1 - \alpha) = (1 - \tau/TTRT)$. That is, as the deadlines become larger, the worst-case achievable utilization approaches the available utilization of the network.

2. In Reference 2, it was shown that for a system in which all relative deadlines are equal to the corresponding message periods $(D_i = P_i)$, a worst-case achievable utilization of $\frac{1}{3}(1 - \alpha)$ can be achieved. That result can be seen as a special case of Equation 3.8: if D_{min} = two TTBT, we have $[D_{min}/TTRT] = 2$ and $U^* = \frac{1}{3}(1 - \alpha)$.

3. TTRT clearly has an impact on the worst-case achievable utilization. From Figure 3.3, we see that when D_{min} = 50 ms, the case of TTRT = 5 ms gives a higher worst-case achievable utilization than the other plotted values of TTRT. When D_{min} = 125 ms, the case of TTRT = 10 ms gives a higher U^* than the other plotted values of TTRT. This observation provides motivation for maximizing U^* by properly selecting TTRT once D_{min} is given.

3.1.4.2 Selection of Target Token Rotation Time

Recall that TTRT, the target token rotation time, determines the expected value of the token rotation time. In contrast to the synchronous bandwidths of individual nodes, TTRT is common to all the nodes and should be kept constant during run time. As we observed in the last section, TTRT has an impact on the worst-case achievable utilization. Thus, we would like to choose TTRT in an optimal fashion, so

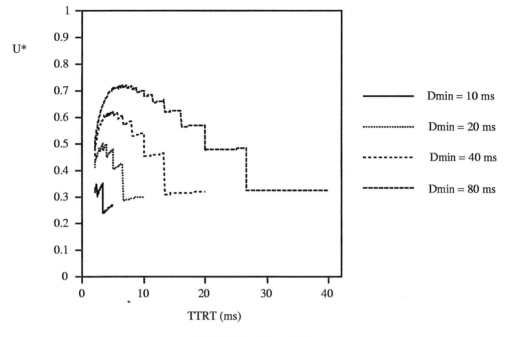

FIGURE 3.4 U^* vs. TTRT.

that the worst-case achievable utilization U^* is maximized. This will increase the amount of real-time traffic that can be supported by the network.

The optimal value of TTRT has been derived in Reference 13 and is given by

$$TTRT = \frac{D_{min}}{\dfrac{-3 + \sqrt{9 + \dfrac{8 D_{min}}{\tau}}}{2}}. \qquad (3.9)$$

The impact of an appropriate selection of TTRT is further evident from Figure 3.4, which uses Equation 3.9 to show U^* vs. TTRT for several different values of D_{min}. As with Figure 3.3, τ is taken to be 1 ms. From Figure 3.4 the following observations can be made:

1. The curves in Figure 3.4 verify the prediction of the optimal TTRT value given by Equation 3.9. For example, consider the case of D_{min} = 40 ms. By Equation 3.9, the optimal value of TTRT is 5 ms. The curve clearly indicates that at TTRT = 5 ms the worst-case achievable utilization is maximized. Similar observations can be made for the other cases.
2. As indicated in Equation 3.9, the optimal TTRT is a function of D_{min}. This coincides with the expectations from the observations of Figure 3.3. A general trend is that as D_{min} increases, the optimal TTRT increases. For example, the optimal values of TTRT are approximately 2.5 ms for D_{min} = 10 ms, 4 ms for D_{min} = 20 ms, 5 ms for D_{min} 40 ms, and 6.67 ms for D_{min} = 80 ms.
3. The choice of TTRT has a large effect on the worst-case achievable utilization U^*. Consider the case of D_{min} = 40 ms shown in Figure 3.4. If TTRT is too small (say, TTRT = 2 ms), U^* can be as low as 45%. If TTRT is too large (say TTRT = 15 ms), U^* can be as low as 31%. However, when the optimal value of TTRT is used (that is, TTRT = 5 ms), U^* is 62%. This is an improvement of 17 and 31%, respectively, over the previous two cases.

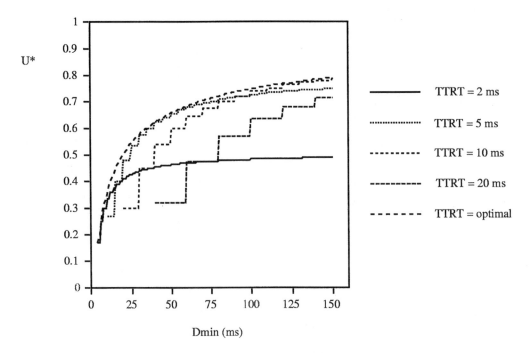

FIGURE 3.5 U^* vs. $D_{\min}$ with optimal TTRT.

The effect of choosing an optimal value for TTRT is also shown in Figure 3.5. Figure 3.5 shows the worst-case achievable utilization vs. $D_{\min}$ when an optimal value of TTRT is used. For ease of comparison, the earlier results of Figure 3.3 are also shown in Figure 3.5.

3.1.4.3 Buffer Requirements

Messages in a network can be lost if there is insufficient buffer space at either the sending or the receiving node. To avoid such message loss, we need to study the buffer space requirements.

There are two buffers for synchronous messages on each node. One buffer is for messages waiting to be transmitted to other nodes, and is called the *send buffer*. The other buffer is for messages that have been received from other nodes and are waiting to be processed by the host. This buffer is called *receive buffer*.

We consider the send buffer first.

- *Case of $D_i \leq P_i$.* In this case, a message must be transmitted within the period it arrives. At most, one message will be waiting. Hence the send buffer need only accommodate one message.

- *Case of $D_i > P_i$.* In this case, a message may wait as long as D_i time without violating its deadline constraint. At node i, messages from stream S_i arrive every P_i time, requesting transmission. During the D_i time following the arrival of a message, there will be further $[D_i/P_i]$ message arrivals. Thus, the send buffer may potentially have to accommodate $[D_i/P_i] + 1$ messages. When the deadline is very large (10 or 100 times larger than the period, for example, which can occur in voice transmission), this might become a problem in practice due to excessive buffer requirements.

  However, when the synchronous bandwidths are allocated as in Equation 3.7, the impact of D_i on required buffer size is limited: it can be shown that if Equation 3.7 is used, the waiting time of a message from stream S_i is bounded by $\min(D_i, P_i + 2\,TTRT)$.[14] Thus, the maximum number of messages from S_i that could be queued at node i is no more than $[(\min(D_i, P_i + 2TTRT))\,/\,P_i]$. This allows us to reduce the required size of the send buffer. Let BS_i denote the send buffer size at node i. The send buffer will never overflow if

$$BS_i \geq \left\lceil \frac{\min\left(D_i, P_i + 2\,\text{TTRT}\right)}{P_i} \right\rceil B_i, \tag{3.10}$$

where B_i is the number of bytes in a message of stream S_i.

An interesting observation is that if $D_i \geq P_i + 2\,\text{TTRT}$, then

$$BS_i \geq \left\lceil \frac{P_i + 2\,\text{TTRT}}{P_i} \right\rceil B_i. \tag{3.11}$$

That is, the send buffer requirements for stream S_i are independent of the deadline D_i. As mentioned earlier, one would expect that increasing the deadline of a stream could result in increased buffer requirements. When the scheme in Equation 3.7 is used, however, the buffer requirements are not affected by increasing the deadline once the deadline reaches a certain point.

Now let us consider the receive buffer size. Suppose that node j is the destination node of messages from nodes $j_1, \ldots, j_k$. This means that messages from streams $S_{j1}, \ldots, S_{jk}$ are being sent to node j. The size of the receive buffer at node j depends not only on the message traffic from these streams, but also on the speed at which the host of node j is able to process incoming messages from other nodes. The host at node j has to be able to keep pace with the incoming messages; otherwise the receive buffer at node j can overflow. When a message from stream S_{ji} arrives at node j, we assume that the host at node j can process the message within P_{ji} time.

Assuming that the host is fast enough, then it can be shown that the number of messages from S_i that could be queued at the destination node is bounded by $[(\min(D_i, P_i + 2\,\text{TTRT})) / P_i] + 1$. With this bound, we can derive the space requirements for the receive buffer at node j. Let BR_j denote the receive buffer size at node j. The receive buffer will never overflow if

$$BR_j \geq \sum_{i=1}^{k} \left\lceil \frac{\min\left(D_{j_i}, P_{j_i} + 2\,\text{TTRT}\right)}{P_{j_i}} + 1 \right\rceil B_{j_i}. \tag{3.12}$$

As in the case of the send buffer, we can observe from Equation 3.12 that if the deadlines are large, the required size of the receive buffer will not grow as the deadlines increase.

3.1.5 Final Remarks

In this section, we introduced the architecture of FDDI and discussed its fault-tolerant capability. We presented a methodology for the use of FDDI networks for real-time applications. In particular, we considered methods of selecting the network parameters and schedulability testing. The parameter selection methods are compatible with current standards. The synchronous bandwidth allocation method has the advantage of only using information local to a node. This means that modifications in the characteristics of synchronous message streams, or the creation of new synchronous message streams, can be handled locally without reinitializing the network. Schedulability testing that determines whether the time constraints of messages will be met were presented. They are simple to implement and computationally efficient.

The materials presented in this section complement much of the published work on FDDI. Since the early 1980s, extensive research has been done on the timed token protocol and its use in FDDI networks.[4,11,16,21] The papers by Ross,[18] Iyer and Joshi,[9] and others[15,22,23] provided comprehensive discussions on the timed token protocol and its use in the FDDI. Ulm[24] discussed the protocol performance with respect to parameters such as the channel capacity, the network cable length, and the number of stations. Dykeman and Bux[7] developed a procedure for estimating the maximum total throughput of asynchronous messages when using single and multiple asynchronous priority levels. The analysis done by Pang and Tobagi[17] gives insight into the relationship between the bandwidth allocated to each class of traffic and the timing parameters. Valenzo et. al.[25] concentrated on the asynchronous throughput and the average token rotation time when the asynchronous traffic is heavy. The performance of the timed token ring depends on both the network load and the system parameters. A study on FDDT by Jam[10] suggests that a value of 8 ms for TTRT is desirable as it can achieve 80% utilization on all configurations and results in around 100 ms maximum access delay on typical rings. Further studies have been carried out by Sankar and Yang[20] to consider the influence of the TTRT on the performance of varying FDDI ring configurations.

References

1. FDDI Station Management Protocol (SMT), ANSI Standard X3T9.5/84-89, X3T9/92-067, August 1992.
2. G. Agrawal, B. Chen, and W. Zhao, Local synchronous capacity allocation schemes for guaranteeing message deadlines with the timed token protocol, *Proc. IEEE Infocom'93*, 186–193, 1993.
3. G. Agrawal, S. Kamat, and W. Zhao, Architectural support for FDDI-based reconfigurable networks, presented at Workshop on Architectures for Real-Time Applications (WARTA), 1994.
4. B. Albert and A. P. Jayasumana, *FDDI and FDDI-II Architecture, Protocols, and Performance*, Artech House, Boston, 1994.
5. B. Chen, H. Li, and W. Zhao. Some timing properties of timed token medium access control protocol, *Proceedings of International Conference on Communication Technology*, June 1994, 1416–1419.
6. B. Chen, S. Kamart, and W. Zhao. Fault-tolerant real-time communication in FDDI-based networks, *Proc. IEEE Real-Time Systems Symposium*, 141–151, December 1995.
7. D. Dykeman and W. Bux, Analysis and tuning of the FDDI media access control protocol, *IEEE J. Selected Areas Commun.*, 6, (6), July 1988.
8. R. M. Grow, A timed token protocol for local area networks, *Proc. Electro/82, Token Access Protocols*, May 1982.
9. V. Iyer and S. P. Joshi, FDDIs 100 M-bps protocol improves on 802.5 Spec's 4-M-bps limit, *Electrical Design News*, 151–160, May 2, 1985.
10. R. Jam, Performance analysis of FDDI token ring networks: effect of parameters and guidelines for setting TTRT, *IEEE LTS*, 16–22, May 1991.
11. R. Jam, *FDDI Handbook — High-Speed Networking Using Fiber and Other Media*, Addison-Wesley, Reading, MA, 1994.
12. S. Kamat, G. Agrawal, and W. Zhao, On available bandwidth in fddi-based reconfigurable networks, *Proc. IEEE Infocom '94*, 1994.
13. N. Malcolm and W. Zhao, Guaranteeing synchronous messages with arbitrary deadline constraints in an FDDI network, *Proc. IEEE Conference on Local Computer Networks*, 186–195, Sept. 1993.
14. N. Malcolm, Hard Real-Time Communication in High Speed Networks, Ph.D. thesis, Department of Computer Science, Texas A&M University, College Station, 1994.
15. J. Mccool, FDDI — getting to the inside of the ring, *Data Commun.*, 185–192, March 1988.
16. A. Mills, *Understanding FDDI*, Prentice-Hall, Englewood Cliffs, NJ, 1995.
17. J. Pang and F. A. Tobagi, Throughput analysis of a timer controlled token passing protocol under heavy load, *IEEE Trans. Commun.*, 37, (7), 694–702, July 1989.
18. F. E. Ross, An overview of FDDI: the fiber distributed data interface, *IEEE J. Selected Areas Commun.*, 7, 1043–1051, September 1989.
19. K. C. Sevcik and M. J. Johnson, Cycle Time Properties of the FDDI Token Ring Protocol, *IEEE Trans. Software Eng.*, SE-13, (3), 376–385, 1987.

20. R. Sankar and Y. Y. Yang, Performance analysis of FDDI, *Proc. IEEE Conference on Local Computer Networks*, Minneapolis, MN, October 10–12, 1989, 328–332.

21. A. Shah and G. Ramakrishnan, FDDI: A High Speed Network, Prentice-Hall, Englewood Cliffs, NJ, 1994.

22. R. Southard, Fibre optics: A winning technology for LANs, *Electronics*, 111–114, February 1988.

23. W. Stallings, *Computer Communication Standards*, Vol 2: *Local Area Network Standards*, Howard W. Sams & Co., 1987.

24. J. N. Ulm, A timed token ring local area network and its performance characteristics, *Proc. IEEE Conference on Local Computer Networks*, 50–56, February 1982.

25. A. Valenzano, P. Montuschi, and L. Ciminiera, Some properties of timed token medium access protocols, *IEEE Trans. on Software Eng.*, 16, (8), August 1990.

26. Q. Zheng and K. C. Shin, Synchronous bandwidth allocation in FDDI networks, *Proc. ACM Multimedia '93*, 31–38, August 1993.

3.2 Multiple Access Communications Networks

Izhak Rubin

3.2.1 Introduction

Modern computer communications networks, particularly Local Area Networks (LANs) and Metropolitan Area Networks (MANs), employ multiple access communications methods to share their communications resources. A multiple access communications channel is a network system whose communications media are shared among distributed stations (terminals, computers, users). The stations are distributed in the sense that there exists no relatively low cost and low delay mechanism for a single controlling station to gain access to the status of all stations. If such a mechanism exists, the resulting sharing mechanism is identified as a multiplexing system.

The procedure used to share a multiple access communications medium is the multiple access algorithm. The latter provides for the control, coordination, and supervision of the sharing of the system communications resources among the distributed stations which transport information across the underlying multiple access communications network system.

In the following, we present a categorization of the various medium access control (MAC) methods employed for the sharing of multiple access communications channel systems. We demonstrate these schemes by considering applications to many different classes of computer communication networks, including wireline and wireless local and MANs, satellite communications networks, and local area optical communications networks.

3.2.2 Features of Medium Access Control Systems

Typically employed topologies for shared medium communications networks are shown in Figure 3.6, and are characterized as follows:[1]

a. A **star** topology, under which each station can directly access a single central node, is shown in Figure 3.6a. A switching star network results when the star node provides store and forward buffering and switching functions, while a broadcast star network involves the employment of the star node as an unbuffered repeater which reflects all the incoming signals into all outgoing lines.

A wired star configuration is shown in Figure 3.6a. In Figure 3.6a2, we show a wireless cell which employs radio channels for the mobile terminals to communicate with a central base station. The terminals use a multiple access algorithm to gain access to the shared (mobile to base-station, reverse access) radio channel(s), while the base station employs a multiplexing scheme for the transmission of its messages to the mobile terminals across the (base station to terminals, forward access) communications channels.

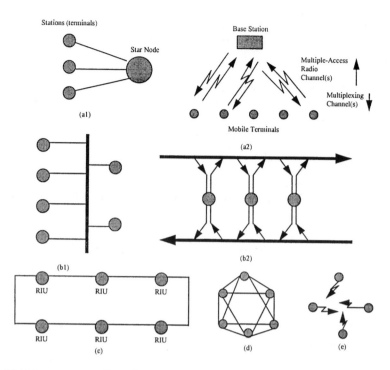

FIGURE 3.6 Multiple access network topologies: (a) star, (b) bus, (c) ring, (d) mesh, and (e) broadcast radio net.

b. A **bus** topology, under which stations are connected to a bus backbone channel, is shown in Figure 3.6b. Under a logical bus topology, the message transmissions of each station's are broadcasted to all network stations. Under a physical bus implementation, the station is passively connected to the bus, so that the failure of a station does not interfere with the operation of the bus network. The station is then able to sense and receive all transmissions that cross its interface with the bus; however, it is not able to strip information off the bus or to properly overwrite information passed along the bus; it can transmit its own messages across the bus when assigned to do so by the employed MAC protocol.

It is noted that when fiber links are used, due to the unidirectional nature of the channel, to achieve full station-to-station connectivity, a bus network implementation necessitates the use of two buses. As depicted in Figure 3.6b2, a common approach is to employ two separate counterdirectional buses. Stations can then access each fiber bus through the use of a proper read (sense) tap followed by a write tap.

c. A **ring** topology, under which the stations are connected by point-to-point links typically in a closed loop topology, is shown in Figure 3.6c. In a physical ring network implementation, each station connects to the ring through an active interface, so that transmissions across the ring pass through and are delayed in the register of the ring interface units (RIUs) they traverse. To increase network reliability, such implementations include special circuits for ensuring a rapid elimination of a failed RIU, to prevent such a failure from leading to a prolonged overall network failure. Due to the use of an active interface, the station is now able to strip characters or messages it receives off the medium, as well as to overwrite onto symbols and messages transmitted across the medium, when the latter pass its interface.

An active interface into the medium further enables the station to amplify the information it passes along, thus leading to considerable reduction of the insertion loss. This is of particular importance in interfacing a fiber-optic medium, whereby a passive interface causes a distinct insertion loss, thus leading to a significant limitation in the number of stations that can be passively connected to a fiber bus, without the incorporation of optical amplifiers.

For fiber-optic links, electrical amplification and MAC processing at the active interface involve double conversion: optical-to-electronic of the received signal and electronic-to-optical of the amplified transmitted signal. As a result, the access burst rate of the station needs to be selected so that it is compatible with the electronics processing rate at the station interface, so that no rate mismatch exists at the electronics/optics interfaces between the potentially very high transmission burst rate across the optical channel and the limited processing rate capacity of the electronically based VLSI access processor at the medium interface unit of the station.

   d. A **mesh** topology, under which the stations are connected by point-to-point links in a more diverse and spatially distributed (mesh) topology (Figure 3.6d). To traverse a mesh topology, switching nodes are required. Through the use of cross-connect switches, multiple embedded multiple access subnetworks are constructed. These subnetworks can be dynamically reconfigured (by adjusting the cross-connect matrices at the switches) to adapt to variations in network loading conditions and in interconnection patterns, and to allow the support of private virtual networks. Such an architecture, as demonstrated by the SMARTNet optical network, is presented in Section 3.2.8.

   e. A **broadcast** multiaccess radio net, as shown in Figure 3.6e provides, by the physical features of the radio propagation across the shared medium, a direct link between any pair of nodes (assuming the terrain topography to induce no blockages in the node-to-node line-of-sight clearances). A packet transmitted by a single station (node) will be received directly by all other stations. Note that this broadcast property is physically achieved also by wireline two-way bus network systems shown in Figure 3.6b1 and 3.6b2. In turn, the other network topologies shown in Figure 3.6 require signal repeating or retransmission protocols to be used to endow them with the broadcast property.

   In considering general high-speed local and metropolitan area networks, the following frame distribution topologies can be distinguished for basic implementations: broadcast (logical bus) topology and mesh (switching) topology.

A. Broadcast (Logical Bus) Topology

Under such a distribution method, station messages are routed through the use of a broadcasting method. Since each message frame (the MAC level Protocol Data Unit) contains addressing information, it is copied automatically from the medium by the intended destination station or stations, so that MAC routing is automatically achieved. Typically bus, ring, or broadcast star topologies are used to simplify the characterization of the broadcasting path.

   The communications link is set up as a bus (for a physical bus implementation), or as a point-to-point link (for a physical ring implementation). The corresponding Station Interface Unit (SIU) acts as a passive or active MAC repeater. A passive MAC node does not interfere with ongoing transmissions along the bus, while being able to copy the messages transmitted across the bus system.

   An active MAC station interface unit operates in one out of two possible modes:

   Repeat mode, whereby it performs as a repeater, serving to repeat the frame it receives.

   Nonrepeat mode, under which the SIU is not repeating the information it receives from the medium. In the latter case, under a logical bus configuration, the SIU is also typically set to be in a stripping mode, stripping from the medium the information it receives. During this time, the SIU is able to transmit messages across the medium, provided it gains permission to do so from the underlying MAC protocol. Note that if the SIU is provided with a store-and-forward capability, it can store all the information it receives while being in a nonrepeat mode, and retransmit this information, if so desired, at a subsequent time.

   A logical bus topology can also be associated with an active interface of the station onto the fiber bus/buses. This is, for example, the case for the IEEE 802.6 Distributed Queue Dual Bus (DQDB) implementation.[1] Under DQDB, the station can overwrite each passing bus bit through optical to electrical conversion, an electrical "or" write operation and an electrical to optical reconversion. The active interface of the station can provide it with the capacity to also strip bits, and thus message frames, off the bus.

   MAC procedures for logical bus configurations can also be characterized by the following features relating to the method used for removing message frames and the constraints imposed upon the number of simultaneous transmissions carried along the medium. As to the latter feature, we differentiate between the following implementations.

Frame transmission initiations in relation to the number of simultaneous transmissions along the logical bus are as follows:

a. **Single message frame transmission across the medium.** A single MAC-frame transmission is permitted across the medium at any time instant; thus, no more than one station can initiate a transmission onto the medium at any given time instant, and no other station can initiate a transmission until this later transmission is removed from the medium.

This is the technique employed by the IEEE 802.5 token-ring LAN under a late token release mode, whereby a station holding the token does not release it until it receives its own frame (following the latter's full circulation around the ring). This simplifies the operation of the protocol and provides the station with the ability to review its acknowledged message frame prior to releasing the token, thus enabling it to immediately retransmit its MAC message frame if the latter is determined to have not been properly received by the destination.

The latter token ring LAN uses twisted-pair or coaxial media at transmission rates of 4 to 16 Mbps. At 10 Mbps, the transmission time of a 1000 bit frame is equal to 100 μs, which is longer than the typical propagation time across the medium when the overall LAN length is shorter than 20 km, considering a propagation rate of about 5 μs/km.

In turn, when considering a 100 Mbps LAN or MAN fiber-optic-based system, such as the FDDI token ring or the DQDB reservation bus systems, which can span longer distances of around 100 km, the corresponding frame transmission time and network-wide propagation delay are equal to 10 and 500 μs, respectively. For a corresponding MAN system which operates at a channel rate of 1 Gbps, the propagation delay of 500 μs is much larger than the frame transmission time of 1 μs. Thus, under such high transmission rate conditions, each message transmission occupies only a small physical length of the logical bus network medium. It is therefore not efficient, from message delay and channel bandwidth MAC utilization considerations, to provide for only a single transmission at a time across the logical bus medium. The following mode of operation is thus preferred. (The following mode, b.1, can also be used for the token ring network under the early token release option, which can lead to performance improvements at higher data rate levels.)

b. **Multiple simultaneous frame** transmissions along the logical bus medium. While permitting multiple frame transmissions across the medium, we can consider two different procedures as it relates to whether single or multiple simultaneous message frame transmission initiations are allowed.

b.1 **A single transmission initiation** is allowed at any instant of time. When a fiber-optic-based token ring MAC scheme, such as FDDI, is used, the station releases the token immediately following the transmission of the station message, rather than waiting to receive fully its own message prior to releasing the token. In this manner, multiple simultaneous transmissions can take place across the bus, allowing for a better utilization of the bus spatial bandwidth resources.

A slotted access scheme is often used for logical bus linear topologies. Under such a scheme, a bus controller is responsible for generating successive time slots within recurring time cycles. A slot propagates along the unidirectional fiber bus as an idle slot until captured by a busy station; the slot is then designated as busy and is used for carrying the inserted message segment. A busy station senses the medium and captures an available idle slot which it then uses to transmit its own message segment. For reasons of fairness, a busy station may be allowed to transmit only a single segment (or a limited maximum number of segments) during each time cycle. In this manner, such a MAC scheme strives to schedule busy station accesses onto the medium such that one will follow the other as soon as possible, so that a message train is generated utilizing efficiently the medium bandwidth and space (length) dimensions. However, note that no more than a single station is allowed at any time to initiate transmissions onto the unidirectional medium; transmission initiations occur in an order that matches the positional location of the stations along the fiber. As a result, the throughput capacity of such networks is limited by the data rate of the shared medium .

b.2 **Multiple simultaneous frame transmission initiations** are included in the MAC protocol. In using such a MAC scheduling feature, multiple stations are permitted to initiate frame transmissions at the same time, accessing the medium at sufficiently distant physical locations, so that multiple transmissions can propagate simultaneously in time along the space dimension of the shared medium. The underlying MAC algorithm needs to ensure that these simultaneous transmissions do not cause any frame overlaps (collisions).

In a ring system, such a procedure can involve the proper use of multiple tokens or ring buffer insertions at each station's interface. In a logical bus operation, when a slotted channel structure is employed, such an operation can be implemented through the designation of slots for use by a proper station or group of stations, in accordance with various system requests and interconnectivity conditions. For example, in the DQDB MAN, stations indicate their requests for channel slots, so that propagating idle slots along each fiber bus can be identified with a proper station to which they are assigned; in this manner, multiple simultaneous frame transmission initiations and ongoing message propagations can take place. Similarly, when TDMA and demand-assigned circuit or packet-switched TDMA MAC schemes are used.

Further enhancements in bandwidth and space utilization can be achieved by incorporating in the MAC scheme an appropriate message frame removal method, as indicated in the following.

Frame Removal Method

When considering a logical bus network with active SIUs, the removal of frames from the logical bus system can be in accordance with the following methods:

a. **Source Removal.** Considering loop topologies, under a source removal method the source station is responsible for the removal of its own transmitted frames. This is, as noted above, the scheme employed by the IEEE 802.5 and fiber-based FDDI token ring systems. Such a MAC feature permits the source station (following the transmission of a frame) to receive an immediate acknowledgment from its destination station (which is appended as a frame trailer) or to identify immediately a no-response status when the destination station is not operative.

b. **Destination Removal.** Under such a scheme, a station, upon identifying a passing frame destined for itself, removes this frame from the medium.

Under such a removal policy, a frame is not broadcasted across the overall length of medium, but just occupies a space segment of the medium stretching between the source and destination stations. Such a method can lead to a more complex MAC protocol and management scheme. Improvement in delay and throughput levels can, however, be realized through spatial reuse, particularly when a noticeable fraction of the system traffic flows among stations that are closely located with respect to each other's position across the network medium.

To apply such a scheme to a token ring system, multiple tokens are allowed and are properly distributed across the ring. When two closely located stations are communicating, other tokens can be used by other stations, located away from the occupied segment(s) to initiate their own nonoverlapping communications paths. Concurrent transmissions are also attained by the use of a slotted access scheme or through the use of a buffer insertion ring architecture, as illustrated later.

When such a scheme is applied to a bus system with actively connected SIUs, which is controlled by a slotted access scheme, the destination station is responsible for stripping the information contained in the slot destined to itself from the bus and releasing the corresponding slot for potential use by a downstream station.

c. **Removal by Supervisory Nodes.** It can be beneficial to employ special supervisory nodes, located across the medium, to remove frames from the medium. In this manner, the frame destination removal process can be implemented in only supervisory nodes, relieving regular nodes of this task.

By using such frame removal supervisory nodes, the system interconnectivity patterns can be divided (statically or dynamically) into a number of modes. Under an extensively divisive mode, the supervisory stations allow communications only between stations that are located between two neighboring supervisory stations to take place. Under a less divisive connectivity mode, the system is divided into longer disjoint communications segments. Under a full broadcast mode, each frame is broadcast to all network stations. Dependent upon the network traffic pattern, time cycles can be defined such that during each of which a specific mode of operation is invoked.

If such supervisory stations (or any station) are operated as store-and-forward buffered switching units, then clearly the distribution processes across the medium segments can be isolated and inter-segment message frames would then be delivered across the logical bus system in a multiple-hop (multi-retransmission) fashion. This is the role played by buffer insertion ring architectures for the specialized ring topology and, in general, by the mesh switching architectures discussed in the following.

B. Mesh (Switching) Topology

Under a mesh (switching) topology, the network topology can be configured as an arbitrary mesh graph. The network nodes provide buffering and switching services. Communications channels are set up as point-to-point links interconnecting the network nodes. Messages are routed across the network through the use of routing algorithms.

Depending on whether the switching node performs store-and-forward or cross-connect switching functions, we distinguish between the following architectures:

Store-and Forward Mesh Switching Architecture: Under a store-and-forward mesh switching archi-tecture, the nodal switches operate in a store-and-forward fashion as packet switches. Thus, each packet received at an incoming port of the switch is examined, and based on its destination address it is queued and forwarded on the appropriate outgoing port. Each point-to-point channel in such a mesh network needs to be efficiently shared among the multitude of messages and connections that are scheduled to traverse it. A statistical multiplexing scheme is selected (and implemented at each switch output module) for dynamically sharing the internodal links.

Cross-Connect Mesh Switching Architecture: In this case, each switching node operates as a cross-connect switch. The latter serves as a circuit (or virtual circuit) switch which transfers the messages belonging to as established circuit (or virtual circuit) from their incoming line and time slot(s) (or logical connection groups for cross connect virtual path switches used by asynchronous transfer mode, ATM, networks) to their outgoing line and time slot(s) (or logical group connections). The cross-connect matrix used by the node to implement this switching function is either preset and kept constant, or it can be readjusted periodically, as traffic characteristics vary, or even dynamically in response to the setup and establishment of end-to-end connections. (See also Section 3.2.8 for such an optical network identified as SMARTNet).

Hybrid Mesh Switching Architecture: The nodal switch can integrate fixed assigned and statistical operations in multiplexing traffic across the mesh topology links. For example, in supporting an inte-grated circuit-switched and packet-switched implementation, a time cycle (time frame) is typically defined, during which a number of slots are allocated to accommodate the established circuits which use this channel, while the remainder of the cycle slots are allocated for the transmission of the packets waiting in the buffers feeding this channel. Frequently, priority-based disciplines must be employed in statistically multiplexing the buffered packets across the packet-switched portion of the shared link, so that packets belonging to different service classes can be guaranteed their desired quality of service, as it relates to their characteristic delay and throughput requirements. For example, voice packets must subscribe to strict end-to-end time delay and delay jitter limits, while video packets induce high through-put support requirements (see Section 3.2.9). ATM network structures that employ combined virtual path and virtual circuit switches serve as another example.

C. Hybrid Logical Bus and Buffered Switching Topologies

High-speed, multigigabit communications networking architectures that cover wider areas often need to combine broadcast (logical bus) and mesh (switching) architectures to yield an efficient, reliable, and responsive integrated services fiber-based network.

In considering logical bus topologies with active SIUs, we noted above that certain nodes can be designated to act as store-and-forward processors and also to serve as frame removal supervisory nodes. Such nodes thus actually operate as MAC bridge gateways. These gateways serve to isolate the segments interconnecting them, each segment operating as an independent logical bus network. The individual segments can operate efficiently when they serve a local community of stations, noting that each segment spans a shorter distance, thus reducing the effect of the end-to-end propagation delay on message performance. To ensure their timely distribution, it can be effective to grant access priority, in each segment, to the intersegment packets that traverse this segment.

3.2.3 Layered Protocols and the MAC Sublayer

In relation to the OSI reference model, the data link layer of multiple access (such as typical LAN and MAN) networks is subdivided into the Medium Access Control (MAC) lower sublayer and the Logical Link Control (LLC) upper sublayer. Services provided by the MAC layer allow the local protocol entity to exchange MAC message frames (which are the MAC sublayer protocol data units) with remote MAC entities.

In considering typical LAN and MAN systems, we note the MAC sublayer to provide the following services:

1. MAC sublayer services provided to the higher layer (such as the LLC sublayer for LAN and MAN systems). LLC Service Data Units (SDUs) are submitted to the MAC sublayer for transmission through proper multiple access medium sharing. In turn, MAC protocol data units (PDUs) received from the medium and destined to the LLC are transferred to the LLC sublayer as proper LLC-SDUs. The underlying LLC-SDUs include source and destination addresses, the data itself, and service class and quality of service parameters.

2. MAC sublayer services are similarly provided to directly connected isochronous and non-isochronous (connection-oriented circuit and packet switched) channel users (CUs) allowing a local CU entity to exchange CU data units with peer CU entities. These are connection-oriented services, whereby after an initial connection setup, the CU is able to access the communications channel directly through the proper mediation of the MAC sublayer. The CU generates and receives its data units through the MAC sublayer over an existing connection, on an isochronous or nonisochronous basis. Such CU-SDUs contain the data itself and possibly service quality parameters; no addressing information is needed, since an established connection is involved. Corresponding services are provided for connectionless flows.

3. MAC sublayer services are provided to the local MAC station management entity, via the local MAC layer management interface. Examples of services include: the opening and closing of an isochronous or nonisochronous connection, its profile, features, and the physical medium it should be transmitted on, and the establishment and disestablishment of the binding between the channel user and the connection end-point identifier. The MAC sublayer requires services from the physical layer that provides for the physical transmission and reception of information bits.

Thus, in submitting information (MAC frames) to the physical layer, the MAC sublayer implements the medium access control algorithm which provides its clients (the LLC or other higher-layer messages) access to the shared medium. In receiving information from the physical layer, the MAC sublayer uses its implemented access control algorithm to select the MAC frames destined to itself and then delivers them to the higher layer protocol entities. MAC layer addresses are used by the MAC layer entity to identify the destination(s) of a MAC frame.

In Figure 3.7, we show the associated layers as implemented by various local and metropolitan area networks, as they relate to the OSI data link and physical layers.

3.2.4 Categorization of Medium Access Control Procedures

A multitude of access control procedures are used by stations to share a multiaccess communications channel or network system. We provide in this section a classification of MAC schemes.

MAC Dimensions

The multiple access communications medium resource can be shared among the network stations through the allocation of a number of key resource dimensions. The assignment space considered is the *T, F, C, S* (Time, Frequency, Code, Space). The allocation of access to the shared medium is provided to

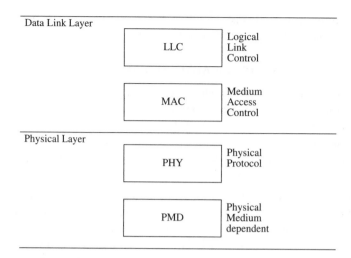

FIGURE 3.7   The MAC sublayer and its interfacing layers across the OSI data link and physical layers.

messages and sessions/calls of stations through the assignment of slots, segments, and cells of the time, frequency, code, and space dimensions.

1. **Time division scheduling.** Stations share a prescribed channel frequency band by having their transmissions scheduled to take place at different segments of time. Typically, only a single message can be transmitted successfully across the designated channel at any instant of time.
2. **Frequency and wavelength division allocation.** The bandwidth of the communications channel is divided into multiple disjoint frequency bands (or wavelength channels for an optical channel) so that a station can be allocated a group, consisting of one or more frequency/wavelength bands, for use in accessing the medium. Multiple time-simultaneous transmissions can take place across the channel, whereby each message transmission occupies a distinct frequency/wavelength channel.
3. **Code division multiple access.** The message of each station is properly encoded so that multiple messages can be transmitted simultaneously in time in a successful manner using a single-band shared communications channel, so that each message transmission is correctly received by its destined station. Typically, orthogonal (or nearly orthogonal) pseudonoise sequences are used to spread segments of a message randomly over a wide frequency band (frequency hopping method) or to time-correlate the message bit stream (direct sequencing method). A message can be encoded by an address-based key sequence that is associated with the identity of the source, the destination, the call/connection, or their proper combination. A wider frequency band is occupied by such code divided signals. In return, a common frequency band can be used by all network stations to carry successfully, simultaneously in time, multiple message transmissions.
4. **Space division multiple access.** Communications channels are shared along their space dimension. For example, this involves the sharing of groups of physically distinct links, or multiple space segments located across a single high-speed logical bus network.

Considering the structure of the access control procedure from the dimensional allocation point of view, note the similarity between space division and frequency/wavelength division methods, in that they induce a channel selection algorithm that provides an allocation of distinct frequency/wavelength or physical channels to a user or to a group of users. In turn, once a user has been assigned such a frequency/wavelength band, the sharing of this band can be controlled in accordance with a time division and/or a code division MAC method. Thus, under a combined use of these dimensions, the medium access control algorithm serves to schedule the transmission of a message by an active station by specifying the selected channel (in a frequency/wavelength division or space division manner) and subsequently specify the time slot(s) and/or multiple access codes to be used, in accordance with the employed time division and/or code division methods.

MAC Categories

In Figure 3.8, we show our categorization of medium access control procedures over the above-defined (T, F, C, S) assignment space. Three classes of access control policies are identified: fixed assignment (FA), Demand-Assignment (DA), and Random-Access (RA). Within each class, we identify the signaling (SIG) component and the information transmission (IT) method used. Note that within each class, circuit-switching as well as packet-switching mechanisms (under connectionless and/or connection-oriented modes) can be used, in isolation or in an integrated fashion.

Under a Fixed Assignment (FA) scheme, a station is dedicated, over the (T, F, C, S) space, a communications channel resource which it can permanently use for accessing the channel. Corresponding access control procedures thus include Time Division Multiple Access (TDMA), Frequency Division Multiple Access (FDMA), Wavelength Division Multiple Access (WDMA), Code Division Multiple Access (CDMA), and Space Division Multiple Access (SDMA) schemes. A corresponding signaling procedure is implemented to ensure that station transmission boundaries are well recognized by the participating stations, along the respective (T, F, C, S) dimensions.

Medium resources (along each dimension) can be assigned on a fixed basis to specified sessions/connections, to a station or to a group of stations. When allocated to a connection/call, such schemes provide the basis for the establishment of isochronous circuits and can lead to efficient sharing of the medium bandwidth resources for the support of many voice, video, high-speed data, and real-time session connections, when a steady traffic stream of information is generated. Effective channel sharing can also result when such channel resources are dedicated for the exclusive use by a station or a group of stations, provided the latter generate steady traffic streams which can efficiently utilize the dedicated resources at an acceptable quality of service level.

For example, under a TDMA procedure, a station is allocated a fixed number of slots during each frame. Under a Packet-Switched TDMA (PS-TDMA) operation, the station is accessing the shared medium by multiplexing its packets into the allocated slots. In turn, under a Circuit-Switched TDMA (CS-TDMA) operation, the station uses its allocated time slots to establish circuits. A circuit consists of a fixed number of the station time slots allocated in each time frame (e.g., a single slot per frame). Connections are assigned by the station available circuits at the requested rate. The messages generated by a connection are then transmitted in the time slots belonging to the circuit allocated to this connection.

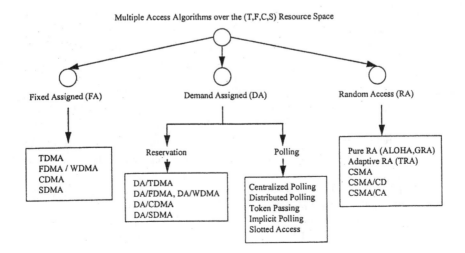

FIGURE 3.8 Classes of medium access control schemes with illustrative access procedures noted for each class.

Under a Random Access (RA) scheme, stations contend for accessing the communications channel in accordance with an algorithm that can lead to time-simultaneous (overlapping, colliding) transmissions by several stations across the same frequency band, causing at times the retransmission of certain packets.

Under fixed random access schemes, a station transmits its message frame across the channel at a random time, in a slotted or unslotted fashion, without coordinating its access with other stations. A station that detects (through channel sensing, or through nonreceipt of a positive acknowledgment from its destination) its transmission to collide with another transmission will retransmit its message after a properly computed random retransmission delay (whose duration may depend upon the estimated channel state of congestion). Under an adaptive channel-sensing random access algorithm, a ready station first senses the channel to gain certain channel state information, and then uses this information to schedule its message for transmission in accordance with the underlying protocol, again without undertaking full access coordination with all other network stations. Various fixed (such as ALOHA[2] and Group Random Access[3]) and adaptive channel-sensing random access algorithms such as CSMA,[4] CSMA/CD, DGRA,[5] Tree-Random-Access collision resolution based,[6] and others[7-9] can be invoked.

In general, random access schemes are not well suited as information transmission methods for governing the access of information messages onto a very high speed fiber communications channel, unless provisions are made to limit considerably the extent of message overlaps due to collisions, or to limit the use of this access technique to certain low-throughput messaging classes.

A random access procedure is typically employed for supporting packet-switching services. At higher normalized throughput levels, high packet delay variances can occur. A proper flow control regulation mechanism must be employed to guarantee maximum delay limits for packets associated with isochronous and real-time-based services.

Nonchannel-sensing random access procedures yield low channel utilization levels. In turn, the efficiency of a channel-sensing random access procedure critically depends on the timeliness of the sensed channel occupancy information, and thus upon the overhead durations associated with sensing the channel. For high-speed logical bus systems, when channel-sensing random-access schemes are used, performance degradation is caused by the very high ratio of the channel propagation delay to the packet transmission time. An active station is required to obtain the channel state prior to initiating a transmission. However, since transmission of a station must propagate across the total logical bus medium before its existence can be sensed by all network stations, and since stations make uncoordinated access decisions under a random access scheme, these decisions can be based on stale channel state information; as a result, this can lead to the occurrence of excessive message collisions.

The degrading effects of the propagation delay to message transmission time ratio can be somewhat reduced by the following approaches:

1. Limit the overall propagation delay by dividing the network into multiple tiers or segments, as illustrated by the double-tier LAN and MAN architecture presented in Reference 10. The network stations are divided into groups. A shared backbone medium can be used to provide interconnection between groups (on a broadcast global basis using repeaters as gateways, or on a store-and-forward bridged/routing basis), using a noncontention (non-random-access)-based MAC algorithm. A random access MAC strategy can be used for controlling the access of messages across the shared group (segment) medium, which now spans a much shorter distance and thus involves considerably reduced propagation delays. Similarly, this is done by interconnecting LAN (such as reduced distance-span ethernet broadcast domain workgroup) segments by routers, bridges, or switches (e.g., by ethernet switches).

2. While the propagation delay depends on the conducting medium and is independent of the transmission speed, the packet transmission time is reduced as the transmission speed is decreased. The MAC message frame transmission time can be increased by using FDMA/WDMA or SDMA methods to replace the single-band high-bandwidth medium with a group of shared multiple channels, each of lower bandwidth.[11] Clearly, this will provide an improvement only if the under-

lying services can be efficiently supported over such a reduced bandwidth channel when a random access (or other MAC) scheme is used.

3. Random access schemes can serve, as part of other MAC architectures, as effective procedures for transmitting signaling packets for demand-assigned schemes. Random access schemes can also be used in conjunction with CDMA and Spread Spectrum Multiple Access (SSMA) procedures, over high-speed channels.

Random access protocols are further presented and discussed in Section 3.2.6.

Under a Demand Assigned (DA) scheme, a signaling procedure is implemented to allow certain network entities to be informed about the transmission and networking needs and demands of the network stations. Once these network entities are informed, a centralized or distributed algorithm is used to allocate to the demanding stations communications resource segments over the (T, F, C, S) assignment space.

The specific methods used in establishing the signaling channel and in controlling the access of individual demanding stations onto this channel characterize the features of the established demand assignment scheme.

The signaling channel can be established as an out-of-band or in-band fixed channel to which communications resources are dedicated, as is the case for many DA/TDMA, DA/FDMA, or DA/WDMA schemes. In turn, a more dynamic and adaptive procedure can be used to announce the establishment of a signaling channel at certain appropriate times, as is the case for the many polling, implicit polling, slotted, and packet-train-type access methodologies.

The schemes used to provide for the access of stations onto the signaling channel, are divided by us into two categories: polling and reservation procedures.

Under a reservation scheme, it is up to the individual station to generate a reservation packet and transmit it across the signaling (reservation) channel to inform the network about its needs for communications resources. The station can identify its requirements, as they occur or in advance, in accordance with the type of service required (isochronous vs. asynchronous, for example) and the precedence level of its messages and sessions/connections. Reservation and assignment messages are often transmitted over an in-band or out-of-band multiple access channel, so that a MAC procedure must be implemented to control the access of these MAC-signaling messages.

Under a polling procedure, it is the responsibility of the network system to query the network stations so that it can find out their transmission needs, currently or in advance. Polling and polling-response messages can be transmitted over an in-band or out-of-band multiple access channel, so that a MAC procedure must be implemented to control the access of these MAC-signaling messages.

Under an implicit polling procedure, stations are granted access into the medium (one at a time, or in a concurrent noninterfering fashion) when certain network state conditions occur; such conditions can be deduced directly by each station without the need to require the station to capture an explicit polling message prior to access initiation. Such schemes are illustrated by slotted bus systems whereby the arrival of an idle slot at a station serves as an implicit polling message, and by buffer-insertion schemes whereby a station is permitted to access the ring if its ring buffer is empty. Additional constraints are typically imposed on the station in utilizing these implicit polling messages for access to the channel to ensure a fair allocation of communications network bandwidth resources. Such constraints often guarantee that a station can transmit a quota of packets within a properly defined cycle.

A multitude of access policies can be adopted in implementing integrated services polling and reservation algorithms. Note that polling techniques are used in implementing the IEEE 802.4 token bus and IEEE 802.5 token ring LANs, the FDDI fiber LAN and many other fiber bus and ring network systems. Reservation techniques are employed by the IEEE 802.6 DQDB MAN (whereby a slotted positional priority implicit polling access procedure is used to provide access to reservation bits,[12]) by demand-assigned TDMA, FDMA, and WDMA systems, and many other multiple access network systems. Integrated reservation and polling schemes are also used. See Section 3.2.5 for further discussion of polling and implicit polling methods.

3.2.5   Polling-Based Multiple Access Networks

In this section, we describe a number of multiple access networks whose medium access control architectures are based on polling procedures.

Token Ring Local Area Network

The token ring network is a local area network whose physical, MAC, and link layer protocol structures are based on the IEEE 802.5 standard. The Ring Interface Units (RIUs) are connected by point-to-point directional links which operate at data rates of 4 or 16 Mbps. A distributed polling mechanism, known as a token passing, protocol is used as the medium access control scheme. A polling packet, known as a token, is used to poll the stations (through their RIUs). When a station receives the token (buffering it for a certain number of bit times to allow for the station to identify the packet as a token message and to react to its reception), it either passes the token along to its downstream neighboring station (when it has no message waiting for transmission across the ring network) or it seizes the token. A captured token is thus removed from the ring, so that only a single station can hold a token at a time. The station holding the token is then allowed to transmit its ready messages. The dwell time of this station on the ring is limited through the setting of the station's token holding timer. When the latter expires, or when the station finishes transmitting its frames, whichever occurs first, the station is required to generate a new token and pass it to its downstream neighboring station. The transmitted frames are broadcast to all stations on the ring: they fully circulate the ring and are removed from the ring by their source station. Such a procedure is also known as a source removal mechanism.

The token ring protocol permits a message priority-based access control operation. For this purpose, the token contains a priority field. A token of a certain priority level can be captured by only those stations that wish to transmit across the medium a message of a priority level equal or higher than the token's priority level. Furthermore, stations can make a reservation for the issue of a token of a higher priority level by setting a priority request tag in a reservation field contained in circulating frames (such tags can also be marked in the reservation field of a busy token, which is a token that has been captured, but not yet removed from the ring, and tagged as busy by the station that has seized it). Upon its release of a new idle token, the releasing station marks the new token at a priority level that is the highest of all those levels included in the recently received reservations. At a later time, when no station is interested in using a token of such a high priority level, this station is responsible for downgrading the priority of the token. In setting priority-based timeout levels for the token holding timer, different timeout values can be selected for different priority levels.

The throughput capacity attainable by a polling scheme, including a token passing distributed polling mechanism such as the token ring LAN, is dependent upon the walk time parameter. The system walk time is equal to the total time it takes for the polling message (the token) to circulate around the network (the ring) when no station is active. For a token ring network, the walk time is thus calculated as the sum of the transmission times, propagation delay, and ring buffer delays incurred by the token in circulating across an idle ring. Clearly, the throughput inefficiency of the network operation is proportional to the ratio between the walk time and the overall time occupied by message transmissions during a single rotation of the token around the ring. The maximum achievable throughput level of the network, also known as the network throughput capacity, is denoted as $L(C)$ (bits/s); the normalized throughput capacity index is set to $s(C) = L(C)/R$, where R (bits/s) denotes the data rate of each network link, noting that $0 \leq s(C) \leq 1$. Thus, if $S(C) = 0.8$, the network permits a maximum throughput level which is equal to 80% of the link data rate. To assess the network throughput capacity level, we assume the network to be highly loaded so that all stations have frames ready for transmission across the ring. In this case, each cycle (representing a single circulation of the token around the ring) has an average length of $E(C) = NK(F/R) + W$, where N denotes the number of ring stations (RIUs), K denotes the maximum number of frames that a station is allowed to transmit during a single visit of the token, F is the average length of the frame (in bits), so that F/R represents the average frame transmission time across a ring link, while W denotes the average duration (in seconds) of the token walk time. The network normalized throughput capacity index is thus given by $s(C) = NK(F/R)/E(C)$. Clearly, a higher throughput level is attained as the walk time duration (W) is reduced. Note that $W = R(p)L + (T/R) + N(M/R)$, where $R(p)$ denotes the propagation delay across the medium (typically equal to 5 μm/km for wired links), L (km) is the distance spanned by the ring network so that $R(p)L$ represents the overall propagation delay around the ring; T is the token length, so that T/R represents the token transmission time around the ring; M denotes the number of bit time delays incurred at the RIUs interface buffer, so that $N(M/R)$ expresses the overall delay incurred by a frame in being delayed at each interface around the ring.

The delay throughput performance behavior exhibited by polling systems, such as the distributed polling token-passing scheme of the token ring LAN, follows the behavior of a single server queueing system in which the server dynamically moves from one station to the next, staying (for a limited time) only at those stations which have messages requiring service (i.e., transmission across the shared medium). This is a stable operation for which message delays increase with the overall loading on the network, as long as the latter is lower than the throughput capacity level $L(C)$. As the loading approaches the latter capacity level, message delays rapidly increase and buffer overflows can be incurred.

The FDDI Network

The Fiber Data Distribution Interface (FDDI) network also employs a token-passing access control algorithm. It also employs a ring topology. Two counterrotating fiber-optic rings are used (one of which is in a standby mode) so that upon the failure of a fiber segment the other ring is employed to provide for a closed-loop topology. The communications link operates at a data rate of 100 Mbps. The network can span a looped distance of up to 200 km.

As described above for the token ring network, upon the receipt of a token, an idle station passes it to its downstream station after an interface delay. If the station is busy, it will capture the token and transmit its frames until its token holding timer timeouts. As for the token ring network, a source removal mechanism is used, so that the transmitted frames are broadcasted to all stations on the ring. They fully circulate the ring and are removed from the ring by their source station. The timeout mechanism and the priority support procedure used by FDDI are different from those employed by the token ring network and are described below. When a station terminates its dwell time on the medium, it immediately releases the token. This is identified as an early token release operation. For lower-speed token ring implementations, a late token release operation can be selected. Under the latter, the token is released by the station only after its has received all of its transmitted messages and removed them from the ring. Such an operation leads to throughput performance degradations when the network uses higher-speed links since then the station holds the token for an extra time which includes as a component the ring propagation delay. The latter can be long relative to the frame's transmission time. As a result, higher-speed token ring networks are generally set to operate in the early token release mode.

The FDDI MAC scheme distinguishes between two key FDDI service types: synchronous and asynchronous. Up to eight priority levels can be selected for asynchronous services. To date, most commonly employed FDDI adapter cards implement mostly only the asynchronous priority 1 service.

The access of frames onto the medium is controlled by a Timed Token Rotation (TTR) protocol. Each station continuously records the time elapsed since it has last received the token (denoted as the token rotation time). An initialization procedure is used to select a Target Token Rotation Time (TTRT), through a bidding process whereby each station bids for a Token Rotation Time (TRT) and the minimum such time is selected. As noted above, two classes of service are defined: (1) synchronous, under which a station can capture the token whenever it has synchronous frames to transmit, and (2) asynchronous, which permits a station to capture a token only if the current TRT is lower than the established TTRT. To support multiple priority levels for asynchronous frames, additional time thresholds are defined for each priority level. In this manner, a message of a certain priority level is allowed to be transmitted by its station, when the latter captures the token, only if the time difference between the time this station has already used (at this ring access) for transmitting higher-priority messages, and the time since the token last visited this station is higher than the corresponding time threshold associated with the underlying message priority level. This priority-based access protocol is similar to the one used for the IEEE 802.4 token bus LAN system.

Using this procedure, stations can request and establish guaranteed bandwidth and response time for synchronous frames. A guaranteed maximum cycle latency-based response time is established for the ring, since the arrival time between two successive tokens at a station can be shown to not exceed the value of $2 \times$ TTRT.

As a polling scheme, the throughput performance of the FDDI network is limited by the ring walk (W) time. The ring throughput is thus proportional to 1 W/TTRT. While lower TTRT values (such as 4 to 8 ms) yield lower guaranteed cycle response times (token intervisit times lower than 8 to 16 ms), higher TTRT values need to be selected to provide for better bandwidth utilization under higher load conditions. The ring latency varies from a small value of 0.081 ms for a 50-station, 10-km LAN, to a value of 0.808 ms for a 500-station, 100-km LAN. Using a TTRT value of 50 ms, for a LAN that supports 75 stations and 30 km of fiber, and having a ring latency $W = 0.25$ ms, a maximum utilization of 99.5% can be achieved.[13]

To provide messages the desired delay performance behavior across the FDDI ring network, it is important to calibrate the FDDI network so that acceptable levels of queueing delays are incurred at the station access queues for each service class.[14] This can be achieved by proper selection of the network MAC parameters, such as the TTRT level, timeout threshold levels when multipriority asynchronous services are used, and the station synchronous bandwidth allocation level when a station FDDI adapter card is set to provide also an FDDI synchronous service. The latter service is effective in guaranteeing quality of service support to real time streams, such as voice, compressed video, sensor data, and high-priority critical message processes which require strictly limited network delay jitter levels.[14] Note that when a token is received by a station which provides an FDDI synchronous service, the station is permitted to transmit its frames which receive such a service (for a limited time which is equal to a guaranteed fraction of the TTRT) independently of the currently measured TRT. When no such messages are queued at the station at the arrival of the token, the token immediately starts to serve messages which receive asynchronous service so that the network bandwidth is dynamically shared among all classes of service.

The delay throughput performance features of the FDDI networks follow the characteristics of a distributed polling scheme discussed above.

Implicit Polling Schemes

Under an implicit polling multiple access mechanism, the network stations monitor the shared medium and are then granted access to it by identifying proper status conditions or tags. To illustrate such structures, we consider slotted channel, register insertion, positional priority, and collision avoidance access protocols.

Under a slotted channel access protocol, the shared medium link(s) are time-shared through the generation of time slots. Each time slot contains a header which identifies it as busy or idle. In addition, time slots may also be reserved to connections, so that a circuit-switched mode can be integrated with the packet-switched multiple-access mode described in the following. To regulate station access rates, to assign time circuits, and to achieve a "fair" allocation of channel resources, the time slots are normally grouped into recurring time frames (cycles). A ready station, with packets to transmit across the medium, monitors the medium. When this station identifies an idle slot which it is allowed to capture, it marks it to be in a busy state and inserts a single segment into this slot. Clearly, in transmitting a packet, a station must break it into multiple segments, whereby the maximum length of a segment is selected such that it fits into a single slot. The packet segments must then be assembled into the original packet at the destination station.

To be able to insert packets into moving slots, the station must actively interface the medium by inserting an active buffer into the channel. A common configuration for such a network is the slotted ring topology. A slotted access protocol can also be used in sharing a linear "logical bus" topology, with active station interfaces. The later configuration is used by the Distributed Queue Dual-Bus (DQDB) MAN defined by the IEEE 802.6 standard. The latter uses a fiber-optic-based dual-bus configuration so that each station is connected to two counterdirectional buses.

To regulate the maximum level of bandwidth allocated to each station, in accordance with the class of service provided to the station, and to control the fair allocation of network resources among stations which receive the same class of service, the access algorithm can limit (statically or dynamically) the number of slots which can be captured by a station during each frame. For the DQDB network, a reservation subchannel is established for stations to insert reservation tags requesting for slots to be used for the transmission of their packets. A station is allowed to capture an idle slot which passes its interface only if it has satisfied all the earlier requests it has received signifying slot reservations made by other stations. The DQDB network also integrates a circuit-switched mode through the establishment of isochronous circuits as part of the call setup procedure. A frame header is used to identify the slots which belong to dedicated time circuits.

Under a register insertion configuration, the medium interface card of each station includes a register (buffer) which is actively inserted into the medium. Each packet is again broken down into segments. A station is permitted to insert its segment(s) into the medium when its register contains no in-transit segments (thus deferring the transmission of its own segments until no in-transit segments are passing by) or when the gap between in-transit segments resident in its register is sufficiently long. In-transit packets arriving at the station interface when the station is in the process of transmitting its own segment are delayed in the register. To avoid register overflows, its size is set to be equal to at least the maximum segment length. The IBM Metaring/Orbit[15] network[1] is an example of such a network system which employs a ring topology as well as the destination removal spatial reuse features presented below.

At higher speeds, to further increase the throughput efficiency of the shared medium network, a destination removal mechanism is used. This leads to spatial reuse since different space segments of the network medium can be used simultaneously in time by different source–destination pairs. For example, for a spatial-reuse slotted channel network (such as slotted ring- or bus-based topologies), once a segment has reached its destination station, the latter marks the slot as idle allowing it to be reused by subsequent stations it visits. Similarly, the use of an actively inserted buffer (as performed by the register insertion ring network) allows for operational isolation of the network links, providing spatial reuse features.

To assess the increase in throughput achieved by spatial reuse ring networks, assume the traffic matrix to be uniform (so that the same traffic loading level is assumed between any source–destination pair of stations). Also assume that the system employs two counterrotating rings, so that a station transmits its segment across the ring that offers the shortest path to the destination. Clearly, the maximum path length is equal to half the ring length, while the average path length is equal to one fourth of the ring length. Hence, an average of four source–destination station pairs time-simultaneously communicate across the dual ring network. As a result, the normalized throughput capacity achieved by the spatial reuse dual ring network is equal to 400% (across each one of the rings), as compared with a utilization capacity of 100% (per ring) realized when a source removal mechanism is used. Hence, such spatial reuse methods can lead to substantial throughput gains, particularly when the network links operate at high speeds. They are thus especially important when used in ultrahigh-speed optical networks, as we note in Section 3.2.8.

Positional-Priority and Collision Avoidance Schemes

Under a hub-polling (centralized polling) positional priority scheme, a central station (such as a computer controller) polls the network stations (such as terminals) in an order which is dictated by their physical position in the network (such as their location on a ring network with respect to the central station), or by following a service order table.[16] Stations located in a higher position in the ordering list are granted higher opportunities for access to the network shared medium. For example, considering a ring network, the polling cycle starts with the controller polling station 1 (the one allocated highest access priority). Subsequently station 2 is polled. If the latter is found to be idle, station 1 is polled again.

In a similar manner, a terminal priority–based distributed implicit polling system is implemented. Following a frame start tag, station 1 is allowed to transmit its packet across the shared medium. A ready station 2 must wait a single slot of duration T (which is sufficiently long to allow station 1 to start transmitting and for its transmission to propagate throughout the network so that all stations monitoring the shared channel can determine that this station is in the process of transmission) before it can determine whether it is allowed to transmit its packet (or segment). If this slot is determined by station 2 to be idle, it can immediately transmit its packet (segment). Similarly, station i must wait for (i–1) idle slots (giving a chance to the i–1 higher-priority terminals to initiate their transmissions) following the end of a previous transmission (or following a frame start tag when the channel has been idle) before it can transmit its packet across the shared communications channel.

Such a multiple access mechanism is also known as a collision avoidance scheme. It has been implemented as part of a MAC protocol for high-speed back-end LANs (which support a relatively small number of nodes) such as HyperChannel. It has also been implemented by wireless packet radio networks, such as the TACFIRE field artillery military nets. In assessing the delay throughput behavior of such an implicit polling distributed control mechanism, we note that the throughput efficiency of the scheme depends critically upon the monitoring slot duration T, while the message delay behavior depends upon the terminal priority level.[17] At lower loading levels, when the number of network nodes is not too large, acceptable message delays are incurred by all terminals. In turn, at higher loading levels, only higher-priority terminals will manage to attain timely access to the network while lower-priority ones will be effectively blocked from entering the shared medium.

Probing Schemes

When a network consists of a large number of terminals, each generating traffic in a low duty cycle manner, the polling process can become highly inefficient in that it will require relatively high bandwidth and will occupy the channel with unproductive polling message transmissions for long periods of time. This is induced by the need to poll a large number of stations when only a few of them will actually have a packet to transmit.

A probing[18] scheme can then be employed to increase the efficiency of the polling process. For this purpose, rather than polling individual stations, groups of stations are polled. The responding individual stations are then identified through the use of a collision resolution algorithm. For example, the following Tree-Random-Access[6,18] algorithm can be employed. Following an idle state, the first selected polling group consists of all the net stations. A group polling message is then broadcast to all stations. All stations belonging to this group which have a packet to transmit then respond. If multiple stations respond, a collision will occur. This will be recognized by the controller which will subsequently subdivide the latest group into two equal subgroups. The process will proceed with the transmission of a subgroup polling message. By using such a binary search algorithm, all currently active stations are eventually identified. At this point, the probing phase has been completed and the service phase is initiated. All stations that have been determined to be active are then allocated medium resources for the transmission of their messages and streams.

Note that this procedure is similar to a reservation scheme in that the channel use temporally alternates between a signaling period (which is used to identify user requests for network support) and a service period (during which the requesting stations are provided channel resources for the transmission of their messages). Under a reservation scheme, the stations themselves initiate the transmission of their requests during the signaling periods. For this purpose, the stations may use a random access algorithm, or other access methods. When the number of network stations is not too large, dedicated minislots for the transmission of reservation tags can be used.

3.2.6   Random Access Protocols

In this section, we describe a number of random access protocols which are commonly used by many wireline and wireless networks. Random access protocols are used for networks which require a distributed control multiple access scheme, avoiding the need for a controller station which distributes (statically or dynamically) medium resources to active network stations. This results in a survivable operation which avoids the need for investment of significant resources into the establishment and operation of a signaling subnetwork. This is of particular interest when the shared communications medium supports a relatively large number of terminals each operating at a low duty cycle. When a station has just one or a few packets it needs to transmit in a timely manner (at low delay levels) on infrequent occasions, it is not effective to allocate to the station a fixed resource of the network (as performed by a fixed assignment scheme). It is also not effective to go through a signaling procedure to identify the communications needs of the station prior to allocating it a resource for the transport of its few packets (as performed by demand-assigned polling and reservation schemes). As a consequence, for many network systems, particularly for wireless networks, it is effective to use random access techniques for the transport of infrequently generated station packets; or, when active connections must be sustained, to use random access procedures for the multiaccess transport of signaling packets.

The key differences among the random access methods described below are reflected by the method used in performing shared medium status monitoring. When stations use full-duplex radios and the network is characterized by a broadcast communications medium (so that every transmission propagates to all stations), each station receives the transmissions generated by the transmitting station, including the transmitting station itself. Stations can then rapidly assess whether their own transmission is successful (through data comparison or energy detection). This is the situation for many LAN implementations. In turn, when stations are equipped with half-duplex transceivers (so that they need to turn around their radios to transition between a reception mode and a transmission mode), and/or when a fully broadcast channel is not available (as is the case for mobile radio nets for which topographical conditions lead to the masking of certain stations, so that certain pairs of stations do not have a line-of-sight based direct communications link) the transmitting station cannot automatically determine the status of its transmission. The station must then rely on the receipt of a positive acknowledgment packet from the destination station.

ALOHA Multiple Access

Under the ALOHA random-access method, network stations do not monitor the status of the shared communications channel. When a ready station receives a packet, it transmits it across the channel at any time (under an Unslotted ALOHA scheme) or at the start of time slot (under a Slotted ALOHA algorithm, where the length of a slot is set equal to the maximum MAC frame length). If two or more stations transmit packets (frames) at the same time (or at overlapping times), the corresponding receivers will not usually be able to receive the involved packets correctly, resulting in a destructive collision. (Under communications channel capture conditions, the stronger signal may capture the receiver and may be received correctly, while the weaker signals may be rejected.) When a station determines its transmitted packet to collide, it then schedules this packet for retransmission after a random retransmission delay. The latter delay can be selected at random from an interval whose length is dynamically determined based on the estimated level of congestion existing across the shared communications channel. Under a binary exponential back-off algorithm, each station adapts this retransmission interval on its own, by doubling its span each time its packet experiences an additional collision.

The throughput capacity of an unslotted ALOHA algorithm is equal to $s(C) = 1/(2e) = 18.4\%$, while that of a slotted ALOHA scheme is equal to $s(C) = 1/e = 36.8\%$. The remainder of the shared channel's used bandwidth is occupied by original and retransmitted colliding packets. In effect, to reduce the packet delay level and the delay variance (jitter), the loading on the medium must be reduced significantly below the throughput capacity level. Hence, the throughput efficiency of ALOHA channels is normally much lower than that attainable by fixed assigned and demand-assigned methods. The random access network system is, however, more robust to station failures, and is much simpler to implement, not requiring the use of complex signaling subnetworks.

The ALOHA shared communications channel exhibits a bistable system behavior. Two distinctly different local equilibrium points of operation are noted. Under sufficiently low loading levels the system state resides at the first point, yielding acceptable delay throughput behavior. In turn, under high loading levels, the system can transition to operate around the second point. Loading fluctuations around this point can lead to very high packet delays and diminishing throughput levels. Thus, under high loading bursts the system can experience a high level of collisions which in turn lead to further retransmissions and collisions causing the system to produce very few successful transmissions. To correct this unstable behavior of the random access multiaccess channel, flow control mechanisms are frequently used. The latter regulate admission of new packets into the shared medium at times during which the network is congested. Of course, this in turn induces an increase in the packet-blocking probability or in the delay of the packet at its station buffer.

Carrier Sense Multiple Access (CSMA)

Under the CSMA random access method, network stations monitor the status of the shared communications channel to determine if the channel is busy (carrying one or more transmissions) or is idle. A station must listen to the channel before it schedules its packet for transmission across the channel.

If a ready station senses the channel to be busy, it will avoid transmitting its packet. It then either (under the nonpersistent CSMA algorithm) takes a random delay and subsequently remonitors the channel, or (under the 1-persistent CSMA algorithm), it keeps persisting on monitoring the channel until it becomes idle.

Once the channel is sensed to be idle, the station proceeds to transmit its packet. If this station is the only one transmitting its packet at this time, a successful transmission results. Otherwise, the packet transmission normally results in a destructive collision. Once the station has determined its packet to collide, it schedules its packet for retransmission after a random retransmission delay.

Many local and regional area packet radio multiaccess networks supporting stationary and mobile stations, including those using half-duplex radio transceivers, have been designed to use a CSMA protocol. The performance efficiency of CSMA networks is determined by the acquisition delay index $a = t(a)/T(P)$, where $T(P)$ denotes the average packet transmission time while $t(a)$ denotes the system acquisition time delay. The latter is defined as the time elapsed from the instant that the ready station initiates its transmission (following the termination of the last activity on the channel) to the instant that the transmission of the packet has propagated to all network stations so that the latter can sense a transmitting station and thus avoid initiating their own transmissions. The acquisition delay $t(a)$ includes as components the network end-to-end propagation delay, the radio turnaround time, the radio detection time for channel busy-to-idle transition time, the radio attack time (time to build up the radio's output power), various packet preamble times, and other components. As a result, for half-duplex radios, the network acquisition delay $t(a)$ may assume a relatively large value. The efficiency of the operation is, however, determined by the factor "a" which is given as the ratio of $t(a)$ and the packet transmission time $T(P)$; since once the station has acquired the channel and is the only one currently active on the channel, after a period of length $t(a)$, it can proceed with an uninterrupted transmission of its full packet, for a period of duration $T(P)$.

Clearly, this is a more efficient mechanism for packet radio networks which operate at lower channel transmission rates. As the transmission rate increases, $T(P)$ decreases causing the index "a" to increase, so that the delay throughput efficiency of the operation degrades. A CSMA network will attain good delay throughput performance behavior for acquisition delay index levels lower than about 0.05. For index levels around or higher than 0.2, the CSMA network exhibits a throughput level which is lower than that obtained by a slotted (or even Unslotted) ALOHA multiaccess net. Under such conditions, the channel-sensing mechanism is relatively ineffective since the window of vulnerability for collisions ($t(a)$) is now relatively too long. It is thus highly inefficient to use a CSMA mechanism for higher data rate channels, as well as for channels which induce relatively long propagation delays (such as a satellite communication network) or for systems which include other mechanisms which contribute to an increase in the value of the acquisition delay index.

As for the ALOHA scheme, a CSMA network exhibits a bistable behavior. Thus, under loading bursts the channel can enter a mode under which the number of collisions is excessive so that further loading of the channel results in diminishing throughput levels and higher packet delays. A flow control–based mechanism can be used to stabilize the behavior of the CSMA dynamic network system.

CSMA/CD Local Area Networks

A local area network that operates by using a CSMA/CD access control algorithm incorporates into the CSMA multiple access scheme described above the capability to perform Collision Detection (CD). The station access module uses a full-duplex radio and appends to its CSMA-based sensing mechanism a CD operation. Once the ready station has determined the channel to be idle, it proceeds to transmit its packet across the shared channel while at the same time it is listening to the channel to determine whether its transmission has resulted in a collision. In the latter case, once the station has determined its packet to be involved in a process of collision, the station will immediately abort transmission. In this manner, colliding stations will occupy the channel with their colliding transmissions only for a limited period of time — the collision detection $T(CD)$ time. Clearly, if $T(CD) < T(P)$, where $T(P)$ denotes the transmission time of a packet (MAC frame) across the medium, the CSMA/CD operation leads to improved delay throughput performance over that of a CSMA operation.

The ethernet LAN developed by Xerox Corporation is a bus-based network operating at a channel data rate of 10 Mbps and using a 1-persistent CSMA/CD medium access control algorithm. The physical, MAC, and link layers of such a CSMA/CD network are defined by the IEEE 802.3 standard (and a corresponding ISO standard). Ethernet nets can employ different media types: twisted pair, coaxial cable, fiber-optic line, as well as radio links (in the case of an ethernet-based wireless LAN system). The configuration of the ethernet LAN is that of a logical bus so that a frame transmission by a station propagates across the bus medium to all other stations. In turn, the physical layout of the medium can assume a bus or a star topology. Under the latter configuration, all stations are connected to a central hub node at which point the access lines are connected by a reflecting repeater module so that a transmission received from an access line is repeated into all other lines. Typical ethernet layouts are limited in their geographic span to an overall distance of about 500 to 2000 m.

The delay throughput efficiency of the CSMA/CD network operation is determined by the acquisition time delay $t(a)$ and index "a," as for the CSMA network. In addition, the network efficiency also depends on the CD time $T(CD)$. It is noted that $T(CD)$ can be as long as twice the propagation delay across the overall span of the bus medium, plus the time required by the station to establish the occurrence of a collision. To ensure that a collision is reliably detected, the power levels of received packets must be sufficiently high. This also serves as a factor limiting the length of the bus and of the distance at which repeaters must be placed. As a result, network stations are frequently configured so that stations belonging to a single workgroup, assuming a sufficiently short span, are attached to a single ethernet segment. Multiple ethernet segments are interconnected by gateways or routers. The latter act as store-and-forward switches which serve to isolate the CSMA/CD multiaccess operation of each segment.

As for the random access mechanisms discussed above, burst loads applied to the CSMA/CD network can cause large delay throughput degradations. In supporting application streams which require limited packet delay jitters, it is thus required to prevent the bus from being excessively loaded. Many implementations thus plan the net's offered traffic loading levels to be no higher than 40% (or 4 Mbps, for a 10-Mbps ethernet operation). Also note that as for other random access schemes, large loading variations can induce an unstable behavior. A flow control mechanism must then be employed to regulate the maximum loading level of the network.

When shorter bus spans are used, or when shorter access link distances are employed in a star configuration, it is possible to operate this network at higher channel data rates. Under such conditions, an ethernet operation at a data rate of 100 Mbps (known as fast ethernet) has been implemented. Through the use of multiple access fibers, the access rate is being extended to a Gbps range (leading to gigabit ethernet systems).

3.2.7 Multiple Access Schemes for Wireless Networks

Under a cellular wireless network architecture, the geographic area is divided into cells. Each cell contains a central base station. The mobile terminals communicate with the base station controlling the cell in which they reside. The terminals use the reverse traffic and signaling channel(s) to transmit their messages to the base station. The base station multiplexes the messages it wishes to send to the cell mobile terminals across the forward traffic and signaling channels.

First-generation cellular wireless networks are designed to carry voice connections employing a circuit switching method. Analog communications signals are used to transport the voice information. The underlying signaling subnetwork is used to carry the connection setup, termination, and handover signaling messages. The voice circuits are allocated through the use of a reservation-based demand-assigned/FDMA scheme. A ready mobile uses an allocated signaling channel for its handset to transmit to its cell base station a request message for the establishment of a voice circuit. If a frequency channel is available for the allocation of such a circuit (traffic channel), the base station will make this allocation by signaling the requesting handset.

Second-generation cellular wireless networks use digital communications channels. A circuit switching method is still employed, with the primary service providing for the accommodation of voice connections. Circuits are formed across the shared radio medium in each cell by using either a TDMA access control scheme (through the periodic allocation of time slots to form an established circuit, as performed by the European GSM and the U.S. IS-54 standards) or by employing a CDMA procedure (through the allocation of a code sequence to a connection's circuit, as carried out by the IS-95 standard). A signaling subnetwork is established. Reverse Signaling Channels (RSCs) are multiple access channels which are used by the mobile terminals to transmit to their cell base station their channel request packets (for a mobile originating call), as well as for the transmission of paging response packets (by those mobiles which terminate calls). Forward Signaling Channels (FSCs) are multiplexing channels configured for the transmission of packets from the base station to the mobiles. Such packets include channel allocation messages (which are sent in response to received channel request packets) and paging packets (which are broadcast in the underlying location area in which the destination mobile may reside).

For accessing the reverse signaling channels, a random access algorithm such as the ALOHA scheme is frequently employed. For TDMA systems, time circuit(s) are allocated for the random access transmission of signaling packets. For CDMA systems, codes are allocated for signaling channels; each code channel is time shared through the use of a random access scheme (employing, for example, a slotted ALOHA multiple access algorithm).

Paging and channel allocation packets are multiplexed (in a time shared fashion) across the forward signaling channels. To reduce battery consumption at the mobile terminals, a slotted mode operation can be invoked. In this case, the cell mobiles are divided into groups, and paging messages destined to a mobile belonging to a certain group are transmitted within that group's allocated channels (time slots).[19] In this manner, an idle mobile handset needs to be activated for the purpose of listening to its FSC only during its group's allocated slots.

The IEEE 802.11 protocol defines the multiple access operation of a wireless LAN that provides data transport services. This involves an operation over a distance span of about 1 km at a data rate of 1 to a few Mbps. The standard defines two configurations: (1) an ad-hoc wireless LAN configuration involving a number of stations which are within communications range of each other; (2) an infrastructure wireless LAN that uses access point nodes to which the wireless terminals transmit their data packets, and that serves as a gateway to a backbone network.

Consequently, the IEEE 802.11 MAC scheme includes a distributed access protocol and a centralized access protocol. Asynchronous and time-bounded service classes are defined. The point coordination function (PCF) involves the use of a polling-based access control protocol for implementing the centralized access scheme. The Distributed Coordination Function (DCF) employs a contention-based algorithm, the CSMA/CA (CSMA with collision avoidance) protocol. In the following we discuss the latter MAC protocol.

Under the CSMA/CA scheme, a station wishing to transmit its packet across the shared wireless medium senses first the channel, as performed by a CSMA scheme. However, if the channel is sensed to be idle, the station must first wait for a minimum period of time, identified as an InterFrame Space (IFS), before it can initiate its access time calculation. To provide three levels of access priorities, the scheme defines three IFS periods: the shortest one is the Short IFS (SIFS), followed by the PCF IFS (PIFS), and then by the (longest period) DCF IFS (DIFS). Stations that have high-priority data packets to transmit, such as Acknowledgment (ACK) or Clear To Send (CTS) packets are permitted to wait for a period equal to SIFS before they access the channel. In this manner, as these stations are at that time the only ones to have such high-priority packets ready for transmission (since they follow a single successful transmission after a sufficiently short period), these stations are provided contentionless access to the channel for the transmission of these packets. In turn, regular data packets are transmitted at a time which is calculated only after the DIFS period has elapsed. At the later time, the station waits for a random backoff period, and if the channel is still idle at that time, the station transmits its packet.

This operation allows for three types of station-to-station interactions to take place: (1) random data access through the use of the DIFS periods, as described above; (2) a Request To Send (RTS) followed by a clear to send mode, under which the sender transmits a RTS packet which is followed by a contentionless response (after a SIFS period) from the destination of an ACK packet, and subsequent contentionless transmissions of the data and ACK packets; (3) a regular contention-based access of a data packet followed by a contentionless access of the responding ACK packet.

Third-generation cellular wireless networks are planned to provide for the support of both voice and data services. These networks employ packet-switching principles. Many of the multiple access schemes described in previous sections can be employed to provide for the sharing of the mobile-to-base-station multiple access reverse communications channels. In particular, random access, polling (or implicit polling), and reservation protocols can be effectively implemented. For example, the following versions of a reservation method have been investigated. Under the Packet Reservation Multiple Access (PRMA)[20] or random access burst reservation, procedure, a random access mechanism is used for the mobile to reserve a time circuit for the transmission of a burst (which consists of a number of consecutively generated packets). For voice bursts, a random access algorithm is used to govern the transmission of the first packet of the burst across the shared channel (by randomly selecting an unused slot, noting that, in each frame, the base station notifies all terminals which slots are unoccupied). If this packet's transmission is successful, its terminal keeps the captured slot's position in each subsequent frame until the termination of the burst's activity. Otherwise, the voice packet is discarded and the next voice packet is transmitted in the same manner. In selecting the parameters and capacity levels of such a multiaccess network, it is necessary to ensure that connections receive acceptable throughput levels and that the packet discard probability is not higher than a prescribed level ensuring an acceptable voice quality performance.[21]

For connection-oriented packet switching network implementations which also provide data packet transport, including wireless ATM networks, it is necessary to avoid frequent occurrences of packet discards, to reduce the rate of packet retransmissions, and to lower the frequency of occurrence of out-of-order packet deliveries. Reverse- and forward-signaling channels are established to provide for the establishment of connections and for the allocation of virtual circuits. Channel resources can then be allocated to established connections in accordance with the statistical features of such connections. For example, real-time connections can be accommodated through the periodic allocation of time slots (or frequency/code resources) while bursty sources can be supported by the allocation of such resources only for the limited duration of a burst activity (See also Section 3.2.9.) For this purpose, a mechanism must then be employed to identify the start and end times of burst activities. For example, the signaling channels can be used by the mobiles to signal the start of burst activity; while the in-band channel is used to detect the end of activity (directly and/or through the use of tagging marks).

3.2.8 Multiple Access Methods for Spatial-Reuse Ultrahigh-Speed Optical Communications Networks

As we have seen above, token ring and FDDI LAs use token-passing methods to access a shared medium. Furthermore, these networks use a source removal procedure so that each station is responsible for removing its own transmitted frames from the ring. In this manner, each transmitted frame circulates the ring and is then removed by its source station. As a result, the throughput capacity of such networks is limited to a value which is not higher than the ring's channel data rate.

In turn, as observed in Section 3.2.5 (in connection with implicit polling-based multiple access networks), the bandwidth utilization of shared medium networks, with particular applications to local and metropolitan area networks, operating at high speed, using generally fiber-optic links, can be significantly upgraded through the employment of spatial-reuse methods. For example, consider a ring network which consists of two counterrotating unidirectional fiber-optic-link-based ring topologies. Each terminal (station) is connected to both rings through ring interface units (RIUs). The bandwidth resources of the rings are shared among the active stations. Assume now that a destination removal method is employed. Thus, when a frame reaches its destination node (RIU), it is removed from the ring by the latter. The network communications resources occupied by this frame (such as a time slot) can then be made available to source nodes located downstream to the removing station (as well as to the removing station itself).

A source station which has a frame to transmit selects the ring which offers the shortest path to the destination node. In this manner, the length of a path will be no longer than one half the ring length. If we assume a uniform traffic matrix (so that traffic flows between the network terminal nodes are symmetrically distributed), the average path length will be equal to one fourth the ring length. As a result, using such a destination removal technique, we conclude that an average of four source–destination flows will be carried simultaneously in time by each one of the two rings. Therefore, the throughput capacity of such a spatial reuse network can reach a level that is equal to 400% the ring channel data rate, for each one of the two rings.

Spatial-reuse ring networks can be used to carry traffic on a circuit-switched or packet-switched (connectionless or connection-oriented) basis. For packet-switched network implementations, the following two distributed implicit polling multiple access methods have been used. Under a slotted ring operation, time slots are generated and circulated across the ring. Each time slot contains a header which identifies it as either empty or busy. A station that senses an empty time slot can mark it as busy and insert a segment (a MAC frame, which, for example, can carry an ATM cell) into the slot. The destination station will remove the segment from its slot, and will designate the slot as idle so that it can be reused by itself or other stations. Under a buffer insertion ring multiple access scheme, each station inserts a buffer into the ring. A busy station will defer the transmission of its packet to packets currently being received from an upstream station. In turn, when its buffer is detected to be empty, the station will insert its segment into the ring. The buffer capacity is set equal to a maximum size packet so that a packet received by a station while it is in a process of transmission of its own packet can be stored in the station's inserted buffer. Each packet is removed from the ring by its destination node.

As noted above, the DQDB MAN is a slotted dual-bus network which employs active station interfaces. To achieve spatial bandwidth reuse gains of the bus bandwidth resources, frame removal stations can be positioned across the bus. These stations serve to remove frames that have already reached their destinations so that their slots can be reused by downstream stations. Note that such stations must use an inserted buffer of sufficient capacity to permit reading of the frame destination node(s) so that they can determine whether a received frame has already reached its destination(s).

In designing optical communications networks, it is desirable to avoid the need for executing store-and-forward switching and operations at the network nodes. Such operations are undesirable due to the large differences existing between the optical communications rates across the fiber-optic links and the electronic processing speeds at the nodes, as well as due to the difficulty involved in performing intelligent buffering and switching operations in the optical domain. Networks that avoid such operations (at least to a certain extent) are known as all-optical networks.

Systems commonly used for the implementation of a shared medium local area all-optical network employ a star topology. At the center of the star, an optical coupler is used. The coupler serves to repeat the frame transmissions received across any of its incoming fiber links to all of its outgoing fiber links. The coupler can be operated in a passive transmissive mode or in an active mode (using optical amplification to compensate for power losses incurred due to the distribution of the received optical signal across all output links). Each station is then connected by a fiber link to the star coupler, so that the stations share a multiple access communications channel. A multiple-access algorithm must then be employed to control the sharing of this channel among all active stations. Typical schemes employed for this purpose use random access, polling, and reservation methods. Furthermore, multiple wavelengths can be employed so that a WDMA component can be integrated into the multiple access scheme. The wavelengths can be statically assigned or dynamically allocated. To illustrate the latter case, consider a reservation-based DA/WDMA procedure. Under a connection-oriented operation, wavelengths are assigned to source–destination nodes involved in a connection. Both transmitter and receiver wavelength assignment-based configurations can be used: the transmitter or receiver (or both) operating wavelengths can be selected dynamically to accommodate a configured connection.

Due to its broadcast feature, the star-configured optical network does not offer any spatial- or wavelength-reuse advantages. In turn, the meshed ring SMARTNet (Scalable multimedia adaptable meshed ring terabit) optical network introduced by Rubin[22] capitalizes on the integrated use of spatial-reuse and wavelength-reuse methods. In the following, we illustrate the architecture of the SMARTNet configuration. Two counterrotating fibers make up the ring periphery. The stations access the network through their RIU, which are actively connected to the peripheral fiber rings. The fiber link resources are shared through the use of a wavelength division scheme. Multiple wavelengths are used and each station has access to all wavelengths (or to a limited number). The rings are divided into segments. Each segment provides access to multiple stations through their connected RIUs. Each segment is connected to its neighboring two segments through the use of wavelength cross-connect routers. Such a router switches messages received at multiple wavelengths from its incoming ports to its outgoing ports. No store-and-forward switching operations are performed. The switching matrix is preset (on a static or adjustable basis) so that messages arriving across an incoming link at a certain wavelength are always immediately switched to a prescribed outgoing link (at the same wavelength; although wavelength translations can also take place). The wavelength routers are also connected (to certain other wavelength routers) by chord links (each consisting of a pair of counterdirectional fiber links). The routers and their chord links form the chord graph topology.

For each configuration of the chord graph, for each assignment of wavelengths, and for each switching table configuration of the routers, the SMARTNet topology can be divided into wavelength graphs. Each wavelength graph represents a subnetwork topology and is associated with a specific wavelength; all stations accessing this subnetwork (which is set here to be a ring or a bus topology) can communicate to each other across this subnetwork by using the associated wavelength. Within each subnetwork, destination-removal spatial-reuse techniques are used. Among the subnetworks, wavelengths can be reused by a number of subnetworks which do not share links.

As discussed above for the spatial-reuse ring networks, a SMARTNet system can be used to support connection-oriented circuit-switching and packet-switching (including ATM) architectures as well as connectionless packet-switching networking modes. For packet-switching networks, the links (with each wavelength channel) can be time-shared by using slotted or buffer insertion implicit polling schemes or through the use of other multiple access methods.

As a network which internally employs cross-connect switches, no random delays or packet discards are incurred within the network. Such a performance behavior is advantageous for the effective support of multimedia services, including voice and video streams (which require low network delay jitter levels) and data flows (which prefer no network packet discards, avoiding the need for packet and segment retransmission and reordering operations). An integrated circuit- and packet-switched operation is also readily implemented through the allocation of time, wavelength, and subnetwork channels. Such allocations can also be used to support and isolate virtual subnetworks (or virtual private networks) and their associated user communities.

The SMARTNet configuration illustrated in Figure 3.9 consists of six wavelength routers; each router has a nodal degree of four, being connected to four incoming and outgoing pairs of counterdirectional links. The chord topology is such that every router is connected to its second-neighboring router (rather than to its neighboring router). Analysis shows[23] the network to require the use of a minimum of five wavelengths. Five different groups of wavelength graphs are noted. Each group includes three wavelength graph subnetworks (rings) which reuse the same wavelength. Hence, a wavelength reuse factor of 3 is attained. Each ring subnetwork offers a spatial reuse gain of 2.4 (rather then 4, due to the nonuniform distribution of the traffic across the wavelength graph). Hence, this SMARTNet configuration provides a throughput capacity level of $3 \times 2.4 = 720\%$ of the data rate across each wavelength channel, multiplied

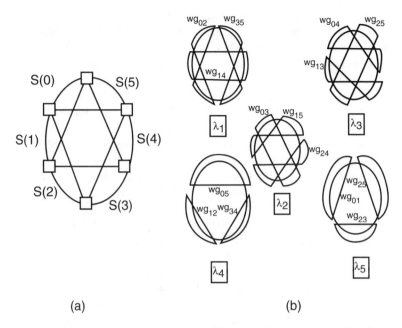

FIGURE 3.9 SMARTNet all-optical network (a) the network layout; nodes represent wavelength routers (b) groups of wavelength graphs; five groups are defined, and each group contains three wavelength graphs.

by the number of wavelengths used (equal to at least five, and can be selected to be any integral multiple of five to yield the same throughput efficiency factor) and by the number of parallel fiber links employed (two for the basic configuration discussed above).

It can be shown that an upper bound on the throughput gain achieved by a SMARTNet configuration which employs wavelength routers of degree four is equal to 800%. Hence, the illustrated configuration, which requires only six wavelength routers and a minimum of only five wavelengths, yields a throughput gain efficiency which reaches 90% of the theoretical upper bound. Clearly, higher throughput gains can be obtained when the router nodes are permitted to support higher nodal degrees (thus increasing the number of ports sustained by each router). Furthermore, it can be shown that such a network structure offers a throughput performance level that is effectively equal to that obtained when store-and-forward switching nodes are employed for such network configurations.

It is also possible to connect the SMARTNet terminals directly to the cross-connect switches (wavelength routers), offering a more scalable configuration. Such a network system has been recently studied as well[24] demonstrating significant throughput utilization gains. Issues investigated for the latter network configuration include the optimal topological layout, methods to achieve an highly efficient operation under the use of a reduced number of identifiers (such as wavelengths), and design approaches to guarantee a high level of survivability.

3.2.9 Quality of Service–Based Integrated Services Multiple Access Communications Networks

Many multiple access networks are designed to support multimedia applications. This requires the sharing of multiaccess channel resources among multiple classes of traffic processes, which can include voice, data and video streams. As a result, it is necessary to design integrated services multiple access methods that offer an array of service classes.

In following the ATM categorization of service classes, integrated services multiple access networks may offer the following services:[25,26]

1. Continuous Bit Rate (CBR) services support applications that generate data at a steady rate and that demand tight delay jitter levels and packet loss rates. Circuit emulation applications require such a service.
2. Real-time Variable Bit Rate (rt-VBR) services support traffic streams that exhibit wide output rate fluctuations and require tight delay jitter levels and packet loss rates. Applications include variable-rate encoded voice and video processes, as well as bursty data flows, which require timely packet deliveries.
3. Non-real-time Variable Bit Rate (nrt-VBR) services support traffic streams that exhibit wide output rate fluctuations and require limited packet loss rates. Included are loss-sensitive batch and inter-active data applications.
4. Available Bit Rate (ABR) service may offer a minimum bandwidth guarantee, while an Unspecified Bit Rate (UBR) service provides no Quality of Service (QoS) guarantees at all. They are appropriate for many regular data applications.

CBR and VBR services offer the user guaranteed QoS levels. For a connection-oriented network, such as an ATM network, the user is required to undergo a Call Admission Control (CAC) process prior to the establishment of the requested connection. For this purpose, the user specifies the call's traffic descriptor (which includes its peak rate, average rate, and maximum burst duration), and its requested QoS levels (including maximum packet delay, peak delay jitter, and packet loss rate, noting that for a nrt-VBR connection only the desired packet loss rate is specified). The CAC manager must check the network resources (through consultation with the network switching nodes along the selected route of the flow) to determine whether the call can be admitted or must be blocked. If admitted, each switching node along the flow route must guarantee the provision of the call's specified QoS levels. For this purpose, the node employs at each one of its buffers scheduling (weighted general processor sharing) algorithms which order the queued packets for service in accordance with their desired (and guaranteed) QoS levels.

For connectionless networking systems,[26] such as TCP/IP networks (including typical LANs and the Internet), similar class-based queueing and scheduling (weighted fair queueing) mechanisms are being implemented. Flow type of service identifiers are used by each node (router) to allocate, to the multitude of flows traversing it, the node input and output buffering, queueing, and processing resources.

In turn, ABR (aside from its minimum rate requirement, if any) and UBR services are recognized as best-effort services. Packet flows are allocated the residual capacities available at the network nodes. A node may reserve an overall bandwidth for the support of a best-effort class to ensure acceptable performance levels; yet, no per-flow reservations are made. Furthermore, a congestion control mechanism[25] may be used to control the rate at which ABR sources drive the network, to reduce the occurrence of packet losses in the network.

In designing an integrated services multiple access communications channel that offers such service classes, we note the following. The class-based queueing and scheduling mechanisms implemented at the network nodes employ statistical multiplexing procedures. The packets sharing the (input or output) service resource at the switch node are stored at a common buffer. The underlying processor-sharing algorithm can then examine the contents of this buffer and order its packets for service in a desirable fashion. If residual bandwidth (service rate) is currently available (after guaranteed service rates are allocated to active flows), it can immediately be allocated to guaranteed service flows which are able to use it, with the leftover bandwidth allocated to best-effort flows. In turn, for a multiple access system, the channel resources are shared by geographically distributed stations. As a result, the QoS-based sharing mechanism may not be able to capitalize rapidly on statistical fluctuations in loading and subsequent variations in channel resource availability. This is particularly true for satellite communications networks due to the underlying long round-trip propagation delay. Yet, it is also a critical factor for many wireless and wireline local and metropolitan area networks due to the underlying signaling delays and bandwidth utilizations involved in implementing the signaling structure required to dynamically share the channel's resources among the different service types. Furthermore, when networks are operated at high speeds, the propagation delays involved in the dynamic bandwidth reallocation process become a limiting performance factor.

Hence, QoS-based integrated service multiple access methods may adopt the following approaches:

1. For CBR applications, it may be necessary to use a fixed assignment scheme to allocate channel resources to a connection, or to invoke the use of a call level demand assignment scheme. The latter uses the parameters specified by the user during a CAC phase to allocate fixed MAC channel resources for the duration of the call.

2. For VBR connections, a call-level demand assignment scheme may have to be employed if the underlying propagation delays and burst start signaling latencies are long. Due to the statistical nature of the traffic generated by the source, the CAC manager will employ statistical multiplexing techniques to calculate the channel resources to be allocated to the user to satisfy its requested QoS levels. Thus, a connection may be established at a data rate whose range lies between the average and peak rates of the call , so that the requested delay and loss QoS requirements are met. Otherwise, the system may use a burst-level demand assignment MAC algorithm. Under such a scheme, channel resources are assigned to the call/flow at the start of each identified burst. Consider the flow activity to consist of a recurring cycles, each consisting of a burst period followed by an inactivity period. Assume that the user can signal the network MAC controller (or all other channel stations in a distributed implementation) in a timely fashion, the end and start times of the underlying bursts. The network can then allocate MAC resources to the user for the support of its bursts in accordance with the required QoS levels. Burst continuation/end signaling tags can be embedded in currently supported burst packets in continuously informing the network controller about the needs of an active user For rt-VBR connections, a call-level demand-assignment or fixed assignment scheme may have to be employed if the underlying propagation delays and burst start signaling latencies are too long.

3. For best-effort applications, as well as for certain VBR flows, a packet-level access algorithm may be employed. Involved are two key MAC protocols:
   a. Random access algorithms are effective for the transmission of short packets by a large number of low-duty-cycle burst users, at low channel utilization levels.
   b. Packet level demand assignment schemes may be effective as well. Under a packet level reservation scheme, the active user signals the network when it has one or more packets to transmit. The network (in a centralized or distributed fashion) then allocates the user resources (such as time slots, frequency bands, CDMA code, and space segments) for the transmission of its packets. Reservation packets can be transmitted across a signaling channel using, for example, a random access algorithm, leading to a reservation ALOHA scheme.

Defining Terms

Carrier sense multiple access (CSMA): A random access scheme that requires each station to listen to (sense) the shared medium prior to the initiation of a packet transmission; if a station determines the medium to be busy, it will not transmit its packet.

Carrier sense multiple access with collision avoidance (CSMA/CA): A CSMA procedure under which each station uses the sensed channel state and an *a priori* agreed scheduling policy to determine whether it is permitted, at the underlying instant of time, to transmit its packet across the shared medium; the scheduling function can be selected so that collision events are reduced or avoided. To implement priorities, a station with a higher-priority message can access the channel by waiting a shorter duration (than lower-priority stations) following the end of the most recent transmission.

Carrier sense multiple access with collision detection (CSMA/CD): A CSMA procedure under which each station is also equipped with a mechanism that enables it to determine if its ongoing packet transmission is in the process of colliding with other packets; when a station detects its transmission to collide, it immediately acts to stop the remainder of its transmission.

Code division multiple access (CDMA): A multiple access procedure under which different messages are encoded by using different (orthogonal) code words while sharing the same time and frequency resources.

Demand assignment multiple access (DAMA): A class of multiple access schemes which allocate dynamically (rather than statically) a resource of the shared medium to an active station; polling and reservation policies are protocols used by DAMA systems; reservation-based TDMA and FDMA systems are also known as DA (demand assigned)/TDMA and DA/FDMA systems.

Fixed assignment multiple access scheme: A multiple access procedure that allocates to each station a resource of the communications medium on a permanent dedicated basis.

Frequency division multiple access (FDMA): A multiple access method under which different stations are allocated different frequency bands for the transmission of their messages.

Integrated services multiple access network: A multiple access network that provides multiple services for the support of multimedia applications; both best-effort services and guaranteed services are supported; the latter can be provided requested QoS performance objectives.

Multiple access algorithm: A procedure for sharing a communications resource (such as a communications medium) among distributed end users (stations).

Multiplexing scheme: A system that allows multiple co-located stations (or message streams) to share a communications medium.

Polling scheme: A multiple access method under which the system controller polls the network stations to determine their information transmission (or medium resource) requirements; active stations are then allowed to use the shared medium for a prescribed period of time.

Random access algorithm: A multiple access procedure which allows active stations to transmit their packets across the shared medium without first coordinating their selected access time (or other selected medium resource) among themselves; this can thus result in multiple stations contending for access to the medium at overlapping time periods (or other overlapping resource sets), which may lead to destructive collisions.

Reservation scheme: A multiple access method under which a station uses a signaling channel to request a resource of the shared medium when it becomes active; also known as demand assigned scheme.

Time division multiple access (TDMA): A multiple access scheme under which a station is assigned dedicated distinct time slots within (specified or recurring) time frames for the purpose of transmitting its messages; different stations are allocated different time periods for the transmission of their messages.

Token-passing scheme: A polling multiple access procedure which employs a distributed control access mechanism; a token (query) message is passed among the network stations; a station that captures the token is permitted to transmit its messages across the medium for a specified period of time.

References

1. Rubin, I. and J.E. Baker, Medium access control for high speed local and metropolitan area networks, *Proceedings IEEE*, Special Issue on High Speed Networks, 78 (1), 168-203, January 1990. Also included in Chapter 2 of *Advances in Local and Metropolitan Area Networks*, W. Stallings, Ed., IEEE Computer Society, New York, 1994, 96–131.
2. Abramson, N., The ALOHA system-another alternative for computer communications, in *Computer Communications Networks*, N. Abramson and F. Kuo, Eds., Prentice-Hall, Englewood Cliffs, N.J., 1973.
3. Rubin, I., Group random-access disciplines for multi-access broadcast channels, *IEEE Trans. In. Theor.*, September 1978.
4. L. Kleinrock and F. A. Tobagi, Packet switching in radio channels. Part I-Carrier sense multiple access models and their throughput-delay characteristics, *IEEE Trans. Commun.*, Vol. 23, pp. 1417–1433, 1975.

5. I. Rubin, "Synchronous and Carrier Sense Asynchronous Dynamic Group Random Access Schemes for Multiple-Access Communications," *IEEE Trans. Commun.*, 31 (9), 1063–1077, 1983.

6. Capetanakis, J. I., Tree algorithm for packet broadcast channels, *IEEE Trans. In. Theor.*, 25 (5), 505–515, September 1979.

7. Sachs, S.R., Alternative local area network access protocols, *IEEE Commun. Mag.*, 26 (3), 25–45, 1988.

8. Maxemchuk, N.F., A Variation on CSMA/CD that yields movable TDM slots in integrated voice/data local networks, *Bell Syst. Tech. J.*, 61 (7), 1982.

9. Stallings, W., *Networking Standards: A Guide to OSI, ISDN, LAN, and MAN Standards*, Addison-Wesley, Reading, MA, 1993.

10. Rubin, I. and Z. Tsai, Performance of double-tier access control schemes using a polling backbone for metropolitan and interconnected communication networks, *IEEE Trans. Spec. Top. Commun.*, December 1987.

11. Chlamtac, I. and A. Ganz, Frequency-time controlled multichannel networks for high-speed communication, *IEEE Trans. Commun.* 36, 430–440, 1988.

12. Newman, R.M. and J.L. Hullett, Distributed queueing: a fast and efficient packet access protocol for QPSX, *Proc. ICCC'86*, Munich, 1986.

13. Ross, F.E., FDDI—a tutorial, *IEEE Commun. Mag.*, 24 (5), 10–17, 1986.

14. Shah, A., D. Staddon, I. Rubin, and A. Ratkovic, Multimedia over FDDI, *Proceedings of IEEE Local Computer Networks Conference*, Minneapolis, Sept. 1992. Also in the *Multimedia Networking Handbook*, J. P. Cavanagh, Ed., Auerbach, New York, 1995, Chap. 6.4.

15. I. Cidon and Y. Ofek, METARING—A Ring with Fairness and Spatial Reuse, Research Report, IBM, T. J. Watson Research Center, 1989.

16. Baker, J.E. and I. Rubin, Polling with a general service order table, *IEEE Trans. Commun.*, 35 (3), 283–288, 1987.

17. Rubin, I. and J.E. Baker, Performance analysis for a terminal priority contentionless access algorithm for multiple access communications, *IEEE Trans.Commun.*, 34 (6), 569–575, 1986.

18. Hayes, J. F., *Modeling and Analysis of Computer Communications Networks*, Plenum Press, New York, 1984.

19. Rubin, I. and C. W. Choi, Delay analysis for forward signaling channels in wireless cellular networks, *Proceedings IEEE INFOCOM'96 Conference*, San Francisco, March 1996.

20. Goodman, D. J., R.A. Valenzuela, K.T. Gayliard, and B. Ramamurthi, Packet reservation multiple access for local wireless communications, *IEEE Trans. Commun.*, 37, 885–890, 1989.

21. Rubin, I., and S. Shambayati, Performance evaluation of a reservation access scheme for packetized wireless systems with call control and hand-off loading, *Wireless Networks J.*, 1, 147–160, 1995.

22. Rubin, I., SMARTNet: a scalable multi-channel adaptable ring terabit network, presented at the Gigabit Network Workshop, *IEEE INFOCOM'94*, Toronto, Canada, June 1994.

23. Rubin, I., and H.K. Hua, SMARTNet: An all-optical wavelength-division meshed-ring packet switching network, *Proceedings IEEE GLOBECOM'95 Conference*, Singapore, November 1995.

24. Rubin, I., and J. Ling, Survivable all-optical cross-connect meshed-ring communications networks, *Proceedings SPIE Conference*, 3228, 280–291, Dallas, 1997.

25. Schwartz, M., *Broadband Integrated Networks*, Prentice-Hall, Englewood Cliffs, NJ, 1996.

26. Stallings, W., *High Speed Networks: TCP/IP and ATM Design Principles*, Prentice-Hall, Englewood Cliffs, NJ, 1998.

3.3 Digital Subscriber Line Technologies

John (Toby) Jessup

Digital Subscriber Line (DSL) technology is a new digital link technology for use on copper wire telephone lines. Like ISDN, DSL network access products provide both voice and data service, but DSL is a newer technology offering better performance and reduced deployment costs.

This section is divided into six parts. Section 3.3.1 provides a brief glossary of telephony terms relevant to DSL. Section 3.3.2 provides an overview of DSL types and origins. Section 3.3.3 explores standardization efforts. Section 3.3.4 describes DSL equipment. Section 3.3.5 provides an in-depth look at DSL engineering, and Section 3.3.6 looks at DSL service delivery and applications.

3.3.1 Telephone Network Issues and Definitions

Some understanding of telephone networks is needed before delving into the details of DSL. The following definitions will be helpful for readers unfamiliar with telephony terms.

Bridged tap: Bridged taps are branches of copper wire used to extend a local loop to multiple destinations. Historically, carriers have often tapped local loops with new branches when lines get reassigned to new service locations, creating treelike wire topologies. Bridged taps cause signal reflections, and on some circuits the removal of bridged tap segments is required to achieve maximum DSL speeds. Adding additional phone jacks within a residence is another form of bridged tap.

Central office (CO): The central office is the switching equipment center where the copper wire telephone lines (local loops) for a community (a service area) are terminated and connected into the telephone network. The central office contains the telephone switches, electrical power/batteries, and equipment racks needed to provide service.

Digital loop carrier (DLC): LECs may use T1/E1 or fiber lines to create digital loop carrier trunk lines between the CO and a remote terminal location. This strategy reduces the wire pairs needed at the CO and extends the CO service area to more customers. Carriers can also use DLC fiber links and remote terminals to create short local loops for high-speed DSL services (VDSL).

Filters or "splitters" (splitterless designs): DSL access line products provide both data and analog voice service over one copper wire pair by using FDM. Analog voice traffic travels on a 4 kHz passband while the data channel is implemented on higher frequencies. Often, a lowpass filter is needed to divide the signals cleanly. Some low-speed DSL designs can operate without the filters installed, reducing deployment costs by allowing existing telephones and residential station wire to be used without modification (a "splitterless" DSL system).

Load coil/loading: Longer POTS lines include inductive load coils to limit bandwidth to a 4 kHz voice telephony passband, allowing carriers to extend the maximum distance of local loops (beyond 18,000 ft). Load coils must be removed when converting POTS lines to ISDN or DSL because these digital services use frequencies above the POTS 4 kHz voice band.

Local exchange carrier (LEC): This is the company(s) or public entity (PTT) which provides the local telephone service to a community. The LEC maintains and operates the lines and equipment used for the local telephone network. In competitive markets there may be multiple DSL LECs reselling the local loop service originally provided by an incumbent LEC. DSL is for copper wire lines only; it is not a technology for long-distance carriers or others using fiber-optic networks.

Local loop: Network access DSL services are designed to run over the local loop circuit, the single pair of copper wires connecting a home or small business to the local CO. Historically, these wires have supported analog telephone circuits, one pair of wire for each telephone line.

POTS: This acronym stands for Plain Old Telephone Service, the typical analog telephone service used on most copper wire local loops. ISDN and network access DSL services replace the analog POTS service with a digital circuit to support both voice and high-speed data communications.

Remote terminal: In larger urban telephone networks, remote terminals are remote termination points for copper local loops outside the CO (usually an underground vault). The remote terminal is where copper local loops are terminated and multiplexed onto a high-capacity digital trunk line (a digital loop carrier).

Trunk: Trunk lines are high-capacity digital links where telephone circuits are multiplexed together. T3/E3 and T1/E1 lines are commonly used to provide trunks among regional COs or as a link between carrier networks. Some DSL products replace T1/E1 copper wire trunks.

3.3.2 DSL Overview and Origins

DSL, or xDSL as it appears in some publications, is not one technology, but many competing technologies, each providing services over copper telephone wire. In this broader sense, ISDN is just another form of DSL, and actually, the first use of the term was in published ISDN specifications.

DSL products are designed to meet two application needs: DSL products can provide network access services over a single pair of copper wire in place of a POTS local loop, or DSL can be used for channelized trunk services designed to replace T1/E1 lines. Network access products provide both voice and digital data services to a residence or business. Channelized services are used for trunk or DLC replacement, and these products do not provide a POTS voice channel.

3.3.2.1 HDSL

High-speed digital subscriber line (HDSL) was developed in the early 1990s as a channelized trunk replacement technology for copper wire T1/E1 and DLC lines. These older trunk lines using 1544-kHz baseband signaling can cause interference on adjacent digital lines in a copper cable or "binder group" (typically a 50-pair copper wire cable). In addition, T1s are expensive to deploy because they require clean wire pairs (no bridged tap) and repeaters every 6000 feet.

HDSL takes 2B1Q, the 2-bits-per-baud, echo-canceled baseband encoding of ISDN, and applies it to T1 and E1 lines, resulting in a (T1) signal rate of 384 kHz on each wire pair (HDSL, like T1/E1, uses two wire pairs). With the offending higher frequencies removed, carriers can deploy T1/E1 services with less interference to adjacent digital lines and on longer lines, including some lines with bridged tap. HDSL provides the same bit rate and channel configuration as T1 or E1 lines.

HDSL deployment is widespread, and many carriers and private campus networks use HDSL equipment for the copper T1/E1 lines in their telephone networks. HDSL-2 is a newer published standard for T1/E1 trunk service using a single pair of wires. HDSL-2 uses a 768-kHz 2B1Q signal.

3.3.2.2 ADSL

In the early 1990s modem developers began exploring technologies to improve local loop digital performance. Telephone companies, fearing competition from cable companies, were seeking a technology to transmit compressed video signals over local loops. Tests of loop performance had demonstrated that asymmetric channel configurations, where the channel from the CO to the customer is faster than the return channel, yielded substantially higher bit rates and transmission distance on typical local loops, and this configuration offered a promising technology for the distribution of video streams.

However, in service trials asymmetric digital subscriber line (ADSL) did not provide adequate video bit rates on longer local loops, and carriers abandoned their deployment plans. The Internet became popular and developers realized this new technology provided an excellent transport for the less-demanding asymmetric send/receive traffic load typical of Internet usage. Many data applications, such as loading Web pages with a browser program, generate more network-to-user traffic than user-to-network traffic, making these applications well suited to asymmetric allocations of access line capacity.

Numerous companies saw the same opportunity, and development efforts proceeded in parallel. As a result, two competing, incompatible encoding methods have emerged in DSL access line products: Discrete Multi-Tone (DMT) and Carrierless Amplitude Phase-shift (CAP). Both technologies improve and extend older transmission encoding techniques, and the advantages of either approach are disputed among competing manufacturers.

Fixed-rate ADSL products are generally obsolete, having been replaced by more flexible RADSL technology.

3.3.2.3 RADSL

Unlike straight ADSL service, which is configured for fixed send/receive line rates, rate adaptive digital subscriber line (RADSL) services can automatically shift speeds according to variations in line transmission quality, and according to service limits defined by the carrier. RADSL modems shift to a lower speed if line quality is poor, offering dynamic performance optimized for current loop conditions (attenuation, spectral disturbance, etc.). RADSL products also give carriers additional service provisioning options, such as the ability to offer custom service levels (maximum bit rate, adjustable send/receive channel balance, etc.).

RADSL receive channel rates can reach 8 Mb/s, but the actual speed of the line is limited by line length and line quality. Typical (affordable) service rates will be slower for most subscribers.

3.3.2.4 SDSL

Some DSL products provide symmetric send and receive channels (as serial communications lines always have). Business applications, such as a transaction processing or LAN interconnection, may work better over symmetric links. SDSL could also serve as a link for packetized voice applications.

However, configuring symmetric DSL service will result in a reduction of total capacity when compared with an asymmetric configuration of the same line. SDSL products are based on HDSL 2B1Q technology, providing a maximum symmetric data rate of 704 kb/s on a single pair of wires (voice service is provided via a 64 kb/s digital channel). RADSL products with configurable upstream and downstream channels can also simulate SDSL channel configurations (nearly), but only at speeds under 2 Mb/s.

3.3.2.5 IDSL

In a further twist on the DSL label, some carriers market IDSL service, which is just ISDN without the telephone network. IDSL service is data-only, and it is not circuit-switched (unlike ISDN, there is no telephone number dialing). IDSL is not a new technology; it is only a service description of what is essentially a data-only ISDN product.

3.3.2.6 UADSL (or G.lite, UDSL, ADSL lite)

Universal ADSL is an initiative to standardize low-speed ADSL service for consumer markets. The UADSL proposed standard specifies DMT line encoding. UADSL, because of its lower speed, will use simple equipment which is easier to deploy, offering a splitterless service where DSL equipment can be installed by a nontechnical consumer. UADSL modems will be inexpensive retail products installed into a PC (or built in) just as analog modems are today, and all wire connections will use the existing telephone wire without modification.

UADSL will be a fixed-rate service intended for Internet access lines. Speeds will be limited to 1536 kb/s; however, this is more than adequate for Internet access applications today, since typical Internet access performance rarely exceeds about 500 kb/s (given the realities of ISP bandwidth limitations, network latency, and loaded server performance).

3.3.2.7 VDSL

VDSL, which stands for Very-high-speed Digital Subscriber Line, is a term applied to future DSL systems offering bit rates above RADSL speeds. VDSL is often cited as a technology for transmission of high-definition television and multimedia Web traffic at speeds up to 51 Mb/s. The key to achieving these bit rates over copper wire is a shortened local loop, although continuing improvements in line encoding will also contribute to performance gains. VDSL will require remote terminals and new fiber-optic DLC systems. A 1000-ft local loop could provide 51 Mb/s service to a home or business.

3.3.3 DSL Standards and the Marketplace

DSL standards are evolving and likely to change substantially in the months and years ahead. To date, the ITU, ANSI, and IEEE have defined few standards for DSL. Other industry groups involved with DSL standards include the ADSL Forum and the Universal ADSL Working Group (UAWG), an industry-sponsored organization which has proposed the UADSL standard. DMT ADSL has been published as a standard (ANSI T1.413, 1997), and the ITU has endorsed this ANSI standard (V.ADSL, 1998). Extensions of this standard and others will surely follow. An accepted UADSL standard (ITU) is expected in 1998. HDSL and HDSL-2 are also published ANSI standards.

However, a published standard does not imply interoperabilty among competing products. Consumer products will probably converge into a working UADSL standard, but high-speed DSL products may never, because of the higher layer features built into these products. In addition to line encoding differences, data-link layer protocols may differ, along with proprietary features such as management capabilities and product integration into carrier network architectures. As with most technical markets, vendor-specific innovations will continue to exceed the performance levels of published standards.

3.3.4 DSL Equipment Overview

3.3.4.1 Network Access Premises Equipment

DSL service will require new equipment at the subscriber's site. The DSL equipment attached to a local loop or trunk line is often referred to as a modem; however, some documents may use the term ADSL Termination Unit — Remote, or "ATU-R." Regardless of the label, the DSL network access equipment provides, at a minimum, the following functions:

Data channel line encoding (CAP or DMT)
Physical interfaces to premises equipment as required (computers and telephones)
Support for a single POTS voice channel
Data-link layer protocol encapsulation on the data channel (PPP and/or ATM)
Protocol interface for data equipment (via ethernet data-link or installable PC device driver)

In addition, some products may offer other features:

Protocol routing and bridging
Provider-configured service levels (RADSL bandwidth limits, etc.)
Management interface for support personnel (SNMP, HTTP, telnet, etc.)
Security features (encryption, filtering, etc.)

Network access DSL modems can be deployed in one of two ways at a subscriber site: as an external device inserted into the existing premises telephone wire, like an external analog modem would be, or as an internal device implemented as a circuit card or feature built into a PC. External equipment is required for high-speed DSL products because filters must be added to the line to ensure a clean 4 kHz signal to the telephones. An external DSL modem provides separate ports for the connection of telephone line and data communication equipment (ethernet for the data, a standard jack for the telephone line). Filtering may be integrated into the modem unit, or microfilters may be required on each telephone station line.

External units are suitable for business applications where data communications equipment is already present, and higher-speed (higher-cost) service is desired. But for residential applications, splitterless designs offer easier deployment and lower costs. UADSL and other low-speed services will not require a filter component in the customer site equipment, eliminating the need for an external modem. Conversion from POTS to UADSL service will only require adding a UADSL modem card to the PC (or buying a PC with the interface built-in). All existing POTS devices (phones, fax, etc.) will operate as before, and the residence wiring will be unmodified.

3.3.4.2 Central Site DSL Data Communications Equipment

DSL network access systems serving a community of subscribers require specialized equipment at a central site (a CO or a remote terminal vault) to terminate DSL local loop pairs. A DSL Access Multiplexer (DSLAM) provides data channel multiplexing and network interconnection for many DSL access lines. The DSLAM (also called an ATU-C in some documents) includes modem cards to terminate DSL data channels, along with connections into an ATM data network or to other data transmission services, such as trunk lines and ethernet (Figure 3.10). The DSLAM is the interconnection point between DSL and these other data networks.

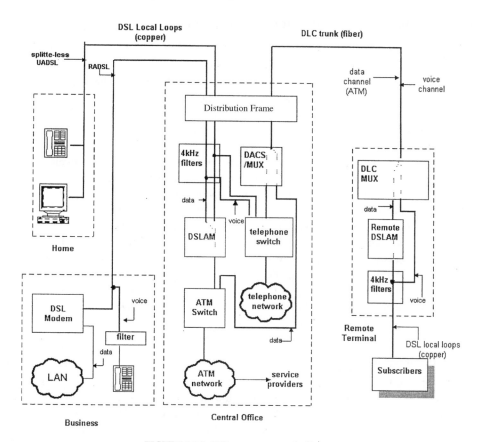

FIGURE 3.10 DSL access network diagram.

Some DSLAM systems perform modem pooling, assigning a modem to a subscriber data line only during line activity (the modem switching delay is transparent to users). Pooling reduces the carrier's equipment and CO engineering costs, making service more affordable.

Voice band filters are also used at the central site. The DSL 4-kHz voice signal is terminated onto existing voice network interfaces (just as POTS lines would be). The data channels are directed into the DSLAM (Figure 3.10).

Manufacturers may offer a variety of other features in a DSLAM, including management, routing, and service-level controls. Finally, it is likely that telephone switch manufacturers will add DSLAM features to future products, and upgraded telephone switches may eventually eliminate the need for separate DSLAM systems.

3.3.4.3 Access Line Voice Services

All DSL access line products provide both voice and data service over a single-pair local loop. Unlike ISDN, DSL separates the data network and telephone network into two separate transmission bands. The voice service connection is circuit-switched POTS, ensuring compatibility with existing devices (like POTS, DSL provides telephone service during power outages). Analog telephones equipment will work without modification (although higher-speed services will require filters). Use of the telephone has no effect on the permanent data channel.

Voice service quality may suffer on DSL lines longer than 18,000 ft. Since the load coils have been removed, telephone signal clarity will be impaired at this distance. Filters can be used to remove high-frequency noise from the signal.

3.3.5   DSL Engineering Issues and Applications

3.3.5.1   Local Loop Engineering Issues in Public Networks

The local loops used in typical metropolitan phone systems vary widely in line length and other aspects. Within each CO service area, some subscribers live close to the CO while others are located on the periphery of the service area. Some loops may have bridged tap segments and some may use DLC systems. Loops may share cable binder groups with other digital services, such as T1, ISDN, or other DSL lines. These factors and others described below can limit DSL performance and reach.

DSL manufacturers strive to offer the highest performance possible on the widest range of local loop conditions. Lower-speed products, such as UADSL, can provide consistent service throughout a CO service area, while higher-speed DSL products yield a range of performance levels determined by loop conditions.

The maximum recommended loop length for metropolitan phone systems is 18,000 ft (5.5 km). However, longer loops do exist in many systems, and on these longer lines DSL services may be limited. Although bridged taps are generally acceptable on a DSL loop (at lower speeds), DSL, like ISDN, requires an unloaded local loop, since DSL must use frequencies higher than 4 kHz for the data channel band.

Carriers deploying DSL service will offer low-speed service to all customers, while positioning higher-speed services for use by businesses and high-demand Internet users. Services over long lines and existing DLC systems will require re-engineering of the local loop, raising deployment costs. In many service areas, older DLC systems designed originally for voice services may delay DSL deployment, since these systems lack the bandwidth needed to transmit both voice and data. DLC systems supporting DSL will need fiber-optic links and new termination equipment (a remote DSLAM).

3.3.5.2   DSL Line Encoding

Current DSL products use one of three encoding methods, 2B1Q, DMT, or CAP, to transmit bits over copper wires (Table 3.1). 2B1Q baseband signaling is suitable for symmetric services, such as HDSL and SDSL. However, because access line DSL products use FDM to reserve a 4-kHz POTS voice channel, one of two modulated passband encoding methods, either CAP or DMT, must be used in access line products (baseband signals cannot work in an FDM system).

TABLE 3.1   Comparison of Copper Wire Transmission Methods

	Frequency, kHz	Max. Bit Rate, kb/s @ Length, ft	Bit Rate kb/s @ Max. Length, ft	Encoding
Ethernet (10BASE-T)	0–20,000	10,000 @ 384	10,000 @ 384	Manchester
ISDN/ISDL	0–80	160 @ 18,000	160 @ 18,000	2B1Q
HDSL (4 wire)	0–384	2048 @ 12,000	2048 @ 12,000	2B1Q
SDSL (2 wire + POTS)	0–384	704 @ 12,000	704 @ 12,000	2B1Q
CAP RADSL (send)	240–1409	7000 @ 6000	1536 @ 18,000	CAP
CAP RADSL (receive)	35–191	640 @ 6000	64 @ 18,000	CAP
DMT RADSL (EC)	25–1104	8000 @ 6000	1536 @ 18,000	DMT
UDSL (send)	140–550	1536 @ 18,000	1536 @ 18,000	DMT
UDSL (receive)	20–130	384 @ 18,000	384 @ 18,000	DMT

Note: Specific vendor claims of maximum data rate and line reach vary, and actual performance achieved depends on loop quality, including adjacent signal disturbance; these figures are only a rough guideline.

3.3.5.2.1   CAP Encoding

CAP is a high-speed version of QAM, the trellis-encoded carrier modulation commonly used in analog modems. However, due to the higher-frequency range available on an unloaded loop, CAP can use faster signaling and larger constellation patterns to send higher bit rates. RADSL CAP modems shift speeds on the line by adjusting the size of the QAM constellation pattern (the number of bits per QAM signal).

CAP systems use FDM to separate send/receive data channels, placing the customer-to-network channel on a lower band (35 to 191 kHz) and the network-to-customer channel on a higher band (240 to 1409 kHz).

CAP has not been accepted as a DSL standard, although a majority of vendors and carriers remain committed to the technology. CAP is an older, more proven technology than DMT, and some would argue it is better suited to the spectral realities of typical telephone networks. CAP is simpler to implement in modem designs, so products may cost less (DMT requires more complex DSP chip sets). CAP also has a power consumption advantage over DMT, operating with 1 to 2 W of line power while current DMT systems require 8 W.

3.3.5.2.2 DMT Encoding

Unlike CAP, which is a single-carrier transmission scheme, DMT divides the data channel frequency range into 256 4-kHz carrier subchannels, each using QAM signaling. DMT spreads a string of bits (a transmission symbol) across the available subchannels, allocating some to each, according to the current QAM capacity of each subchannel. QAM constellations are adjusted on each channel, with lower frequency subchannels typically supporting larger patterns (up to 15 bits). Echo cancellation or FDM can be used to separate send and receive signals (the T1.413 ADSL standard leaves this choice to the manufacturer).

DMT and CAP offer similar speed and reach capabilities; however, DMT offers better RADSL rate adaption behavior. If a line is experiencing interference, DMT can disable or adjust the troublesome subchannels, providing dynamic fine adjustment of the line rate. CAP RADSL products must stop transmission and retrain the line using a smaller QAM pattern, resulting in a coarser adjustment to the transmission rate.

DMT is patented and licensed by Texas Instruments (formerly by Amati Corporation).

3.3.5.3 Digital Line Performance in Typical Networks

DSL is performance limited by the quality of the local loop infrastructure. Factors limiting performance include:

Local loop length
Wire gauge (or gauges) used in the local loop
Bridged taps
Channel bandwidth symmetry
Send/receive channel separation (either by FDM or echo cancellation)
Adjacent digital lines in a cable binder group
Environmental impairments, such as AM radio and impulse noise

Generally, the challenge for DSL modem manufacturers and service providers is to minimize signal corruption due to transmission loss (attenuation) and electrical disturbance by adjacent signals (cross talk). We will look at attenuation first, followed by cross talk disturbance caused by DSL itself, and then cross talk caused by other digital services.

3.3.5.3.1 Attenuation and Frequency

For any copper wire network, signal attenuation of higher frequencies increases as wire length increases, so digital transmissions on longer lines tend to use lower frequencies. Ethernet 10BASE-T, a short-distance copper network, rides on a 20,000 kHz baseband carrier, but only for 328 ft. T1 lines operate at 1544 kHz over 6000 ft, while HDSL offers the same data rate (1536 kb/s) at a much lower 384-kHz bandwidth, resulting in significantly longer reach (12,000 ft) and reduced cross talk. A 2B1Q ISDN BRI signaling at 80 kHz can operate over 18,000 ft of 24 AWG copper wire (see Table 3.1).

In addition to wire length, the thickness (gauge) of wire determines attenuation loss. Telephone systems use 22, 24 and 26 American wire gauge (AWG) wire. 24 AWG (0.5 mm) is quite common, and most published specifications are based on 24 AWG. Performance and reach is best on 22 gauge, while loops built with 26 gauge wire (or combinations of wire gauge) will see reduced DSL loop lengths and bit rates.

Echo-canceled DMT uses frequencies up to 1104 kHz, giving DMT a slight loop length advantage over CAP systems which typically use an upper pass band channel extending to 1409 kHz (Figure 3.11). On

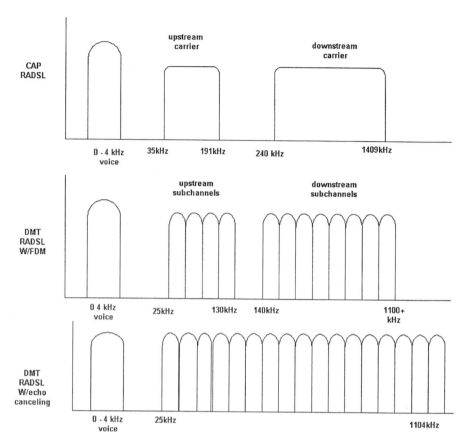

FIGURE 3.11 DSL bandwith allocation.

longer loops, RADSL systems reduce the upper-frequency content of the signal to maintain communication at a lower bit rate.

3.3.5.3.2 Cross Talk (NEXT/FEXT) among DSL Circuits

In addition to the effects of attenuation, DSL transmissions are also impaired by adjacent signal cross talk disturbance. Cross talk is present in all twisted-pair copper wire media, particularly in cable ends where the close proximity of untwisted conductors (in connectors and on distribution frames) can induce bit errors on adjacent wires. Near-end cross talk (NEXT) is the most common problem, occurring when a strong transmitting signal disturbs a weaker (attenuated) receive signal operating in the same frequency on an adjacent wire. Far-end cross talk (FEXT) occurs when a strong transmitting signal disturbs a weaker signal at the opposite end of the cable.

One way to eliminate NEXT is to use FDM to place send and receive channels on separate frequencies. All CAP and some DMT systems do this. However, FDM reduces available bandwidth because unused guard bands must be allocated between the sending and receiving channels. The guard bands may shift the upper channel toward higher frequencies, potentially reducing loop reach. FEXT is still present in FDM systems, but FEXT disturbance is less common than NEXT. Some DMT systems use echo cancellation, and these systems, while providing improved loop reach by using lower frequencies, may be susceptible to self-NEXT disturbance.

3.3.5.3.3 Asymmetry Improves Performance

Using asymmetric send and receive channels also reduces cross talk disturbance, giving ADSL and RADSL a significant bit rate and loop length advantage over symmetric systems such as HDSL and SDSL. In telephone systems, cross talk disturbance is itself asymmetrical: it is a condition localized to the CO or remote terminal end of the circuit (where all the copper wires come together) and is less of a problem on the customer side of the wire. Asymmetric DSL takes advantage of this condition by reducing the bit rate of the attenuated (and therefore CO cross talk-susceptible) customer-to-network channel. In addition, the customer-to-network channel is placed on lower-frequency bands where attenuation is reduced (see Figure 3.11).

At the subscriber end of the circuit the opposite conditions exist: cross talk is minimal since there are few noisy wires sharing the termination point. So, the arriving high bit rate (and high frequency) network-to-customer signal, although attenuated by transmission over the loop, is not further impaired by cross talk disturbance.

Asymmetric DSL services minimize the spectral interference most likely to occur in telephone networks: NEXT disturbance of attenuated customer-to-network signals on the CO side of the circuit. The result is an optimized system offering the highest (asymmetric) bit rate possible.

3.3.5.4 Cross Talk Caused by Other Digital Services

Like all digital signals, DSL is also disturbed by inductive energy radiating from adjacent wires running other digital signals, such as T1 and ISDN. Telephone carriers have dealt with this problem for years, taking care to spread digital services among binder groups to avoid spectral disturbance. DSL is just another digital service with similar network design and provisioning parameters.

Generally, baseband transmitters, such as T1/E1 and ISDN (and HDSL/SDSL), which radiate energy across all frequencies below the signaling rate, create more problems in a binder group (combining DSL with T1 circuits in a binder group is not recommended). Modulated carrier signals such as DMT and CAP offer a more benign spectral content which is easier to deploy. However, DMT and CAP signals can still corrupt each other when combined in a common system.

3.3.5.5 Disturbance by RF Signals

Both DMT and CAP DSL systems share bandwidth with the AM radio band (550 to 1400 kHz), and under some conditions, these signals can also disturb DSL signals. DMT RADSL modems handle this interference well, automatically reducing usage of the conflicting frequency subchannels.

3.3.5.6 Other Transmission Impairments

Phase distortion increases with wire length, and both CAP and DMT can tolerate moderate phase distortion (through error correction). Bridged taps cause signal reflections which can cause bit errors, particularly on longer lines where attenuation is present. Impulse noise, such as caused by a telephone going off-hook, is present in all telephone systems. DSL systems can be susceptible to disturbance by excessive impulse noise.

3.3.5.7 Future (VDSL) Engineering Issues

VDSL systems may finally bring DSL back into the video transmission market originally envisioned for ADSL. Although currently an emerging technology, these systems designed for video and multimedia Web services show much promise. Manufacturers are beginning to make DSLAM equipment designed for field installation in remote terminal vaults linked by DLC fiber trunks. VDSL service will require significant engineering investment to make short length loops available across a service area and, in an era of deregulated competition, carriers are likely to proceed cautiously, looking for proven market success before deploying.

3.3.6 DSL Service Delivery and Applications

DSL provides users with a dedicated link to a specific service provider, such as an ISP. DSL data service is an always-on connection: there is no dialing or call setup delay as would occur with an analog modem or ISDN connection. A DSL subscriber is permanently attached to a service provider, functioning much like a workstation in a LAN environment. The service provider site is usually linked into the carrier's regional data network through an ATM access line (Figure 3.12).

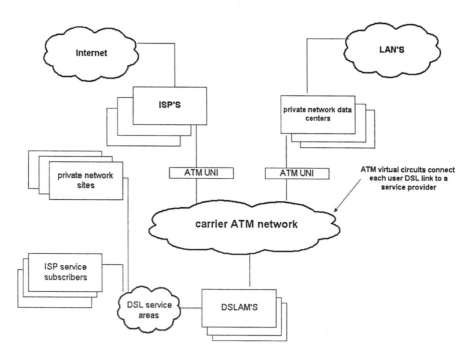

FIGURE 3.12 DSL applications.

The main competitor of DSL in residential markets is cable modems. CATV-based data channels offer higher bit rates (27 Mb/s), but they use shared media, placing numerous subscribers on the same channel bus, similarly to a shared media ethernet LAN. Congestion is possible, and the cable itself becomes a single point of failure (as it is with cable television). New wiring will be needed to connect telephones to the cable box. Splitterless DSL services will not require new wiring, and all DSL systems use a reliable star topology (the local loop network) to provide each subscriber with guaranteed bit rates over a secure path.

3.3.6.1 Data-Link Layer Protocols

Like any point-to-point data channel, DSL end systems must support compatible data-link layer protocols to allow interconnection. Some manufacturers are using PPP, while others are combining PPP/AAL5 and ATM. Extending ATM over DSL adds a virtual circuit capability in the remote DSL equipment (the ATU-R modem becomes an ATM UNI), providing numerous benefits to service providers, including end-to-end Quality of Service (QoS) controls, management, and efficient multiplexing of user traffic in a cell switching network.

Carrier networks built with ATM and DSL can provide either a data-link layer (bridged) connection or routed (IP, IPX) connection to users. Most DSL modems and ATM UNI devices can support either function.

3.3.6.2 DSL Applications

Consumer Internet access is the most common use of DSL services today. ISPs serving a DSL community are each linked into the local carrier DSL network via an ATM data link (Figure 3.12).

Private WAN networks are another opportunity for DSL. Companies can build DSL/ATM networks linking branch offices or home workers. Branch sites and telecommuters use DSL while the central site data center uses ATM. When DSL service becomes ubiquitous over a metropolitan area, DSL could replace or be integrated with existing private WAN services such as frame relay and leased lines.

DSL also has applications in private campus networks. A business or government campus, university, or residential community can use DSL over its existing premises wire. Voice channels are tied into the PBX system as before, and DSL data channels are linked into a central data center. In this way, DSL can significantly reduce the premises wiring expense in a campus environment where existing telephone wire can now be used to support a fast data channel in addition to telephony. Separate LAN wiring is eliminated, and cheaper EIA Category 3 wiring and connectors can be used (standard telephone station wire).

3.3.6.3   Service Switching

A dedicated, always-on link to an ISP is suitable for Internet applications of DSL, but telecommuting and private business applications may require some form of service switching to allow a user to reach multiple service providers, such as both an ISP and the servers within a private company network. And some users of DSL may require support for multiple protocols, not just IP.

In a private IP-only DSL network, IP routing can connect user stations to multiple services within the private network. A proxy server or firewall link to the Internet (perhaps provided by the DSL carrier) can be added to allow DSL users to reach both internal servers and external Internet hosts by using standard IP DNS or LAN directory services.

Virtual private network (VPN) tunneling protocols such as L2TP and PPTP provide another IP-based service switching option. Home users permanently linked to an ISP can use a VPN (implemented in the PC or in the DSL modem) to establish a private remote LAN link into their company network when telecommuting. Tunneling adds framing overhead to the data stream, and security may be a concern, but this option may be the simplest approach to implement.

Both PPP and ATM AAL-5 protocols can carry non-IP traffic across a DSL/ATM network, including Novell IPX, Appletalk, and IBM (LLC, NetBIOS) traffic; however, some modem products may not support these protocols. Multiprotocol features will most likely show up in high-end products designed for business networks (such as branch office routers).

3.3.6.4   The Future of DSL

For decades a digital "last mile" connection to telephone networks has proved elusive. ISDN has gained some degree of success, but mostly in markets dominated by PTT carriers, such as in Europe and Japan. In the U.S. there has been less deployment of ISDN because the switch upgrade costs are high and other data services are priced competitively.

DSL deployment makes business sense for carriers in competitive markets. The service can be introduced one loop at a time, reducing the up-front capital risk of conversion. But more importantly, DSL preserves existing analog voice services via FDM, making deployment and support significantly easier. DSL removes data traffic from the circuit-switched telephone network, reducing the network congestion caused by lengthy data telephone calls.

For users, DSL is attractive because it is fast, inexpensive, and the connection is always on. Data services will be much more reliable because today's computer applications and protocols are inherently designed for a persistent data link. Dial-up service often requires routers performing complex protocol spoofing, and DSL eliminates this major weakness in ISDN products.

DSL will fulfill the promise of ISDN by offering evolution instead of revolution. The analog voice network is preserved and data services are given a separate, always-on high-speed channel. It offers an attractive service which carriers can easily deploy in today's competitive marketplace.

3.4   SONET and SDH

Chris B. Autry and Henry L. Owen

3.4.1   Introduction

The synchronous optical network (SONET) and the synchronous digital hierarchy (SDH) are digital transmission standards. SDH is the international version of SONET, which is used in North America. The two major differences between SONET and SDH are the terminology and the basic line rates used [Autry and Owen, 1997]. First, an overview of SONET is presented, which is then followed by an overview of SDH.

SONET is a physical transmission vehicle that is capable of transmissions in the gigabit range. SONET is defined by a set of electrical as well as optical standards. SONET is intended to be the transmission means over the next several decades in the same manner that T1 technology has been the basic transport mechanism. Since SONET uses the capability of already installed fiber-optic cable, eliminates complexity, and reduces equipment functionality requirements, local and interexchange carriers have incentive to install SONET over T3. Immediate savings in operational cost, as well as preparing for higher bandwidth applications, justify this installation. The technological step forward provided by SONET allows the realization of a new generation of high bandwidth services in a more economical manner than has previously existed [Davidson and Muller, 1991; Sexton and Reid, 1992; Minoli, 1993].

At least two problems with the plesiochronous digital hierarchy (PDH) transmission system motivated the development of SONET. The first is the complexity involved in signal access. In accessing a given 1.544 Mb/s T1 from a 44.736 Mb/s T3, the entire T3 signal must be demultiplexed in order to access the given T1 of interest. The second problem with the existing PDH system is the complexity of network management. It has proved not to be a simple task to measure network performance, respond to network failures, or manage remote network equipment from control centers. SONET was created to solve these as well as other problems by [Davidson and Muller, 1991]:

- Unifying North American and international standards;
- Including intelligence in the multiplexers for protection switching, administration, operations, and maintenance;
- Making multivendor networks possible;
- Using a base rate that is able to accommodate existing T1 and T3 rates;
- Using a synchronous network to simplify multiplexing and demultiplexing for easy signal access;
- Supporting continually increasing optical bit rates;
- Including sufficient overhead channels and functions to support facility maintenance.

3.4.2   SONET Frame

The basic building block in SONET is the synchronous transport signal level-1 (STS-1). It is transported as a 51.840 Mb/s serial transmission using an optical carrier level-1 (OC-1) optical signal. Even though SONET frames are physically transmitted serially, one normally finds it easier to think of the frames in terms of bytes. In fact, hardware implementations of SONET systems do the majority of the processing in terms of bytes, not bits. The grouping of a specified set of bytes in the STS-1 is called a frame. An STS-1 frame consists of 6480 bits, which is 810 bytes. The 810 bytes are typically represented as 90 columns by 9 rows, as shown in Figure 3.13. The bytes are numbered from left to right across the top row until 90 is reached, then continues on the second row again left to right starting with byte 91. In a given byte, the most significant bit is transmitted first. This group of bytes (the STS-1 frame) is transmitted in 125 μs so that 8000 frames occur per second.

3.4.3   SONET Data Rates

Higher data rates are transported in SONET by synchronously multiplexing N lower-level modules together. Table 3.2 lists valid values of N and the corresponding data rates. SONET standards define both optical and electrical signals. Optical carrier level-N (OC-N) and synchronous transport signal level-N (STS-N) correspond to the optical and electrical transmissions, respectively, of the same data rate. The maximum value of N is limited by the requirement that each individual STS-1 is allocated only one 8-bit identification value and this value must be unique.

TABLE 3.2   Sonet Optical Carrier and Data Rates

Optical Carrier	Data Rate (Mb/s)
OC-1	51.84
OC-3	155.52
OC-12	622.08
OC-24	1244.16
OC-48	2488.32
OC-192	9953.28

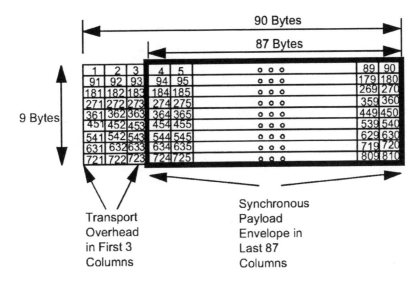

FIGURE 3.13 STS-1 frame.

3.4.4 Physical, Path, Line, and Section Layers

As shown in Figure 3.13, the STS-1 frame structure has two parts, the transport overhead and the synchronous payload envelope (SPE). Starting with a payload (for example, DS1, DS3, etc.), which is to be transported by SONET, the payload is first mapped into the SPE. This operation is defined to be the path layer and is accomplished using path terminating equipment. Associated with the path layer are some additional bytes, named the path overhead (POH) bytes, which are also placed into the SPE. After the formation of the SPE, the SPE is placed into the frame along with some additional overhead bytes, which are named the line overhead (LOH) bytes. The LOH bytes are used to provide information for line protection and maintenance purposes. This LOH is created and used by line terminating equipment such as OC-N to OC-M multiplexers. The next layer is defined as the section layer. It is used to transport the STS-N frame over a physical medium. Associated with this layer are the section overhead (SOH) bytes. These bytes are used for framing, section error monitoring, and section level equipment communications. The physical layer is the final layer and transports bits serially as either optical or electrical entities.

3.4.5 Overhead Byte Definitions

There are four types of overhead bytes: section, line, path, and virtual tributary (VT). The first three columns in the SONET STS-1 frame are where the section and the line overhead bytes are always located. Table 3.3 gives a brief summary of each of the overhead bytes in an STS-1 frame. The SONET overhead bytes are defined in Table 3.3 using the byte numbers from Figure 3.13.

The framing bytes A1 and A2 are used by SONET hardware to identify the start of the STS-1 frame. Bytes H1 and H2 indicate the start of the SPE, and H3 is the negative justification opportunity byte. Bit interleaved parity (BIP) codes are used to detect errors. Bytes D1 to D3 form a 192-kb/s data communication channel (DCC), and bytes D4 to D12 comprise a 576-kb/s DCC. DCCs are used for alarms, maintenance, control, monitoring, and administration needs. An order wire provides voice communication between maintenance personnel. Each identification byte is used to identify a connection at the specified layer. The growth bytes are available to support services and technologies of the future. Additional details on the overhead bytes may be found in ANSI [1991].

TABLE 3.3   SONET Section, Line, Path, and VT Overhead Byte Allocations

Byte No.	Name	Type	Description of Usage
1	A1	Section	Framing
2	A2	Section	Framing
3	C1	Section	An 8-bit STS-1 identifier
91	B1	Section	Bit interleaved parity-8 (BIP-8) code over section
92	E1	Section	Order wire
93	F1	Section	User channel
181	D1	Section	1 byte of 192-kb/s data communications channel
182	D2	Section	1 byte of 192-kb/s data communications channel
183	D3	Section	1 byte of 192-kb/s data communications channel
271	H1	Line	Pointer (most significant byte)
272	H2	Line	Pointer (least significant byte)
273	H3	Line	Negative justification opportunity byte
361	B2	Line	Bit interleaved parity-8 (BIP-8) code over line
362	K1	Line	Automatic protection switching (most significant byte)
363	K2	Line	Automatic protection switching (least significant byte)
451	D4	Line	1 byte of 576-kb/s data communications channel
452	D5	Line	1 byte of 576-kb/s data communications channel
453	D6	Line	1 byte of 576-kb/s data communications channel
541	D7	Line	1 byte of 576-kb/s data communications channel
542	D8	Line	1 byte of 576-kb/s data communications channel
543	D9	Line	1 byte of 576-kb/s data communications channel
631	D10	Line	1 byte of 576-kb/s data communications channel
632	D11	Line	1 byte of 576-kb/s data communications channel
633	D12	Line	1 byte of 576-kb/s data communications channel
721	Z1	Line	Growth
722	Z2	Line	Growth
723	E2	Line	Order wire
In SPE	J1	Path	Repeating 64-byte identification
In SPE	B3	Path	Bit interleaved parity-8 (BIP-8) code over path
In SPE	C2	Path	Payload type identification
In SPE	G1	Path	Path status including far end block error (FEBE)
In SPE	F2	Path	User channel
In SPE	H4	Path	Multiframe frame count
In SPE	Z3	Path	Growth
In SPE	Z4	Path	Growth
In SPE	Z5	Path	Growth
In SPE	V5	VT	Error checking, signal label, and path status
In SPE	J2	VT	Path trace
In SPE	Z6	VT	Path growth
In SPE	Z7	VT	Path growth

Sources: ANSI [1991] and Bellcore [1994].

TABLE 3.4   H1 and H2 Byte Contents*

Bit Number and Usage in Byte H1									Bit Number and Usage in Byte H2						
1	2	3	4	5	6	7	8	9	10	11	12	13	14	15	16
N	N	N	N	*	*	I	D	I	D	I	D	I	D	I	D

Source: Bellcore [1994].

3.4.6   Synchronous Payload Envelope and Pointers

In addition to the transport overhead, the STS-1 frame contains the synchronous payload envelope (SPE) as previously mentioned. The SPE consists of 783 bytes and may be thought of as 87 columns by 9 rows. Two columns (30 and 59) are not used for carrying payload; instead they contain "fixed stuff" bytes. Another column contains STS-1 POH. The remaining 756 bytes are used for carrying payload. The SPE may have its first byte anywhere inside the 87 columns by 9 rows of the SPE area in the STS-1 frame and in fact can move around in this area. The overhead bytes H1 and H2 indicate the starting location. These two bytes contain a 10-bit binary value between 0 and 782 that corresponds to the location in the STS-1 frame in which the first payload byte is located. The contents of the H1 and H2 pointer bytes are shown in Table 3.4. Bits 1 through 4 (new data flag) contain bits that signal that a new pointer value is to be used, allowing for a complete change in payload. Bits 6 and 7 are not defined for use. Bits 7, 9, 11, 13, and 15 are defined to be increment bits, while bits 8, 10, 12, 14, and 16 are defined to be decrement bits.

In a SONET network element, there is a receive side clock and a transmit side clock that are not always at the exact same frequency. An objective in SONET is to recover and to distribute highly accurate clocks; however, this is not always possible in all operating scenarios. For example in a failure mode, a clock may be derived from a less accurate source to allow network operation to continue. In such situations it is possible that the STS-1 frame rate is faster or slower than the SPE rate. In this situation it is necessary to transmit one extra byte in the negative justification opportunity byte or one fewer byte (known as a positive stuff byte) in a given STS-1 frame to accommodate the SPE. In a positive justification action, the normally used byte immediately after the H3 overhead byte is not used to carry payload. In a negative justification action, the H3 overhead byte, which normally does not carry payload, is used to carry a payload byte.

The corresponding 10-bit pointer value is used to indicate which one of these two cases is occurring in the same STS-1 frame by the increment or decrement indication inside the 10-bit pointer value. A positive stuff byte usage in this frame is indicated by inverting the bits 7, 9, 11, 13, and 15 (the I bits), which denotes that the positive stuff opportunity byte does not contain a payload byte and that all bytes of the payload may now be found 1 byte later in the SPE area of the STS frame. The pointer value must be incremented by one in this case. An example of this is shown in Figure 3.14.

Conversely, if the negative justification byte opportunity is used the bits 8, 10, 12, 14, and 16 (the D bits) are inverted, which denotes that the negative stuff opportunity byte does contain payload data and that all bytes in the SPE may now be found 1 byte earlier. The pointer value is decremented by one. An example of this is shown in Figure 3.15.

The physical byte locations and the corresponding pointer values that range from 0 to 782 are shown in Figure 3.16. Note that when the pointer value is 522, the payload fits into the SPE area of a single frame. Conceptually, this is the simplest situation to visualize. When the pointer value is 523, the first byte of the payload is found 1 byte later, and consequently the last byte of the payload is not contained in this same frame but instead is found in the following frame. This is conceptually more difficult to visualize. As shown in Figure 3.16, if the pointer value is 0, the first byte of the payload is found in the byte immediately after the H3 byte. The last byte of the payload is found in the byte before the H1 byte in the following frame.

3.4.7 Virtual Tributaries

Four different sizes of payloads, which are named virtual tributaries (VT), fit into the SPE. They are the VT1.5 which is 1.728 Mb/s, VT2 which is 2.304 Mb/s, VT3 which is 3.456 Mb/s, and last the VT6 which is 6.912 Mb/s. A VT requires a 500 s structure (4 STS-1 frames) for transmission. These required four frames are called a superframe or multiframe. The bytes that make up the four frames of a given VT include the special bytes named V1, V2, V3, and V4. The V1 and V2 bytes comprise the VT pointer that indicates the alignment of the VT within the allocated VT bytes independent of the other VTs in the same STS-1. The pointer bits contained within the two VT pointer bytes (named the V1 and V2 bytes) are defined similarly to the STS-1 pointer bytes shown in Table 3.4. The definitions of the locations for the VT pointer values are shown in Figure 3.17. These VT pointer bytes (V1 and V2) point to the first byte of the VT payload, a byte named the V5 byte. The V5 byte contains overhead information about the payload contained within the VT. The V5 byte includes error checking in the form of a 2-bit bit interleaved parity (BIP-2), a far end block error (FEBE) bit, path signal identification, and a downstream equipment error indication. The V3 byte is allocated as the negative justification opportunity byte for

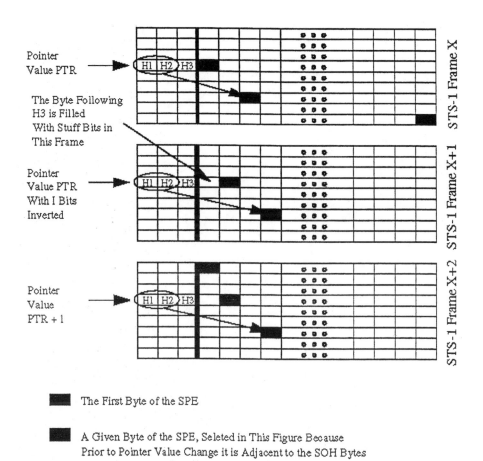

Pointer
Value PTR

The Byte Following
H3 is Filled
With Stuff Bits in
This Frame

Pointer
Value PTR
With I Bits
Inverted

Pointer
Value
PTR + 1

■   The First Byte of the SPE

■   A Given Byte of the SPE, Seleted in This Figure Because
    Prior to Pointer Value Change it is Adjacent to the SOH Bytes

■   A Given Byte Of the SPE, Selected in This Figure Because Prior to
    Pointer Value Change it is the Last Byte of the STS-1 Frame.

FIGURE 3.14 Positive pointer activity. (*Source:* ANSI, 1991.)

the VT. The byte immediately after the V3 byte is the positive justification byte opportunity. The V4 byte is an undefined byte. The remaining bytes in a VT are used for the various mappings of DS1, DS2, E1, etc. into the various VTs.

In a given STS-1 frame, a single VT1.5 occupies 3 columns out of 87. It is possible to have 28 VT1.5s located in an STS-1 frame. A single VT2 occupies 4 columns; it is possible to have 21 VT2s in an STS-1 frame. A VT3 occupies 6 columns; thus 14 total may be accommodated in a single STS-1 frame. A VT6 occupies 12 columns; thus, 7 total are possible in an STS-1 frame. Mixes of these VTs are allowed by dividing the SPE into seven VT groups. Each of these groups requires 12 of the 87 columns with two of the remaining three columns containing fixed stuff and the remaining column containing POH. The SPE is constructed by taking 1 byte from group number one, the second byte from group number two, and so on. A given group may be constructed out of 4 VT1.5s, 3 VT2s, 2 VT3s, or a single VT6. This approach allows mixing various payloads in the same SPE. An example SPE constructed out of a mixture of VT1.5s, VT2s, VT3s, and VT6s is shown in Figure 3.18.

3.4.8   Scrambling

A scrambling function is used on all bytes of the STS-1 frame except for the A1, A2, and C1 bytes. This scrambling function is used to prevent long strings of ones or zeros from occurring. A generating

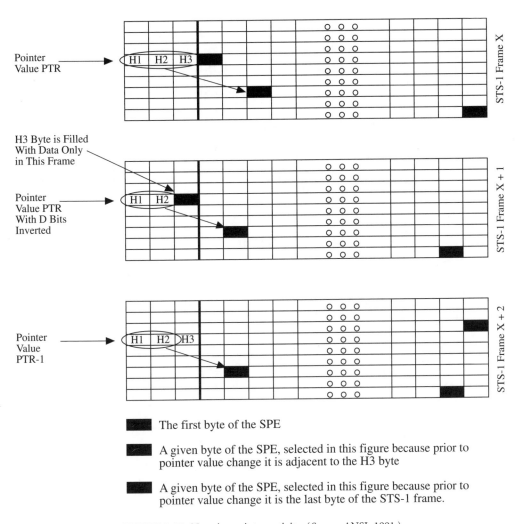

FIGURE 3.15 Negative pointer activity. (*Source:* ANSI, 1991.)

FIGURE 3.16 Pointer value locations in the STS-1 frame. (*Source:* ANSI, 1991.)

VT1.5

V1	78	79
80	81	82
83		
86		
89		
92		
95		
98		
101	102	103

VT2

V1	105	106	107
108	109	110	111
112			
116			
120			
124			
128			
132			
136	137	138	139

VT3

V1	159	160	161	162	163
164	165	166	167	168	169
170					
176					
182					
188					
194					
200					
206	207	208	209	210	211

VT6

V1	321	322	323	324	325	326	327	328	329	330	331
332	333	334	335	336	337	338	339	340	341	342	343
344											
356											
368											
380											
392											
404											
416	417	418	419	420	421	422	423	424	425	426	427

Frame 1 of 4

VT1.5

V2	0	1
2	3	4
5		
8		
11		
14		
17		
20		
23	24	25

VT2

V2	0	1	2
3	4	5	6
7			
11			
15			
19			
23			
27			
31	32	33	34

VT3

V2	0	1	2	3	4
5	6	7	8	9	10
11					
17					
23					
29					
35					
41					
47	48	49	50	51	52

VT6

V2	0	1	2	3	4	5	6	7	8	9	10
11											
23											
35											
47											
59											
71											
83											
95	96	97	98	99	100	101	102	103	104	105	106

Frame 2 of 4

VT1.5

V3	26	27
28	29	30
31		
34		
37		
40		
43		
46		
49	50	51

VT2

V3	35	36	37
38	39	40	41
42			
46			
50			
54			
58			
62			
66	67	68	69

VT3

V3	53	54	55	56	57
58	59	60	61	62	63
64					
70					
76					
82					
88					
94					
90	91	92	93	94	95

VT6

V3	107	108	109	110	111	112	113	114	115	116	117
118	119	120	121	122	123	124	125	126	127	128	129
130											
142											
154											
166											
178											
190											
202	203	204	205	206	207	208	209	210	211	212	213

Frame 3 of 4

VT1.5

V4	52	53
54	55	56
57		
60		
63		
66		
69		
72		
75	76	77

VT2

V4	70	71	72
73	74	75	76
77			
81			
85			
89			
93			
97			
101	102	103	104

VT3

V4	106	107	108	109	110
111	112	113	114	115	116
117					
123					
129					
135					
141					
147					
153	154	155	156	157	158

VT6

V4	214	215	216	217	218	219	220	221	222	223	224
225	226	227	228	229	230	231	232	233	234	235	236
237											
249											
261											
273											
285											
297											
309	310	311	312	313	314	315	316	317	318	319	320

Frame 4 of 4

FIGURE 3.17 VT pointer values: V1 and V2 contain the VT pointer; V3 is the negative justification byte opportunity, and V4 is unused. (*Source:* ANSI, 1991.)

polynomial of $1 + x^6 + x^7$ is used in each frame. The scrambler is reset to all ones on the most significant bit of the byte following the C1 byte. A functional diagram of a scrambler is shown in Figure 3.19.

3.4.9 Synchronous Transport Signal Level-N (STS-N)

An STS-N is formed by byte interleaving the STS-1 signals that comprise the STS-N signal. It may be thought of as $N \times 810$ bytes or as an $N \times 90$ column $\times$ 9 row structure. The transport overhead of the individual signals that make up the STS-N is frame aligned before interleaving. The SPEs, which are contained in the STS-N, are not required to be aligned because each STS-1 will have a unique payload pointer to indicate the position of each of the SPEs.

3.4.10 STS Concatenation

A concatenated STS (STS-Nc) is a number of STS-1s which are kept together. Certain services such as asynchronous transport mode (ATM) payloads may use this super-rate payload capability. The multiples of the STS-1 rate are mapped into an STS-Nc SPE. The STS-Nc is multiplexed, switched, and transported as a single unit. The SPE of an STS-Nc consists of $N \times 783$ bytes, which may be thought of as an $N \times 87$ column $\times$ 9 row structure. Only one set of STS POH is actually used in the STS-Nc; the POH always

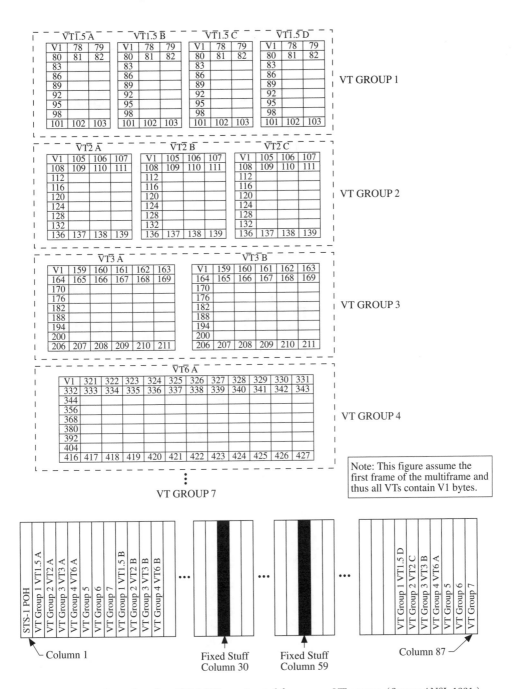

FIGURE 3.18 Examples of an STS-1 SPE constructed from seven VT groups. (*Source:* ANSI, 1991.)

appears in the first of the N STS-1s that make up the STS-Nc. Mappings for STS-3c and STS-12c have been defined [Bellcore, 1994].

3.4.11 Transmission of ATM in SONET

The STS-3c, STS-12c, etc. may be used to transport ATM cells. The H4 byte contains a number in the lower order 6 bits which indicates the number of bytes between the H4 byte and the first byte of the "first" ATM

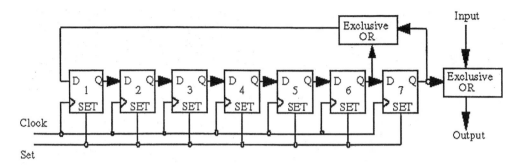

FIGURE 3.19 Scrambler functionality. (*Source:* ANSI, 1991.)

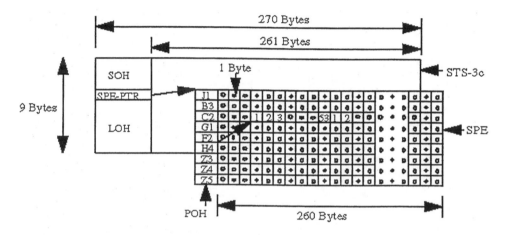

SOH Section Overhead
LOH line Overhead
POH Path Overhead
SPE Synchronous Payload Envelope

FIGURE 3.20 ATM mapping in STS-3c. (*Source:* ANSI, 1991.)

Flag 0x7E	Address 0xFF	Control 0x03	Protocol 16 bits	Information Payload such as an IP datagram	FCS 16 or 32 bits	Flag 0x7E

FIGURE 3.21 Point-to-point protocol frame.

cell, which is contained in the SPE [ANSI, 1991]. The ATM cell payload is scrambled using a generator polynomial of $1 + x^{43}$. This is done to prevent the payload information from providing false A1A2 framing values. Figure 3.20 shows the ATM mapping.

3.4.12 Point-to-Point Protocol over SONET

Similar to ATM over SONET, Point-to-Point Protocol (PPP) over SONET maps directly into the SPE of a concatenated STS frame. In contrast to the 53-byte cell used in ATM, PPP uses an variable-length, HDLC-like frame, as shown in Figure 3.21. PPP over SONET has recently gained popularity for carrying

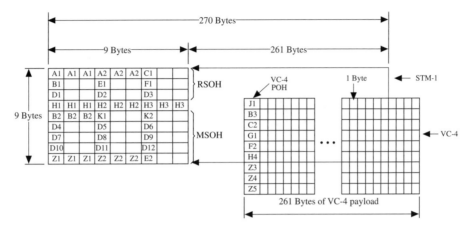

FIGURE 3.22 STM-1 frame structure with VC-4 payload. (*Source:* ITTU-T G.709, 1993.)

IP datagrams between backbone routers with high-speed OC-3 and OC-12 interfaces. IP over SONET using PPP is more efficient than using IP over ATM over SONET. The PPP overhead tax is only 2% as compared with the ATM overhead tax of approximately 25% when the distribution of packet sizes is taken into account [Manchester et al, 1998].

As shown in Figure 3.21, PPP uses the flag sequence (0x7E) to delineate the beginning and end of a frame. Furthermore, 0x7Es are used for interframe fill. For transparency, an escape sequence (0x7D) is defined. When the PPP frame contains a 0x7E byte, the transmitter converts each 0x7E payload byte into a two-byte sequence of 0x7D5E. The receiver converts 0x7D5E sequence back into a single 7E payload byte. Similarly, a 0x7D in the payload is translated into a 2-byte sequence of 0x7D5D. This method prevents a 0x7E or 0x7D in the PPP frame from being interpreted as the flag sequence or the escape sequence, respectively.

The relevant standards are Internet Engineering Task Force (IETF) RFC 1619 "PPP Over SONET/SDH," RFC 1661 "The Point-to-Point Protocol (PPP)," and RFC 1662 "PPP in HDLC-like Framing." One problem with the standards is the omission of a payload scrambling requirement. Since the SONET payload is concatenated instead of multiplexed, a malicious user could defeat the standard SONET scrambler and cause operational problems such as loss of signal and loss of frame. As a result, using a self-synchronous scrambler with a generator polynomial of $x^{43} + 1$ is recommended for scrambling the entire PPP frame plus the interframe fill [Manchester et al., 1998].

3.4.13 Synchronous Digital Hierarchy

The Synchronous Digital Hierarchy (SDH) is a set of international, digital transmission standards. SONET uses a basic line rate of 51.84 Mb/s, and SDH uses a basic line rate of 155.52 Mb/s, which is exactly three times the SONET basic rate [Omidayr and Aldridge, 1993; Sexton et al., 1993]. The compatibility between SDH and SONET allows for internetworking at the administrative unit-4 (AU-4) level [ITU-T G.708, 1993].

3.4.14 Frame Structure

The synchronous transport module-1 (STM-1) frame consists of 2430 bytes (19,440 bits) and is represented as a 9-row × 270-column structure as depicted in Figure 3.22. The frame is transmitted serially a bit at a time from left-to-right and from top-to-bottom. The frame consists of section overhead, AU pointers, and payload. An STM-1 frame is transmitted every 125 μs, and thus has a bit rate of 155.52 Mb/s. This rate is defined as the basic rate in SDH [ITU-T G.708, 1993].

TABLE 3.5 SDH Bit Rates

Synchronous Transport Module	Line Rate (Mb/s)
STM-1	155.52
STM-4	622.08
STM-16	2488.32
STM-N	$N \times 155.52$

Source: ITU-T G.707, 1993.

3.4.15   Synchronous Transfer Module-N (STM-N)

An STM-N signal is formed by byte-interleaving the individual STM-1 signals. The rate of an STM-N is N times the rate of an STM-1. Currently N is only defined for the values of 1, 4, 16, and 64. Table 3.5 shows the bit rates for the serial transmission of the corresponding STM.

3.4.16   Administration Unit (AU), Tributary Unit (TU), and Virtual Container (VC) Definitions

A container-n (C-n: n = 11, 12, 2, 3, or 4) as shown in Figure 3.23 contains the client information to be transported and forms the payload of the corresponding virtual container-n (VC-n). A VC-n (n = 11, 12, 3, or 4) is a structure consisting of payload from a container plus the addition of some path overhead information. A given VC is either classified as a lower-order VC or a higher-order VC. The VC-11, VC-12, and VC-2 are lower-order VCs. The VC-4 is a higher-order VC. In ITU-T G.708 [1993], VC-3s are classified only as higher-order VCs. However, as pointed out in Sexton and Reid [1992] the VC-3 really has a dual role as a lower-order VC when used as the payload of a tributary unit-3 and as a higher-order VC when used as the payload of an AU-3.

A tributary unit-n (TU-n: n = 11, 12, 2, or 3) contains a lower-order VC and a TU pointer. The TU pointer indicates the start of the lower-order VC relative to the start of the supporting higher-order VC. An administrative unit-n (AU-n: n = 3, 4) consists of a higher-order VC and a AU pointer. The AU pointer indicates the start of the higher-order VC relative to the start of the STM-N frame. A tributary unit group-n (TUG-n: n = 2,3) is a collection of one or more tributary units. A TUG-2 consists of one TU-2, three TU-12s, or four TU-11s. A TUG-3 contains one TU-3 or seven TUG-2s. Furthermore, an administrative unit group-n (AUG-n) is a collection of one or more administrative units. An AUG consists of either one AU-4 or three AU-3s [Asatani et al., 1990] [ITU-T G.708, 1993].

3.4.17   Administrative Unit Pointer Mechanism

The AU pointer indicates the start of the higher-order VC relative to the start of the STM-N frame. Thus, the pointer mechanism allows the VC to float inside the frame. When the incoming clock is running faster than the outgoing clock at a node in the SDH network, a finite-size buffer inside the equipment at that SDH node will begin to fill up. Thus, the fill of the buffer will eventually reach an upper threshold and cause a negative justification pointer action. A negative justification pointer action prevents overflow by transmitting a data byte instead of a stuff byte in the negative justification opportunity, the H3 byte. Therefore, the buffer fill is reduced. A negative justification pointer action inverts the decrement (D) bits in the AU pointer (H1 and H2 as shown in Table 3.4 and Figure 3.22) and reads a data byte from the buffer into the H3 byte. Normally, the negative justification opportunity carries a stuff (dummy information). When the outgoing clock is running faster than the incoming clock at a node, the buffer will start emptying. To prevent underflow, the buffer fill will eventually reach a lower threshold that results in a positive justification pointer action. A positive justification pointer action prevents underflow by sending a stuff byte instead of a data byte in the positive justification opportunity, byte 0 of the VC. Thus, the buffer fill will rise. To perform a positive justification pointer action, the increment (I) bits in the AU pointer (H1 and H2 as shown in Table 3.4 and Figure 3.22) are inverted, and a stuff byte is transmitted in the positive justification opportunity. Normally, the positive justification opportunity contains data from the buffer. Consecutive pointer actions must be at least four frames apart (500 s).

3.4.18   TU-3 Pointer Mechanism

When a VC-4 contains three VC-3s, the TU-3 pointer (H1 and H2 as shown in Figure 3.24) indicates the offset of the VC-3 relative to the start of the supporting VC-4. The pointer mechanism works similar to the AU pointer mechanism described above. Each TU-3 has an independent pointer (H1 and H2), negative justification opportunity (H3), and positive justification opportunity (VC byte 0). When TUG-2s

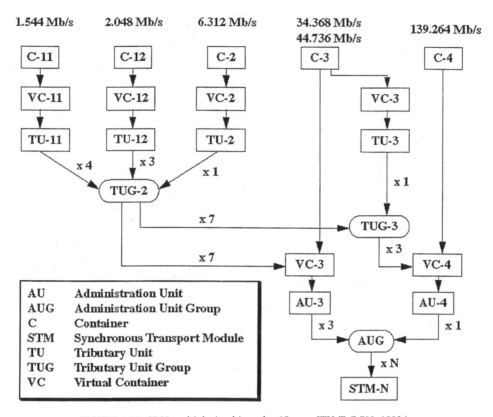

FIGURE 3.23 SDH multiplexing hierarchy. (*Source:* ITU-T G.709, 1993.)

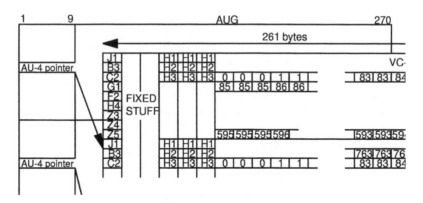

FIGURE 3.24 TU-3 pointer offset numbering. (*Source:* ITU-T G.709, 1993.)

are carried inside a VC-4, the TU-3 pointer is not used and has a null pointer indication (NPI) value [ITU-T G.709, 1993].

3.4.19 TU-1/TU-2 Pointer Mechanism

The TU pointer indicates the start of the lower-order VC relative to the start of the associated higher-order VC. The TU pointer mechanism works essentially the same way as the AU pointer mechanism and is only used in the floating mode. The negative justification opportunity is the V3 byte, and the positive justification opportunity is the VC byte after V3 in the TU multiframe. Bytes V1 and V2 as shown in Figure 3.25 comprise

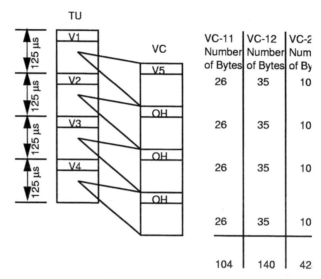

	VC-11 Number of Bytes	VC-12 Number of Bytes	VC-2 Num of By
	26	35	10
	26	35	10
	26	35	10
	26	35	10
	104	140	42

TU   Tributary Unit
VC   Virtual Container
V1   TU Pointer Byte1
V2   TU Pointer Byte 2
V3   Negative Justification Byte
V4   Reserved

FIGURE 3.25 Virtual container mapping in multiframe tributary unit. (*Source:* ITU-T G.709, 1993.)

the TU pointer that points to the first byte in the VC, the V5 byte. Consecutive pointer actions must be separated by at least four multiframes (2 ms) [ITU-T G.709, 1993].

3.4.20 Section Overhead

Figure 3.22 shows the section overhead (SOH) of an STM-1 signal. As shown, the SOH is divided into the regenerator section overhead (RSOH) and the multiplex section overhead (MSOH). Equipment that performs regeneration terminates the RSOH. Regenerators pass the MSOH unmodified. Multiplexing equipment terminates the MSOH as well as the RSOH [ITU-T G.708, 1993]. In SONET, the RSOH is called the section overhead, and the MSOH is termed the line overhead. The name and the purpose of each of the overhead bytes may be found in ITU-T G708 and G709 [1993].

3.4.21 Mapping of Tributaries

Both synchronous and asynchronous tributaries are allowed in SDH. These mappings are fully defined in ITU-T G.709 [1993]. Asynchronous mappings of payloads into TUs involve the use of justification bits. Synchronous tributary mappings do not use justification bits in the mapping process. The asynchronous mapping of a 139,264 kb/s signal is accomplished by dividing each of the nine rows in the STM-1 frame into blocks of 13 bytes. Each of these 13-byte blocks contains a "W," "X," "Y," or "Z" byte followed by 12 bytes that contain bits from the 139,264 kb/s signal. One of the nine rows from the STM-1 frame is shown in Figure 3.26 (excluding the first nine overhead bytes of the STM-1 frame). The W byte is defined to contain 8 data bits from the 139,264 kb/s signal. The Z byte contains 6 or 7 data bits from the signal depending on the usage of the S-bit negative justification opportunity. The S-bit negative justification opportunity, which is used based upon the fill of an incoming 139,264 kb/s buffer, has a nominal justification ratio of 2/9. Hence, since each VC-4 row contains an independent negative bit justification opportunity, each STM-1 frame nominally contains two used S bits. The X and Y overhead bytes contain 0 data bits from the signal. The Y byte contains fixed stuff, while the X byte encodes the S-bit usage in each VC-4 row. .

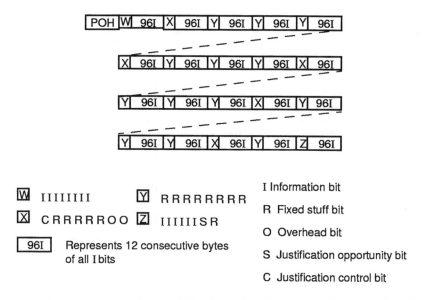

FIGURE 3.26 Asynchronous mapping of 139,264 kb/s tributary into VC-4 (note only one row of the nine row VC-4 container structure is shown. (*Source:* ITU-T G.709, 1993.)

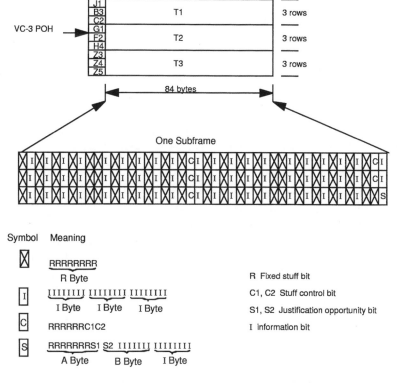

FIGURE 3.27 Asynchronous mapping of 34,368 kb/s tributary into a VC-3. (*Source:* ITU-T G.709, 1993.)

The asynchronous mapping of a 34,368 kb/s signal into a VC-3 is shown in Figure 3.27. The STM-1 frame (excluding the overhead) is broken up into three subframes (T1, T2, and T3). One of these subframes is shown in detail in Figure 3.27. Each VC-3 subframe spans three rows in the STM-1 frame.

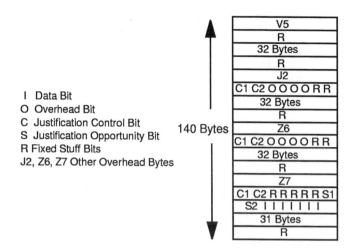

FIGURE 3.28 Mapping of asynchronous 2048 kb/s tributary into VC-12. (*Source:* ITU-T G.709, 1993.)

There are five types of bytes in the subframe. The first type is an I byte that contains 8 bits from the 34,368 kb/s signal. In Figure 3.27, the letter I, with a box around it, represents a group of three I bytes. The R overhead bytes, represented by a box with an X inside, each contain 8 fixed stuff R bits and hence no signal data bits. The A byte contains 0 or 1 signal data bits depending on the usage of the S1 negative bit justification opportunity. The B byte contains 7 or 8 signal data bits depending on the usage of the S2 positive bit justification opportunity. In Figure 3.27, the letter "S," with a box around it, represents a group of one A byte, one B byte and one I byte. The S1 and S2 bit justification opportunities are either used or not used based upon the fill in the incoming 34,368 kb/s buffer. Finally, the C overhead bytes, which contain no signal data bits, encode the S1 and S2 bit usage in each subframe. Since each STM-1 frame contains three VC-3 subframes, each VC-3 has three independent positive/negative bit justification opportunities in each STM-1 frame.

The VC-12 structure shown in Figure 3.28 consists of 1023 data bits, 6 justification control bits, 2 justification bits (S1 and S2), and 8 overhead communications channel bits. The remaining bits are fixed stuff (R) bits. The O bits are reserved for future overhead communications purposes. For a nominal 2048 kb/s signal, the S2 justification bit is always used for user data, and the S1 justification bit is a stuff bit. In the event that the synchronizer node's reference clock is too fast, S2 bit justification opportunities will occasionally not be used for user data. Conversely, a slow synchronizer clock will result in S2 and occasionally S1 being used for user data

There are four types of bytes in the 1544 kb/s signal mapping shown in Figure 3.29. The first type is a normal 8-bit data byte. The second type is an Rx or Ri overhead byte that contains 1 signal data bit. The third type is a purely overhead byte, which contains no signal data bits. The final type is an Rs byte containing 0 (an unused S2 positive bit justification opportunity), 1 (a normally used S2 bit and a normally unused S1 negative bit justification opportunity), or 2 signal data bits (both S1 and S2 bits are used for user data). The S1 and S2 bit justification opportunities are either used or not used based upon the fill of the synchronizer buffer.

3.4.22 Mapping of Asynchronous Transfer Mode (ATM) Cells

The 53-byte ATM cell mapping into SDH is accomplished by aligning the byte structure of every cell with the byte structure of the VC. Legal VC structures for transport of ATM cells include but are not limited to VC-4 (as shown in Figure 3.30) and VC-4-Xc. The ATM cell contains a header error control field that is used as a cell alignment word to determine cell delineation. This method uses the correlation between the 32 header bits to be protected by the header error control field and the control bit of the 8 bit header error

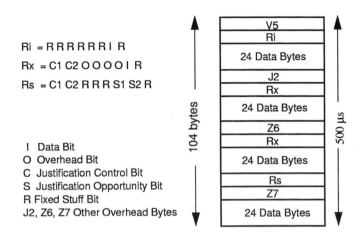

FIGURE 3.29 Asynchronous mapping of 1544 kb/s tributary into VC-11. (*Source:* ITU-T G.709, 1993.)

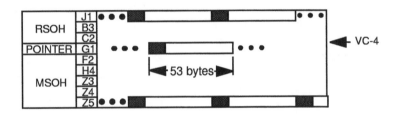

FIGURE 3.30 ATM cells mapping into a VC-4. (*Source:* ITU-T G.709, 1993.)

TABLE 3.6 Standards

Standard	Title
ANSI T1.101	Synchronization Interface Standards for Digital Networks
ANSI T1.102	Digital Hierarchy-Electrical Interfaces
ANSI T1.105	Digital Hierarchy-Optical Interface Rates and Formats Specifications (SONET)
ANSI T1.107	Digital Hierarchy-Formats Specifications
ANSI T1.210	Operations, administration, maintenance, and provisioning (OAM&P) — Principles of functions, architectures, and protocols for telecommunications management network (TMN) interfaces
ITU-T G.707	Synchronous Digital Hierarchy Bit Rates
ITU-T G.708	Network Node Interface for the Synchronous Digital Hierarchy
ITU-T G.709	Synchronous Multiplexing Structure
ITU-T G.781	Structure of Recommendations on Equipment for the Synchronous Digital Hierarchy (SDH) Equipment
ITU-T G.782	Types and General Characteristics of Synchronous Digital Hierarchy Equipment
ITU-T G.783	Characteristics of Synchronous Digital Hierarchy (SDH) Equipment Functional Blocks
ITU-T G.784	Synchronous Digital Hierarchy (SDH) Management
ITU-T G.823	The Control of Jitter and Wander within Digital Networks Which Are Based on the 2048 kb/s Hierarchy
ITU-T G.824	The Control of Jitter and Wander within Digital Networks Which Are Based on the 1544 kb/s Hierarchy
ITU-T G.825	The Control of Jitter and Wander within Digital Networks Which Are Based upon the Synchronous Digital Hierarchy
ITU-T G.911	Parameters and Calculation Methodologies for Reliability and Availability of Fibre Optic Systems
ITU-T G.955	Digital Line Systems Based on the 1,544 kb/s and the 2,048 kb/s Hierarchy on Optical Fibre Cables
ITU-T G.957	Optical Interfaces for Equipments and Systems Relating to the Synchronous Digital Hierarchy
ITU-T G.958	Digital Line Systems Based on the Synchronous Digital Hierarchy for Use on Optical Fibre Cables

control field. A shortened cyclic code with the generator polynomial $g(x) = x^8 + x^2 + x + 1$ is used where the remainder is then added to the fixed pattern 01010101 in order to improve the cell delineation performance. This is similar to conventional frame alignment recovery where the alignment word is not fixed but varies from cell to cell. The H4 byte is not used to indicate cell offset [ITU-T G.709, 1993].

3.4.23   Standards

Some of the standards that are relevant to SONET and SDH are shown in Table 3.6.

Acknowledgments

The texts taken from the ITU material have been reproduced with the prior authorization of the ITU as copyright holder. The sole responsibility for selecting extracts can in no way be attributed to the ITU. The complete volumes of the ITU material can be obtained from International Telecommunications Union, General Secretariat-Sales and Marketing Service, Place des Nations–CH–1211, Geneva 20, Switzerland.

References

ANSI T1.105.1991. Digital Hierarchy Optical Interface Rates and Formats Specifications (SONET), American National Standards Institute, New York.

Asatani, K., Harrison, K., and Ballart, R. 1990. CCITT standardization of network node interface of synchronous digital hierarchy, *IEEE Commun. Mag.*, 28 (8), 15–20.

Autry, C. B. and Owen, H. L. 1997. Synchronous optical network (SONET), in *The Communications Handbook*, CRC Press, Boca Raton, FL, Chap. 39.

Autry, C. B. and Owen, H. L. Synchronous digital hierarchy (SDH), *The Communications Handbook*, Chapter 40, CRC and IEEE Press.

Bellcore 1994. Synchronous Optical Network (SONET) Transport Systems: Common Generic Criterion, GR-253-CORE, Bell Communications Research, Piscataway, NJ.

Davidson, R. and Muller, N. 1991. *The Guide to SONET*, Telecom. Library Inc., New York, NY.

ITU-T Recommendation G.707. 1993. Synchronous Digital Hierarchy Bit Rates, Geneva, Switzerland.

ITU-T Recommendation G.708. 1993. Network Node Interface for the Synchronous Digital Hierarchy, Geneva, Switzerland.

ITU-T Recommendation G.709. 1993. Synchronous Multiplexing Structure, Geneva, Switzerland.

Manchester, J., Anderson, J., Doshi, B., et al. 1998. IP over SONET, *IEEE Commun. Mag.* 36 (5): 136-142.

Minoli, D. 1993. *Enterprise Networking*, Artech House Inc., Norwood, MA.

Omidayr, C. and Aldridge, A. 1993. Introduction to SONET/SDH. *IEEE Commun. Mag.* 31 (9): 30-33.

Sexton, M. and Reid, A. 1992. *Transmission Networking: SONET and the Synchronous Digital Hierarchy*, Artech House Inc., Norwood, MA.

Sexton, M. and Reid, A. 1997. *Broadland Networking ATM, SDH and SONET*, Artech House Inc., Norwood, MA.

Sexton, M., Roverano, F., DeCremiers, F. 1993. SDH Architecture and Standards. *ALCATEL Electrical Commun.*, 4th Qtr.: 299-312.

Simpson, W. 1994. IETF RFC 1619, "PPP over SONET/SDH."

Simpson, W. 1994. IETF RFC 1661, "The Point-to-Point Protocol (PPP)."

Simpson, W. 1994. IETF RFC 1662, "PPP in HDLC-like Framing."

Wu, T. 1992. *Fiber Network Service Survivability*, Artech House Inc., Norwood, MA.

Further Information

A good and very high level introduction to SONET may be found in *The Guide to SONET* [Davidson and Muller, 1991]. A more advanced treatment of SONET is *Transmission Networking: SONET and the Synchronous Digital Hierarchy* [Sexton and Reid, 1992]. Survivability of SONET networks is examined in *Fiber Network Service Survivability* [Wu, 1992]. An excellent chapter on SONET, an explanation of clock distribution in synchronous networks, and how SONET fits into the big picture of networks in general is contained in *Enterprise Networking* [Minoli, 1993]. A single technical document that contains a majority of the details of SONET is "Synchronous Optical Network (SONET) Transport Systems: Common Generic Criterion" GR-253-CORE, which may be obtained from Bellcore [1994].

An introductory source on SDH is the September 1993 *IEEE Communications Magazine* [Omidayr and Aldridge, 1993] that features SDH/SONET. For more details the book *Broadband Networking ATM SDH and SONET* [Sexton and Reid, 1997] is an excellent source of information. Finally, the details of SDH are best obtained directly from the ITU standards.

3.5   Wireless and Mobile Networks

Zygmunt Haas

3.5.1   The State of the Art in Wireless and Mobile Networks

The currently offered commercial wireless services can be classified based on two criteria: mobility support and technology. The mobility classification is strongly related to the coverage of the corresponding system; i.e., the higher the speed of the mobile user, the larger the coverage of the system. Some representative examples of mobile services are shown in Figure 3.31.

Paging, traditionally a one-way receive-only service, utilizing, for example, the FLEX™ protocol, has recently been offered as a full two-way service through the ReFLEX™ paging protocol.[*] Undoubtedly, this expansion of paging technology is responsible for the renewed growth of the paging industry. It is predicted that the two-way paging service will experience significant growth in the next few years; prognosis of between 50 to 100% per year in the next 2 to 3 years is quite common.

By far, the biggest market for wireless services in the U.S. is that of cellular telephones. Although it has experienced some saturation, its growth is still considerable. For instance, it was anticipated that by the year 2000 there would be between 60 and 90 million cellular telephones in the U.S. This increase is from *the* about 40 million cellular telephone users in 1998, most of whom still used phones that operate based on the analog (first-generation) Advanced Mobile Phone System (AMPS) technology. Forecast for the penetration of the PCS services (in addition to the cellular telephones) is also impressive — between 15 to 23 million subscribers by the year 2000.

The microcellular technology (see Section 3.5.2) is still limited to a local environment. However, with the growth in the cellular systems and the introduction of personal communications services, it is anticipated that the demand for cellular services will continue to grow, forcing the systems into cell sizes to increase the network capacity.

Today's U.S. cellular market is composed of 306 Metropolitan Statistical Areas (MSA 1 to 306) and 428 Rural Service Areas (RSA 307 to 734). As two service providers, the *wireline* and the *nonwireline* providers, are defined for each area, there are total of 1468 cellular networks covering the U.S. The geographic coverage in the U.S. is nearly ubiquitous. Standards that ensure "seamless" interconnection of these networks were defined (e.g., TIA IS-41), allowing users to continue to receive service automatically while roaming among these networks.

Due to the quite often underestimated growth of the wireless business, with the radio spectrum being limited due to technical and political considerations, schemes to increase network capacity have been continuously proposed and studied. Extensions to the original (first-generation) cellular systems based on the AMPS technology have been incorporated; for example, the Narrowband AMPS (NAMPS). A further increase in network capacity requires more radical change to the air protocols. As such, schemes based on digital technology (still mainly supporting voice with limited data capabilities) have been standardized, e.g., the TDMA-based GSM, the TDMA-based IS-136, and the CDMA-based IS-95. Currently, these second-generation systems are still islands of connectivity, requiring dual-use phones[**] to

---

[*]FLEX and ReFLEX are trademarks or registered trademarks of Motorola, Inc.

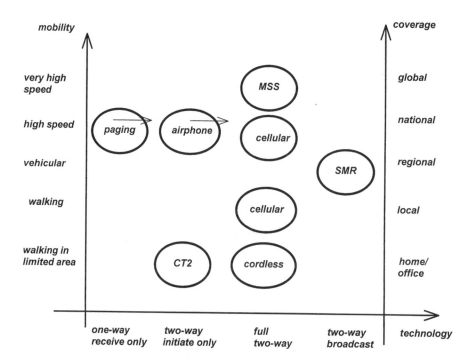

FIGURE 3.31 Examples of some currently offered wireless services.

obtain more than local coverage. It is not clear how the digital technology will converge in the future, given these two, so different air protocols.

The future promises to bring third-generation systems. Although the precise definition of a third-generation system is not unique, it is generally envisioned that a third-generation system will support multimedia communication, will integrate services that are today provided by separate networks, and will incorporate new features commensurate with the "anytime, anywhere" motto.

In parallel with the high-mobility systems, some companies pursue deployment of lower-mobility systems, examples of which are the Personal Access Communications System (PACS) and the Japanese Personal Handyphone System (PHS). These systems may not be suitable to support high-mobility environments, but are satisfactory for limited-coverage systems with low-mobility users.

In addition, there is limited specialized data market in the U.S. It is offered through the Cellular Digital Packet Data (CDPD) service and through two wide-area data networks: (RAM) Mobile Data (based on the Mobitex protocol) and (ARDIS).

In the international arena, there are multiple cellular standard. Older (first-generation) systems include Total Access Cellular System (TACS), for example, in the U.K.; Japanese TACS (JTACS) in Japan; Nordic Mobile Telephone (NMT) in the Scandinavian countries. Second-generation systems include Global System for Mobile (GSM) communications with extended coverage in western Europe.

Globally, the Mobile Satellite Systems (MSS) promise to provide wide-range coverage throughout the globe by portable equipment; examples include Iridium (Motorola) and Teledesic (Microsoft/McCaw).

Three acronyms are associated with the third-generation wireless networks: UPT, FPLMTS, and PCS. Definitions of the three are as follows:

Universal Personal Telecommunication (UPT) enables access to telecommunication services while allowing personal mobility. It enables each UPT user to participate in a user-defined set of subscribed services and to initiate and receive calls on the basis of a unique, personal, network-transparent UPT

"Dual use phones support a new digital air protocol in the area of coverage, while they revert to the AMPS system when the digital service is not available.

Number across multiple networks at any terminal, fixed or mobile, irrespective of geographical location, limited only by terminal and network capabilities and restrictions imposed by the network operator. [CCITT SG-I; F.851 — UPT Service Principles and Operational Provisions — Version 5, COM-I88, Geneva, May 28 to June 7, 1991]

Future Public Land Mobile Telecommunications Systems (FPLMTS) are systems which will provide telecommunications services to mobile or stationary users by means of one or more radio links. This mobility will be unrestricted in terms of location within the radio coverage area. FPLMTS will extend the telecommunication services of the fixed network to those users over wide geographic areas, subject to constraints imposed by spectrum allocation and radio propagation and, in addition, will support a range of services particular to mobile radio systems. [CCIR* Draft Recommendation 687 and Reports 1153 and 1155].

Personal Communications Service (PCS) is an extension and integration of existing and future wireless and wired communication networking features and capabilities, ultimately allowing communication with a *person*, regardless of his or her location. PCS features include unified access, personalized service profile, personal phone number, support for personal and terminal mobility, improved digital communication, etc.

Both UPT and PCS concepts address the following issues: management of users' profiles, wireless access, personal mobility, and terminal mobility.

3.5.2 The Enabling Technologies

In this section, we revisit some of the basic schemes used in wireless systems: Frequency Division Multiple Access (FMDA), Time Division Multiple Access (TDMA), Code Division Multiple Access (CDMA), the cellular principle (and the associated topic of channel assignment strategies), and handoff schemes.

3.5.2.1 Frequency Division Multiple Access

In FDMA, the total spectrum is divided into channels, which are assigned to users for the duration of a call; i.e., during the call in progress, a channel is dedicated to a pair of users. When the call terminates, the channel can be reassigned to another pair of users. FDMA is used in nearly all first-generation radio systems and many second-generation systems as well. AMPS is an example of an FDMA-based system, in which of the total 832 full-duplex channels, each unidirectional channel is 30 kHz wide and the two directions are separated by exactly 45 MHz. (Communication from the base station to a subscriber is called a forward link and, from a subscriber to its base station, a reverse link.) Of these channels, 42 are control channels and the remaining 790 are voice channels.

To estimate the required number of channels to support some user populations, the design takes into the account the required quality of service, the average call duration, the traffic intensity, and the call activity factor. The quality of service is usually the percentage of blocked and dropped calls. Erlang-B formula (blocked calls cleared) is routinely used to perform these calculations:

$$E(\Gamma, N) = \frac{\Gamma^N/N!}{\sum_{i=0}^{N} \Gamma^i/i!},$$

where Γ is the total offered load, N is the number of channels, and $E(\Gamma, N)$ is the probability of a call being blocked. For example, the Public Switched Telephone Network (PSTN) system is designed for 1% blocked calls, while the cellular phone system is designed for 2% blocking probability. The emerging future PCS is expected to provide 1% blocking.

---

*CCIR=Consultative Committee for International Radio

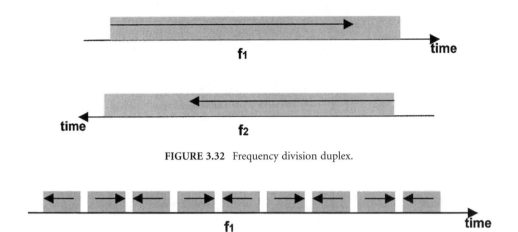

FIGURE 3.32 Frequency division duplex.

FIGURE 3.33 Time division duplex.

The design of channelized access (such as FDMA) usually relies on the trunking efficiency, also called channel group efficiency. Due to the trunking efficiency, a larger pool of available channels can serve larger than the proportional increase in the user population with the same quality of service (and the channel utilization is higher). For example, consider two cases: a pool of 15 channels and a pool of 45 channels. When designed for 1% blocking probability, the 15-channel pool can support, on the average, 8 calls (at 53% occupancy), while the 45-channel pool can support, on the average, 33 calls (at 73% occupancy).

3.5.2.2 Time Division Multiple Access (TDMA)

TDMA allows multiple users to share the same frequency band by multiplexing their transmissions in time; i.e., time is divided into non-overlapping-in-time slots. These time slots are assigned to calls. In TDMA, the signaling rate is equal to the sum of all the data rates of all the multiplexed transmissions. Thus, the bandwidth of the frequency band needs to be wide enough to accommodate this aggregated rate.

In practice, the total spectrum is divided into frequency channels (FDM) and each frequency channel is further divided in time by TDMA. Thus, the access scheme is often termed FDM/TDMA. Practically, to avoid overlapping between transmissions in adjacent slots, some guard time is included in every slot. The overlap can be created by imperfect synchronization or uncompensated differences in delays among mobiles at different distances from the base station.

TDMA is used in many second-generation systems, such as IS-54/136, GSM, or PHS. For example, in IS-54/136, the original 30 kHz AMPS channels are subdivided into six time slots each. Since the data rate of each slot is 8 kbps, the total data rate in a 30-kHz channel is 48 kbps. To accommodate 8 kbps coded speech and overhead, 16 kbps per user is required. Thus, each user uses two slots per frame (1 and 4, 2 and 5, 3 and 6). Consequently, the capacity of IS-54/136 is three times that of AMPS. Future improvements in voice coding are expected to bring down the coded speech to 4 kbps, which (with its overhead) can be accommodated in a single slot per frame. Consequently, IS-54/136 will improve the capacity of AMPS by six times.

In the context of multiplexing schemes, the term *duplexing* refers to the way that the two directions of a connection are multiplexed in frequency and time. In particular, frequency division duplexing (FDD) assigns a separate frequency channel to each direction of a connection (Figure 3.32). In Time Division Duplexing (TDD), a single frequency channel is used alternatively in time by the two directions, as shown in Figure 3.33.

3.5.2.3 Code Division Multiple Access (CDMA)

CDMA allows multiple users to share the same frequency band at the same time by multiplexing their transmissions in the code space. In other words, different transmissions are encoded with orthogonal codes and thus can coexist at the same time on the same frequency band. As the crosscorrelation between any two codes is very low (ideally zero), the receiver can retrieve the transmissions by correlating the received signal with the appropriate replica of the code.

CDMA is implemented through the use of *spread spectrum* techniques. Developed initially for military applications, spread spectrum spreads the power of a signal over a bandwidth that is considerably larger than the signal bandwidth. The spreading can be done, for example, by multiplying the signal in time with the orthogonal code — Direct Sequence (DS) — or by hopping in frequency among multitude of carriers — Frequency Hopping (FH).

The features of the CDMA technique, useful for military communications are as follows:

- The resulting spectral density is considerably smaller than the original one; this can be used to hide the signal.
- After decoding, the power density of a narrowband interferer (intentional or not) is very small; this can be used for antijamming protection.
- After decoding, because of small crosscorrelation between different codes, transmission encoded with different code appears as noise; this can be used to multiplex a number of transmissions on the same channel (i.e., multiple access scheme).

The noiselike interference among the different CDMA transmissions limits the number of users that can concurrently use the same CDMA "channel." The required quality of service determines this number of users. (Note, in CDMA, a "channel" is defined as a spectral bandwidth that is shared among many users.)

Multiple Access with Spread Spectrum—Transmissions from other users (with orthogonal codes) are seen as noise. A receiver needs to know the code that a transmission was encoded with, to be able to decode the message. (Thus, some limited security is provided by the spread spectrum technique.) To decode the signal, for example, the received signal is compared with the phases of the replica of the PN code, with which the signal was originally encoded.

Processing Gain (PG) of CDMA is defined as:

$$\text{Processing Gain} = \frac{\text{Bandwidth of the spread signal}}{\text{Data rate of the original signal}}$$

The PG indicates on the amount of improvement in the signal-to-interference ratio (SIR) resulting from spreading the signal bandwidth. For example, assume that the original pulse duration is n times the chip duration. Thus, the PG equals n. For example, if the chip bit rate is 1 Gbps and the signal is 10 Mbps, the PG = 100 or 20 dB.

Advantages of CDMA are as follows:

- Flexible system design can accommodate variable traffic load with temporal quality of service degradation.
- Statistical multiplexing takes advantage of idle connections (such as speed activity factor).
- Interference is based on the average, rather than peak, energy level.
- Good discrimination from interfering transmissions is provided.
- No channel assignment (reassignment) is required for reuse of 1.
- *Soft handoffs* are supported.
- Performance in a multipath environment is improved.
- Coexistence with other wireless technology (such as microwave — especially with broadband-CDMA) is possible.
- Some limited degree of security is provided by the spread spectrum technique.

Among the disadvantages of CDMA are

- Code (PN sequence) synchronization and tracking is required.

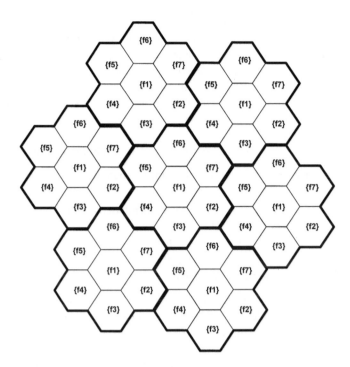

FIGURE 3.34 Reuse pattern of seven.

- (Adaptive) power control to eliminate the "near–far problem" is required.
- Potential interference problems exist (especially in narrowband-CDMA) due to sharing the broad spectrum of CDMA.

3.5.2.4 The Cellular Principle

The initial mobile phone systems were based on a single base station, which served a large coverage area (Mobile Telephone Service, St. Louis, 1946; Improved Mobile Telephone Service, 1965; and New York City Mobile Phone Service, 1976). Due to the limited radio spectrum, these systems could not meet subscribers' demands. Cellular technology, which was invented at Bell Laboratories in 1946, increases the network capacity. It relies on the concept of transmission *concurrency.* Concurrency is created by reusing channels in different cells: *channel reuse.* This allows increase in total capacity of the system (i.e., the number of supported users.)

The total coverage area is divided into cells. In each cell, only a subset of all the channels is available. All the channels are partitioned into sets, which are assigned to cells. The same set is assigned to two cells that are geographically distant "enough," so that the interference between the co-channel cells is very small. Different channel assignments are possible. In particular, assignments can be based on Fixed Channel Assignment (FCA) or Dynamic Channel Assignment (DCA). FCA can be based on different repetitious patterns, e.g., three- or seven-cell patterns (Figure 3.34). Additionally, the transmission powers of the mobiles (and the base stations) are lower than in a noncellular system.

3.5.2.4.1 Advantages and Disadvantages of Cellular Systems
Among the advantages of cellular systems are

- More capacity due to the spectral reuse;
- Lower transmission power due to the smaller mobile/base station separation;
- More "predictable" propagation environment due to a smaller transmitter/receiver distance;
- More robust system; service is interrupted only in the area of a base station failure.

Disadvantages of a cellular network, as compared with a special mobile radio system are

- Need for more infrastructure; i.e., more base stations;
- Need for a network to interconnect the base stations;
- Residual interference from co-channel cells;
- Handoffs procedures required to provide seamless service;
- "Hot spots" in user concentration, especially in the FCA schemes.

Although for ease of analysis and design cells are modeled as hexagons to describe continuous coverage, the actual coverage of the radiation pattern is highly irregular and is influenced by various effects, such as terrain topology, anthropogenic structures, atmospheric conditions, etc. The cell size can vary from $\frac{1}{2}$ mile in metropolitan areas to 10 miles in rural areas.

The purpose of the channel assignment schemes is to allocate channels to mobile/base station connections, in such a way that the channel reuse is maximized, subject to some maximal co-channel interference constraints. The simplest scheme is the Fixed Channel Allocation (FCA); channels are permanently assigned to cells, based on a repetitious reuse pattern. FCA does not solve the "hot spots" problem.

In pure Dynamic Channel Allocation (DCA) schemes, there is no preassignment of channels to cells; reuse of channels is dictated by already assigned channels in other cells. DCA can allocate more capacity to areas (cells) with higher demand.

Most DCA schemes (and FCA) assume worst-case conditions; i.e., the mobile is located on the circumference of a cell. Thus channels are allocated to *cells*, rather than to *mobiles* (traffic-bounded schemes).

Other DCA schemes were proposed (see, for example, Haas et al. [1997]) that rely on interference measurements. These schemes can "maximally" pack the channel allocations.

3.5.2.4.2 *How the Reuse Factor Is Determined?*
Definitions:

D = minimum distance between centers of co-channel cells
R = radius of a cell
C = the power of a carrier frequency
C/I = carrier-to-interference ratio (determines the relative impairment of the interfering signal)
(C/I) min = the minimum acceptable C/I to satisfy some quality of service (e.g., some acceptable speech quality)
r = propagation exponent

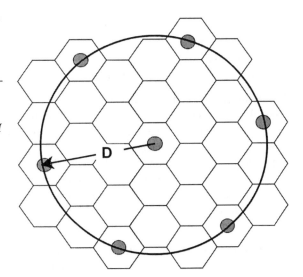

- The number of cells in a repetitive pattern is termed *reuse factor, N.*
- In hexagonal cell pattern, only certain values of N are possible:

$$N = i^2 + i \cdot j + j^2, \quad \text{where } i, j = 0,1,2,3,\dots$$

- Thus, possible values of N are 1, 3, 4, 7, 9, 12, 13, 16, 19, 21, ...
- Also, from other geometric considerations it follows that

$$\left(\frac{D}{R}\right) = \sqrt{3 \cdot N}$$

- From the hexagonal geometry, the "worst case" C/I ratio can be calculated using the following approximation:

$$\left(\frac{C}{I}\right) \approx \frac{1}{6}\left(\frac{D}{R}\right)^{r}$$

- Thus, given the minimal requirements on C/I, the D/R ratio can be determined. Given the D/R ratio, the reuse factor can be determined. For instance,

System	(C/I) min	(D/R)	N
AMPS	~18 dB	~4.6	7
GSM	~11 dB	~3.0	4
IS-54	~16 dB	~3.9	7
CDMA	~−15 dB	~0.7	1

3.5.2.4.3   Improving the Capacity and Reducing the Interference in Cellular Systems

There is close correspondence between the network capacity (expressed by N) and the interference conditions (expressed by C/I). Capacity can be increased and interference can be reduced by any of the following three schemes:

- Cell sectoring
- Cell splitting
- Cell sizing (microcellular networks)

Cell sectoring reduces the interference by reducing the number of co-channel interferers that each cell is exposed to. For example, for 60° sectorization, only one interferer is present, compared with six in omnidirectional antennas (Figure 3.52). But, cell sectorization also splits the channel sets into smaller groups, reducing the *trunking efficiency.*

Cell splitting allows creation of smaller cells. Thus, the same number of channels is used for smaller area. For the same probability of blocking, more users could be allocated.

Advantages of cell splitting include more capacity and requiring only local redesign of the system. Among the disadvantages of cell splitting are an increased number of handoffs, increased interference levels, and requirement for more infrastructure (base stations).

3.5.2.4.4   Cell Sizing — Microcellular/Picocellular Systems

In cell sizing, to allow more capacity, the size of the cells is scaled down. Since the quality of service (C/I) depends only on the ratio (D/R), the performance (i.e., interference level) is unaffected by the scaling. However, the same number of channels can now be used in a smaller area (i.e., larger user density), increasing the total number of concurrent users. The increase is as α^{-2}, where α is the scaling factor. Smaller cells also imply less transmitted power — thus smaller and lighter handsets are possible. However, smaller cells also imply larger handoff rate. As an example, consider the following two cases.

Case I: Cell radius = 1mile
                Number of cells = 32
                Number of channels = 336
                Reuse factor = 7
                →48 channels per cell
                →1536 concurrent calls

Case II: Cell radius = 0.5 mile (= 1/2)
                Number of cells = 128
                Number of channels = 336
                Reuse factor = 7
                →48 channels per cell
                →6144 concurrent calls

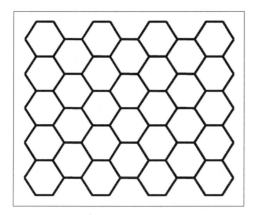
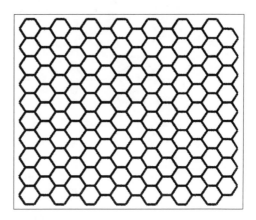

3.5.2.5   Handoff, Handover, or Automatic Link Transfer (ALT)

Handoff (U.S. Cellular Standards) or handover (CCITT/CCIR Standards) or ALT (ANSI T1P1) is a procedure of changing the mobile-terminal-to-fixed-network-port binding, from one port to another port. For example, in a cellular network and on the physical layer, the cellular telephone is the terminal and the ports are the base stations.

In general, the handoff procedure provides means for improving the quality (e.g., RF signal strength) of the received signal while the mobile terminal moves or when reception conditions change. In this case, the trigger for handoff is the RF signal strength falling below some threshold, and when there is another base station that can serve the mobile with a more strongly received RF signal.

There are two types of handoff schemes: Mobile Assisted Handoff (MAHO), in which the mobile actively participates in the handoff decision by providing measurements, and network initiated handoff, in which the decision is made solely by the network measurements.

Internal handoff refers to channel reassignment procedure — it is not a "real" handoff. *Soft handoff* is a handoff in which the new binding is completed before the old binding is torn down; it reduces the probability of call dropping, also called the "make-before-break" handoff.

Handoff Strategies—Common handoff performance metrics include (see also, Pollini [1996]):

- *Call blocking probability:* The probability of a new call being blocked
- *Call dropping probability:* The probability that, due to a handoff, a call is terminated
- *Call completion probability:* The probability that an admitted call is not dropped before it terminates
- *Probability of unsuccessful handoff:* The probability that a handoff is executed while the reception conditions are inadequate
- *Handoff blocking probability:* The probability that a handoff cannot be successfully completed
- *Handoff probability:* The probability that a handoff occurs before call termination
- *Rate of handoff:* The number of handoffs per unit time
- *Interruption duration:* The duration of time in which a mobile is not connected to any network port
- *Handoff delay:* The distance between the locations at which the handoff is executed and the location at which is should be performed

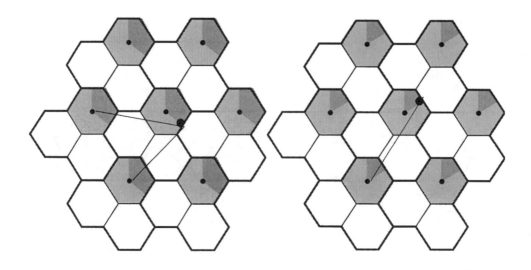

3-sector cells   **6-sector cells**

FIGURE 3.35   Sectorization and interference for $N = 3$.

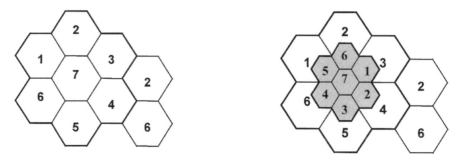

FIGURE 3.36   Cell splitting.

Measurement of the signal strength is the primary parameter for handoff decision. To eliminate the rapid fluctuations in the RF signal strength (i.e., fast fading), the measured signal is averaged over some short time. Another possible parameter for the handoff decision is the Bit Error Rate (BER). The following are some examples of handoff algorithms (refer to Figure 3.37 for an understanding of how the different schemes affect the instant of handoff).

- *Relative Signal Strength:* Handoff will occur at L_1. Note that this approach leads to "ping-ponging" and that it may lead to unnecessary handoff
- *Relative Signal Strength with Threshold:* Depends on the value of the threshold. At Th_1 handoff occurs at L_1 and at Th_2 handoff occurs at L_2. The ping-ponging effect is still possible.
- *Relative Signal Strength with Hysteresis:* The handoff will occur at L_3 for hysteresis H. This eliminates the ping-pong effect, but may lead to unnecessary handoff.
- *Relative Signal Strength with Hysteresis and Threshold:* The handoff will occur at point L_3 if the threshold is Th_1 or Th_2 and will occur at L_4 for threshold Th_3.
- *Prediction Techniques:* In which the level of the received signal is predicted from previous measurements.

Some studies evaluate handoff algorithms in combination with the channel assignment schemes. In particular, it is possible to dedicate some number of "guard channels" to ensure lower dropping than blocking probability.

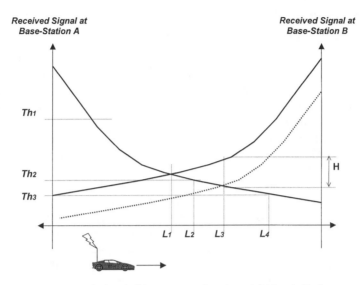

FIGURE 3.37 The handoff instance as a function of the handoff scheme.

The above handoff algorithms can be evaluated based on statistical analysis of the signal strength. For example, the relative signal strength can be expressed as:

$$BS_0 \text{ to } BS_1: \text{ if } Y_1 > Y_0 \quad \text{and} \quad BS_1 \text{ to } BS_0: \text{ if } Y_0 > Y_1,$$

where Y_i is the averaged signal power of base station i (BS_i). Similarly, the relative signal strength with hysteresis can be formulated as:

$$BS_0 \text{ to } BS_1: \text{ if } Y_1 > Y_0 + H \quad \text{and} \quad BS_1 \text{ to } BS_0: \text{ if } Y_0 > Y_1 + H$$

and the relative signal strength with hysteresis and threshold as:

$$S_0 \text{ to } BS_1: \text{ if } \left(Y_1 > Y_0 + H\right)_1 \left(Y_0 < Th\right) \quad \text{and} \quad BS_1 \text{ to } BS_0: \text{ if } \left(Y_0 > Y_1 + H\right)_1 \left(Y_1 < Th\right)$$

Of course, there are possible variations as well. For instance, consider a scheme in which the hysteresis is used only when the signal strength at the current station is strong enough (above the threshold). Otherwise, no hysteresis is employed. Further modification involves two thresholds, Th_L, Th_U, in which case no handoff occurs if the signal is above the upper threshold, Th_U.

Direction-biased handoff algorithms have also been considered, which give preference to a handoff to base stations in the direction of the mobile movement. Let set A include the base stations that the mobile is approaching and set B include the base stations that the mobile is moving away from. Also, two new hysteresis values are defined, encouraging, H_e, and discouraging, H_d. The relations between these values is $H_d \le H \le H_e$. If the current serving base station is BS_s, then handoff to the base stations $BS_i \in \mathbf{B}$ occurs when:

$$BS_s \text{ to } BS_i: \text{ if } \left(Y_i > Y_s + H\right) \text{ and } BS_S \notin \mathbf{B} \quad \text{or} \quad \text{if } \left(Y_i > Y_s + H_d\right) \text{ and } BS_S \notin \mathbf{A}$$

and handoff to the base station occurs when:

$$BS_s \text{ to } BS_i: \text{ if } \left(Y_i > Y_s + H_e\right) \text{ and } BS_S \notin \mathbf{B} \quad \text{or} \quad \text{if } \left(Y_i > Y_s + H\right) \text{ and } BS_S \notin \mathbf{A}$$

3.5.3   Some Common Air Interfaces: AMPS, IS-54/136, IS-95, GSM

3.5.3.1   Advanced Mobile Phone System (AMPS)

The AMPS is a first generation system with nearly ubiquitous coverage in the U.S. It is based on FDMA/FDD access, with channels of 30 kHz. Forward (downlink) channels from the base station to the mobile are in the frequency range of 869 to 894 MHz, while reverse (uplink) channels from the mobile to the base station occupy the 824 to 849 MHz frequency range. Channels are paired, with forward/reverse channel separation of 45 MHz.

AMPS channels are of two types: Voice Channels (VC) and Control Channels (CC). The Forward Control Channels (FCCs) are of broadcast type and are used for subscriber paging, voice channel assignments, and for other control functions. The paired Reverse Control Channels (RCCs) are randomly accessed with "sensing" information provided by corresponding FCCs. VCs, forward or reverse, are dedicated channels assigned to a single call throughout its duration.

The total number of channels is 416 per provider (333 in the original AMPS version before spectrum augmentation by the FCC in 1987), out of which 21 channels are used as control channels and 395 as voice channels. Voice channels are FM modulated with ±12 kHz peak deviation. Control channels are FSK modulated with deviation of ±8 kHz and signaling rate of 10 kbps. AMPS frequency reuse is based on seven repetitious patterns, with fixed channel assignment (FCA) scheme. Coverage of an AMPS cell is generally in the range of 2 to 25 km. Figure 3.38 depicts the frequency assignments to the two service providers (A, non-wireline and B, wireline) and Table 3.7 shows the frequency allocations to these channels. Note that channel numbers 800 to 900 are not assigned.

The configuration of AMPS hardware is shown in Figure 3.39 and, in general terms, consists of the base station, the base station controller, and the mobile switching center* (MSC). The MSC serves the function of a central office (CO) in a switched telephone network. In AMPS, MSC is also responsible for assignment of voice channels to incoming or outgoing calls. The call set up process is shown in Figures 3.40 and 3.41, for mobile initiated (outgoing) and network initiated (incoming) calls, respectively.

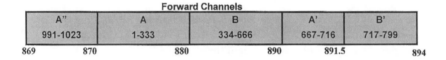

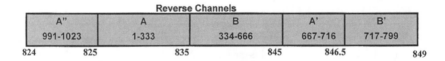

FIGURE 3.38 AMPS frequencies assignments.

TABLE 3.7   AMPS Channels Assignments

Channel	Channel Number	Center Frequency (MHz)
Reverse channels	$1 \leq N \leq 799$	$0.030 \cdot N + 825.0$
	$991 \leq N \leq 1023$	$0.030 \cdot (N - 1023) + 825.0$
Forward channels	$1 \leq N \leq 799$	$0.030 \cdot N + 870.0$
	$991 \leq N \leq 1023$	$0.030 \cdot (N - 1023) + 870.0$

---

*MSC is sometimes also referred to as MTSO (mobile telephone switching office).

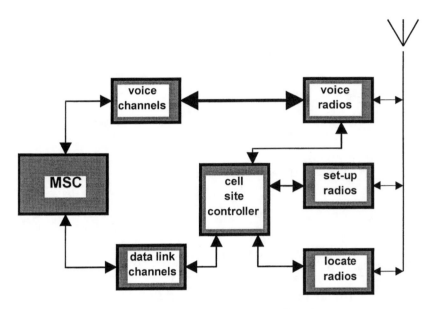

FIGURE 3.39 AMPS base station configuration.

3.5.3.1.1 Detailed AMPS Channel Description

The Reverse Control Channel (RCC) in AMPS can transmit up to six words in a single frame (Figure 3.42). The data field in each word is 48 bits long and encoded with (48;36;5) BCH block code. Each data field is repeated five times for reliability purposes (majority voting is employed at the receiver after detection). Each frame is delimited by a Bit Sync (BS) dotting sequence of 30 bits (101010...1010) and followed by the word sync (WS) of 11 bits (11100010010). The Digital Color Code (DCC) is used to distinguish transmissions in co-channel cells. Thus, a frame with a single word contains 288 bits and corresponds to a data rate of 271 bps (overhead excluded) at the 10 kbps signaling rate.

The AMPS Forward Control Channel (FCC) is a broadcast type of channel. Two words, each 40 bits long, are transmitted in each frame (Word A and Word B). Each word is encoded with the (40;28;5) BCH block code and is transmitted five type (for reliability purposes) and the replicas of the two words are interleaved. Each frame starts with 10 bits/field (1010101010), followed by the word sync (WS) pattern of 11 bits (11100010010). A special feature of the FCC is that it provides information about the status (idle or busy) of its corresponding RCC, though the "busy/idle" bits that are inserted every tenth bit in the frame. This allows mobiles to "sense" the status of the RCC before accessing it, thus reducing the probability of access collision. The "busy/idle" bits are depicted in Figure 3.43 as ↑.

Total frame size is thus 463 bits and at 10 kbps signaling rate, the data rate (excluded overhead) is 1.2095 kbps.

3.5.3.1.2 AMPS Signaling Formats

AMPS uses a number of signaling tones and a special audio quenching scheme to transmit (in-band) signaling information. Supervisory tone (SAT) is an indication that the connection is still "alive." It is continuously transmitted in both directions of any VC that carries voice information. It can be generated at one of the three following frequencies: 5970, 6000, or 6030 Hz. SAT is detected at the mobile on the FVC and reproduced on the corresponding RVC. If the SAT is missing for 1 s, the connection is declared lost and the call is dropped.

The Signaling Tone (ST) is a 10-kbps sequence of alternating ones and zeros with duration of 200 ms. It coexists with SAT and signals "on/off hook" events.

Blank-and-Burst encoding is a scheme that allows fast transmission of control information of a VC. It is implemented by turning off the voice (FM) transmission for a typical period of 100 ms and replacing it with an FSK, 10 kbps NRZ Manchester encoded data with deviation of ±8 kHz. The data are encoded

MSC	BAS	Mobile
	• Continuously transmits the setup data on the FCC	
		• Scans and locks on FCC • Initializes call • Seizes RCC • Sends service request
	• Forwards service request	
• Selects a VC • Sends channel assignment to BAS		
	• Forwards channel assignment to the mobile (on FCC)	
		• Tunes transmitter/receiver to the assigned VC • Transmits SAT on the RVC
	• Detects SAT • Sends confirmation message to MSC	
• Completes call through the PSTN		

FIGURE 3.40 The setup process for a mobile initiated call.

MSC	BAS	Mobile
	• Continuously transmits setup data on FCC	
		• Scans and locks on the strongest FCC
• Incoming call is received • Returns audible ring to the caller • Sends paging message to the cells		
	• Reformats the paging message • Sends the paging message on the FCC	
		• Detects the page • Seizes RCC • Sends service request
	• Forwards service request to MSC	
• Selects VC • Sends channel assignment to the BAS		
	• Forwards channel assignment to the mobile (on FCC)	
		• Tunes transmitter/receiver to the assigned VC • Transmits SAT on the RVC
	• Detects SAT • Sends alert to mobile on FVC	
		• Alerts user • Sends ST on RVC
	• Detects ST	
		• User answers • Stops ST on RVC
	• Detects absence of ST • Sends answer message to MSC	
• Receives answer message • Stops audible ring to the caller • Completes connection through the PSTN		

FIGURE 3.41 The setup process for a network initiated call.

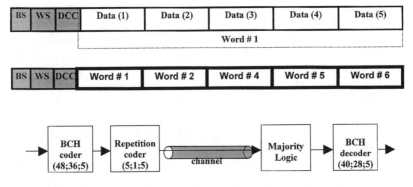

FIGURE 3.42 RCC frame structure and its coding/encoding scheme.

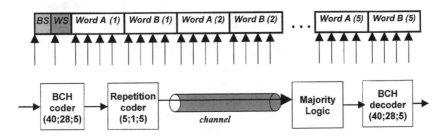

FIGURE 3.43 FCC frame structure and its coding/encoding scheme.

with (40;28) BCH code on the FVC and with (48;36) BCH on the RVC. The resulting data rates are 271 bps and 662 to 703 bps, on the FVC and the RVC, respectively. The blank-and-burst encoding is used to exchange control information (e.g., commands) between the mobile and the base-station, while a call is in progress (i.e., the mobile and the base station are tuned to the VC). Examples of such commands include change power level, handoff, and directed retry.

3.5.3.1.3 AMPS Mobile Power Levels

The power of the mobile is controlled by the base station through change power level commands. As the mobile moves closer to the base station and depending on the radio propagation conditions, the mobile's transmission power may need to be reduced accordingly. AMPS defined power setting in increments of 4 dB, whereas the maximum power depends on type of the mobile equipment. For example, for mobile class I, there are eight different power levels: +6, +2, –2, –6, –10, –14, –18, and –22 dBW. For Class III mobile, there are only six power settings: 2, –6, –10, –14, –18, and –22 dBW. Table 3.8 outlines the acceptable power ranges for the three mobile types.

TABLE 3.8 AMPS Power Ranges for Different Mobile Types

Mobile Type	Maximum Power	Minimum Power
Class I (mobile)	+6 dBW (4 W)	–22 dBW
Class II (transportable)	+2 dBW (1.6 W)	–22 dBW
Class III (portable)	–2 dBW (0.6 W)	–22 dBW

3.5.3.2 IS-54/136 — the U.S. Digital Cellular System

IS-54/136 is the second-generation North American Digital Cellular standard. Its major objective is to increase the network capacity and system performance. However, because of the widespread use of AMPS, a standard that would provide an easy migration path was very desirable. As such, IS-54 preserves the structure of the control channels and the spectral frequency channelization of AMPS voice channels.

Thus, the frequencies of channel width are the same as in AMPS and the operation is still FDD based. However, the IS-54 carriers for voice transmission are divided into six time slots, allowing ultimate six times improvement in capacity (currently, two time slots are assigned per user, leading to an increase of three times in capacity, as compared with AMPS). The signaling rate on the voice channels is 48.6 kbps, which with two slots per user results in 16.2 kbps per user, out of which 13 kbps is used for compressed voice, and the rest is overhead. This compatibility in control channels allows implementation of dual-mode systems, which switch to AMPS in areas where IS-54 is not yet supported. As a second migration step IS-136 provides its own structure to control channels, with time division multiplexing of control channels. Figure 3.44 depicts the IS-54/136 frame structure. Note that voice channel pairing remains the same as in AMPS, with 45 MHz separation between the two directions.

The structure of a IS-54/136 forward traffic (voice) channel (i.e., the format of a TDMA slot) is presented in Figure 3.45. Its starts with 28 synchronization bits, used for timing synchronization, equalization, and time slot identification. Sync is followed by 12 bits of the slow associated control channel, which allows conveying of control information between the base station and the mobile at a relatively low data rate. The digitized and compressed voice bits are included as two blocks of 130 bits each. Alternatively, the DATA field can contain digital information. Between the two DATA blocks, there are 12 bits of the coded digital verification color code (CDVCC), which serves a similar role to the SAT tone in AMPS. In particular, it allows distinguishing between the reuse instances of the same carrier in co-channel cells. CDVCC is encoded in shorted hamming code (12;8;3).The slot ends with 1 reserved bit (RSVD) and 11 bits of the coded digital control channel locator (CDL).

The distinguishing feature of the reverse digital traffic channel is that it contains two new fields: the Guard Time (G) and the Ramp Time (R) (Figure 3.46). The G field is necessary because the reverse frequency channel is shared among a number of mobiles (three, in the current implementation), whose distance to the receiver (at the base station) is usually not equal. Thus, measures need to be taken to ensure that transmissions from the mobiles assigned adjacent slots (on the same frequency channel) do not overlap at the receiver. The G field, equivalent to 6 bits in duration allows for such a synchronization. The R field, also equivalent to 6 bits in duration, is used to power up the transmitter at the mobile (i.e., switch the RF on). The DATA is split into three blocks; one of 16 bits and two of 122 bits, for the total of 260 bits per slot. The 28 synchronization bits are used to synchronize the local oscillator to the transmitter clock. The function of the SACCH and the CDVCC are as in the forward traffic channel.

The slot structure of the forward control channel in IS-136 is shown in Figure 3.47. The main new features here are the Shared Channel Feedback (SCF) and the coded superframe phase (CFSP). In particular, the SCF is used to indicate the status of the reverse control channel (which is a random access channel), very similar to the function of the busy/idle bits in AMPS.

Finally, the slot structure of the reverse control channel in IS-136 is presented in Figure 3.48. Two formats are possible: the abbreviated time slot format is used by a mobile during handing off to another channel and includes an additional Abbreviated Guard (AG) time field. Additional synchronization fields are present (e.g., preamble (PREAM) and sync+) to allow fast synchronization on short bursts transmitted over the control channel by mobile stations. The PREAM field is also used for power control purposes.

IS-54/136 provides for two types of schemes for communication of control information on traffic channels: the slow Associated Control Channel (SACCH) and the fast associated control channel (FACCH). In each slot (of traffic channel) 12 bits are reserved for the SACCH to convey 66-bit-long messages encoded with rate $\frac{1}{2}$ convolutional code. (The 66 bits include 48 bits of actual data, in addition to 16 bits of cyclic redundancy code, CRC, 1 continuation bit, and 1 reserved bit). Thus, a control message requires 132 bits to transmit, which corresponds to 11 slots or 220 ms at the 218 bps data rate of the SACCH.

For faster transmission of control messages, FACCH is used, in which the messages are sent by temporarily interrupting the transmission of the data; i.e., a control message is inserted into the DATA field of a traffic slot. This is similar to the blank-and-burst scheme in AMPS. Messages, 65 bits in length, are encoded in rate $\frac{1}{4}$ convolutional code to produce the 260 bits, which are inserted into a single slot.

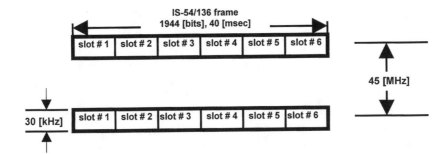

FIGURE 3.44 IS-54/136 frame structure (voice).

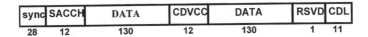

FIGURE 3.45 The forward digital traffic TDMA slot.

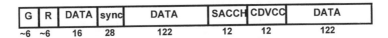

FIGURE 3.46 The reverse digital traffic TDMA slot.

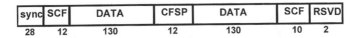

FIGURE 3.47 The forward digital control TDMA slot.

- normal slot

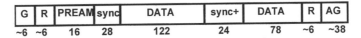

- abbreviated slot

FIGURE 3.48 The reverse digital control TDMA slot.

Examples of IS-54/136 control messages are shown in Table 3.9.

TABLE 3.9 IS IS-54/136 Control Messages

- Forward channels
- Alert with info (Caller ID)
- Measurement order (MAHO)
- Flash with info (Message Waiting)
- Handoff
- Power control
- Reverse channels
- Connect
- Acknowledgment
- Channel quality measurements
- Flash with info

3.5.3.3 IS-95: CDMA Digital Cellular Standard

The CDMA technique, promoted by Qualcomm, resulted in Interim Standard 95, the IS-95 standard. IS-95, similarly to IS-54, is compatible with the existing AMPS frequency assignment. (Frequency assignment for the reverse band is 824 to 849 MHz and for the forward band is 869 to 894 MHz. Channels are paired, with 45 MHz separation.) Thus, systems can be designed to be dual mode. (Qualcomm offered CDMA/AMPS dual-mode phone from 1994.) Also, IS-95 is compatible with IS-41.

IS-95 is based on reuse of one and, thus, avoids frequency planning — a big advantage in planning of a cellular system. The channels are created within 1.25-MHz bands (each direction); i.e., 10% of the available cellular spectrum. Thus, incremental deployment is possible. Practically, a 270-kHz guard band is required for AMPS carriers on each end of the CDMA spectrum. The modulation and spreading is different from the forward and the reverse links in IS-95.

The maximum user data rate is 9.6 kbps and is spread to a channel chip rate of 1.2288 Mchip/s, which corresponds to spreading factor of 128. The speech coder used in IS-95 is the Qualcomm 9600 Code Excited Linear Prediction (QCELP) coder. It reduces the data rate during periods of silence to 1200 kbps. Thus, depending on the voice activity, the coder output rate is 1.2, 2.4, 4.8, or 9.6 kbps. An improved coder, QCELP13, was introduced by Qualcomm in 1995. By increasing the convolutional code data rates, the effective information data rates were increased by 50%.

3.5.3.3.1 IS-95 — Forward Link

The forward channel (Figure 3.49) consists of:

- Pilot channel: Allows to acquire timing information by the mobile, provides phase reference for the coherent demodulation process, and provides means for signal strength comparison for the purpose of handoff determination
- Synchronization channel: A broadcast channel for synchronization messages (at 1200 bps)
- Up to seven paging channels: A broadcast channel for control information and paging messages (at 9.6, 4.8, and 2.4 kbps)
- Up to 63 Forward Traffic Channels (FTC): For user data (at 9.6, 4.8, 2.4, and 1.2 kbps)

The BAS transmits to mobiles on the forward channel with the spreading codes being mobile specific. Mobiles in a cell are assigned different Walsh spreading functions, providing excellent signal separation. Scrambling by a PN sequence of length 2^{15} chips provides further separation of users with the same Walsh functions in different cells (co-code users). It also improves the wideband spectral characteristics of the signals.

A pilot code is continuously being transmitted on one of the channels (channel 0) at a higher power level than the user channels. It allows coherent detection and channel propagation estimation.

Transmission on the forward channel includes:

- Data stream is encoded with rate $\frac{1}{2}$ rate convolutional code
- Interleaved
- Spread by one of the 64 (orthogonal) Walsh functions
- Scrambled using PN sequence 2^{15} chips long

The speech coder detects idle periods in speech and uses repetition to keep the symbol rate at 19.2 kbps.

Long PN sequences are $2^{42} - 1$ chips long. There are two types of long code: public and private. The public type is created by using the mobile ESN number. The private sequence is created by a private procedure. Calls are initiated using the public long code. After authentication, the private long code is used.

Open loops and fast closed loops are used to control power and eliminate the "near–far" problem. The power control information is sent on the forward channel at 800 bps, which is taken from the data channel. Power control is performed by power control commands sent to the mobiles every 1.25 ms on the forward channel. The commands specify 1-bit operation: increase or decrease by 1 dB.

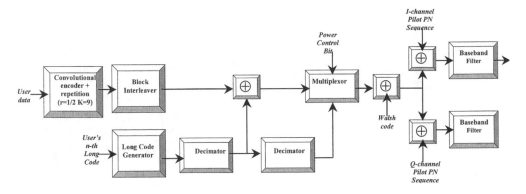

FIGURE 3.49 Forward channel in IS-95.

Soft handoff diversity is used, in which the mobile combines predetected signals from two base stations much in the same way as in the multipath case.

3.5.3.3.2 IS-95 — Reverse Link

Reverse channel is noncoherently detected, is asynchronous, and is power-controlled by the forward channel commands (Figure 3.50). Three-finger RAKE receivers are used to reduce the effect of multipath.

Reverse Channel spreading consists of:

- Data stream encoded with rate $\frac{1}{3}$ rate convolutional code
- Interleaved
- Six symbols are mapped into one of the 64 (orthogonal) Walsh functions (this mapping is not for spreading, but for modulation purposes)
- Spread by user-specific (with period of $2^{42} - 1$) and base-specific (period of 2^{15}) codes; this results in the rate of 1.2288 Mchips

3.5.3.4 Global System for Mobile Communications (GSM)

GSM is a second-generation, digital, TDMA-based, cellular standard with extensive coverage in Europe, including seamless coverage across European countries. Its total RF spectrum of about 50 MHz is located in the following bands: 890 to 915 MHz (reverse link) and 935 to 960 MHz (forward link). In all, 124 RF carriers (62 for each one of the two service providers), each of 200 kHz width, occupy the above spectrum. The carrier signaling rate is about 270.8 kbps, GMSK modulated. The offset of the forward/reverse links pairing is fixed 45 MHz. Each RF carrier is further divided into eight time slots per frame (TDMA), creating eight physical channels per carrier. There is no preassignment of any channels to a specific or exclusive use.

Physical layer slot structure can support both voice and data. Speech is encoded at 13 kbps. There are four error-control mechanisms: $\frac{1}{2}$ rate convolutional code, cyclic redundancy code, variable-depth interleaving, and flexible automatic retransmission request scheme.

Logical channels (i.e., virtual circuits) are realized on top of physical channels. Different logical channels are specified:

- BCCH — Broadcast Control Channel
- FCCH — Frequency Correction Channel
- SCH — Synchronization Channel
- CCCH — Common Control Channel
- TCH — Traffic Channel
- SACCH — Slow Associated Control Channel

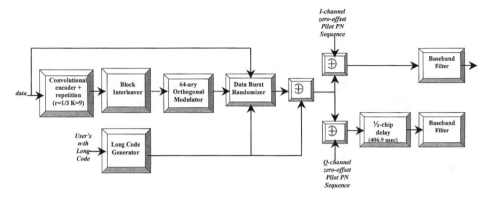

FIGURE 3.50 Reverse channel in IS–95.

- FACCH — Fast Associated Control Channel
- DCCH — Dedicated Control Channel

The top system view of GSM is depicted in Figure 3.51. It is very similar to the first-generation cellular networks, with the clear exception that MSCs can be directly connected one to another, without the interconnection of PSTN. The structure of the GSM base station system is depicted in Figure 3.52.

GSM has a very elaborate hierarchy of its frame structure, shown in Figure 3.53, with clock periods of up to nearly 3.5 h defined by the hyperframe structure. At the bottom of this hierarchy is the basic TDMA slot, whereas information within a slot is referred to as a *burst*. Several types of bursts are available:

- Normal burst
- Frequency correction burst
- Synchronization burst
- Access burst
- Dummy burst

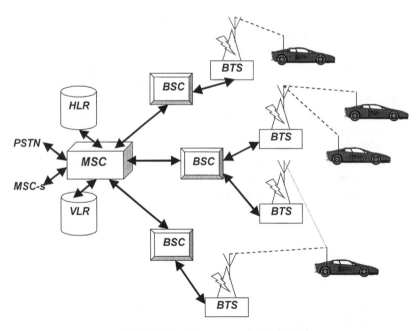

FIGURE 3.51 Top system view of GSM.

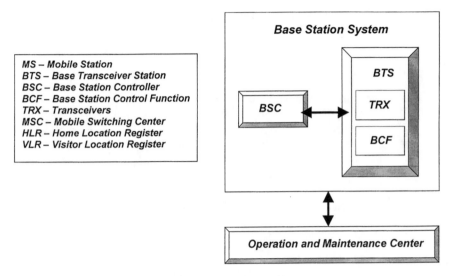

FIGURE 3.52 GSM base station system.

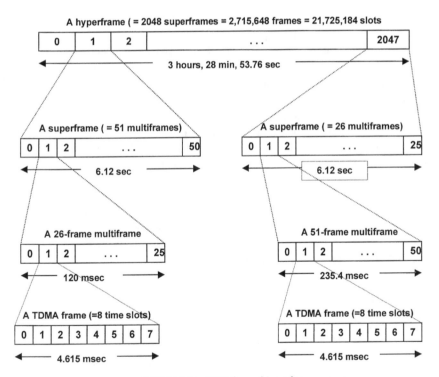

FIGURE 3.53 GSM frame hierarchy.

An example of the slot with a normal burst is shown in Figure 3.54. The Tail Bits (TB) allow synchronization of transmissions from mobiles located at different distances from the base station. The *stealing flag* bits indicate whether the DATA portion of the slot is used for transmission of user data or for control information. Finally, the training sequence allows determination of the equalizer settings. The guard times allow for timing errors and enough time for equalization on fast changing channels. Fault detection is also provided.

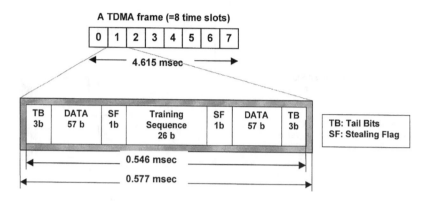

FIGURE 3.54 The frame structure of a normal burst.

The equivalent full-rate single-channel data rates in GSM is:

$$\frac{114 \,[\text{bits}/\text{frame}] \times 24 \,[\text{frames}/\text{multiframe}]}{120 \,[\text{ms}/\text{mulitframe}]} = 22.8 \text{ kbps}$$

as each multiframe consists of 24 traffic frames + one SACCH frame (#12) + one IDLE frame (#25). Data rate of a half-rate channel is 11.4 kbps.

The goal of the handoff and channel reassignment scheme is to operate on the best-quality RF carrier at any time and at any location. To achieve this, in GSM a mobile continuously scans and measures the RF quality of adjacent channels (including the currently serving cell). The mobile reports the strongest six channels. If (a) the RF signal quality falls below some threshold for a period of 5 s, or (b) the RF signal quality from another cell is better than some threshold for a period of 5 s, then the mobile will be handed off to the other cell. However, there will be no handoff because (b) is within less than 15 s of the previous handoff. Channel measurements are a combination of RF signal measurements, together with BER measurements, as bad channels cannot be always identified by RF measurements. And, because of forward error control, handoff indication based solely on BER measurements may give too short a notice, reducing the time for successful completion of the handoff procedure.

Load leveling is performed in GSM; i.e., spreading the assigned channels among carriers, cells, and areas. GSM also provides "holding call" procedure that allows reconnection of "dropped calls."

GSM spans most of the ISO/OSI model layers and it is compatible with signaling system 7. As an example, consider the GSM security features, which include:

- Identity authentication (authentication)
- Identity confidentiality (privacy)
- Signaling data confidentiality (concealing information)
- User data confidentiality (privacy)

Implementation of these security features is facilitated by a mobile-assigned Subscriber Identity Module (SIM) at the subscription time. SIM includes:

- IMSI — International Mobile Subscriber Identity
- Individual Subscriber Authentication Key
- Authentication Algorithm

To authenticate a service request, the following procedure is performed:

- Mobile transmitting its temporary IMSI (TIMSI);
- Network replying with a random-generated number (RAND);

- Mobile computing the signed response (SRES) using the authentication algorithm, the key (which is a function of the frame number) and RAND, and sending the SRES back to the network;
- Network comparing the SRES from the mobile with its computed SRES. This results in either authentication or denial.

Privacy is provided by a Temporary Mobile Subscription Identity (TMSI), valid during its binding to a VLR, and computed after the authentication procedure.

3.5.4 Mobility Management in Wireless Cellular Networks

In this section, we describe how mobility management is implemented in the AMPS network. Although some details are significantly different in other networks (in particular, the second-generation networks), the basic principles are relatively similar to these of AMPS.

The primary function of mobility management is the ability to track mobile users. As a user roams throughout the coverage of a cellular system, its location (i.e., the base station that a mobile is currently in its coverage) needs to be identified when a call has been initiated from the fixed network (or another mobile user), with the user being its destination. The user tracking function is currently supported through a number of processes, chief of which are the registration and the routing functions. These processes are implemented through signaling system 7 (SS7).

The basic interconnection of a mobile to the fixed Public Switched Telephone Network (PSTN) is described in Figure 3.55. The communication between a mobile and the base-station (BAS) is through four types of channels: reverse control channel, forward control channel, reverse voice channel, and forward voice channel. These channels are implemented based on one of the Common Air Interfaces (CAIs), described above. A BAS will serve many mobiles in its coverage area. The BAS is, in turn, connected to an MSC through two types of connections, voice circuits and data links, which can be carried over terrestrial or microwave links. An MSC may serve many BASs (Figure 3.56).

MSC controls the following functions with the assistance of SS7: call setup, call transfer, billing, interaction with PSTN, etc. In particular, MSC utilizes the SS7 signaling network to validate location and to deliver calls for roamers. To do so, it uses three databases: Home Location Register (HLR), Visitor Location Register (VLR), and Authentication Center (AUC). HLR, VLR, and AUC may or may not be located in close proximity to the MSC.

A mobile is associated with the HLR of its MSC; the association of mobile with an MSC is determined by the prefix of the mobile telephone number. The HLR is used to register the current location of a mobile. It is updated through the registration process. VLR is used to record the roamer in a specific system. It is updated through the registration process. AUC is used authenticate a roamer to the network to ensure that the user is eligible to receive the requested services.

There are three main functions associated with mobility management: *roaming, registration,* and *routing,* which use these three databases. Roaming allows user to move from the coverage of its home system (home MSC) to the coverage of another system (visitor MSC) and continue to receive the requested services. Registration is a process through with a roamer notifies the currently serving MSC (visitor MSC) about its presence in the MSC coverage area. The visitor MSC updates the VLR and forwards the registration to the home MSC, which updates the HLR. Finally, routing is a process by which the system selects a route through the network to interconnect the party with the called party.

Upon receipt of a mobile user's registration, the MSC determines whether the user is a roamer or a local subscriber. If it is a roamer, the visited MSC requests verification of the subscriber information (MIN-Mobile Identification Number, and ESN, Electronic Serial Number) from the user's HLR. (Each HLR maintains the MIN and ESN information for all its subscribers.) The roamer's supplied information (MIN + ESN) are compared with the information at the home AUC. If it does not match, the visitor MSC is instructed to deny service to the user. If the roamer's information matches the HLR record, the HLR is updated with the identity of the mobile's currently visited MSC and the *customer profile* is returned to the visitor MSC. (The *customer profile* contains features such as call forwarding, call waiting, conference calling, etc.)

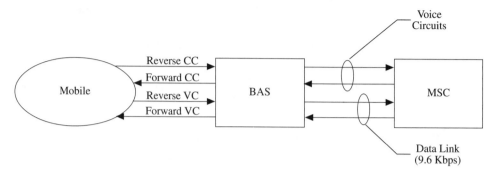

FIGURE 3.55 Mobile/BAS/MSC interconnection.

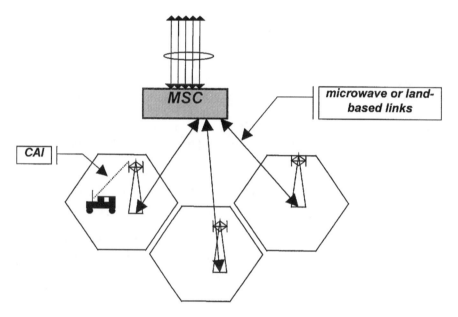

FIGURE 3.56 BAS/MSC interconnection.

Upon receipt of registration confirmation, VLR stores the customer profile (and possibly some additional antifraud information). This concludes the registration process. VLR is a dynamic database and continues to keep information about the current roamers in the MSCs coverage. Billing is done by the visited MSC, at a rate which depends whether the mobile is a roamer, or not.

After the initial registration, the mobile periodically sends the MIN + ESN information to the visitor MSC over control channels, in response to the registration command. This feature is called *autonomous registration,* and is performed every 5 to 10 min. A mobile determines whether it is a roamer by comparing its stored station ID (SID) with the advertised SID in its current the coverage area. (SID identifies the coverage area of an MSC.) The registration process is done exclusively over the SS7 signaling network and not over the PSTN lines.

Call routing is done by the SS7 system routing the call to the user's home MSC, which checks the HLR for the subscriber's current location. If the subscriber is a roamer, the home MSC routes the call to the visitor MSC. To route call to the visitor MSC, the home MSC send a *route request* to the visitor MSC, which returns the *temporary directory number* (TDN). The home MSC uses then the TDN to forward the call to the visitor MSC. Upon receiving a call connection request, the MSC pages a subscriber throughout its coverage area and alerts the subscriber of a connection request. Note, as the MSC does

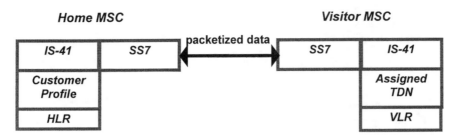

FIGURE 3.57 Intersystem handoffs.

not know the actual location of the mobile in its coverage area, pages are issued to all the base stations in the MSCs coverage. If the called party does not answer the call, the visitor MSC returns a *redirection* to the home MSC to reroute the call (such as to voice mail or another phone number).

To accommodate intersystem handoffs, IS-41 is used for communicating information between MSCs (Figure 3.57). This allows uninterrupted service to the roaming mobiles.

3.5.5   Mobility in the Internet (Mobile IP)

The Transmission Control Protocol/Internet Protocol (TCP/IP) protocol suite is the most extensively used protocol for computer communications. The IP is based on the notion of "best-effort delivery" of data-grams, which are routed through connectionless mode between routers of different networks. The IP addresses are the basis for this routing process; each network is identified by an IP address, which is a prefix to the IP address of individual machines residing on the particular network. Thus, to route between machines, the datagram needs to be forwarded from the source machine network to its gateway. Then, the datagram will hop from a router to a router on the path between the source and the destination gateway and finally arrive at the network identified by the destination IP address. From the destination network gateway, the datagram will be delivered locally to the destination machine.

The TCP adds reliability to the "best-effort delivery" IP protocol. In particular, it ensures sequenced and errorless delivery of a stream of data from an application to its peer application running of remote machines (or locally on the same machine).

Both TCP and IP were designed for stationary networks that seldom (if ever) change their point of attachment to the Internet. As a result, the IP addresses for machines serve two purposes: they specify the identity of a machine and they determine its point of attachment (as the prefix of a machine's IP address constitutes the IP address of the attached to network). Thus IP address determine the routing of a datagram in the Internet.

For a mobile machine, changing its point of attachment to the Internet means that a route to the host cannot be determined from its IP address. Since IP addresses also represent the identity of a machine, they cannot be arbitrarily changed without the need to notify all the machines that may communicate with the mobile machine in question. Thus, clearly, a modification to the IP is required to support mobility in the IP networks. The difficulty in making any changes to the IP is that the resulting protocol must be backward compatible; only machines that care to implement the mobility support need to be affected.

The Internet Engineering Task Force (IETF) proposed a solution to the mobility in the Internet in its RFC 2002 document. In what follows, the main features of the RFC 2002 are described. First, however, we define the term *triangular routing*.

Based on the traditional IP routing mechanism, datagrams destined to a *mobile host* (MH) will end up in the gateway of its *home network* (i.e., the network whose IP address matches the prefix of the mobile host). In triangular routing, when a mobile is away from home, its datagrams are captured by a process termed *home agent* (HA), which runs in the mobile's home network. The HA, which masquerades as the away-from-home mobile, encapsulates the captured datagram as a new datagram with destination address

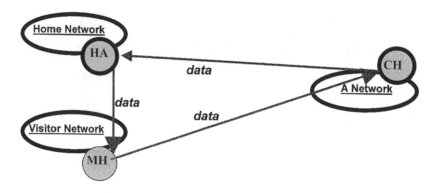

FIGURE 3.58 Triangular routing.

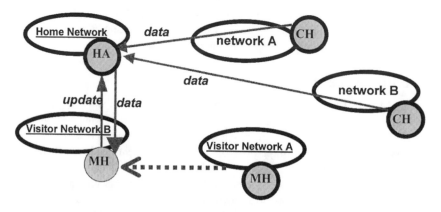

FIGURE 3.59 The registration process.

equal to the IP address of a process running on the *visitor network* and termed *foreign agent* (FA). The FA, upon receipt of the encapsulated datagrams, decapsulates them and delivers the datagram locally to the visitor mobile host. Because of this routing, datagrams from any fixed host to a mobile host always go through the home network and are then redirected to the visitor network. However, datagrams from the mobile host to a fixed host, referred to here as a *corresponding host* (CH), travel directly through the Internet without any redirection. Thus, traffic between a MH and a CH create a "triangular" shape, as is indicated in Figure 3.58.

Through another process, termed the *registration* process, the HA and the FA learn about the current *point-of-attachment* of the mobile to the Internet — this is referred to as the *care-of-address* (COA) and is usually identified as the IP address of an FA (see Figure 3.59).

3.5.5.1 IP Mobility Support

The goals of the IP mobility support protocol were to limit the number of administrative messages sent between the MN and the agents, and to keep to a minimum the number of duplicating functionality messages on mobile links. Obviously, these goals were motivated by the fact that the communication between the mobile and the rest of the network will be carried over a wireless medium.

The following entities are defined in the RFC 2002:

- Mobile Node (MN): A host or a router capable of changing its attachment point from one network/subnetwork to another.
- Home Agent (HA): A router resident on MN home network and forwarding traffic destined to the MN through encapsulation. HA also maintains current mobility bindings for the MN.
- Foreign Agent (FA): A router on the visited network assisting the MN.

- Care-of-Address (COA): An IP address either of a FA or temporarily assigned to a MN (RFC 1541, R. Droms, Oct. 1993). COA represents one end of a tunnel (the other end is the HA), allowing tunneling of traffic to the MN.

The following functions are defined in the RFC 2002:

- Agent Discovery (AD): The process by which a newly attached MN learns about the identity of HA or FA. AD is performed either through agents advertising their availability or through MN solicitation.
- Registration (RG): The process through which the away-from-home MN registers its COA with its HA. RG can be performed directly with the HA or through the FA, which forwards the registration information to the HA.
- Encapsulation (EP): Also known as tunneling, the process allows forwarding of intercepted datagrams by the HA to the COA. This is accomplished through enveloping the intercepted datagrams into another IP datagram.
- Decapsulation (DP): The reverse process of EP. Performed at the COA, the original datagrams are extracted from the enveloped datagrams.

The following requirements are defined in the RFC 2002:

- The MN shall be able to communicate with other nodes when reconnected to the Internet at different points of attachment. The MN may be disconnected from the Internet while changing points of attachment.
- The MN shall continue to use its own IP address (i.e., home address).
- The MN shall be able to communicate with nodes that do not implement the IP mobility support.
- No changes are required in hosts or routers that do not serve any mobility functions.
- MN shall support registration authentication.

The assumptions underlying the protocol operation are:

- The IP mobility support protocol does not constrain the MN IP address assignment and the address will be used regardless of the MN point of attachment.
- The frequency with which the MN changes its points of attachment will be at most once per second.
- IP datagrams routing is based on the destination address and not on the source address.

The registration process creates mobility bindings and is done through UDP registration messages. Registration can be performed with or without the assistance of an FA. When the registration is done without FA, the MN registers its COA (which was acquired dynamically on the visitor network) with its HA. The two options are depicted in Figure 3.60.

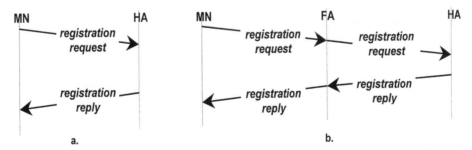

FIGURE 3.60 RFC 2002 registration process; directly with the HA (a) and through FA(b).

The mobile, especially wireless, environment is potentially quite vulnerable to security attacks, such as eavesdropping, active reply attacks, or impersonating another entity. RFC 2002 defined security measures to ensure that appropriate messages are authenticated. It does not provide a privacy (encryption) recommendation. Authentication is achieved through the mobility security association (MSA) tables, which are maintained by each MN and each mobility agent. The tables are indexed by IP addresses. MSA between a MN and its HA should be set up by the HN administration and will allow one to authenticate registration messages with the HA. HAs and MHs must be capable of using the authentication procedure; the default is keyed MD5 (RFC 1321, April 1992), with 128-bit keys, used in "suffix + prefix" mode. Because of the key management problem, not all FA-related messages are authenticated. For commercial applications, messages between FA and HA may need to be authenticated (e.g., billing, service provision).

To ensure location privacy, an MH may establish a tunnel to its HA, forcing the MH datagrams to be routed through the HA and, thus, appear as sourced from the HN. Protection against reply attacks is provided in RFC 2002 by the identification field in the registration request, which is used to ensure the currency of the request. Two methods are outlined: *nonces* (Figure 3.61) and *time-stamps* (Figure 3.62).

When nonces are used, the lower 32 bits in the registration reply are used to match it with the registration request. The HA issues a new nonce in the registration reply to be used in the next registration request. The low- and the high-order 32 bits are different from the previously used numbers. The nonce protocol is self-synchronizing — rejection due to invalid nonce is equipped with a new nonce.

When time stamps are used, the clocks at the two corresponding node must be adequately synchronized. The lower-order 32 bits are used to match the request with the reply.

Forwarding datagrams to the MN is done by encapsulation and can be implemented either as IP-in-IP (Figure 3.63) or as Minimal Encapsulation (ME) (Figure 3.64). In the IP-in-IP scheme, the time-to-live (TTL) in the outer header is set to the original datagram. At decapsulation, the TTL is decreased by one in the IP header. ME can be used for nonfragmented datagrams and an MH learns about ME capability of the FA from the agent's advertisement.

RFC 2002 allows for use of *mobile routers* (MR); i.e., routers that are themselves capable of mobility. In other words, the mobile network itself, together with the router, can be mobile with respect to the fixed network. Examples of such networks include a network on a train, where a mobile user may move between the cars of the train, and the whole network is mobile with respect to the fixed network along the train path. Routing to an MH served by a MR is through double tunneling: once between the HAMH and the MR and second between the HAMR and the FAMR. The packet route is, thus, CH → HA_{MH} → HA_{MR} → FA_{MR} → MR → MH. This is depicted in Figure 3.65.

3.5.5.2 Protocol Stack of Mobile IP

In Mobile IP, the datagrams are forwarded using the standard IP tunneling technique. Thus, the protocol stack for user datagrams is as shown in Figure 3.66. Control datagrams, such as for the purpose of registration, for example, will be composed with the UDP protocol. Thus, the protocol stack for the control datagram is as depicted in Figure 3.67.

3.5.6 Cellular Digital Packet Data (CDPD)

CDPC is an intermediate technology that allows communication of digital (packetized) data over the existing AMPS infrastructure (Figure 3.68). As mentioned previously, the design of AMPS, being a first-generation system, was geared exclusively toward supporting voice communications through analog (FM) modulation. As such, it has no provisions for digital data transmission. Although digital data can be transmitted over AMPS through the use of wireless modems (and cellular point-to-point data link control, such as MNP 5 or 10), such communication is not well suited for bursty computer data, leading to high charges for transmission of small amounts of data (due to lengthy setup costs) and to long periods of time when there is no activity. Furthermore, depending on the propagation conditions, the achievable data rates are rather low.

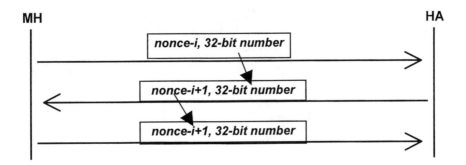

FIGURE 3.61 Use of nonces in the registration process.

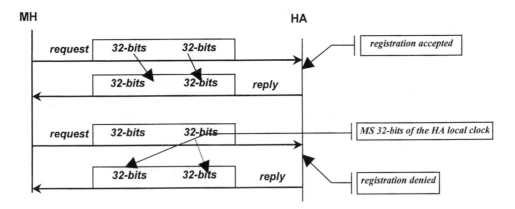

FIGURE 3.62 Use of time stamps in the registration process.

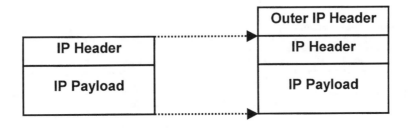

FIGURE 3.63 IP-in-IP encapsulation.

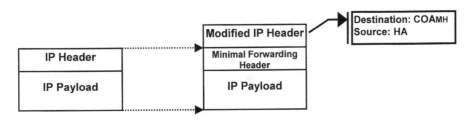

FIGURE 3.64 Minimal encapsulation.

Most, if not all, of the second-generation systems, being digital in nature, allow support for transmission of digital data with data rates of 9.6 and 19.2 kbps. However, these systems are still not extensively deployed, creating "islands" on coverage. In contrast, AMPS has nearly ubiquitous coverage in North America.

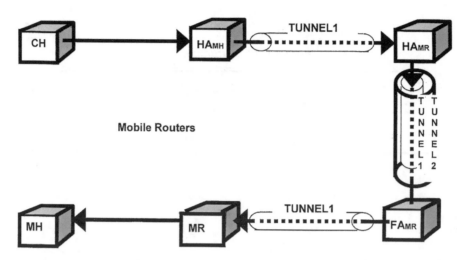

FIGURE 3.65 Mobile router.

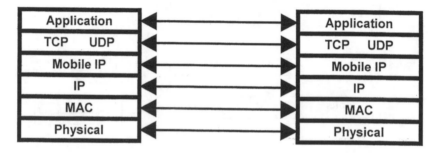

FIGURE 3.66 Forwarded datagrams.

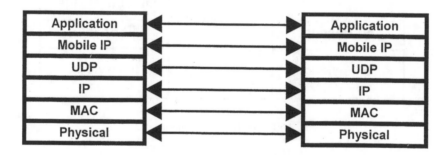

FIGURE 3.67 Control datagrams.

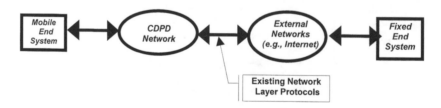

FIGURE 3.68 General flow of traffic across CDPD network.

A number of wireless data networks exist in the U.S. Two well-known examples are the ARDIS and the RAM mobile data networks. As with the second-generation cellular networks, the wireless data networks do not enjoy extended coverage. Rather, the systems are targeted at the mobile travelers market, providing coverage to places such as airports or convention centers.

CDPD was introduced, on the one hand, to allow digital packetized transmission, and, on the other hand, to support extensive AMPS coverage. It is done by forming an adjunct network that uses the same AMPS frequencies, in the periods of time when these channels are not in use by AMPS (i.e., CDPD utilizes the unused wireless capacity by the voice customers to transmit packet bursts on idle, unoccupied, AMPS channels.

The initial CDPD consortium was supported by eight cellular providers: AMCI, Bell Atlantic Mobile, Contel Cellular, GTE Mobile Communications, McCaw, NYNEX, PacTel Cellular, and Southwestern Bell Mobile Systems. CDPD products are available today from many manufacturers. The coverage of the system is already significant.

The goal of CDPD is to provide ubiquitous service as a commercial public mobile data communication network with data rates of 19.2 kbps. The CDPD service is connectionless or connection oriented. For connectionless service, the supported protocols are connectionless network protocol (ISO 8473) or IP. Connection-oriented services are provided by transport protocol class 4 (TP4) or TCP.

CDPD architecture is depicted in Figure 3.69 and the corresponding protocol stack in Figure 3.70.

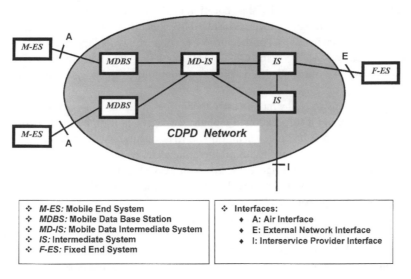

FIGURE 3.69 CDPD network architecture.

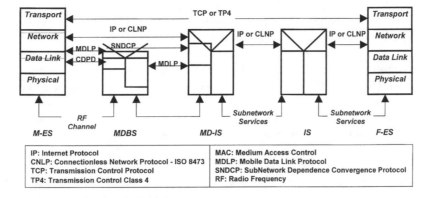

FIGURE 3.70 CDPD protocol architecture.

3.5.7   Summary and Concluding Remarks

In the last decade, the communication and computing paradigm has been shifting from individual users and stationary personal computing to groupware with support for mobile access. While most of today's mobile systems are still analog and voice oriented (the first generation), the perception of multimedia communication, in general, and data, in particular, is becoming increasingly important in mobile networks. As such, the first-generation systems (e.g., AMPS) are being replaced by second-generation digital systems (e.g., IS-136, IS-95, GSM). Thus, the technological basis to accommodate the shift toward multimedia support is being established. Furthermore, it is envisioned that, with proliferation of the mobile technology, the number of mobile users will increase dramatically. Also, decreasing hardware prices and increasing availability of mobile-centric applications and services are expected to stimulate the mobile and wireless multimedia market.

The usable RF spectrum is limited by technical and political considerations. To follow the increase in demand for wireless communications, the cellular technology that is based on concurrency is being used in most wireless networks. The cellular principle allows reuse of the RF spectrum and, thus, increases the spectrum utilization. In particular, smaller cells imply better reuse and more capacity. Thus, micro- and picocells, combined with macrocells, are expected to be the prevalent technology for future wireless networks

In cellular first- and second-generation networks, mobility management is supported by two databases — Home Location Registry (HLR) and Visitor Location Registry (VLR) — and is implemented through the SS7 signaling network. The major processes involved in cellular mobility management are registration, call delivery, and intersystem handoffs. For the purpose of supporting mobility management, the optimal location tracking strategy depends on the Call-To-Mobility Ratio (CTMR). For large CTMR, frequent mobile location update is recommended. Small CTMR favors a lower update rate and more search-based user location tracking.

The vision of ubiquitous access by mobile users to land-based resources is a major driver behind the research, standardization, and incorporation of mobility management features into the existing networking protocols. Although there is no unique definition of what constitutes a third-generation system, some attributes are generally accepted in the technical literature as representative of the future generation of wireless and mobile networks. For instance, the vision of PCS advocates profile-driven communication by an individual, allowing communication with *anyone, anywhere, anytime, and in any format.* In particular, PCS supports both personal and terminal mobility.

In the data communication area, the TCP/IP protocol is the most extensively used protocol suite for local- and wide-area connectivity. Undoubtedly, the *Internet* and its services (e.g., www) contribute to the proliferation of TCP/IP. However, the IP and TCP protocols were designed for stationary networks, routers, and computers, and cannot support mobility without proper modifications. In particular, IP addresses are used for both equipment identification and route determination (i.e., IP addresses serve as *location* identifiers). The IETF WG Mobility Support (RFC 2002) was introduced to allow mobility in the TCP/IP-based machines. It is based on two special entities, home agent and foreign agent, and on the notion of care-of-address. The main processes in IETF-WG Mobile IP are agent advertisement, agent discovery, registration, encapsulation, forwarding, and decapsulation. Routing in IETF Mobile IP is triangular. The newer version of the IP protocol, the IPV6, provides an opportunity to improve the IP mobility support protocol by eliminating the inefficient triangular routing present in the current version of IP (i.e., IPV4). The IETF RFC 2002 also includes extensive security provisions based authentication of corresponding entities.

Above the IP layer, the performance of the TCP protocol is affected by mobility through incorrectly interpreting datagram loss during handoffs as network congestion. This results in engagement of retransmission back-off and congestion-control mechanisms, leading to substantial throughput loss. Extensions and modifications to the TCP protocol were proposed to cure the above abnormality. Other proposed TCP modifications deal with the special characteristics and requirements of the mobile/wireless communication environment (such as asymmetrical design to offload processing from the mobile device).

CDPD is also based on the IP protocol (along with CLNP). Although the strategies are different, there are many similarities between mobile IP and CDPD. While CDPD supports a full protocol stack, addressing lower-layer issues as well, the IETF mobile IP scales considerably better with the number of networks.

It is expected that mobile and wireless networks will become more and more an indispensable part of our lives, allowing remote access to multimedia information and applications. Many of the applications will be location-dependent services, assisting users in their everyday tasks, as well as in providing services to travelers and mobile workers.

References

D. Bertsekas and R. Gallager, *Data Networks*, 2nd ed., Prentice-Hall, Englewood Cliffs, NJ, 1992.

G. Calhoun, *Digital Cellular Radio*, Artech House, Norwood, MA, 1988.

G. Calhoun, *Wireless Access and the Local Telephone Network*, Artech House, Norwood, MA, 1992.

K. Feher, *Wireless Digital Communications — Modulation and Spread Spectrum Applications*, Prentice-Hall, Englewood Cliffs, NJ, 1995.

M. J. Feuerstein and T. S. Rappaport, *Wireless Personal Communications*, Kluwer Academic Publishers, Boston, 1993.

V. K. Garg and J. E. Wilkes, *Wireless and Personal Communications Systems*, Prentice-Hall, Englewood Cliffs, NJ, 1996.

Z. J. Hass et al., Simulation results of the capacity of cellular systems, *Trans. Meh. Technol.*, 46 (4), 1997.

G. C. Hess, *Land-Mobile Radio System Engineering*, Artech House, Norwood, MA, 1993.

J. M. Holtzman and D. J. Goodman, Eds., *Wireless Communications — Future Directions*, Kluwer Academic Publishers, Boston, 1993.

J. M. Holtzman and D. J. Goodman, Eds., *Wireless and Mobile Communications*, Kluwer Academic Publishers, Boston, 1994.

J. Holtzman, Ed., *Wireless Information Networks — Architecture, resources, Management, and Mobile Data*, Kluwer Academic Publishers, Boston, 1996.

W. C. Jakes, Ed., *Microwave Mobile Communications*, IEEE Press, New York, 1974.

W. C. Y. Lee, *Mobile Communications Design Fundamentals*, 2nd ed., John Wiley & Sons, New York, 1993.

E. A. Lee and D. G. Messerschmitt, *Digital Communications*, 2nd ed., Kluwer Academic Publishers, Boston, 1994.

R. C. V. Macario, *Personal and Mobile Radio Systems*, Peter Peregrinus, London, 1991.

R. V. V. Macario, *Cellular Radio Principles and Design*, McGraw-Hill, New York, 1993.

A. Mehrotra, *Cellular Radio: Analog and Digital Systems*, Artech House, Norwood, MA, 1994.

A. Mehrotra, *Cellular Radio Performance Engineering*, Artech House, Norwood, MA, 1994.

M. Mouly and M. B. Pautet, *The GSM System for Mobile Communications*, Michel Mouly and Mary Bernadette Poutet Press, Palaiseau, France, 1992.

S. Nanda and D. J. Goodman, Eds., *Third Generation Wireless Information Networks*, Kluwer Academic Publishers, Boston, 1992.

J. D. Parsons and J. G. Gardiner, *Mobile Communication Systems*, Halsted Press, Glasgow, 1989.

J. N. Pelton, *Wireless and Satellite Telecommunications*, Prentice-Hall, Englewood Cliffs, NJ, 1995.

G. P. Pollini, Trends in handover design, *IEEE Commun. Mag.*, 34 (3), 1996.

J. G. Proakis, *Digital Communications*, 3rd ed., McGraw-Hill, New York, 1995.

T. S. Rappaport, Ed., *Cellular Radio and Personal Communications*, IEEE Press, New York, 1995.

T. S. Rappaport, *Wireless Communications — Principles and Practice*, Prentice-Hall, Englewood Cliffs, NJ, 1996.

D. W. E. Rees, *Satellite Communications*, John Wiley & Sons, New York, 1990.

R. Schneiderman, *Wireless Personal Communications — The Future of Talk*, IEEE Press, New York, 1994.

G. Varrall and R. Belcher, *Data over Radio*, Quantum Publishing, Denver, 1992.

A. J. Viterbi, *CDMA — Principles of Spread Spectrum Communication*, Addison-Wesley, Reading, MA, 1995.

J. Walker, Ed., *Mobile Information Systems,* Artech House, Norwood, MA, 1990.

D. J. Withers, *Radio Spectrum Management,* Peter Peregrinus, London, 1991.

M. D. Yacoub, *Foundations of Mobile Radio Engineering,* CRC Press, Boca Raton, FL, 1993.

H. E. Young, *Wireless Basics,* Intertec, Chicago, 1992.

4

Video Communications

Dan Minoli, Editor
Teleport Communications Group

Emma Minoli
Red Hill Consulting

Larry Sookchand
Teleport Communications Group

4.1 An Overview of Digital Video and Multimedia Technologies and Applications*

4.1.1 Introduction

Interest in high-quality video-based digital communications for business as well as for entertainment applications goes back several years. At one end, digital methods are now planned to be deployed for the support of a new generation of television known as advanced TV (ATV) or digital TV (DVT), although there are several approaches to the matter (terrestrial broadcast, satellite broadcast, and cable TV distribution). At the other end, there is significant interest in delivering desktop business video applications via the Internet protocol (IP) suite of protocols, and over the corporate intranet, the Internet, or other affiliated extranets. In addition to traditional room-based videoconferencing, vertically integrated applications, such as distance learning and telemedicine, are also receiving renewed attention of late. And, PC-based multimedia (e.g., video from digital disks) is becoming ever more popular. Furthermore, as

---

*Portions of this material were synthesized from the numerous articles, books, and papers the senior author has published in the recent past on the topic of digital video.

mobile communications become a basic part of modern life, the next step is to support mobile multimedia applications; in fact, already there are evolving standards and equipment to support handheld (or laptop)-based video telephony.

Over the recent past, a major change in the area of TV distribution has occurred. While in some countries wireless video is equivalent with the reception of terrestrial analog TV signals via a rooftop antenna, in other countries TV programs are received via satellite, cable, or even multichannel microwave distribution system (MMDS) links. The advent of DTV transforms the traditional TV channel to a data transmission medium that supports large data rates (e.g., upwards of 38 Mbps) at very low bit error rates ($<10^{-11}$). In this situation, the term *television* loses part of its original meaning: digital television is no longer restricted to transmitting sound and images; instead, it becomes a data broadcasting mechanism that is fully transparent to the content.[1]

This section explores issues related to digital video in the area of trends, standards, applications, and challenges. Portions of the discussion focus on support of video over the intranet and Internet, and the intrinsic need for quality of service (QoS) in these networks. More specifically, the following topics are addressed:

1. The emergence of video-based applications, including DTV broadcast
2. ITU-T H.320/H.323
3. ISO MPEG family, including MPEG-1, MPEG-2, and MPEG-4
4. ISO MPEG-2 (details)
5. QoS issues and video over the Internet

There are a number of driving forces for video-based digital services. First and foremost, there is the demand for higher quality in the video and audio spectrum; secondarily, one finds an array of technological developments, that, in fact, is spurring innovation and deployment. No longer will barely discernible talking heads with unsynchronized lips and video, plagued by annoying encoding delays and jumping or exploding heads, be tolerated by the business user community for videoconferencing applications — 15 years of this is enough. Fortunately there are new technologies entering the field at the commercial level that have the potential to address problems that have been holding back the deployment of video services for years. These technologies span both high-performance Digital Signal Processing (DSP), as well as broadband transmission. Until now, transmission of business video has had to be accomplished using Integrated Services Digital Network (ISDN) basic rate service at 128 kbps, or, at most, at 384 kbps (by employing expensive inverse multiplexers to merge several ISDN channels into one aggregate channel). Now broadband services, both at the data link layer, e.g., Asynchronous Transfer Mode (ATM), as well as at the network layer (e.g., via high-performance "gigarouters" or IP switches), are becoming available. Figure 4.1 depicts the typical business video environment that can be found at this time.

In general, digital video applications now emerging can be classified in three categories:

1. Corporate/institutional video transmission (classical systems, as well as the newer H.323-based systems);
2. Stored digital video for multimedia applications;
3. Entertainment video (video broadcast, including post-production distribution).

In corporate/institutional video transmission, the turn-of-the-decade news in video relates to the development of QoS-based networks that support video transmission, along with the introduction of IP/Local-Area Network (LAN)-based (specifically, H.323) coders for desktop applications. Standardization and the use of ubiquitously available IP networks will go a long way to foster the proliferation of these video services. These developments drive the general need for QoS-based communications. At this time, most discussions about QoS have an enterprise network point of reference; however, since many corporate networks use the Internet in some fashion, this translates ultimately into a requirement on the Internet itself. In reference to stored digital video, the turn-of-the-decade news relates to the availability of DVD (digital video disk or, according to some, digital versatile disk), along with the supporting digital

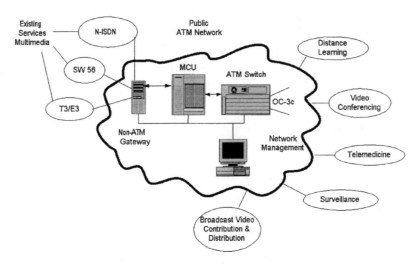

FIGURE 4.1 Typical business video environment (MCU = multipoint control unit).

encoding methods. The news in entertainment video is the plan now under way to broadcast DTV commercially, particularly in terrestrial networks, but also in cable and satellite applications.

In order to facilitate ubiquitous communications, standards are required. The video space is no different. High-quality digital encoding schemes such as International Organization for Standardization/International Electrotechnical Commission (ISO/IEC) MPEG-2 (Motion Picture Expert Group-2) and MPEG-4 are evolving. The first such standard in the family, the ISO MPEG-1 (ISO 11172-2, 1993), is a "high-quality" audiovisual coding standard supporting the storage and retrieval of multimedia information on a CD-ROM up to about 1.5 Mbps. ISO MPEG-2 (ISO 13818-2) followed in 1995; the standard addresses broadcast TV applications. MPEG-4 is a new flexible low-bandwidth standard now under development; the advanced capabilities of this standard enable it to support a number of evolving applications, ranging from wireless videophones to Internet multimedia presentations, broadcast TV, and DVD. MPEG-4 enables robust video transmission over noisy communications channels (such as wireless video links). Naturally, a standard that supports these diverse functionalities and associated applications turns out by necessity to be fairly complex. MPEG-4 reached *Committee Draft* standard status in November 1997; it was expected to be an official *International Standard* in January 1999.* [*Note:* Originally another standard, MPEG-3, was designed to support high-definition TV (HDTV) with sampling dimensions up to 1920 × 1080 at 30 fps. This standard has now been abandoned since, with a little tweaking, MPEG-2 and MPEG-1 work well for HDTV at rate of 20 and 40 Mbps. Hence, HDTV is now part of the MPEG-2 High-1440 Level specification (discussed in Section 4.3).]

For corporate video applications, there has been a set of standards going back a number of years, in the form of traditional ITU-T H.261/H.262 recommendations (under the H.320 umbrella). Recently two new recommendations have emerged for the support of IP-based multimedia (H.323) and for mobile IP-based telephony (H.324). Applications include stand-alone videophones, PC-based multimedia applications, inexpensive voice/data modems, World Wide Web (www) browsers with live video, video-only security cameras, and others.[2]

On the entertainment side, in late 1996, the Federal Communications Commission (FCC) adopted a DTV transmission standard for terrestrial broadcast in the U.S. Since the mid-1980s, many people have closely followed the intense efforts directed at developing a North American DTV standard. These efforts have included standards for digital program origination (known as Production Standards, and typically developed within the Society of Motion Picture and Televisions Engineers, SMPTE), and the recently publicized DTV Terrestrial Broadcasting Standard (the DTV transmission standard was developed by a private-sector consortium known as the Grand Alliance, under the supervision of the FCC Advisory Committee on Advanced Television Services).* As a consequence of this interindustry cooperation (spanning broadcasters, manufacturers of both professional and consumer electronics, computing, motion picture film, and telecommunications industries), the standards that have emerged are cost-effective and flexible.[3] The baseline DTV Terrestrial Broadcasting Standard encompasses both digital SDTV (standard definition TV, i.e., 525-line-based digital television), and digital HDTV. Although it took a lot of effort for the DTV standard to be developed and approved, in the view of many that actually now seems the easier part of the job, in light of the challenges that now lie ahead for the entire industry. After all, a single analog video format having just one resolution and one frame rate is now planned to be retired in favor of multiplicity of digital video formats varying in both resolution and frame rate. The FCC has mandated deadlines at specific time instances in the future for deployment of digital services.[4]

The sections that follow provide details on the digital video themes that have been introduced thus far in this summary view of the industry.

4.1.2 Evolving Use of Digital Video in Corporate and Entertainment Environments — QoS Support

4.1.2.1 Background on Corporate/Institutional Video Transmission

Significant gains in automation and productivity have been seen in North America in the past decade. These gains have been made possible by the plethora of application software that has emerged. The near-ubiquitous penetration of "standardized" PC operating systems, such as Microsoft Windows on the client side and Windows NT and UNIX on the server side, friendly graphic user interfaces (GUIs) along with network GUIs (specifically, WWW browsers), and the transmission control protocol/Internet protocol (TCP/IP) apparatus, have made this software and productivity revolution possible. The 1990s saw widespread introduction of client/server and Web-based intranet/Internet systems in the corporate environment.

As we move forward, a whole gamut of *new applications* are evolving and are being deployed to reach the next plateau in business support tools, and to rein in ensuing additional productivity gains. Multimedia, desktop videoconferencing, voice over data networks, and Computer Telephony Integration (CTI), to list a few, from both and intranet and Internet perspective, are expected to play an increasingly important role in the corporate landscape during the next decade.

In the enterprise network supporting these informatics applications, there has been the deployment of new LAN technologies (such as switched and/or 100 Mbps ethernet). At the Wide-Area Network (WAN) level, networks have shifted in two directions (in comparison with, for example, a decade ago):

1. The movement away from Time Division Multiplexing (TDM) transmission facilities of fixed bandwidth, particularly when focusing at the whole (nationwide) network (rather than just the access tail, which continues to be based on traditional telephony facilities, such as DS1, DS3 lines, or OC-3).

2. The movement away from dedicated point-to-point lines, which become impractical as the number of interconnected sites increases, and toward the use of packet technology (whether at the network layer in the form of IP-routed systems or at the data link layer, such as frame relay and ATM).

Hence, many Fortune 500 companies now use packet network services (e.g., frame relay, ATM, and IP) for corporate data, intranet, and electronic commerce applications. It follows that there is interest in addressing the question of services and media integration. Integration has found reasonable effectiveness in the frame relay context for the support of small office/home office (SoHo) locations. In

---

*The standards are formally documented by the ATSC (Advanced TV Standards Committee).

the end, an integrated network has a great deal of appeal for transmission efficiencies and technology/network management reasons. The two trends noted above have both benefits and drawbacks. In general, the benefits have weighed on the side of economics and scalability; the drawbacks have weighed on the side of performance and QoS. The goal, therefore, is to achieve integration and at the same time support QoS.

Video/multimedia applications, as well as more traditional but mission-critical applications, require enterprise networks that support QoS. Appropriate local-area, campus, wide-area, and international communication infrastructures will be needed to support this move to digital video and other broadband applications, including QoS support. Video streams have markedly different requirements compared with traditional data flows that supported applications such as E-mail, word processing, and financial analysis. The need for QoS is currently driven by three factors:

1. Support of voice over *packet networks* (intranets and Internet) with ensuing statistical gains and lower cost. There has been an interest going back at least to the mid-1970s (if not earlier) in utilizing integrated networks that support all of the organization's media, because of the efficiencies in transport and management that would result.[5-8]
2. Support of desktop video over the enterprise network and the Internet. Newly developed standards such as ITU-T H.323 are expected to see penetration in corporate LANs in the next couple of years. H.320-based systems are also in place at this time. The H.320/H.323 standards make video on LANs possible at the technology and interoperability level; however, as soon as more than a few users fire up conferences (particularly multipoint conferences/multicasts), the network may be severely impacted, as well as the quality of the conference.
3. Support of priority-based data applications over the intranet and Internet.

QoS includes guaranteed bandwidth-on-demand (minimum, average, and peak), as well as predictable (small) end-to-end delay, delay variation (jitter), and unit (block, cell, frame) loss. QoS-enabled networks are needed not only for time-sensitive applications, such as voice and video, but also to support data applications, which, as networks become more congested and more integrated (both at the corporate level and at the Internet level), need to receive a guaranteed level of performance.

There is now new significant industry work under way to address QoS in data networks. There have been successful as well as unsuccessful attempts to address the integration/QoS issue over the years (Table 4.1). Efforts in ISDN and ATM have also been aimed at voice support in general and at multimedia in particular. Efforts on the data side have included support of voice in LANs (such as IEEE 802.9 and FDDI II) and enhancements to routers, IP (e.g., IPv6 and RSVP), and network-layer handling of packets (such as multiprotocol label switching, MPLS, multiprotocols over ATM, MPOA, and tag switching/net-flows).

Key mechanisms showing promise now include ATM user-to-network interface (UNI) 4.0, IPv6, resource reservation protocol (RSVP), along with the integrated services architecture (ISA), real-time transport protocol (RTP), and network layer switching/MPLS. At the same time, existing services such as frame relay may or may not be upgraded to support QoS, which could turn out to be a problem.

The QoS issue has to be addressed at an end-to-end level, i.e., on a LAN–WAN (Internet)–LAN basis, as well as at multiple layers of the communication protocol suite (specifically, data link layer and network layer). From an end-to-end perspective, one has to look at the various LAN technologies and understand their QoS support; then, one has to do the same for the WAN technologies. On the protocol layer view, one could look at the problem as being addressed end-to-end, regardless of the underlying subnetworks, *if the QoS reservation protocol is end-to-end (end system to end system) and if every element at the layer under discussion (e.g., routers if one is looking at the IP/RSVP level) supports the QoS fulfillment mechanism.* When considering the discrete subnetworks, LAN technologies tend to have the issue of QoS driven by protocol considerations (e.g., contention) issues rather than by bandwidth (and, hence, cost) issues. For WANs this is exactly the opposite: the QoS support in protocols may exist, but the bandwidth is expensive, so the QoS fulfillment has to be done in a cost-effective manner.

TABLE 4.1   Support of QoS on Existing, Evolving, or Obsolete Technologies

Technology	Kind	Integration Support	QoS Support	Success
100 Mbps ethernet (shared/switched)	STDM, LAN	Yes	Some, based on reduced congestion	Prevalent
ATM LAN	STDM, LAN	Yes	Yes	Very limited deployment
ATM WAN	STDM, WAN	Yes	Yes	Medium deployment now
Ethernet (shared)	STDM, LAN	Yes, with some effort	Minimal	Ubiquitous
Ethernet (switched)	STDM, LAN	Yes	Some, based on reduced congestion	Ubiquitous
FDDI II	TDM-like in circuit channel, LAN	Yes	Yes, but fixed channels	No deployment
Frame relay	STDM, WAN	Possible now	Limited: via engineering, no intrinsic support (FRF.13 just begins to address the issue)	Wide-scale deployment
Frame relay on ISDN	STDM, WAN	Possible	Limited: via engineering, no intrinsic support	No deployment
IEEE 802.1q and.1p	STDM LAN	Possible	Yes	Not yet deployed
IEEE 802.9	TDM-like in circuit channel, LAN	Yes	Yes, but fixed 64 kbps channels	No deployment
IP	STDM, LAN/WAN	Yes	Limited; perhaps "gigarouters" have more punch and can lower delay and delay variation compared to existing routers	Ubiquitous
IPv6	STDM, LAN/WAN	Yes	Possible to some degree if vendors implement feature	Not yet deployed
ISDN 2B+D/23B+D/H0/H11	TDM, WAN	Yes	Yes, but fixed bandwidth	Limited deployment for integration purposes, mostly for Internet access
ISDN Packet on B	TDM/STDM, WAN	Yes	Unlikely, depends on packet switch	No deployment
ISDN Packet on D	TDM/STDM, WAN	Yes	Unlikely, depends on packet switch	No deployment
Network layer switching (flavors)	STDM, LAN/WAN	Yes	Some, based on the fact that a more efficient treatment of IP is possible	Limited deployment
Packet	STDM, WAN	Yes	Unlikely, depends on packet switch	Limited deployment now
RSVP with IP	STDM, LAN/WAN	Yes	Yes, but it is only a reservation mechanism; IP and (perhaps) ATM needed	Only now being deployed
RTP	STDM, LAN/WAN	Yes	Yes	Limited deployment

Fortunately, as this book was written, router vendors such as Cisco System were starting to build products to support QoS, in order to facilitate integrated voice/data/video networking at the corporate branch level, as well as over the intranet and Internet. Leading-edge vendors are including RSVP capabilities in their routers. They are also modifying their routers to accommodate better buffer management to deliver predictable QoS in a standard IP environment.

In summary, to carry video on enterprise networks, QoS issues, both at the protocol and at the technology level, need to be addressed. This topic is revisited in Section 4.5.

TABLE 4.2   Examples of DVC Products and Players

Company	Type	Name	Type of Connection
8 × 8, Inc.	Codec chip camera	VCP	10 Mbps ethernet
		Impression camera	10 Mbps ethernet
Accord Video	Video terminal	10 Mbps ethernet	
Telecommunications	MCU	10 Mbps ethernet	
Connectix	Camera	Quickcam	10 Mbps ethernet
Digital Semiconductor	Video terminal	10 Mbps ethernet	
	Video terminal	25 Mbps ATM	
First Virtual Corp.	Video terminal	PictureTel PCS 50	25 Mbps ATM/ISDN
Kenwood USA	PCMCIA card	Conference card	10 Mbps ethernet
Lucent Technologies L2W-323	Gateway	RADVision	10 Mbps ethernet
Microsoft software	Conferencing	NetMeeting	Internet
Picture Tel gateway	Video terminal	LiveLAN	10 Mbps ethernet
	PicTel Live Gateway	10 Mbps ethernet	
RADVision, Inc.	Gateway	10 Mbps ethernet	
VCON, Inc.	Video terminal	Armada Cruiser	10 Mbps ethernet
VDOnet Corp.	10 Mbps ethernet		
RVideoserver Gateway	MCU	H.323	10 Mbps ethernet
	H.323	10 Mbps ethernet/ISDN	
Vista Imaging	Digital Camera	ViCam	10 Mbps ethernet
Zydacron, Inc.	Video terminal	Z250	10 Mbps ethernet/ISDN

4.1.2.2 Business Video and Multimedia Applications

Observers claim that the new software and hardware that have entered the corporate landscape in the past decade are changing the way people do work in the corporation. At the root of this claim is the introduction of graphic and video technology in a variety of enterprise settings. Business video takes the form of PC-based conferencing, multimedia, video-server-based computer-based training, reception of digitized broadcast video on a PC window, and imaging-based document management/workflow systems.[9] The field of desktop video communications (DVC) is emerging.[10] (Table 4.2[11]).

In addition to networked multimedia per se, a variety of vendors and suppliers are pursuing the objective of bringing real-time two-way video to the corporate desktop. For example, multimedia conferencing enables designers in remote locations to review and/or work cooperatively on the same project using PCs that incorporate text, graphics, audio, visual, and tactile (touch-screen) capabilities. An increasing number of companies are now utilizing videoconferencing as part of their normal business practices, even though most of the applications are still in the conference-room-to-conference-room arrangement, rather than being desktop based. Such desktop conferencing systems, however, are now beginning to appear. These systems utilize PCs and high-end workstations that employ ISDN, ATM, or IP services.

Proponents make the case that videoconferencing on the desktop is only a matter of time. They quote well-known "benefits" that are direct carryovers from traditional videoconferencing: (1) reduced travel expenses; (2) more effective use of time; and (3) the ability to connect dispersed workgroups. Conference-room cost from $15,000 to $50,000, while desktop systems can be as inexpensive as $200.

During the 1990s there were a battery of new standards approved by the ITU-T for LAN/intranet/Internet-based multimedia, that promised to enable interworking among products and lower prices. The key new umbrella standard is H.323 (Table 4.3[12] and Figure 4.2), which is discussed in more detail in Section 4.2. Figure 4.3 depicts the key standards that relate to corporate digital video. Once standards are widely supported, a technology may see rapid introduction: industry forecasts predict over 6 million DVC worldwide shipments by 2001 in the business market and 14 million shipments in the residential market.[13]

As noted in the previous section, video and multimedia have two important characteristics that impact the type of transmission technology that can be employed at the LAN or WAN level for its delivery[14-28]:

TABLE 4.3   Some Milestones for Multimedia

1982	Introduction of compact disk — consumer audio
1984	Introduction of CD-ROMs; Macintosh GUIs
1986	Initial CD-I (Compact Disk Interactive) specification; Microsoft Windows
1987	DVI (digital video interactive) technology announced
1988	Erasable optical disks; initial ATM standards
1990	MPC (multimedia PC) standard; IMA (Interactive Multimedia Association) Compatibility Project; commercial multimedia applications
1992	ATM-based LAN development (155 Mbps to the desktop); FDDI 100 Mbps connections possible for less than $1000
1994	Wide-area ATM networks; networked multimedia systems and applications
1995	New low-speed voice compression standards (e.g., G.729, G.723.1, G.729A) and other interoperability standards for desktop multimedia are developed; also, DVD launched, H.323 developed
1998	Increased penetration of multimedia in corporate America for "mission-critical" applications; by then, IP had become ubiquitous in intranets and in the Internet; multimedia over LANs sees penetration

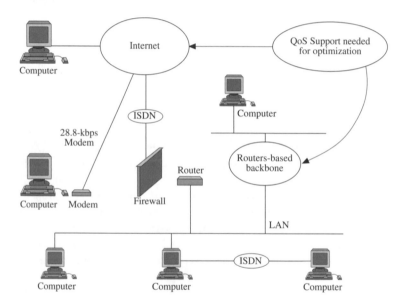

FIGURE 4.2   LAN/Intranet/Internet based multimedia.

1. The packetized video must be delivered with low and predictable end-to-end delay. Additionally, the delay variation must be small. Usually one cannot use store-and-forward methods except for non-real-time video, unless (a) the network is absolutely well-tuned or (b) the data rate is very small (e.g., 2×64 kbps in ITU-T H.261) or (c) the network supports QoS, as discussed in the previous section. Traditional protocols, not to say routers, do not have the ultimate performance capabilities for supporting hundreds of simultaneous corporate users using the enterprise backbone or LAN for videoconferencing applications.

2. Simple digitization of a video signal can yield from 140 Mbps for traditional full-motion NTSC (National Television Standards Committee) video to 1 Gbps for HDTV. Recently developed digital compression algorithms and supporting hardware reduce these values by about 100-fold or better. Compression is an economical method for storage and transmission of video in limited-bandwidth environments, including an organization's enterprise network. Compression methods include ISO MPEG schemes and ITU-T H.261/H.263, in addition to vendor proprietary methods.[14] However, the data rate is still fairly high, as follows:

   • Desktop videoconferencing applications using H.320/H.261 ITU-T standards produce from 128 to 768 kbps per user; the newer H.263 provides video support at 28.8 kbps or higher.

Multimedia Terminal Standards and Interworking
• H.320 (Narrowband terminals on N-ISDN)
• H.321 (Narrowband terminals on B-ISDN)
• H.322 (Narrowband terminals on Guaranteed QoS LAN)
• H.323 (Narrowband terminals on Non-Guaranteed QoS LAN)
• H.324 (Low bit rate terminals on GSTN)
• H.310 (broadband MPEG-2 & narrowband terminals on B-ISDN)
• ATM Forum Video On Demand
• ATM Forum Phase 2 Terminal Interworking
• ATM Forum VBR MPEG-2 Over ATM
• ATM Forum H.320 Over AAL5

FIGURE 4.3 Key standards relating to corporate digital video.

- Entertainment video and distance-learning applications using MPEG-1/MPEG-2 standards produce 1.5 or 6 Mbps, respectively, per user.
- Multimedia applications, using MPEG-1 or proprietary methods (for examples the Intel Indeo, DVI, and Wavelets), produce about 1.5 Mbps per user.

Note: Until the mid-1990s a number of vendors used Motion Joint Photographic Expert Group (JPEG), which is an adaptation from still photography — this method typically requires from 10 to 30 Mbps; other broadcast-level digitization schemes used in the recent past required 45 Mbps.

This brief discussion on data rates should make it immediately clear that traditional *shared* ethernet LANs operating at 10 Mbps over a traditional router-based network of low bandwidth is marginally adequate for the support of video in the corporation. Studies indicate that unless the video quality on PCs approaches the quality to which users are accustomed to on their TV sets or VCRs, that the deployment of video in the organization will be negatively impacted. This baseline requirement leads to the realization that one needs signal bandwidth of 1.5 to 6 Mbps *per user* for some business applications (lower resolution will be acceptable for other applications). A traditional ethernet would then be able to support only one to four users per segment.

An increasing number of vendors and users have taken the approach of microsegmenting existing ethernets to support a single user per segment, and then using ethernet switches (layer 2 switches such as the Cisco Catalyst 5000) to achieve the desired networkwide connectivity (in addition to routers to interconnect IP subnetworks). Although these switches are fairly inexpensive (e.g., $500 per port or less), and are readily available from dozens of vendors, an enterprisewide video architecture based on such an approach can be somewhat limiting. A 100-Mbps ethernet is another approach developers are taking to address the problem: 100 Mbps could support from 15 to 60 video users at 6 or 1.5 Mbps, respectively, unless one used switched ethernet. Gigabit ethernet systems are also appearing. A high-speed switch

- Standards (H.310, late 1998)
- Compatible.with 320/323/321/3 22/324
- Video over IP over ATM
- MPOA, IP Switching, …
- RSVP to QoS Mapping
- ATM Access Technology
- ATM Service Ubiquity
- End-User's Affordability

FIGURE 4.4   ATM-based multipoint conferencing solution challenges.

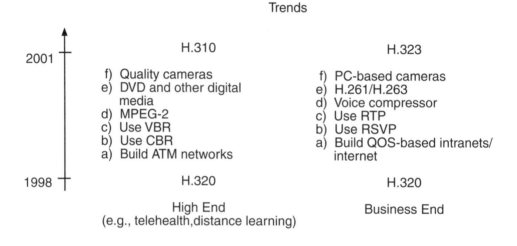

Trends

2001

H.310

f) Quality cameras
e) DVD and other digital
   media
d) MPEG-2
c) Use VBR
b) Use CBR
a) Build ATM networks

H.323

f) PC-based cameras
e) H.261/H.263
d) Voice compressor
c) Use RTP
b) Use RSVP
a) Build QOS-based intranets/
   internet

1998

H.320

High End
(e.g., telehealth,distance learning)

H.320

Business End

FIGURE 4.5   Challenges for videoconferencing.

would still be required to support segment-to-segment or segment-to-server connectivity. By contrast, ATM *dedicates* appropriate bandwidth resources to *each* user as a matter of course, up to the maximum line card speed or the speed of the switch's backplane.* The issue, however, is more obvious in the WAN: a 100-Mbps ethernet infrastructure in two buildings with an interbuilding router backbone operating at T1 rates will be totally limited by the bottleneck in bandwidth. It is imperative that bandwidth and QoS issues be addressed in LAN/WAN networks, while accounting for bandwidth demands on existing legacy applications before video-based technology can emerge.

4.1.2.2.1 Videoconferencing

Figures 4.4 and 4.5 summarize the challenges being faced to make videoconferencing in general, and multipoint videoconferencing in particular, a reality. The following elements impact on network connection options for videoconferencing services currently supported:

1. Point-to-point connection
2. Bridging (i.e., point-to-multipoint, multipoint-to-multipoint)
3. Inverse muxing (i.e., one video channel obtained via multiple network connections)

In existing networks, the provisioning of point-to-point connections is the network responsibility. Bridging and inverse muxing are roles of the user's equipment with current videoconferencing service.

---

*These observations relate to the campus/building network. This discussion does not imply that ATM to the desktop is superior to other technologies, such as 100-Mbps ethernet, only that bandwidth and QoS (e.g., frame delay, frame delay variation, frame loss) must be accounted for.

New services, particularly ATM, can be utilized to place some of this functionality in the network. The LAN (of adequate bandwidth) will become a major enabling technology for desktop videoconferencing. A growing number of products are being developed for the LAN; however, the issue about network capacity and QoS on a legacy LAN remains, as addressed earlier; in addition, concerns about the ability of routers are real.[29]

4.1.2.2.2 Multimedia

Multimedia, as a general field, is a technology that is based on the multisensory nature of people and the ability of computers to store, manipulate, and display information such as video, graphics, audio, and text. Multimedia has been enabled by the synergistic confluence of the PC, the television, and the optical file server. Broadband communications networks are another key technical driver. There are now many practical multimedia business applications, including presentation development, kiosks, computer-based training, preparation of business presentations, online magazines, computer-based training (CBT), and desktop videoconferencing.[28]

Industry observers agree that multimedia is one of the key technologies influencing how people will use computers over the next few years. Nearly all new PCs now support multimedia. Corporations are examining the possibility of putting multimedia to work for them, to support their transition to a competitive business posture in the context of the global economy of the new decade. Multimedia is not a single technology, but a class of technologies and applications that span two (voice and data) or more (voice, data, video, and graphics) media. Multimedia can operate delivering from as little as 56 kbps of information to a user to as much as 155, 622, and even higher. The 1.5 to 6.0 Mbps data rate per user is a basic range that is being designed by multimedia technology developers, for example, for educational purposes. MPEG-1 products operating at the 1 Mbps rate have seen good penetration. Table 4.3 summarizes some important multimedia milestones achieved during the past decade.

Initial multimedia applications were confined to the desktop, where all required information resides in a PC-attached CD-ROM videodisk. Recent enterprise networking history has shown that *stand-alone islands* are untenable over time, even in the case of traditional business applications. Companywide connectivity is expected in the next few years in regard to multimedia and digital video. Desktop videoconferencing, access to remote libraries of video or multimedia material, access to archived multimedia corporate records, and downloading of server-based multimedia instructions are just some examples of desirable networked multimedia applications, requiring both local- and wide-area connectivity. Beyond the desktop-based platforms, multimedia requires high-capacity digital networks to provide real-time services such as retrieval, messaging, conversation, and distribution; here too, QoS-enabled networks are needed. In fact, what has held back broadband applications in general, and multimedia in particular, has been the lack of adequate bandwidth and QoS, not only at the WAN level, but also at the local level. Communications technology of adequate capacity, QoS, and flexibility is a critical factor that will be required to enable multimedia to migrate from dedicated desktop systems to more efficient distributed systems, making more, preferably most, employees of a corporation actual users of multimedia applications. As noted, the 1990s burgeoned with new high-quality, high-speed digital services at the local- and wide-area level.

At the LAN level 100-Mbps ethernet and gigabit ethernet come into play. For WAN applications, ATM technology is the underpinning technology for high-end networked multimedia applications, while services such as ISDN and other "fast packet" services can support midrange applications. However, one may not want to run ATM natively because of the changes required, and because these changes would be required throughout many elements of the corporate network. Preferably, one wants to be able to run IP over ATM, and so only upgrade a few backbone routers to ATM. Several approaches have evolved to facilitate this hybrid network architecture.

4.1.2.3 Stored Digital Video

DVD technology is now entering the market at the commercial and consumer level. DVD is the "next generation" of optical disk storage technology. It is essentially a bigger, faster CD that can hold video as well as audio and computer data. DVD aims to encompass home entertainment, computers, and business

TABLE 4.4   DVD Features

- Over 2 hours of high-quality digital video
- Support for widescreen movies and regular or widescreen TVs (16:9 and 4:3 aspect ratios)
- Up to eight tracks of digital audio (for multiple languages), each with up to eight surround channels
- Up to 32 subtitle/karaoke tracks
- Automatic "seamless" branching of video (for multiple story lines or ratings on one disk)
- Up to nine camera angles (different viewpoints can be selected during playback)
- Menus and simple interactive features (for games, quizzes, etc.)
- "Instant" rewind and fast forward, including search to title, chapter, or track
- Durability (no wear from playing, only from physical damage)
- Not susceptible to magnetic fields, resistant to heat
- Compact size (easy to handle and store, players can be portable)

information with a single digital format, eventually replacing audio CD, videotape, laser disk, CD-ROM, and perhaps even video game cartridges. DVD has widespread support from all major electronics companies, all major computer hardware companies, and most major movie and music studios.[30] DVD disks may be used in servers or jukeboxes, and may serve corporate video applications such as training. Likely these servers will be networked.

It is important to understand the difference between DVD-video and DVD-ROM. DVD-video (often simply called DVD) holds video programs and is played in a DVD player hooked up to a TV. DVD-ROM holds computer data and is read by a DVD-ROM drive hooked up to a computer. The difference is similar to that between audio CD and CD-ROM. DVD-ROM also includes (future) variations that are recordable one time (DVD-R) or many times (DVD-RAM). DVD-video has the features shown in Table 4.4.

DVD has the capability to support near-studio-quality video and better-than-CD-quality audio. DVD is vastly superior to videotape, and can be better than laser disk. However, quality depends on many production factors. Until compression knowledge and technology improves, we may see DVD programming that is inferior to laser disks. Also, since large amounts of video have already been encoded for VideoCD using MPEG-1, some early DVDs will use that format (which is no better than VHS) instead of higher-quality MPEG-2. DVD video is compressed from studio ITU-R 601 format to MPEG-2 format. MPEG-2 is a "lossy" compression method that removes redundant information (such as sections of the picture that do not change) and information that is not readily perceptible by the human eye. The resulting video, especially when it is complex or changing quickly, may sometimes contain "artifacts" such as blockiness or fuzziness. It depends entirely on the quality of compression and how heavily the video is compressed. At average rates of 3.5 Mbps, artifacts may be occasionally noticeable. Higher rates result in higher quality, with almost no perceptible difference from the original master at rates above 6 Mbps.[30] (This topic is treated at length in Sections 4.3 and 4.4.)

Capacities of DVD-ROM and DVD-video are as follows (for reference, a CD-ROM holds about 0.64 GB — in the list below, SS/DS means single-/double-sided, SL/DL means single-/dual-layer):

- DVD-5 (12 cm, SS/SL): 4.38 GB (4.7 G) of data, over 2 hours of video
- DVD-9 (12 cm, SS/DL): 7.95 GB (8.5 G), about 4 hours
- DVD-10 (12 cm, DS/SL): 8.75 GB (9.4 G), about 4.5 hours
- DVD-18 (12 cm, DS/DL): 15.90 GB (17 G), over 8 hours

Other formats may evolve, e.g.,

- DVD-1? (8 cm, SS/SL): 1.36 (1.4 G), about 0.5 hour
- DVD-2? (8 cm, SS/DL): 2.48 GB (2.7 G), about 1.3 hours
- DVD-3? (8 cm, DS/SL): 2.72 GB (2.9 G), about 1.4 hours
- DVD-4? (8 cm, DS/DL): 4.95 GB (5.3 G), about 2.5 hours

DVD is primarily the work of Toshiba, Philips, and Sony. There were originally two next-generation standards for DVD, one backed by Sony, Philips, and others, and a competing format backed by Toshiba,

Time Warner, and others. A group of computer companies led by IBM insisted that the DVD proponents agree on a single standard. The approved DVD standard was announced in September 1995, avoiding a confusing and costly repeat of the VHS vs. BetaMax videotape battle. The DVD Licensor Consortium now comprises Hitachi, JVC, Matsushita, Mitsubishi, Philips, Pioneer, Sony, Thomson, Time Warner, and Toshiba. As this technology enters the corporate landscape, the need for networking services will increase as soon as there is a desire to store the material at a central server.

4.1.2.4 Broadcast Applications

On February 2, 1997, Tim Russert closed the NBC broadcast of *Meet the Press* with a historic statement that owed much to recent developments in both digital broadcast technology and network politics: "*Meet the Press* made television history. You didn't see or hear it because your set isn't equipped yet, but this was the first network program to be broadcast in digital high definition, or HDTV. It brings you crisp, movie-quality images and CD-quality sound, HDTV sets will be available in late 1998, and eventually this technology will be the industry standard." This test broadcast marked the implementation of television's most significant technological development since its invention. In a medium that has seen little fundamental innovation in the way its message reaches the populace, the historic *Meet the Press* broadcast suggests not only that a new technology is entering the scene, but also that broadcasters grasp its significance and its potential for improving the quality and flexibility of television services. The HDTV revolution has begun and proponents claim it will have far-reaching consequences for broadcasters, manufacturers, and consumers.[3]

With its 16:9 aspect ratio, six-channel audio, and screen sizes in the 36- to 100-in. range, HDTV will offer consumers a "cinematic" entertainment experience in their home. It is projected that, as terrestrial cable and satellite services vie for share in a dynamic and competitive marketplace, receiver prices will drop rapidly, fostering the rapid introduction of this technology. The U.S. broadcast industry has a large archive of high-definition 35mm motion picture films; these film archives have constituted the source material for about 70% of traditional television prime-time programming. This readily available programming source can be retransferred on high-definition systems that are already appearing on the market (it is expected that such retransfered material will provide a great deal of the programming content for the first few years of HDTV). This is similar to the issuing of AAD audio disks long before DVD disks became more of the norm.

The industry has waited many years for digital DTV emergence as a technologically and economically feasible, and certainly preferable, method of television production and broadcast. Packetized digital video (e.g., digitized video carried in ATM cells) is entering not only the corporate world, but the entire broadcasting business, both over the air and over cable.[14] These advances have been called "broadcasting's third revolution; the first two was radio, then television."[31] Realizing the impact that the first two events had, one can appreciate the implied impact of the third. DTV units not only can carry a mix of video, audio, and data services today, but they also provide the mechanism for adding new services in the future without obsoleting receivers already in the field. For example, during a commercial, a "browseable" brochure can also be downloaded, in the "data channel," that can be consulted via a PC to obtain information such as product specifications, available options, dealers, etc.

The video signal originating from a TV studio camera is considered a baseband component. The three color components (red, green, and blue) are distinct signals. Because early NTSC systems needed to maintain compatibility with black-and-white TV sets, the R, G, B color space in video signals was converted to the Y, U, V color space. Y is the luminous information (lightness). U is defined as $U = B - Y$, and V is defined as $V = R - Y$. Less bandwidth needs to be assigned to color difference signals (U and V) than to the luminance signal. If RGB-to-YUV color space conversion is done maintaining the full bandwidth chroma (hue plus saturation) it is called a 4:4:4 sampling. If the conversion is carried out on chroma samples every other pixel, then it is termed a 4:2:2 sampling scheme. The 4:2:2 sampling halves the chroma resolution horizontally, resulting in a 33% saving in bandwidth compared with a 4:4:4 sampling, yet with little perceptible loss in video quality. A 4:2:0 sampling reduces the chroma bandwidth even more, halving the overall bandwidth.[4] (See Figure 4.6.[32])

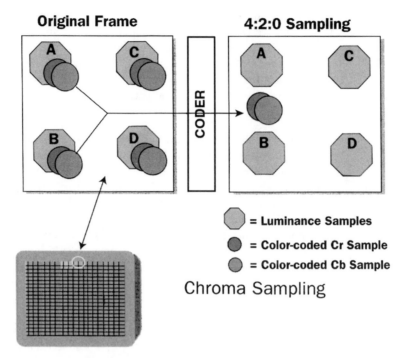

FIGURE 4.6   Chroma sampling. The amount of data encoding needed to reconstruct an accurate image, can also be reduced by various sampling techniques. One of the most efficient of these is chroma sampling. This figure demonstrates how 4:2:0 chroma sampling shares and thus reduces the amount of pixel information required per sample. The scheme takes advantage of the fact that the human eye can detect brightness (L) much better than it can detect color (Cr, Cb). The 4:2:0 sampling mode further reduces the amount of encoded information by multiplexing the Cr and Cb information into one single chrominance channel (Cr). Sampling at an even higher rate of 4:2:2 is also possible and is currently being incorporated into many of the latest codecs.

Digitization has already entered the TV studio, even before HDTV has made its presence felt. Advances in very large scale integration (VLSI) integrated circuits (ICs) have made possible a new class of studio equipment: digital equipment for NTSC video. New production switchers, routers, and tape machines all support digital component 4:2:2 video. Standards such as the ITU-R 601 support high-quality production: ITU-R 601 standard for broadcast video has an active resolution of 710 pixels (picture elements) by 485 lines, plus a 4:2:2 sampling scheme. Also, there are standards for parallel and serial video data interchange between equipment, standards like SMPTE-259D, a 360-Mbps interface between, say, digital tape machines and video routers. Although the vast majority of TV studios still use the analog NTSC equipment, many have begun making the transition to full digital facilities for producing NTSC programs.[4]

Hence, even prior to actual over-the-air digital transmission for end-to-end digital delivery, the broadcast industry is seeing the decline of the analog videotape as the medium for storing video. Videographic designers have been using disks for some time as their primary medium, with videotape reserved for output and occasional input; the trend is now expected to permeate the industry. Vendors of video servers are now emphasizing the advantages of tapeless production and tapeless distribution, including "digital ad insertion." Disk-based storage of video is convenient not only because it is digital and is randomly accessible, but because the information can be transmitted over an (internal) backbone data network, for studio workflow purposes. Digital technology is making it possible to include more functions into video editing systems now reaching the market; these editing systems are available at prices that were unthought of a few years ago.[33]

Uncompressed digital editing has been used during the past few years; but with the introduction of compressed video, the need has arisen to also deal with these newer formats. The 6-Mbps data rate of

MPEG-2 makes it easier to store, download, process, distribute, and archive this material. Video compression is entering the teleproduction industry. Applications include disk-based nonlinear editors, video file servers, and on-air automated playback systems. Soon, similar teleproduction techniques will also enter the mainstream corporation, whether in the public relations department, the training department, the library, the production floor, or the departmentalized server. Improvements in compression technology create a dynamic environment of innovation in the way a company employs video technology. MPEG encoders can now be found for desktop applications, ranging from as little as $500 to $100,000 (high-end real-time systems); many were around $5000 at this writing. We are in a transition era where disk-based compression systems are replacing videotape; the use of nonlinear editors to produce video products continues to increase.[34] The data rates for these applications range from 50 Mbps to 3 Mbps, which relate to compression rates ranging from 4.5:1 to 72:1. [Note that a 2-hour film feature requires about 5 GB.]

Now, the move to DTV will be the catalyst for a total digitization of the broadcast operation. The bottom line is that video studios will have to be redesigned. In the equipment room, a routing switcher (or router), that can have hundreds of video input and output ports, handles all the video in a TV station. Currently this is an analog matrix. At the push of a button, the routing switcher allows easy connectivity among many video cameras, tape machines, and other studio equipment. The production switcher is the main piece of equipment in the control room and is used to handle special effects, such as video fades and wipes, and inserts commercials. In a conventional television studio, NTSC signals are routed on coaxial cables from one piece of equipment to another. With the advent of DTV, the switcher will likely have to be a bona fide multigigabit digital ATM switch.

Digital HDTV and digital SDTV are now supported by industry-developed standards. The standard that is at the base of the end-to-end digitization of TV was developed by the Advanced Television Systems Committee (ATSC)* and "adopted" by the FCC. However, when the FCC ruled on the specifics of the digital transmission parameters for DTV, they left the choice of video "payload" (mix of channels, as well as resolution) up to the broadcasters. The FCC has aimed at leaving choices but no ambiguities in the definition of the digital terrestrial "channel" to be used by broadcasters.[3] Furthermore, SMPTE has delineated for broadcasters, producers, and manufacturers all of the specific parameters for HDTV and SDTV equipment. The NTSC standard defines a video frame as containing a total of 525 interlaced lines, such that all the odd lines are scanned before all the even lines at about a 30-Hz frame rate. Television studio equipment relies on this fixed frame structure for timing and synchronization. The new ATSC standard mandates compressing the video and audio signals as well as using packetized transport for video, audio, and data packets. Clearly, the transition from analog NTSC to the compressed digital ATSC high-definition standard will completely transform how a television studio routes, stores, processes, and transmits the new television signal.[4]

The new ATSC standard** defines four digital television formats; this standard encompass both digital HDTV and SDTV. These formats are described by the number of pixels per line, the number of lines per video frame, the frame repetition rate, the frame structure (interlaced or progressive), and the aspect (width-to-length) ratio, as follows:

1. 1920 × 1080 pixels, 60 interlaced frames per second (also 30 or 24 progressive), 16:9 aspect
2. 1280 × 720 pixels, 60 or 30 or 24 progressive frames per second; 16:9 aspect
3. 704 × 480 pixels, 60 or 30 or 24 progressive frames per second (also 60 interlaced); 16:9 aspect (also 4:3)
4. 640 × 480 pixels, 60 or 30 or 24 progressive frames per second (also 60 interlaced); 4:3 aspect

On the matter of scanning, the ATSC standard includes both the interlaced and the progressive scanned video formats. Interlacing is a technique the camera uses to take two snapshots of a scene within a frame time. During the first scan, it creates one field of video, containing even-numbered lines, and during the

---

*This is a cooperative committee of manufacturers and broadcasters to promote the DTV standard and certify the new equipment.

**See http://www.atsc.org for ATSC Standard Documents A/53 for video and A/52 for audio.

second, it creates another, containing the odd-numbered lines. This technique, which is used in NTSC video, is used to reduce flicker and to provide higher brightness on the TV receiver for a specified frame rate (and bandwidth). Computer-generated video is generally scanned in progressive manner, where one frame of video contains all the lines in their proper order. Equipment providers from the computer side prefer the progressive mode, but would like a lower frame rate. Equipment providers from the TV sets side call for the inclusion of multiple formats, because the use of interlaced formats to be initially more common.

The ATSC DTV transmission standard (ATSC Doc. A/53) specifies the parameters for digital HDTV and SDTV video and audio formats. The ATSC DTV standard defines a transmission system flexible enough to transmit HDTV and/or multiplexed SDTV at different times during the programming day. At the highest resolution level (necessary for true HDTV), the ATSC/SMPTE standards requires a resolution of 1920 (horizontal) × 1080 (vertical, or number of scanning lines) for digital sampling structure. This format can be transmitted digitally at 60 pictures/second interlaced (normal "live" television for sports, etc.) or, alternatively, at 24 or 30 frames per second progressive scan (for 24 fps film-originated material).[3] The same DTV transmission system can support a number of digital 525-line multiplexed SDTV channels (in standard 4:3 aspect ratio or the new 16:9 widescreen aspect ratio). Within the 19.4 Mbps total video "payload" allowed by the DTV standard, broadcasters are free to determine the number of SDTV channels they will offer, by selecting their own individual balance between picture quality and channel data rate.

In fact, as noted above, the standard includes the two HDTV formats: the 1920-pixel-by-1080-line interlaced and the 1280-pixel-by-720-line progressive scan format. Most of the HDTV broadcast equipment now available or emerging is designed for the 1920 × 1080 interlaced video format only (progressive-scanning HDTV cameras and monitors are not yet available); 24 fps film translated to video have a progressive format. The progressively scanned 704-pixel-by-480-line video with 16:9 aspect ratio will probably see some penetration, since some equipment providers (e.g., Panasonic) have announced a line of production, storage, and display products supporting this format. Compared with the NTSC standard, it has a wider picture and eliminates artifacts affecting interlaced video (such as the line crawl that affects some scenes containing slow vertical motion). The ATSC standard also supports SDTV formats (interlaced and 704-pixel-by-480-line or 640-pixel-by-480-line). Given the fact that most of the existing studios support one or the other of these two formats, most local production will probably be of this kind.

The bit stream being transmitted may change according to the nature of the program, e.g., news, followed by an archive segment, followed by a commercial. The ATSC standard recommends that the receiver "seamlessly," and without loss of video, continue to display all these formats in the native format of the television receiver.

The ATSC standard specifies MPEG-2 as the video compression standard. In an end-to-end context, note that the ATSC standard allows controlled coding/decoding because if the video is coded (compressed) and decoded more than a handful of times, the picture quality rapidly degrades, unless a hierarchical (tiered) compression approach is utilized. For example, for video production, the compression can be 4:1 and the coding format be I-frames (intraframes) only.[*] For contribution video, the compression can be 10:1 and use a IPPP coding structure (P being predicted frames). For storage, the compression can be 25:1 with IPIP coding structure. For transmission, the compression can be 50:1 with IPBBBBP coding structure. Figure 4.7 depicts the various bit rates that are achievable with mixed frames. (B is a bidirectionally interpolated frame).

Since the December 1996 FCC ruling on the next generation of DTV broadcast standards, significant strides have been made on three fronts:

1. Consumer Hardware: Various trade shows in recent seasons have seen announcements from many manufacturers of their plans to begin production on widescreen HDTV sets.

---

[*]These terms are discussed in Sections 4.3 and 4.4.

Video Latency and Bit Rate

Mode	Video Latency	Video Bit Rate	
I	Low	8 Mbps	(High)
IP	Low	4.3 Mbps	(Medium)
IBP	Medium	3.1 Mbps	(Medium)
IBBP	Medium	2.7 Mbps	(Low)

FIGURE 4.7 Bit rates achievable with mixed frames.

2. Programming: Several television networks have already publicly expressed their commitment to debut HDTV programming, and several television stations have already begun broadcasting test transmissions of HDTV.

3. Production Hardware: Various trade shows in recent seasons have seen the launch of digital widescreen HDTV acquisition, post-production, and transmission systems for in-the-field production, from major manufacturers. These systems are not just still-on-the-drawing-board prototypes, but will consist of working models, many to be delivered as this book goes to press.

While the advent of DTV is a given, it is important at this time that broadcasters and manufacturers work together and take the steps necessary to ensure that the full promise of digital HDTV is realized. An important first step for broadcasters to undertake will be the dual transition from composite analog to direct-digital component program origination, and from the standard 4:3 aspect ratio to the new 16:9 widescreen format. At this time, direct-digital 4:2:2 SDTV origination offers good video quality with high signal-to-noise ratios. The quality of these programs will be fully appreciated when the signals are digitally manipulated (compressed) for transmission over the DTV channels in the near future; their quality will be appreciated even more when the upconversions to the digital HDTV are required for the next decade or so.

Effects such as instantaneous cut-away, edits, etc. need to be supported. With current production equipment these studio effects can readily be achieved. Production equipment for compressed video also needs to support these capabilities. An intracoded system can do this, since all intracoded frames are independent I-frames, so that altering (or deleting) one frame does not affect others frames in the bit stream. Things are more complex for intercoded systems.

Fortunately, in the past few years video equipment manufacturers have opted to keep expanding the envelope rather than curtail their efforts in light of what the first HDTV incarnation would actually be. For example, Sony and other manufacturers were steadily developing sequential generations of HDTV program origination products, despite the small market of digital broadcasters and program producers. In one example, digital SDTV, Sony has continued its support of the international digital 4:2:2 component video standard that is emerging as the preferred basis for digital 525-line program creation and the associated MPEG-2 digital 4:2:0 transmission format (that is now formally a part of the ATSC DTV transmission standard). A 4:2:2 digital component widescreen camcorder has emerged as an early-entrant system for field acquisition of digital high-end SDTV program material. Sony's subsequent development of an entire digital broadcast product line based upon the recently standardized MPEG 4:2:2 Profile at Main Level brought in a secondary digital "layer" that can constitute the "workhorse" heart of an entire digital broadcasting operation (e.g., Betacam SX is the 4:2:2 recording format associated with that system, and it ushers in a new era of high-performance and highly cost-effective camcorders for digital 4:2:2 applications). The 4:2:2 platform is now supported in anticipation of the crucial role of SDTV within the overall DTV dynamic.[3] A new generation of HDTV studio cameras, an all-digital HDTV camcorder, digital HDTV switchers and multieffects system, and studio monitors have been unveiled recently.

If the bit rate used within the studio is in the range of 200 to 270 Mbps, then the compressed data can be stored in and routed by uncompressed standard definition equipment, that complies with SMPTE 259. Both Sony and Panasonic have proposed systems allowing D1 and D5 tape equipment, currently in wide use, to be used to store compressed data. The 270-Mbps version of SMPTE 259 can be used to store and route up to 200-Mbps data, and the 360-Mbps version can be used for bit rates up to 270 Mbps. But as more equipment supporting HDTV pictures becomes available, digital SDTV will give way to HDTV programming.[4]

The new TV studio must support HDTV as well as SDTV/NTSC equipment. It also must allow compressed operations (like storage and splicing). A high-speed ATM network will route the compressed bit stream and other data around the studio. There will also be transcoders to convert one compressed format into another. In the early stages, video production will be preformed on uncompressed video, but as compressed technology advances, more production will be done on compressed video (splicing, editing, and effects are more complicated in compressed video environments).

Broadcasters are seizing the initiative at this time, in preparing for DTV. CBS station WRAL-TV in North Carolina (officially the first station in the U.S. to transmit digitally), along with five other stations, have been licensed by the FCC to begin experimental digital HDTV broadcasting. Many more are licensed by the FCC to begin experimental digital HDTV broadcasting, and many more have license applications under review. An industry initiative, the *HDTV Model Station* (formally licensed as WHD-TV, and housed in the host station WRC-TV in Washington, D.C.) has created a working laboratory for HDTV equipment. The HDTV Model Station is serving as a proving ground where manufacturers and broadcasters can address technical issues associated with implementation of a simulcast digital DTV/analog NTSC operation. Also, five member stations of the Public Broadcasting Service have formed the Broadcasting Digital Alliance to address the technical and programming issues of DTV cooperatively. To advance the cause of DTV broadcasting, the Public Broadcasting Service has also recently arranged for satellite feeds of the compressed ATSC DTV signal format to allow stations and laboratories (involved in DTV receiver development) to receive this signal. There will also be new operator requirements. For example, there is the matter of reeducating camera persons, directors, set designers, video operators, and post-production specialists in the ways of widescreen and wide-angle origination production. As with any technological innovation, there will be a considerable learning curve.[3]

The FCC approval of the DTV standard means that the timetable for broadcasters to begin the transition to digital television has been initialized. The rollout of DTV will begin in the larger U.S. markets. The FCC has assigned additional channels to broadcasters for digital transmission and has mandated a rapid build-out plan. The four top network affiliates in each of the 30 top markets were to be on the air in November 1998, with all the commercial stations on the air in 2002. At this time the FCC also calls for the return of a second (NTSC) channel by the year 2006. These early services will be a combination of SDTV and HDTV.

4.2   ITU-T H.320/H.323

As noted in Section 4.1, to achieve ubiquity, there is a need for standardization. Today one can call anyone in the world, dial into any modem, or use any Ethernet Network Interface Card and have the desired connectivity. The same is needed in regard to videoconferencing. Fortunately, standards have evolved over the years; unfortunately these standards have been adopted only slowly.

This section addresses the technologies that are evolving to support desktop business video. Codecs receive considerable attention, because they are the basis for the entire video application. Codecs must be compatible (i.e., conforming to industry standards, e.g., ITU-T H.320); be relatively inexpensive (e.g., a $100 chip set); and support reasonable quality.

4.2.1 H.320

ITU-T H.320 is a key group of standards that was developed for traditional videoconferencing applications. ITU-T H.320 is the basis for the majority of standardized "low-bit-rate" videoconferencing systems in use today. H.320 enables a variety of applications such as telehealth, distance learning, and travel-free conferencing. This standard supports data rates from 64 Kbps to 2.048 Mbps. It is intended for traditional "constant-bit-rate" communications channels such as DS1/T1, E1, ISDN primary rate interface, and single or multiple basic rate interface. Figure 4.8 depicts the H.320/1/2 reference model, while Figure 4.9 describes the features of these standards.

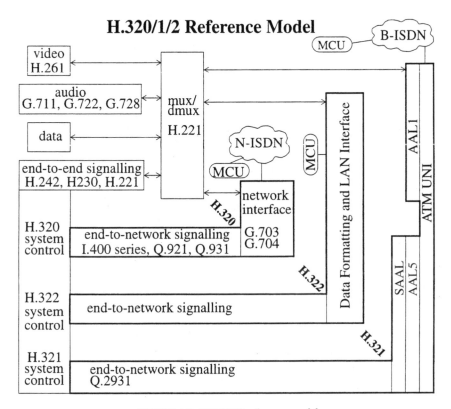

FIGURE 4.8 H.320/1/2 reference model.

- H.3201112 are narrowband visual telephone terminals specifying the same:
  - H.261 video compression
  - G.711, G.722, and 6.728 audio compression
  - H.221, H.230, and H.242 end-to-end signaling including start/stop audio/video, terminal capability exchange, etc.
  - H.221 media multiplexing which overlays on the opened end-to-end channel
- H.320 operates over an N-ISDN network:
  - Px64-kbps channel structure such as BRI, PRI, and H_0
  - Q.931 for end-to-network signaling over the D channel of BRI, PR', H_0
- H.321 operates over a B-ISDN network:
  - ATM VC emulation of Px64-kbps channel.
  - Q.2931 for end-to-network signaling over SAAL/AAL5
- H.322 operates over a guaranteed QoS LAN, e.g., isochronous ethernet, IEEE802.9a:
  - Single LAN segment, no bridges or routers which are not suitable for ensuring QoS
  - H.322 terminals on different segments communicate via N-ISDN
  - End-to-network signaling

FIGURE 4.9 Features of H.320/1/2 standards.

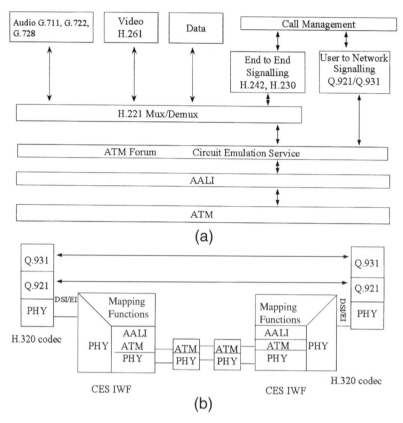

FIGURE 4.10 (a) H.320 protocol reference model when using ATM transport, (b) H.320 signaling reference model when using ATM.

H.320 utilizes the H.261 video encoding specification, which in turn uses the px64 codec. Adoption of standards ensures compatibility. Some vendors have introduced proprietary systems that claim to be "better" (but in fact are not a significant improvement). H.320 operates up to 1.544 Mbps (for U.S. implementations) per user (for comparison, Motion JPEG can go as high as 20 to 45 Mbps per user, and offer significant quality improvements). Field introduction of H.32, however, has been slower than initially expected. Industry groups such Personal Conferencing Workgroup (PCWG) and International Multimedia Teleconferencing Consortium (IMTC) have addressed the interworking details of H.320 and T.120 (whiteboards). A specification for LAN communication support of H.320, specifically H.323, has emerged, addressing the interoperability question. (H.323 is discussed below.)

A number of vendors have looked at the use of H.320 over emulated channels delivered via ATM. ATM offers Circuit Emulation Service, where if the user employs ATM Adaptation Layer 1, the network provides a high-quality cell transfer (low loss and low jitter) to emulate a DS1/T1 channel (in whole or slotted in 64 kbps channels). Figure 4.10 shows the signaling reference model that would be employed in this case.

H.321 adapts H.320 to ATM, but it retains all other capabilities of H.320. Figure 4.11 depicts the H.321 reference model.

4.2.2   H.323

The relatively new H.323 standard provides a foundation for audio, video, and data communications across LAN and IP networks, including the Internet. Specifically, H.323 describes terminals, equipment, and services for multimedia communication over LAN and IP networks that do not provide a guaranteed

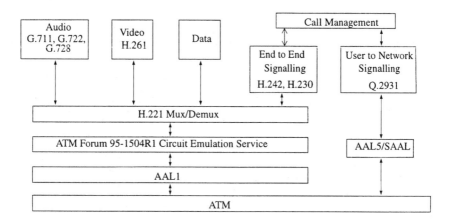

FIGURE 4.11 H.321 protocol reference model.

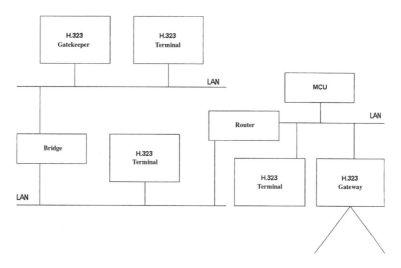

FIGURE 4.12 H.323 environment for video services.

quality of service. H.323 terminals and equipment may carry real-time voice, data, and video, or any combination, including videotelephony. The LAN over which H.323 terminals communicate may be a single segment, or it may be multiple segments with complex intranet topologies.

Figure 4.12 shows the H.323 environment for video services being deployed in the next couple of years. Figure 4.13 identifies some of the key highlights of the environment. Figure 4.14 depicts the H.323 reference model, and Figure 4.15 lists aspects of the reference model.

It should be noted that operation of H.323 terminals over the multiple LAN segments (including the Internet) may actually result in poor performance since the possible means by which QoS might be assured on such types of LANs/internetworks is beyond the scope of the recommendation. H.323 terminals may be integrated into personal computers or implemented in stand-alone devices such as videotelephones. Support for voice is mandatory in the standard, while data and video are optional, but if supported, the ability to use a specified common mode of operation is required, so that all terminals supporting that media type can interwork. Other components in the H.323 series include H.225.0 packet and synchronization, H.245 control, H.261 and H.263 video codecs, G.711, G.722, G.728, G.729, and G.723 audio/speech codecs, and the T.120 series of multimedia communications protocols.

H.323 terminals may be used in multipoint configurations, and may interwork with H.310 terminals on B-ISDN, H.320 terminals on N-ISDN, H.321 terminals on B-ISDN, H.322 terminals on guaranteed

- H.323 operates over a nonguaranteed QoS LAN consisting of a single segment, bridged segments, and/or routed segments
- H.323/H.255.0 extends to H.320IH221 conference connections onto the nonguaranteed QoS LAN environment
- The H.323 protocol stacks are above the transport layer providing such services as reliable and unreliable delivery
- H.323 gateway:
  - H.225.0 ↔ H.221
  - H.245 ↔↔ H.242
  - H.225.0 ↔ Q.931 I Q.2931
  - H.323 terminal video, audio, data transmission format ↔ other terminal video, audio, data transmission format
  - Call setup and clearing on LAN and on other networks
- H.323 gatekeeper:
  - A gatekeeper is optional
  - H.323 terminal registration
  - Admission control, based on bandwidth, authorization, etc.
  - Alias (E. 164 or e-mail style name) to transport address resolution
  - Locating gateway
  - Zone management
  - Call control signaling between H.323 terminals
  - Call authorization
  - Bandwidth reservation
- H.323 terminals:
  - Video compression: H.261 and H.263 (low bit rate)
  - Audio compression: 6711, 6.722, G.723, 6.728, and 6.729
  - Data: T. 120 series

FIGURE 4.13 Highlights at the H.323 environment.

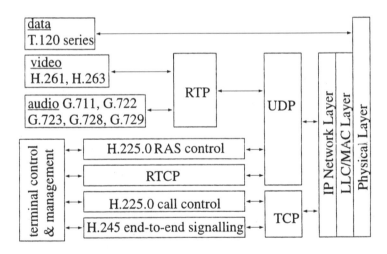

FIGURE 4.14 H.323 reference model.

QoS LANs (e.g., IEEE 802.9), H.324 terminals on the Public Switched Telephone Network (PSTN) and wireless networks (see next section), and V.70 terminals on PSTN.

As a brief history, H.323 work started in May 1995. The standard was adopted by the ITU-T in 1996. Currently, the ITU-T is working on version 2, which supports additional functionality. Completion was expected by the end of 1998 (according to some, ITU-T is now learning to work on "Internet time," that is, to get things out quickly). H.323 components include:

- Terminals
- Gatekeepers
- Gateways (H.323 to H.320/H.324/POTS)

- Transport Address:
  - Identify H.323 terminal, gateway, and gatekeeper
  - An example of IP network transport address: IP address + UDP or TCP port
- H.225.0:
  - Communication protocol between H.323 terminals, gateways, and gatekeeper
  - Based on Q.931 messages of specific information elements
  - RAS (registration, admission, and status) control for an H.323 terminal to signal the gatekeeper
  - Call control for end-to-end call setup between H.323 terminals
  - Option call control via gatekeeper
  - Call setup with other terminal types via an H.323 gateway
- H.245:
  - End-to-end signaling after the end-to-end channel is established by H.225.0
  - Also used in H.310 and H.324
- Real-Time Transport Protocol (RTP)
  - Exchanges audio, video data over unreliable service
  - Carries real-time information such as synchronization information
  - Multiplexes audio and video streams using different transport layer ports (in contrast the H.32011/2 H.221 and the H.324 H.223 use a single stream of media multiplexing)
  - Monitored by an associated real-time control protocol (RTCP)

FIGURE 4.15 Aspects of the reference model.

- MCUs
- Multipoint Controller (MC)
- Multipoint Processor (MP)

The H.323 terminal encompasses two versions:

- Corporate Network (high quality)
- Internet (optimized for low-bandwidth 28.8/33.6 — G.723.1 and H.263)

The terminal may have a built-in multipoint capability for *ad hoc* conferences. Multicast capabilities (specifically, multi-unicast) allow a few people in a call without centralized mixing or switching.

The H.323 gatekeeper supports the following functions:

Address translation
- H.323 alias to transport (IP) address based on terminal registration
- "E-mail-like" names can be supported
- "Phone number-like" names can be supported

Admission control
- Permission to complete call
- Can apply bandwidth limits
- Method to control LAN traffic

Management of gateway
- H.320, H.324, POTS, etc.

Call signaling
- Able to route calls in order to provide supplementary services or to provide multipoint controller functionality

Call management/reporting/logging

An industry migration to H.323 is not going to happen overnight; hence there is a need for interworking between the two systems. Figure 4.16 shows an H.323 gateway. Figure 4.17 lists features of the gateway. H.323 gateways provide global (worldwide) connectivity and interoperability from LAN. For example, interoperability between H.323 and H.320, as well as H.323 to H.324 may be supported. Also, it maps call setup signaling (Q.931 to H.225.0), control (H.242/H.243 to H.245), and media (FEC, multiplex,

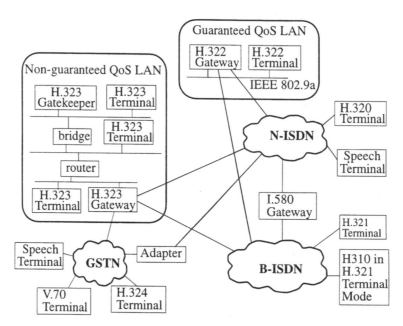

FIGURE 4.16 H.323 gateway.

- H.322 gateway enable interworking of H.323 terminals with
  - H.320 terminals on N-ISDN
  - H.321 terminals and H.310 terminals in H.321 mode, on B-ISDN
  - H.322 terminals on guaranteed QoS LAN via N-ISDN (not via B-ISDN or H.32 gateway)
  - H.324 and V.70 on GSTN (general switched telephone network) but not mentioned in H.324
- H.322 gateway enables interworking of H.322 terminals with
  - H.320 terminals on N-ISDN
  - H.321 terminals and H.310 terminals in H.321 mode, on B-ISDN
  - H.323 terminals on nonguaranteed QoS LAN via N-ISDN
  - Not (yet?) with GSTN terminals
- Interworking of H.324 and H.320
  - Use an interworking adapter between ISDN and GSTN signals; use dual mode (1SDN and GSTN) terminals on the ISDN
  - H.320 on N-ISDN and H.321 on B-ISDN terminals communicate via an AAL1 based L.580 gateway

FIGURE 4.17 Features of the H.323 gateway.

rate matching, audio transcoding, T.123 translation). Related to multipoint functionality, one has the following.

- MC: Multipoint controller portion of a traditional MCU; manages common modes, capabilities.
- MP: Multipoint processor portion of traditional MCU; mixes or switches audio. It is not necessarily co-resident with MC (e.g., MC running multicast conference with each terminal mixing audio).

A traditional MCU supports the following functions (which are not necessarily unique to H.323)[35]:

Media distribution
- Unicast: send media to one terminal (centralized in MP; traditional model)
- Multicast: send to each receiver directly
- Hybrid: some of each

Manage *ad hoc* multipoint calls
- Join, invite, control of conference modes

Traditional MCU applications

Multiprotocol through the utilization of gateways

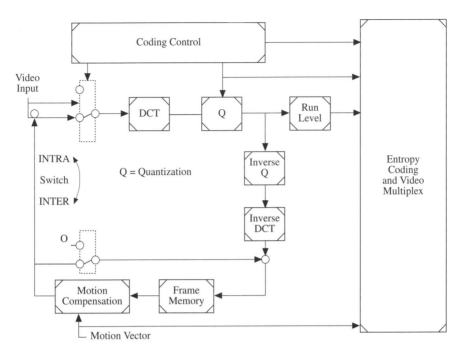

FIGURE 4.18 Approach of the H.263 coder.

The H.263[*] video codec ("Version 2, Video Coding for Low Bit Rate Communication," ITU-T H.263, January 1998) that is used in H.323 has been developed as an improvement over H. 261, which is today's standard video codec for videoconferencing on the ISDN. H.263 can operate at a range of bit rates, but is particularly applicable at rates below 32 kbps where reasonable quality is still possible for video containing limited motion. QCIF (Quarter Common Intermediate Format) resolution (176×144 pixels) is the most common format for these rates. The approach of the H.263 codec is shown in Figure 4.18. The video input is a sequence of digitized pictures (frames) at a rate of 30 frames/s. At QCIF resolution, each picture is divided into 11×9 macroblocks (MB). The MBs consist of 16×16 pixels, that are further subdivided into four 8×8 blocks. The MBs are processed using the apparatus of Figure 4.18. In general, when there is significant activity in one part of the image, the macroblocks corresponding to this area generate more bits than other parts of the image.

Two modes can be selected for each MB, implemented by the switch S shown in the figure: INTRA and INTER (the selection of the path is beyond the scope of the recommendation and is a local implementation issue). In the INTRA mode, no temporal dependency from the previous frames is postulated. The picture is directly coded using the a discrete cosine transform (DCT; the operation of the DTC is discussed in Section 4.3). In the INTER mode, the MB is predicted from the previously reconstructed frame using compensation. The output from the frame memory is identical to the decoded frames at the decoder for error-free transmission. Hence, the same prediction can be formed at the encoder and decoder. Motion compensation implies that the current MB is predicted by a 16×16 block in the previous frame, that is spatially shifted in accordance with a motion vector. In the ideal case, the remaining predication error is negligible, and no more information need be transmitted. In general, however, the remaining prediction error is encoded using a DCT for each 8×8 block. The transform coefficients are

---

[*]Shortly after the original version 1 of H.263 was formally adopted by the ITU-T in March 1996, the ITU-T launched a new effort to further improve H.263 without changing the basic concept of block-based motion compensation. This effort resulted in version 2 of H.263 video coding standard, which is informally known as H.263+ and was adopted by the ITU-T in February 1998. It contains a number of optional feature enhancements which are added to the already existing options in an upward-compatible way.[2]

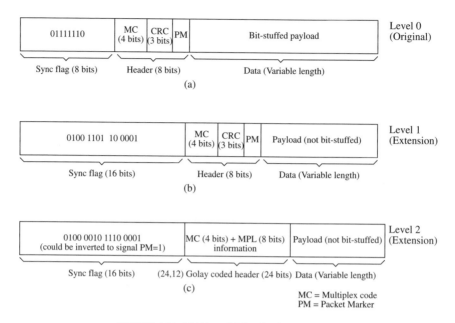

FIGURE 4.19 H.223 multiplex (with extensions).

quantized and encoded as a series of zero runs and quantizer levels. Then, run-level pairs, motion vectors, and mode information are entropy-coded along with the other side information, resulting in variable-length code words that are multiplexed to the video bit stream.[2] Naturally, the INTER mode is preferred if possible, because of its efficiency.

The H.223 multiplex protocol interleaves video, audio, data, and control streams into a combined bit stream, supporting dynamic allocation of bandwidth to the individual channels. Video typically requires the largest portion of the total bit rate. The protocol consists of a lower multiplex layer, that actually mixes the different media streams, and a set of adaptation layers, that perform logical framing, sequence numbering, error detection, and error correction by retransmission as appropriate to each media type.[2] Hence, H.223 is a connection-oriented multiplexer that is designed to mix any number of channels on a circuit-switched network (for low bit rates, however, generally no more than one video, audio, and data channel is used in each direction). H.223 is byte oriented and provides very low multiplex delay.

New capabilities are being added to this protocol. As it existed originally, H.223 was not robust in dealing with transmission errors since it was designed to work with V.34 modems on channels providing low error rates; hence, H.223 was not suitable for wireless applications without some extensions. These extensions have now been included in annexes to the original recommendation. The additions allow a hierarchical, multilevel multiplexing mechanism that allows engineering trade-offs between overhead, complexity, and robustness.

Figure 4.19 depicts the multiplexing structure.[2] The original, default, and simplest level is Level 0. In this V.34-based version packets are variable length and delimited by an 8-bit header containing a multiplex code (MC) identifying the content of the packet. What follows is the payload, which in general can consist of a mix of bits from various sources. The end boundary of the packet is determined by the next appearance of the 8-bit synchronization flag. (Bit stuffing is performed on all data between synchronization flags is to avoid flag misinterpretation.) The major vulnerabilities of H.223 Level 0 arise from the bit stuffing, and from the short and consequently vulnerable synchronization flags and headers. In Level 1 bit stuffing is not performed, and a longer synchronization flag is used. The flag can be simulated by the data, but such simulations are considered not to be problematic. In Level 2, additional robustness is provided by lengthening and adding error protection to the header that describes the contents of the packets. There is also a Level 3; it uses Forward Error Correction (FEC). The packet marker (PM) is a 1-bit field is provided that identifies the length of the payload in bytes. This provides additional redundancy by informing the receiver where the next sync flag can be expected.

There are many control messages that a terminal has to support. Those messages are defined in the H.245 multimedia system control protocol. The control messages are transmitted over a reliable data link layer with error recovery. Recovery via retransmission is accomplished using either the V.42 Link Access Procedure for Modems (LAPM) or Simplified Retransmission Protocol (SRP).

As noted in the introduction of this section, there is interest in supporting wireless multimedia. Indeed, with relatively minor extensions, specifically an error-tracking approach adopted by the ITU-T for H.263 version 2 and with minor extensions of H.245 to include an additional control message video, one now has a multimedia capability for wireless environments.

In 1995 the ITU-T standardized the Algebraic Code Excited Linear Prediction (ACELP) voice algorithms for the coding of speech signals in wide-area networks; ACELP is used for compression rates at or below 16 kbps. ITU-T G.729 (Conjugate Structure ACELP, CS-ACELP) is now an international standard that compresses 64-kbps PCM streams as used in typical voice transmission to as low as 8 kbps. ITU-T G.728 (Low Delay CELP, LD-CELP) is an international standard used to compress voice to 16 kbps. ITU-T G.723.1 compresses voice to rates as low as 5.3 kbps (it also operates at 6.3 kbps). This standard is used in the H.323 recommendation for conferencing over LANs. G.723.1 is considered a good first step, and is best suited for intranets and controlled point-to-point IP-based connections. G.729A, a simplified version of G.729 operates at 8 kbps, and therefore is slightly better in quality than G.723.1.

4.2.3 H.324

ITU-T H.324 is a recently developed recommendation for terminals for low-bit-rate multimedia communication, that may consist of real-time voice, data, and video, or any combination, including videotelephony.[2] The standard makes the assumption that the transmission is based on V.34 modems operating over the widely available general switched telephone network (GSTN) (also known as the PSTN in North America). Developers hold the position that H.324 terminals are likely to play a major role in future multimedia applications. In fact, a gamut of H.324-enabled products has already been developed and is being sold in rapidly increasing numbers. Figure 4.20 illustrates the H.324 model.

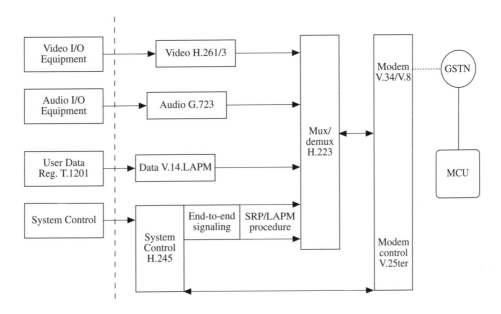

FIGURE 4.20 H.324 model.

In model, only the modem (V.34), multiplex (H.223*), and control protocol (H.245) are mandatory;** the data, video, and audio streams are optional. By design, even the most basic H.324 terminals can interwork with more advanced terminals. During the setup of the session, the terminals negotiate a common set of capabilities used for the connection.

There is interest in using a coding scheme for mobile/wireless applications that has the characteristic of being based on some existing method (to reduce both the standardization interval and the product-development interval), and of requiring low bit rate. H.324 is a good candidate. As noted, H.324 is a complete multimedia terminal operating with V.34/V.8 modems at 28.8 kbps. The system can operate as low as 10 kbps and as high as a few hundred kbps. Hence, it is a good candidate for mobile (wireless) applications (wireless applications are at the low end of the bit rate range). However, a number of extensions are needed to deal with the increased error rate of wireless/mobile channels. In 1994 the ITU-T started an Ad Hoc Group (AHG) to investigate the use of H.324 in mobile environments. This group, which is now part of ITU-T Study Group 16, Question 11, also handles work on error-resilience video coding (the error issue is now embodied in ITU-T Study Group 16, Question 15).

H.324 employs the V.34 modem to transmit (and receive) the bit stream generated by the H.223 multiplex. No data compression or retransmission capabilities are supported. This approach simplifies replacement of the V.34 modem with other connection-oriented transmission systems; in particular, a "wireless interface" with a range of bit rates can be used in place of V.34. This, however, requires certain changes in the setup procedures (for example, for the establishment of a connection). The changes necessary to replace the modem with a wireless interface are identified in Annex C of the H.324 recommendation.

4.2.4   H.310

H.310 is an MPEG-2 audio and video service. Figures 4.21 and 4.22 provide information on H.310. H.310 defines broadband audiovisual (MPEG-2) terminals as:

- Send-Only-Terminals supporting AAL1, AAL5, and AAL1 and 5 (known, respectively, as SOT-1, SOT-5, and SOT-1&5)
- Receive-Only-Terminals supporting AAL1, AAL5, and AAL1 and 5 (known, respectively, as ROT-1, ROT-5, and ROT-1&5)
- Receive-and-Send-Terminals supporting AAL1, AAL5, and AAL1 and 5 (known, respectively, as RAST-1, RAST-5, and RAST-1&5)

For example, ROT-5 terminals can operate in the native communications mode that consists of H.222.1 with 11172-3 Layer II, H.262, H.245 as the audio, video, and end-to-end control protocols, respectively. RAST-5 terminals can operate in the native communications mode that consists of H.222.1 with G.711, H.262, H.245 as the audio, video, and end-to-end control protocols, respectively. H.310 requires RAST terminals to interoperate with H.320/321 terminals. However, RAST-5 needs a gateway for AAL-1 adaptation, H.221 multiplex, and H.242 control message generation in the user's network in order to interwork with these H.320/321 terminals.

To support the conferencing application, the service uses a full MPEG-2 encoder/decoder pair, including stereo audio. The service is optimized for interactive applications where excessive delay is a consideration. Delay is reduced by encoding I-frames only, or I- and P-frames (these are discussed in the Section 4.3). As a result of reduced coding complexity, the operating bandwidth will be higher than encoding I-, B-, and P-frames. The user may be able to have some control over the bandwidth/delay trade-off for various applications and/or bandwidth.

---

*H.223 is the "Multiplexing Protocol for Low Bit Rate Multimedia Communication," adopted in 1996. H.245 is the "Control Protocol for Multimedia Communication" also adopted in 1996.

**Although the recommendation specifies the use of a V.34 modem, extension to other transmission systems is doable.

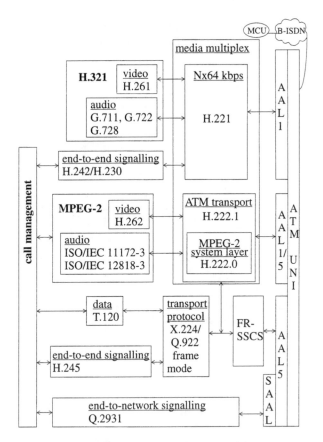

FIGURE 4.21 H.310 reference model.

- H.310, in native mode, is a broadband MPEG-2 terminal operating over a 13-ISDN network:
  - MPEG-2 video compression
  - MPEG-2 audio compression
  - T. 120 series multimedia data
  - H.245 end-to-end signaling
  - Frame mode X.224/Q.922 transport protocol for T.120 and H.245
  - H.222.0: packetize element streams into program or transport streams, clock recovery, etc.
  - H.222.1: transport stream over ATM
  - Q.2931 end-to-network signaling
  - RGT, SOT, RAST terminals over AAL1 and AAL5
- H.310, in H.321 mode, is a narrowband visual telephone terminal operates over a B-ISDN network:
  - H.261 video compression
  - G.711, G.722, and G.728 audio compression
  - H.242/H.230 end-to-end signaling
  - H.221 media multiplexing
  - Q.2931 end-to-network signaling

FIGURE 4.22 Features of the H.310 reference model.

Figure 4.23 shows some early work by the ATM Forum on video distribution. Figure 4.24 offers other details. Signaling is an important aspect of videoconferencing and other services. Figure 4.25 illustrates some of the signaling requirements. The Video-On-Demand (VOD) service defined in af-saa-0049.000 uses a full MPEG-2 decoder including stereo audio, working from a video server. This MPEG-2 decoder needs to interwork with the MPEG-2 encoder functions, up to the full bandwidth using the IBP encoding

Video on Demand
VoD Client H.310/ROT-5 Model

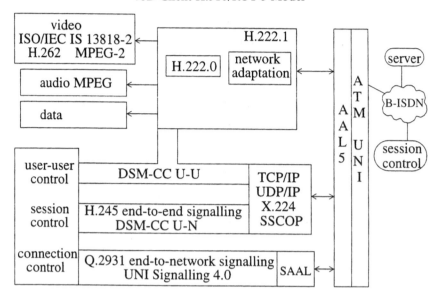

FIGURE 4.23 ATMF models.

- ATM Forum SAA/AMS Phase 1 spec concerning MPEG-2 CBR
- Client is either a PC or set-top terminal based on H.310/ ROT-5 model
- Server is a video server based on H.310/SOT-5 model
- User-to-user control: DSM-CC U-U
- Connection control is either proxy or first-party signaling based on UNI Signaling 4.0
- Network adaptation functions:
  - SPTS (Single Program Transport Stream — one common time base)
  - Default two TS packets encapsulated in eight cells AAL5 CPCSPDU
  - Selectable number of TS packets per PDU in SVC, by signaling

---

FIGURE 4.24 ATMF issues on VOD.

mechanism. The service is targeted at applications that are not fully interactive and can accept coding delays. The objective is to achieve maximum possible video quality at a given bit rate at the expense of delay.

4.3 Basic Compression Concepts and the MPEG Family

4.3.1 Digital Video Compression Overview

Compression algorithms are critical to the viability of digital video, digital video distribution, video-on-demand, multimedia and other video services. This section provides an overview of digital video compression as well as a synopsis of some MPEG-1, MPEG-2, and MPEG-4 principles. The standards discussions is based directly on the ISO standards; such material is included here in order to promulgate the use of open standards; however, developers and other parties working on products of commercial value should refer directly to the original documentation, especially since only some key highlights are included.

Over a dozen video standards/format are available worldwide for TV video (without even counting variant broadcasting schemes beyond the basic NTSC, PAL, and SECAM methods). Table 4.5 summarizes the video standards discussed or alluded to in this section. This section focuses on the higher-end standards

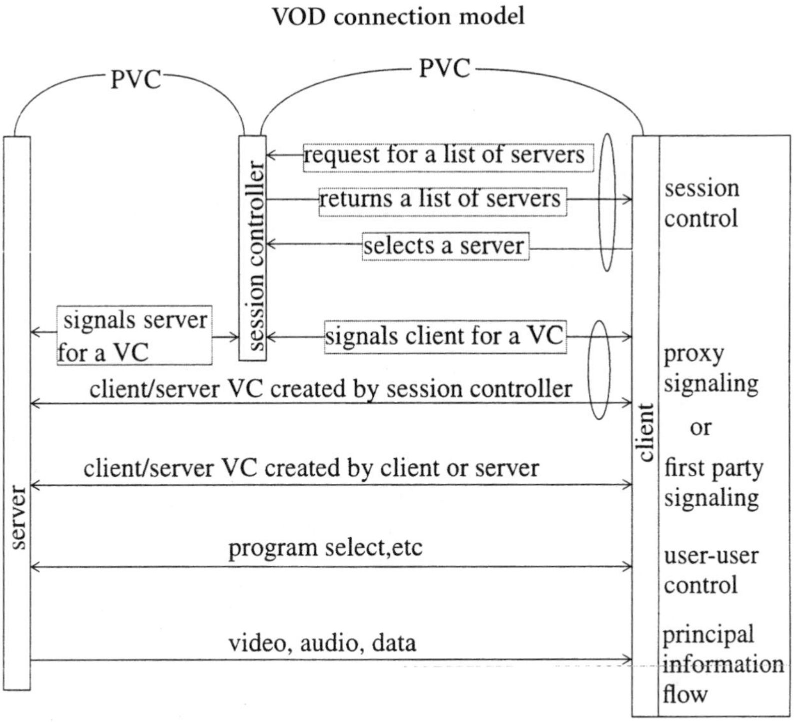

VOD connection model

FIGURE 4.25 Signaling requirements.

TABLE 4.5   Plethora of Video Formats and Coding (in Alphabetical Order)

CD-I	Digital consumer-electronic format
D-1/CCIR 601	Digital production standard
D-2	Digital production standard
DS-3 based	Digital U.S. commercial methods (nonstandard)
DVD-video	Digital consumer electronic format
DVI/Indeo	Early *de facto* digital standards for multimedia
H.261 et al.	Digital videoconferencing format
H.263	Digital videoconferencing format for LAN/IP networks
HDTV	High-definition digital scheme discussed in Section 4.1
JPEG/Motion JPEG	Digital compression format (principally for still video, but also for some video)
MPEG-1	Digital compression full-motion video (low-end entertainment video and multimedia)
MPEG-2	Digital compression full-motion video (high-resolution)
MPEG-4	Robust low-data-rate compression full-motion video
NTSC	Analog U.S. and Japanese format
PAL	Analog European format
SECAM	Analog French and Eastern European format
Vendor-based	Digital vendor-specific videoconferencing formats

(the previous section focused on the lower-end business conferencing standards — however, many of the underlying principles and technologies are similar). In order for digital television to successfully enter the market, it is important that the agreed-upon audio and video compression technique be used on an industry-wide basis. It is also important that the transport approach be also standardized; standards such as MPEG-2 and ATM do exactly that.

4.3.1.1   Compression Methods

Video can be considered a sequence of frames, where each frame is an array of pixels. The goal of a video coding algorithm is to remove redundant information and greatly reduce the data rate. Two types of redundancies exist in video: redundancy within a single frame and redundancy between adjacent frames. There are two classes of compression algorithms: "lossless" algorithms and "lossy" algorithms. Another way of classifying compression algorithms is as entropy coding and source coding.

Lossless compression is one where the entire information contained in the uncompressed message can be faithfully recovered by the decompressor. For example, instead of sending a 100-bit message 0111111111...111111 one could compress it as x0y1, where x and y are octets that take values 0 (base 10) to 255 (base 10). In this case, one would send (00000001)0(01100011)1, which is only 18 bits long, and yet the receiver is still be able to recover the message exactly. Lossless compression algorithms are symmetrical, namely, either the sender or receiver can perform the compression and decompression with the same level of computational complexity and without loss of data integrity. Compression of *data* material, either for transmission or for storage, clearly requires lossless methods. Many hardware and software products implement lossless compression. They typically double or quadruple the storage capacity on a disk (i.e., have a compression ratio of about 2:1 or 4:1), or double the apparent speed of a communication line. These algorithms can also be applied to files that represent voice or image information. Because the redundancy is higher, the compression ratios can be as much as 10:1. However, this is both (1) less effective than the compression obtained with specialized "lossy" techniques and (2) less than the information bandwidth reduction that is sought (typically 100:1 or even 200:1). Lossy compression algorithms do not aim at retaining the entire information, but (just) enough to be adequate for the task at hand. Lossy algorithms result in (slightly) degraded pictures. The advantage of these algorithms is that they can achieve 100:1 or 200:1 compression.

There in another way of looking at compression, namely, source coding vs. entropy coding. Source coding deals with *features* of the source material, and encompasses lossy algorithms. Source coding can be further classified as intraframe and interframe coding. Intraframe coding is used for the first picture of a sequence and for downstream pictures after some major change of scenery. Intraframe coding is used for sequences of similar pictures (even for those including moving objects). Intraframe coding removes only the spatial redundancy within a picture; interframe coding also removes the temporal redundancy between pictures. Entropy coding, on the other hand, achieves compression by using statistical properties of the coded signal, and is, in theory, lossless.

Video can be compressed using lossless or lossy methods. For lossless compression, video can be digitized according to the ITU-R 601 standard; the bit rate is approximately 165 Mbps. Although useful in a number of high-end commercial applications, this data rate is simply too high for user-level applications. Lossy methods, such as MPEG-1 and MPEG-2, are more appropriate for video on demand and digital video distribution.

Two other methods are on the horizon — *fractals* and *wavelets*. The fractal transform uses Mandelbrot's approach of using simple equations to generate natural-looking images in a high level of detail. It is believed by experts to be a good compression scheme, particularly for still images of nature. Based on equations, it can be expanded to sizes even larger than the original, leading to claims of greater compression compared to other schemes. Images are segmented into domains that can be described as squeezed-down, distorted versions of larger parts or "ranges" of the same image.[36] Artifacts include softness and substitution of details by other details typically undetectable in natural scenes. Packages ranging in cost from $500 to $20,000 are available for desktop applications. The wavelet method is also based on mathematical techniques. A wavelet codec transforms a picture into a set of different spatial representations, some of which contain insignificant high frequencies, and one of which contains all the important low-frequency information. This scheme can also compress audio. Artifacts include softness, small random noise, and edge halos. Products based on this approach are also appearing.

4.3.1.2 Traditional Digital Video — Broadcast Quality

As covered briefly in Section 4.1, an image is composed of three elements: a luminance (brightness) element and two chrominance (color) elements. These elements come into play in the digitization process. There are two nearly lossless methods of digitizing television signals: *digital component video* and *digital composite video*.

TABLE 4.6   Summary of the CCIR/ITU-R 601 Digital Video Standard

Coded signals	Y, R-Y, B-Y (luminance and color differences)
Number of samples:	
• Luminance (Y)	858
• Each color difference (R-Y, B-Y)	429
Sampling approach	Orthogonal, line, field, and picture repetitive; R-Y and B-Y samples co-sited with odd Y samples in each line (1st, 3rd, 5th, etc. sample)
Sampling frequency:	
• Luminance (Y)	13.5 MHz
• Each color difference (R-Y, B-Y)	6.75 MHz
Form of coding	Uniformly quantized pulse code modulation (PCM), 8 bits per sample for all three signals (luminance and color differences)
Number of samples per digital active line:	
• Luminance (Y)	720
• Each color difference (R-Y, B-Y)	360
Correspondence between video signal levels and quantization levels:	
• Luminance (Y)	220 quantization levels with the black level corresponding to level 16 and the peak white level corresponding to level 235
• Each color difference (R-Y, B-Y)	224 quantization levels in the center part of the quantization scale with zero signal corresponding to level 128

Digital component video, known as 4:2:2, is a time-multiplexed digital stream of three video signals: luminance (Y), C_r (R-Y), and C_b (B-Y). The 4:2:2 refers to the ratio of sampling rates for each component. This format is also often called D-1, referring to the tape format associated with the digital component recording. CCIR/ITU-R Recommendation 601 was adopted in 1982, after 8 years of study and compromise among European, Japanese, and North American approaches. The standard accommodates equally well NTSC, PAL, and SECAM formats. Typical digital component video encoding systems have utilized the following parameters:

- NTSC: Luminance sampling frequency 13.5 MHz. Sampling frequency for color differences: 6.75 MHz. Pixels: 858 × 525 (about 720 × 484 for the active image area).
- PAL: Luminance sampling frequency of 17.734475 MHz. Sampling frequency for color differences: 8.867236. Pixels: 910 × 525 (about 768 × 484 for the active image area).

At the final stage, the word length for digital image delivery is usually between 8 and 10 bits, but to maintain precision more may be utilized, particularly in the early stages of off-line processing (e.g., 16 bits). Since there is no one single sampling rate to obtain digital video, conversion between digital formats requires transcoding, not only of the formats, but also of the sampling frequencies.

The CCIR/ITU-R 601 standard support both the 525-line, 60 fields/second format and the 625-line, 50 fields/second format. Table 4.6 depicts a summary of the key highlights of the CCIR/ITU-R 601 standard for the 525-line format[37] (MPEG-2, covered later, aims at providing similar quality but at much smaller data rate).

The other encoding method is digital composite video, known as $4f_{sc}$. This format, applicable to NTSC, PAL, and SECAM, also consists of three components: Y, I, and Q. However, I and Q are not multiplexed but quadrature-modulated and summed to the Y component. The result is a single information stream sampled at four times the color subcarrier rate. The term $4f_{sc}$ refers to "4× the frequency of the subcarrier." This format is often called D-2, referring to the associated tape format. Typical digital composite video encoding systems have utilized the following parameters:

1. NTSC: Luminance sampling frequency of 14.31818 MHz (four times the frequency of the NTSC subcarrier)
2. Sampling frequency for color differences: 7.15909
3. Pixels: 910 × 525 (about 768 × 484 for the active image area)

The transmission in digital form of a composite television signal (whether NTSC, PAL, or SECAM, particularly for contribution networks), requires, according to the sampling rates used and the number of bits employed to represent the signal, a rate of 100 to 150 Mbps. In component video (e.g., 4:2:2), the bit rate can reach 216 to 270 Mbps; uncompressed HDTV requires about 1 Gbps.[38] Compression is unquestionably required.

In the U.S., DS3 transmission facilities support about 45 Mbps; in Europe, E3 facilities support about 34 Mbps; also note that DS1 supports 1.544 Mbps and E1 supports 2.048 Mbps. These rates have defined, at the pragmatic level, the boundaries for the commercially available video encoding algorithms. For example, vendor-proprietary methods have been used in the past in the U.S. to encode TV signals at 45 Mbps for remote delivery (e.g., between TV studios in two cities), utilizing commonly available telecommunications carrier services. Several suppliers have developed broadcast-quality DS3 video coders using "relatively mild compression" on 525/60 Hz video. Uncompressed pulse code modulation of NTSC signals would require a minimum sampling rate of 8.4 MHz for a 4.2-MHz bandwidth, and 8 or 9 bits per sample, resulting in an uncompressed rate of 67.2 to 75.6 Mbps. (In reality higher data rates are needed, because, as noted below, the sampling rate is higher.) In Europe, the encoding of interest is 34 Mbps. These network rates clarify why the CCIR published Recommendation 723 that supports standardized video coding at 32 to 45 Mbps for full resolution video/TV signals (720 pixels per line, 483/576 lines per frame, 59.94/50 2:1-interlaced frames per second). However, there is an interest to bring the video data rate down farther, as discussed in the next sections.

4.3.1.3 Compression Algorithms in Common Use

The video compression requirements vary between various applications, digital storage media, transport method (e.g., terrestrial, DBS, etc.), and video programming type (e.g., talk shows vs. sporting events). Nonetheless, it is important that easy interworking and movement between such media be accomplished. Many factors come into play, such as timing, program stream reconstruction, synchronization, demultiplexing/remultiplexing, packeting/repacketizing, and encryption.[39] It has been the objective of the standards study groups to limit the extent of the specifications to a minimum and to define only what it takes to accomplish meaningful interoperability.

The mid-1980s saw the emergence of ITU-Ts Recommendation H.261, supporting video compression and coding at px64 kbps (p = 1, 2, ..., 30). These standards (compared in Section 4.2) are suited for videotelephony and videoconferencing, but are not deemed appropriate for entertainment-quality video-on-demand programming (they could, however, be used in conjunction of digital video in support of these other services, perhaps in support of telecommuting[40]).

The 1980s also saw the emergence of a standard originated by ISO, which was formally adopted at the end of 1992. This standard, ISO/IEC 11172 (also known as MPEG-1), provides video coding for digital storage media with a rate of 2 Mbps or less. H.261 and MPEG-1 standards provide picture quality similar to that obtained with a VCR. Both these standards are characterized by low-bit-rate coding and low spatial resolution. H.261 supports 352 pixels per line, 288 lines per frame, 29.97 noninterlaced frames per second (lower resolution is also supported). MPEG-1 typically supports 352 pixels per line, 240/288 lines per frame, 29.97/25 noninterlaced frames per second. Many useful video and multimedia applications require higher resolution than this, in order to provide an acceptable level of quality to the user. Real time encoding/decoding also introduces delays that increase with decreasing data rates. For example, current technology encoding at 128 kbps may result in unacceptable delays for a quality videoconference.

MPEG-1 was developed as a video compression standard to be used with CD-ROMs. The compression ratio is about 100:1; however, the quality of the picture is marginal for generic broadcast and cable applications of high-action movies and sporting events. MPEG-1 employs a source input format for motion video and associated audio with a data rate up to 1.5 Mbps.

The MPEG standard embodies the concepts of (1) group of frames and (2) interpolated frames (the presence of interpolated frames is optional) (Figure 4.26.) Each group of frames contains a frame that is intraframe coded only, to facilitate random access. There will also be predicted frames. Interpolated

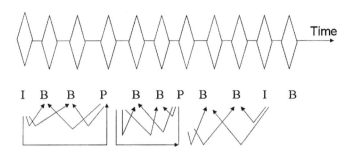

I : Intraframe coded frames (standalone)
P: Predicted Frames (predicted from intra frames of from previously predicted frames

B: Bidirectionally - interpolated frames (form nearest two I or P frames)

FIGURE 4.26 "Group of frames" concept.

frames are formed from adjacent (past and future) "keyframes" (both the stand-alone and predicted frames can be used for interpolation). As seen in Figure 4.26, a group of frames consists of a single stand-alone frame, several predicted frames, and one or more interpolated frames positioned between key-frames. The bandwidth allocated to each type of frame typically conforms to the ratio of 5:3:1 (intraframe coded, predicted, and interpolated). "Future" frames may be employed to predict intermediary frames. (This can be done not only for stored material, but also for real-time material by buffering a few frames for analysis.)

In recent years both ISO and ITU-T have been working on new high-quality video coding standards: full-motion, reasonable-resolution digital video is sought in the 4 to 20 Mbps range. MPEG started its study on the second-phase work (known as MPEG-2) in 1990, with completion in 1995. MPEG-2 (ISO/IEC 13818) supports both full CCIR/ITU-R 601 resolution as well as HDTV. The rates of interest range from 2 to 20 Mbps. As noted, work on MPEG-4 has been undertaken of late.

In a hybrid coder, a type of coder commonly used, an estimate of the next frame to be processed is created from the current frame. The difference between this estimate and the actual value of the variable(s) of interest contained in the next frame, when it arrives at the coder, is encoded by an appropriate mechanism. One of the more common examples of this type of coder is the motion-compensated DCT coder used in MPEG-1 and MPEG-2. Motion compensation capitalizes on the correlation between successive frames of a video sequence. A motion-compensation algorithm assigns a velocity (that is, a speed and direction to a moving object). Constant velocity makes predicting the next frame of video fairly straightforward. Misprediction, however, can cause the loss of two or three frames of video. Even when true motion is not at play in a scene, motion compensation algorithms improve the data rate by seeking a block of identical (or nearly identical) values in the previous frame and spatially close to the block to be encoded; this block is then used to formulate a prediction. Only the information used to find the prediction block is sent to the decoder.[38]

The DCT concentrates the remaining signal into a small number (64 to be precise) of coefficients that can be quantized and efficiently represented. This coder is three to four times as efficient as one that uses no prediction, but it is sensitive to transmission errors and does not permit random access. This coder is utilized (with modifications to facilitate random access) in the MPEG-1 standard; the basic DCT method (see list below) is also used in the JPEG still-image standard (some developers have also used JPEG to encode video, but this is now generally on the decline).

As discussed earlier, encoding/decoding falls into two main types[41]:

Interframe: These encoders/decoders use combinations of key, motion-predicted, and interpolated frames to achieve high compression ratios with low data rate. Examples of these types of algorithms include the various MPEG algorithms.

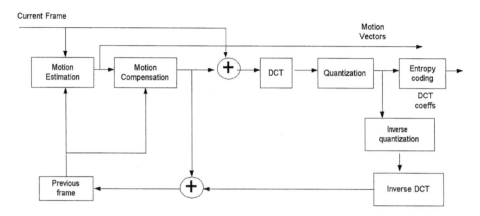

FIGURE 4.27 Typical video coder.

Intraframe: These systems compress every frame (and sometimes every field) of video individually. These algorithms (e.g., motion JPEG, but also MPEG with I-frames) offer the advantage of frame-accurate editability; however, they produce from two to ten times more data than the interframe algorithms.

Coding standards such as H.261and H.263 logically partition the images to be encoded into rows of macroblocks known as groups of blocks (GOBs). From the encoder's point of view, each of the current frames of the video sequence is partitioned into rectangular regions of 16×16 pixels called *macroblocks*. For each macroblock, the motion estimation stage of the video encoder computes the motion vectors that best represent the location in the previous frame where image pixels of similar intensity values occur. (A typical video coder is shown in Figure 4.27.) The motion compensation stage applies these motion vectors to the corresponding macroblocks in the previous frame and computes a motion-compensated frame and the difference with current frame is then computed. The frame representing these differences is called as the *residual frame*. The residual frame represents the information in the current frame that cannot be predicted from the previous image. This residual frame is then coded by the DCT stage. The DCT is typically performed on 8×8 pixel blocks. The quantized DCT coefficients are then coded using variable-length coding (VLC) schemes such as Huffman coding. Hence, for every frame of the video sequence, the video encoder transmits motion vector information, DCT information, and some overhead header information. In order to achieve further coding efficiency, the header and motion vector information are also coded using VLC techniques. These motion-compensated frames of the video sequence are known as *predictive frames* (P-frames) or *interpolated frames* (B-frames, being bidirectionally interpolated). Some frames of the video sequence are coded completely with respect to themselves with no motion compensation, and these are known as *intraframes* (I-frames). Intraframes do not have any motion vector information associated with them.[42]

Until the recent development of chips able to support real-time encoding, the issue of compression algorithm asymmetry was important. (Powerful DSP chips are being developed to support video processing by a number of vendors.) With symmetric algorithms, the compression process requires the same amount of clock time as decompression (playback); asymmetric compression requires considerably more clock time than decompression.[42] In theory this makes the decompression possible on cheap low-end equipment. Recent advances in chip power make this a less important issue, particularly since the suppliers are coming out with a single chip that can be programmed to be an MPEG-1 encoder, MPEG-1 decoder, MPEG-2 encoder, MPEG-2 decoder, a JPEG encoder, a JPEG decoder, or an H.261 encoder/decoder.[43]

A typical MPEG-2 decoder available at this writing supports the following functions:

- MPEG-2 13818 standard for audio, video, and multiplexing
- MPEG audio Layer I and II

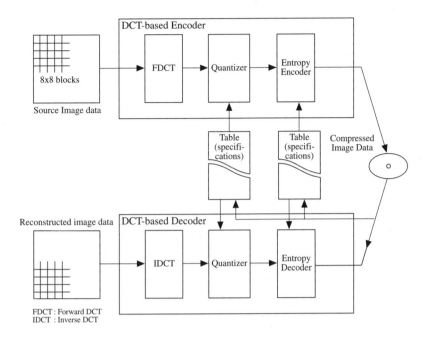

FIGURE 4.28 DTC usage in a JPEG codec.

- I-only, IP, IBP, and IBBP coding
- 0.8 to 15 Mbs bit
- 4:2:0 sampling resolution, 8 bits or 4:2:2 sampling resolution, 10 bits
- Two audio channels
- CCIR 601 and SIF formats
- 5.1 audio channels and support for AC-3
- Multiple input/output ports (A/V switch)

High-end MPEG-2 systems for telemedicine, distance learning, and business video applications went for $10 thousand to $25 thousand at this writing.

4.3.1.4 A Short Discussion of DCT

This section provides a short description of some key DTC features, by discussing them in the context of still picture encoding. Figure 4.28 depicts the key processing steps that are embodied in the DCT-based operation for the case of a single-component[*] (i.e., monochrome) JPEG image sequential codec (the functioning of the DCT in other codecs is similar). Note the forward DCT (FDCT) function and the inverse DCT (IDCT) function.

One can think of DCT-based compression as compression of a stream of 8×8 blocks of gray-scale image samples. Each 8×8 block (represented by 64 point values known as $f(x,y)$, $0 \leq x \leq 7$, $0 \leq y \leq 7$) makes its way through each processing stage, yielding output in compressed form. For progressive-mode codecs, an image buffer is placed between the quantizer and the entropy coding module; this enables the

---

[*]A source image/frame contains three image components (also called colors, spectral bands, or channels), e.g., RGB. Each component consists of an array of samples, to which the DCT can be applied in turn. A sample is expressed as an integer with precision P bits with values in the range $[0, 2^P - 1]$; all samples of all components within the same source image/frame must have the same precision (P can be 8 or 12), but image components may be sampled at different rates compared with each other.

image to be stored and then sent out in multiple scans with follow-up information aimed at successively improving the quality of the received image.

Each 8 × 8 block of source image (frame) samples can be viewed as a 64-point discrete signal that is a function of the two spatial dimensions x and y. At the input to the encoder, these 64 source image samples are cranked through the following equation[44]:

For $0 \leq u \leq 7$, $0 \leq v \leq 7$ calculate the following 64 values:

$$F(u,v) = 0.25 * C(u) * C(v) * \left\{ \sum_{x=0}^{7} \sum_{y=0}^{7} f(x,y) * \cos\left[(2x+1) * u\pi / 16\right] * \cos\left[(2y+1)v\pi / 16\right] \right\}$$

where $C(u) = C(v) = 1/\sqrt{2}$ for u, $v = 0$ and $C(u) = C(v) = 1$ otherwise.

Mathematically, the FDCT takes the input signal and decomposes it into 64 orthogonal basis vector signals. The output of the FDCT is a set of 64 basis signal amplitudes, that are known as "DCT coefficients." The coefficient for the vector (0,0) is called the DC coefficient; all other coefficients are called AC coefficients. The DC coefficient generally contains a significant fraction of the total image energy. Because sample values typically vary slowly from point to point across an image, the FDCT processing achieves data compression by concentrating most of the signal in the lower values of the (u,v) space. For a typical 8 × 8 sample block from a typical source image, many, if not most, of the (u,v) pairs have zero or near-zero coefficients and therefore need not be encoded. At the decoder, the IDCT reverses this processing step. One can use 8-bit or 12-bit source image samples; 12-bit samples, however, require fairly large computational resources for FDCT or IDCT calculations.

In principle, the DCT introduces no loss to the source image samples; it just transforms them to a domain where they can be more efficiently encoded. This means that if the FDCT and IDCT could be computed with perfect accuracy and if the DCT coefficients were not quantized, the original 8 × 8 block could be recovered exactly. But, as seen above, the FDCT (and the IDCT) equations contains transcendental functions (i.e., cosines). Consequently, no finite-time implementation can compute them with perfect accuracy. In fact, a *number* of algorithms have been proposed to compute these values approximately. No single algorithm is found to be optimal for all implementations: an algorithm that runs optimally in software usually does not operate optimally in firmware (say, for a programmable DSP) or in hardware.

Given the finite precision of the DCT inputs and outputs, an interworking challenge arises: coefficients calculated by two different algorithms (say, one the sender and one in the receiver), or even by independently designed implementations of the same FDCT or IDCT algorithm (which differ only minutely in the precision of the cosine terms or intermediate results) will result in slightly different outputs from identical inputs.

Each of the 64 DCT coefficients obtained at the output of the FDCT is then uniformly quantized by utilizing a 64-element quantization table, that must be specified by the application (or user). Each element can take an integer value from 1 to 255 (or 1023), that specifies the step size of the quantizer for its corresponding DCT coefficient. The purpose of quantization is to achieve further compression by discarding information that is not visually significant. Quantization is a lossy process, and is the principal source of lossiness in DCT-based encoders.

When the aim is to compress the image or frame as much as possible but without visible artifacts, each step size is chosen to be the perceptual threshold of human vision. These thresholds are functions of the source image characteristics, display characteristics, and viewing distance.

After the quantization process, the DC coefficient, representing a sort of average of the value of the 64 image samples, is handled separately. Since there is usually high correlation between the DC coefficients of adjacent 8 × 8 blocks, the quantized DC coefficient is encoded differentially, namely, as the difference between the current value and the previous value. In order to facilitate entropy coding, the quantized AC coefficients are ordered into the "zigzag" sequence shown in Figure 4.29.[44] This ordering helps the

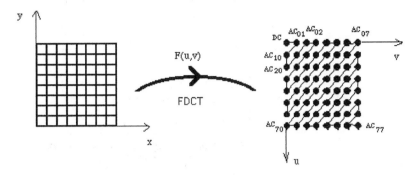

FIGURE 4.29 Ordering of quantized samples for entropy encoding.

entropy coding process by placing low-coordinate coefficients (which are more likely to be nonzero) before high-coordinate coefficients.

The last step for DCT-based encoding is entropy coding itself. This step achieves additional lossless compression by encoding the quantized DCT coefficients more compactly, based on their statistical characteristics. Entropy coding can be viewed as a two-step process. The first step converts the sequence of quantized coefficients (ordered as discussed above) into an intermediate sequence. The second step converts the symbols to a stream in which the symbols no longer have externally identifiable boundaries. Two entropy coding methods are in common use: Huffman coding and arithmetic coding. The sequential codec uses Huffman coding, but codecs with both methods are specified for all modes of operation. Arithmetic coding produces about 10% better compression than Huffman; however, it is more complex. Huffman coding requires that one or more sets of code tables be specified. The same tables used to compress an image are needed to decompress it. Huffman tables may be predefined and used within an application as defaults, or developed specifically for a given image in an initial statistics-gathering pass prior to actual compression.

4.3.2 JPEG and Motion JPEG

The JPEG standard has been developed jointly by both ISO and ITU-T (hence, the nomenclature "joint") for compression of still images. It can compress typical images from $1/10$ to $1/50$ of their uncompressed bit size without visibly affecting image quality. JPEG is the first international digital image compression standard for multilevel continuous-tone still images (both gray scale and color). Some applications to which JPEG addresses itself include: color facsimile, quality newspaper wirephoto transmission, desktop publishing, graphic arts, and medical imaging. Some have used JPEG to support video transmission, (particularly in the medical industry where one wants to be able to adjust the resolution of the transmission in real time, e.g., when viewing the display of a set of X-ray images). JPEG plays a limited role in digital video at this time.

JPEG utilizes a methodology based on the DCT, discussed in the previous section. It is a symmetrical process, with the same complexity for coding and for decoding. JPEG will be an important compression standard for a number of applications since it works relatively well and is already available in the marketplace, as evidenced by vendor support. Many digital video cameras, fax machines, copiers, and scanners now include JPEG chips.

Some vendors also use JPEG methods for encoding of full-motion video, NTSC TV signals in particular. This is known as motion JPEG. While JPEG was not designed for full-motion video, it can accommodate it with some restrictions. For example, audio is not supported in an integrated fashion. One of the limiting factors of this use of the algorithm is that it works independently from frame to frame; hence, it cannot reduce the redundancies that exist between frames. Some view the fact that JPEG performs only intraframe compression as a benefit in the sense that it offers "fast" random access to any frame of the video material. Other full-motion video compression techniques performing interframe compression rely on periodic transmission of a reference frame — if the reference frame is sent every 20 frames one may have to wait as many as 19 frames before the reference frame is received; this would equate to a wait of 20/60 or 0.33 s. With JPEG, one needs to wait only for the time required to decompress one frame, that is, 0.04 s.

Network-based applications using JPEG for full-motion video are not likely to be widely implemented because JPEG is bandwidth intensive. For video material displayable at a PC monitor at medium resolution, i.e., 640 × 480 pixels, 24 bits for color representation, JPEG is required to in order to compress about 1 MB per frame, or 30 MBps (240 Mbps) to a lower value. The downloading, displaying, and manipulation of full-screen video in digital form is a daunting task, even if one were to achieve a 50:1 compression; this is why standards such as MPEG-1 and MPEG-2 are needed for any digital video transmission, except perhaps for desktop multimedia applications.

4.3.3   MPEG-1

As noted, the ISO/IEC/JTC1/SC29/WG11 MPEG working group has produced a specification for coding of combined video and audio information. It is directed at video display as contrasted to still-image display to JPEG addresses. MPEG specifies a decoder and data representation for retrieval of full-motion video information from digital storage media in the 1.5 to 2 Mbps range. Hence, it can be used in an ADSL context. Its principal goal was, however, to support storage of digital video. The specification is composed of three parts: Part 1, Systems; Part 2, Video; and Part 3, Audio. The system part specifies a system coding layer for combining coded video and audio, and provides the capability of also combining private data streams and streams that may be defined at a later date. The specification describes the syntax and semantic rules of the coded data stream.

As mentioned earlier MPEG-1 uses three types of frames: intra (I) picture frames; predicted (P) frames; and bidirectional interpolated (B) frames. I-type frames are compressed using only the information in that frame using the DTC algorithm. An incoming video signal of 1 s will contain at least two I frames. P frames are derived from the preceding I-frames (or from other P-frames) by predicting motion forward in time. P-frames are compressed to approximately 60:1. B frames are derived from the I- and P-frames, based on previous and frame referencing. B frames are required to achieve the low average data rate.[42]

Since MPEG allows coding comparison across multiple frames, it can yield compression ratios of 50:1 to 200:1. The MPEG algorithm is asymmetrical. Namely, it requires more computational complexity (hardware) to compress full-motion video than to decompress it. This is useful for applications where the signal is produced at one source but it is distributed to many. MPEG chips on the market at this writing provide 200:1 compression to yield VHS quality at 1.2 to 1.5 Mbps; they also can provide 50:1 compression for broadcast quality at 6 Mbps.

The MPEG system coding layer specifies a multiplex of elementary streams such as audio and video, with a syntax that includes data fields directly supporting synchronization of the elementary streams. The system data fields also assist in the following tasks[45,46]:

1. Parsing the multiplexed stream after a random access;
2. Managing coded information buffers in the decoders;
3. Identifying the absolute time of the coded information.

The system semantic rules impose some requirements on the decoders; however, the encoding process is not specified in the ISO document and can be implemented in a variety of ways, as long as the resulting data stream meets the system requirements.

Figure 4.30 depicts an MPEG encoder at the functional level. The video encoder receives uncoded digitized pictures called Video Presentation Units (VPUs) at discrete time intervals; similarly, at discrete time intervals, the audio digitizer receives uncoded digitized blocks of audio samples called Audio Presentation Units (APUs). Note that the times of arrival of the VPUs are not necessarily aligned with the times of arrival of the APUs.

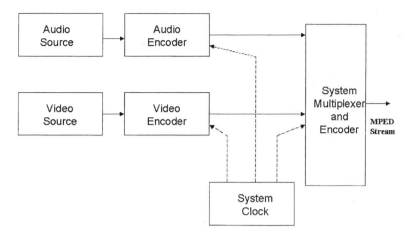

FIGURE 4.30 Generation of an MPEG data stream. *Note:* Since the ISO 11172 specification specifies syntax and semantics of the coded data stream, the encoder model is not part of the specification itself; however, the reference decoder model is specified as part of the semantic definition of the information stream at the system layer.

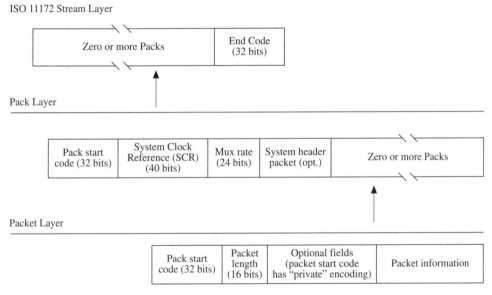

FIGURE 4.31 Simplified view of ISO 11172.

The video and audio encoders, respectively, encode digital video and audio as described in the MPEG Specification Parts 2 and 3, producing coded pictures called video access units (VAUs) and coded audio called audio access units (AAUs). These outputs are referred to generically as elementary streams. The system encoder and multiplexer produce a multiplex stream, referred to as $M(i)$, containing the elementary streams as well as system layer coding described below. Audio is supported from 32 to 384 kbps and can consist of a single channel or two stereo channels.

The MPEG system specification includes a syntax with three coding layers above the layer implicitly defined by the elementary streams. These are (1) the ISO 11172 stream layer, (2) the pack layer, and (3) the packet layer (Figure 4.31).

The ISO 11172 stream layer includes a sequence of packs followed by an end code. The pack layer includes the SCR field, the "mux rate" field, an optional system header packet, and the packet layer. (The "mux rate" field bounds the rate of bytes per second as measured by the current and succeeding SCR values and the number of coded bytes intervening.) The packet layer comprises packets containing information from individual elementary streams (there is information from exactly one elementary stream in each packet). The packet contents and all system layer coding are octet-aligned, although the individual coding elements within elementary streams may not necessarily be octet-aligned. Each packet consists of a packet start code followed by the packet length (in octets, ranging up to $2^{16} - 1$). There are 69 different values of packet start code currently defined: 16 are for video, 32 are for audio, 2 are private, 1 is for padding, and the remainder 18 codes are reserved for future use (of which the Multimedia and Hypermedia Experts Group may take 16). Private data is unrestricted, other than by the syntax and STD model that applies to the entire stream. Three optional fields may be included in the packet: STD buffer size, PTS and DTS. Last, one finds the packet information from the elementary stream. The amount of packet information is limited only by the total available packet length (decremented by the data in the packet header itself) and by constraints imposed by the decoder.

The method of multiplexing the elementary streams (VAUs and AAUs) is not directly specified in MPEG. However, there are some constraints that must be followed by an encoder and multiplexer in order to produce a valid MPEG data stream. For example, it is required that the individual stream buffers must not overflow or underflow. The sizes of the individual stream buffers impose limits on the behavior of the multiplexer. Decoding of the MPEG data stream starting at the beginning of the stream is straightforward since there are no ambiguous bit patterns. Starting decoding operation at random points requires locating pack or packet start codes within the data stream.

An important aspect of MPEG is clock synchronization. Synchronization is a fundamental aspect of communication. The principle may be known to the reader in the context of synchronizing various nodes on a network to enable character, block, or message recovery, as these data entities are transferred from the sender to the receiver. In a multimedia context, synchronization is even more "intimate" in the sense that the various signal objects comprising the combined signal must be stored, retrieved, and transmitted in with precise timing relationships. The basic mechanism to achieve the desired synchronization is, to a large extent, the same in both contexts.

The System Time Clock (STC) is a reference time operating at 90 kHz. The STC is not necessarily phase-locked to the audio or the video sample clocks. The STC produces 33 bit time values (binary values from 0 to $2^{33} - 1$) incremented at 90 kHz. For some (but not necessarily all) VPUs and APUs arriving at the encoder, the value of the STC is determined and stored with the presentation units through the coding, transmission, and decoding processes. These values are called Presentation Time Stamps (PTS). The time stamps, using a reference frequency of 90 kHz and unsigned binary values from 0 to $2^{33} - 1$, allow unique identification of operating time within the information stream over an interval exceeding 24 h.[45,46]

In addition to the PTS, there are two other time stamps associated with the coded information stream itself that are used to ensure synchronization between the various decoders and the information stream source in a decoding system:

1. System Clock References (SCR): These are generated as samples of the STC such that the value of the SCR equals the value of the STC at the time the last octet of the SCR exits the system encoder.
2. Decoding Time Stamps (DTS): These are similar to PST, except that the permutation ordering of pictures in the video coding process is reflected in the DTS values.

To achieve synchronization in multimedia systems that decode multiple video and audio signals originating from a storage or transmission medium, there must be a "time master" in the (decoding) system. MPEG does not specify which entity is the time master. The time master can be (1) any of the decoders, (2) the information stream source, or (3) an external time base; all other entities in the system (decoders and information sources) must slave their timing to the master. If a decoder is taken as the time master, the time when it presents a presentation unit is considered to be the correct time for the use of the other entities. Decoders can implement phase-locked loops or other timing means to assure proper slaving of their operation to the time master. If an information stream source is considered the time master, the SCR values indicate the correct time at the moment that these values are received. The decoders then use this information of what the correct time is to pace their decoding and presentation timing. If the time base is an external entity, all of the decoders and the information sources must slave the timing to the external timing source.

4.3.4 MPEG-2

The compression schemes discussed so far do not produce adequate quality for full-motion video (although JPEG supports high quality, its data rate is too high). MPEG-2 was developed by ISO/IEC/JTC1/SC29/WG11 and is known as ISO/IEC 13118. This is probably going to be the most important video compression standard for digital video and video-on-demand applications. The committee (WG11) consists of largely U.S. and European companies involved in or interested in video and audio compression.[37] As noted earlier, it aims at providing CCIR/ITU-R quality for NTSC, PAL, and SECAM, and also at supporting HDTV quality (this being a relatively newer requirement). MPEG-2 work is now driven by the desire to accomplish global unification of digital TV program generation, editing, storage, retrieval, transport, and display.[39] The standard provides a set of agreed-upon methodologies for audio and video compression, and transport of complex multiplexes of associated data and related data services.

MPEG-2 has been very successful commercially, with significant acceptance in the market, not only for broadcast TV but also for other applications (e.g., distance learning). MPEG-2 has been used in Direct Broadcast Satellite (DBS), DVD, and HDTV. There were plans to develop another compression standard for HDTV (this was referred to as MPEG-3, as noted in Section 4.1); however, planners found that MPEG-2 was suitable (hence it was decided to include HDTV as a separate profile of MPEG-2).

The objectives of newer video coding standards for full-motion video that have been developed recently, specifically MPEG-2, are as follows:

1. Picture quality should be higher than that of the current NTSC/PAL/SECAM broadcast systems.
2. Compression to bit rates in the range of 4 to 6 Mbps for NTSC/PAL/SECAM material and 6 to 10 Mbps for television signals conforming to CCIR Recommendation 601. These target rates are already achievable in experimental systems.
3. The standard(s) need to be flexible enough to allow both high-performance/high-complexity and low-performance/low-complexity (e.g., intraframe mode only operation) codec systems.
4. The standard(s) should take into account existing standards. Compatibility consideration enables smooth migration of new standards while maintaining interoperability among equipment conforming to the old- and new-generation standards.
5. Compatibility should be maintained to the extent possible. There are two types of compatibility: upward/downward compatibility (addressing different picture format sizes) and forward/backward compatibility (addressing different generation standards). A system is upward compatible if a higher-resolution receiver is able to decode material from the signal transmitted by a lower-resolution encoder. A system is downward compatible if a lower-resolution receiver is able to decode material from the signal or part of the signal transmitted by a higher-resolution encoder. A system is forward compatible if the new standard decoder is able to decode material from a signal or part of the signal of an existing standard encoder. A system is backward compatible if an existing standard decoder is able to decode material from the signal or part of the signal of a new standard encoder.

The following four layers come into play in the discussion of MPEG-2 coding:

1. Block: The smallest coding unit in the MPEG algorithm. It consists of 8×8 pixels and can be one of three types: luminance (Y), red chrominance (C_r), and blue chrominance (C_b). The block is the basic unit in intraframe coding.

TABLE 4.7   Initial MPEG-2 Profiles and Levels

	Simple Profile	Main Profile	Next Profile
High Level Type 1: Supports 1152 lines per frame (lpf), 1920 pixels/line (ppl), and 60 fps. This equates with 62.7 M pixels per second (pps) or 60 Mbps.		HDTV (U.S.)	HDTV (European)
High Level Type 2: Supports 1152 lps, 1440 ppl, and 60 fps. This gives 47 Mpps or 60 Mbps.		HDTV (U.S.)	HDTV (European)
Main Level: Supports 576 lpf, 720 ppl, and 30 fps. This gives 10.4 Mpps or 15 Mbps	Cable TV industry	Video broadcasting industry	
Low Level: Supports 288 lpf, 352 ppl, and 30 fps. This gives 2.5 Mpps or 4 Mbps.			

2. Macroblock: The basic coding unit in the MPEG algorithm. It is a 16×16 pixel segment in a frame. Since each chrominance component has one half the vertical and horizontal resolution of the luminance component, a macroblock consists of four Y, one C_r, and one C_b block.

3. Slice: A horizontal strip within a frame. It is the basic processing unit in the MPEG coding scheme. Coding operations on blocks and macroblocks can only be performed when all pixels for a slice are available. A slice is an autonomous unit since coding a slice is done independently of its neighbors. Typically each frame contains 30 slices of 512×16 pixels.

4. Picture: The basic unit of display. It corresponds to a single frame in a video sequence. The spatial dimension of a frame are variable and are determined by the requirements of an application. Typically these dimensions are 512×480 pixels, which is similar to NTSC broadcast quality.

MPEG-2 initially consisted of three *profiles* (simple, main, and next), each further divided into four *levels* (high-level type 1, high-level type 2, main level, and low level) — the simple profile does not support bidirectional B* frames shown in Figure 4.26, eliminating the need to store them (Table 4.7). The levels refer primarily to the resolution of the video produced; for example, the low level refers to a standard image format video with a resolution of 352×240 (also known as source input format, SIF). The main level conforms with the CCIR/ITU-R 601 quality. The high level is for HDTV.

The main level/main profile was standardized in 1993 (with chips available in 1994). Toward the end of 1993, the structure of profiles and levels was changed to reflect likely applications of the standard. The levels were modified to high level, high-1440 level, main level, and low level; the profiles were modified to simple profile, main profile, SNR** scalable profile, Spatially Scalable Profile, and High Profile.[47] A previous profile known as "main + profile" was split into two scalable profiles, anticipating situations of network congestion, whereby the scalable profiles can drop information if it cannot be transported. The "next profile" was renamed "high profile" to reflect the fact that it applies to high-resolution video applications. See Table 4.8.

The MPEG-2 main profile is the one of greatest current interest.[37] It can handle images from the lowest level with MPEG-1 quality, up through broadcast quality at the main level, to HDTV at the high level.

The expectation is that MPEG-2 will play a critical role in most industrial and consumer applications in the foreseeable future. While MPEG-1 was developed for computer applications and, hence, supports progressive scanning only, MPEG-2 is to be used in the TV world and, hence, supports interlaced scanning.

The input to the MPEG-2 encoder is digital component video. The standard covers audio compression, video compression, and transport. In the transport area it defines[39]:

---

*B frames are storage areas where the incoming frame is compared with the preceding frame and used to predict the next one (this, however, requires that the decoder have more than two frames of video memory, thus more than doubling the memory requirement from 1 to 2 MB).

**Note: The acronym SNR is not defined even in the standard which says: "SNR scalability: A type of scalability where the enhancement layer(s) contain only coded refinement data for the DTC coefficients of the base layer."

TABLE 4.8 MPEG-2 Conformance Points

	Simple Profile	Main Profile	SNR Scalable Profile	Spatially Scalable Profile	High Profile
High level		x			x
High-1140 level		x		x	x
Main level	x	x	x		x
Low level		x	x		

Note: x = point of concordance (likely combinations of levels and profiles in actual applications).

1. Program streams: A grouping of audio, video, and data elemental components having a common time relationship, and being generally "associated" for delivery, storage, and playback.
2. Transport streams: A collection of program streams or elementary streams (video, audio, data) that have been multiplexed in a nonspecific relationship for the purpose of transmission.

These efforts at the system layer are aimed at providing a basic data structure; a data structure in this context is viewed as the "semantics and syntax" of a data stream that can serve as a common format for local usage (e.g., storage, edit, etc.) and for broadcast. A number of basic structural elements have been defined and are expected to become part of the system layer syntax. Pivotal to this structure is the fact that the transport stream is based on packet principles. The packets (in the 130 to 192 octet range) contain digital information from a single elementary stream or data type.* Each packet has a header of up to 4 octets that provide information such as packet ID, clear/scramble indication, key (even/odd), and continuity counter.

4.3.5 MPEG-4

MPEG-4 is the newest ISO standard under development by the Motion Picture Expert Group. MPEG-4 is the next audiovisual coding standard that ISO developed after MPEG-1 and MPEG-2. Unlike the previous two standards, which as covered earlier had clear application in mind (e.g., storage or transmission), MPEG-4 is a broad umbrella standard with a number of different technologies aimed at different applications. Initially MPEG-4 was focused at low-bit-rate communications, but later the scope was expanded to include multimedia coding. Interestingly, MPEG-4 is efficient across a range of bit rates, from a few kbps to tens of Mbps.

MPEG-4 developers are currently looking at frame reconstruction from a narrow bandwidth command string, speech and video synthesis, and the use of fractal geometry, computer visualization, and artificial intelligence to build accurate data from minimal frames. Some of the highlights of the standard are[42]:

- Ability to efficiently encode mixed media data such as video, graphics, text, images, audio, and speech, called audiovisual objects (AVOs);
- Ability to create a compelling multimedia presentation by composing these mixed media objects by a compositing script;
- Error resilience to enable robust transmission of compressed data over noisy communication channels;
- Ability to encode arbitrarily shaped video objects;
- Multiplexing and synchronization of the data associated with these objects so that they can be transported over network channels providing a QoS appropriate to the nature of the specific objects;
- Ability to interact with the audiovisual scene generated at the receiver.

---

*One advantage of this packetized approach, besides efficient transport (e.g., over ATM) is the fact that the packets can be encrypted in support of conditional access.

MPEG-4 is a multimedia coding standard; it standardizes tools not only for video coding but also coding audio, graphics, and text. The standard also includes a systems part, that describes how the audio, video, text, and graphics are synchronized. The MPEG-4 Committee has taken the approach of versioning to the standard formation process. Version 1 of the standard is now in CD status; and so is version 2. MPEG-4 version 1 includes a number of useful tools, and version 2 is expected to include some others being developed by the standards body. It is expected that MPEG-4 version 2 will be backward-compatible with MPEG-4 version 1.

Because of the desire for usage in noisy channels, MPEG-4 incorporates several error management features. These mechanisms provide important capabilities such as resynchronization, error detection, data recovery, and error concealment. The mechanisms are

- Video packet resynchronization
- Data partitioning
- Reversible variable-length coding (RVLCs)
- Header extension code (HEC)

When the compressed video data is transmitted over noisy channels, bit and burst errors corrupt the bit stream. A video decoder that is decoding this compromised bit stream will lose synchronization with the encoder (i.e., the decoder is unable to identify the precise location in the image where the current data belong). If rectifying measures are not in place, the quality of the decoded video degrades rapidly and significantly. One way to deal with this problem is for the encoder to insert resynchronization markers in the bit stream. When the decoder detects an error, it can then search for the resynchronization marker and regain resynchronization.

A horizontal row of macroblocks for QCIF images comprises a GOB (in the case of CIF images, each row of macroblocks consists of two GOBs). To provide synchronization resilience, the H.263 encoder has the option of inserting resynchronization markers at the beginning of each GOB. The smallest section to which the error can be isolated and concealed is, therefore, one row of macroblocks. Furthermore, when the resynchronization markers are restricted to be at the beginning of the GOBs, it is only possible for the decoder to isolate the errors to a fixed GOB independent of the image content.

MPEG-2 has a similar (optional) slice resynchronization methodology. MPEG-4 provides an improved method, because the MPEG-4 encoder is not restricted to inserting the resynchronization markers only at the beginning of each row of macroblocks. (H.263 version 2 adopted a resynchronization scheme similar to MPEG-4 in an additional annex.) The encoder has the option of dividing the image into video packets, each made up of an integer number of consecutive macroblocks. These macroblocks can span several rows of macroblocks in the image and can even include partial rows of macroblocks.[42] One simple approach is for the MPEG-4 encoder to insert a resynchronization marker every K bits.[*] Note that when the MPEG-4 encoder inserts the resynchronization markers at uniformly spaced bit intervals, the macroblocks interval between the resynchronization markers is shorter in the high-activity areas and longer in low areas. This implies that in the presence of a short burst of errors, the decoder can quickly localize the error to within a few macroblocks for high-activity areas of the images and retain the image quality in these areas.

After detecting an error in the bit stream and resynchronizing to the next marker, the decoder has isolated the data in error to be in the macroblocks between the two resynchronization markers. The video decoder discards all these macroblocks and replaces the luminance and chrominance of these macroblocks with the luminance and chrominance from the corresponding macroblocks in the previous frame to conceal the errors.

Between two markers, the motion and DCT data for each macroblock are coded together. Consequently, when the decoder detects an error, whether the error occurred in the motion or DCT part, all

---

[*]The spacing of the resynchronization markers in MPEG-4 is recommended based on the bit rates: for 24 kbps it is recommended to insert them at intervals of 480 bits; for bit rates between 25 and 48 kbps, every 736 bits.

the data in the stream has to be discarded: because of the uncertainty of the exact location where the error occurred, the decoder cannot determine that either the motion or DCT data on any macroblocks in the packet is not erroneous.

The data partitioning mode in MPEG-4 partitions the data within a video packet into a motion part and a texture part separated by a unique motion boundary marker (MBM) (the MBM identifies the end of the motion data and the beginning of the DCT data). When an error is detected in the motion section, the decoder flags an error and replaces all the macroblocks in the current packet with skipped blocks until the next resynchronization marker (resynchronization occurs at the next successfully read resynchronization marker). If any subsequent video packets are lost before resynchronization, those packets are replaced by skipped macroblocks as well. When an error is detected in the texture section (and no errors are detected in the motion section), the NMB motion vectors are used to perform motion compensation (NMB is the number of macroblocks in the video packet). The texture part of all the macroblocks is discarded and the decoder resynchronizes to the next resynchronization marker.[42] If an error is not detected in the motion or texture section of the bit stream, and the resynchronization marker is not found at the end of decoding all the macroblocks of the current packet, an error is flagged. In this case, only the texture part of all the macroblocks in the current packet needs to be discarded. Motion compensation can still be applied for the NMB macroblocks since we have higher confidence in the motion vectors since we got the MBM.[42]

The other two methods (reversible variable-length codes and header extension codes) also provide additional error management capabilities for MPEG-4.

4.4 MPEG-2 Standard Details[*]

This international standard was prepared by SC29/WG11, also known as MPEG (Moving Pictures Experts Group). MPEG was formed in 1988 to establish an international standard for the coded representation of moving pictures and associated audio stored on digital storage media. The MPEG-2 standard is published in four parts. *Part 1: Systems* specifies the system coding layer of the MPEG-2. It defines a multiplexed structure for combining audio and video data and means of representing the timing information needed to replay synchronized sequences in real time. *Part 2: Video* specifies the coded representation of video data and the decoding process required to reconstruct pictures. *Part 3: Audio*: specifies the coded representation of audio data. *Part 4: Conformance Testing* was still in preparation at this writing; it will specify the procedures for determining the characteristics of coded bit streams and for testing compliance with the requirements stated in Parts 1, 2, and 3. The specification is fairly complex; the standard is over 400 passes long. (The focus of this section is on the system aspects of the MPEG-2 standards.)

Table 4.9 depicts some of the key terminology and concepts of MPEG-2 (items shown in brackets are specific to one part of the standard only).

4.4.1 Part 1: Systems

The systems part of MPEG-2 addresses the combining of one or more elementary streams of video and audio, as well as other data single or multiple streams that are suitable for storage or transmission. Systems coding follows the syntactic and semantic rules imposed by Part 1 of the MPEG-2 specification and provides information to enable synchronized decoding without either overflow or underflow of decoder buffers over a wide range of retrieval or receipt conditions.

Systems coding is specified in two forms: the *transport stream* and the *program stream*. Each stream is optimized for a different set of applications. Both the transport stream and program stream defined in MPEG-2 provide coding syntax that is necessary and sufficient to synchronize the decoding and presentation of the video and the audio information, while ensuring that coded data buffers in the decoders

---

[*]This section is based directly on ISO/IEC 13818 (Committee Draft, November 1993).

TABLE 4.9   Key MPEG-2 Terms and Concepts

Access unit [System]	A coded representation of a presentation unit: in the case of compressed audio, an access unit is an Audio Access Unit; in the case compressed video, the coded representation of a picture
Bit rate	The rate at which the compressed bit stream is delivered from the storage medium or data link to the input of a decoder
Channel	A digital medium that stores or transports an ISO/IEC 13818 stream
Coded representation	A data element as represented in its encoded form
Compression	Reduction in the number of bits used to represent an item of data
Constant bit rate	Operation where the bitrate is constant from start to finish of the compressed bit stream
Constrained system parameter stream (CSPS) [System]	An ISO/IEC 13818 program stream for which the constraints defined in Part 1 apply
Data element	An item of data as represented before encoding and after decoding
Decoded stream	The decoded reconstruction of a compressed bit stream
Decoder	An embodiment of a decoding process
Decoding (process)	The process defined by MPEG-2 that reads an input coded bit stream and outputs decoded pictures or audio samples
Decoding time stamp (DTS) [System]	A field that may be present in a PES packet header that indicates the time an access unit is decoded in the system target decoder
Digital storage media (DSM)	A digital storage or transmission device or system
Editing	The process by which one or more compressed bit streams are manipulated to produce a new compressed bit stream; conforming edited bit streams must meet the requirements defined in MPEG-2
Elementary stream [System]	A generic term for one of the coded video, coded audio, or other coded bit streams
Encoder	An embodiment of an encoding process
Encoding (process)	A process, not specified in MPEG-2, that reads a stream of input pictures or audio samples and produces a valid coded bit stream as defined in MPEG-2
Entropy coding	Variable-length lossless coding of the digital representation of a single stream to reduce redundancy
Fast forward playback [Video]	The process of displaying a sequence, or parts of a sequence, of pictures in display-order faster than real time
Layer [Video, Systems]	One of the levels in the data hierarchy of the video and system specifications defined in Parts 1 and 2 of the MPEG-2 standard
MPEG-2 (multiplexed) stream [System]	A bit stream composed of zero or more elementary streams combined in the manner defined in Part 1 of the MPEG-2 standard
Packet data [System]	Contiguous bytes of data from an elementary stream present in a packet
Packet identifier (PID) [System]	A unique integer value used to associate elementary streams of a program in a single or multiprogram transport stream
Packet [System]	A packet consists of a header followed by a number of contiguous bytes from an elementary data stream; it is a layer in the system coding syntax described in the MPEG-2 standard
Packetized elementary stream (PES) [System]	A packetized elementary stream consists of PES packets, all of whose payloads consists of data from a single elementary stream, and all of which have the same stream ID
Padding [Audio]	A method to adjust the average length of an audio frame in time to the duration of the corresponding PCM samples, by conditionally adding a slot to the audio frame
PES packet header [System]	The data structure used to convey information about the elementary stream data contained in the PES packet data
Presentation time stamp (PTS) [System]	A field that may be present in a PES packet header that indicates the time that a presentation unit is presented in the system target decoder
Presentation unit (PU) [System]	A decoded audio access unit or a decoded picture
Program-specific information (PSI) [System]	Normative data which is necessary for the demultiplexing of transport streams and the successful regeneration of the program; in some cases, the nonmandatory information table is privately defined
Program [System]	A collection of elementary streams with a common time base
Random access	The process of beginning to read and decode the coded bit stream at an arbitrary point
Side information	Information in the bit stream necessary for controlling the decoder
Source stream	A single nonmultiplexing stream of samples before compression coding

TABLE 4.9   Key MPEG-2 Terms and Concepts (continued)

Start codes [System, Video]	32-bit codes embedded in that coded bit stream that are unique; used for several purposes including identifying some of the layers in the coding syntax
STD input buffer [System]	A first-in first-out buffer at the input of system target decoder for storage of compressed data from elementary streams before decoding
Still picture	A coded still picture consists of a video sequence containing exactly one coded picture which is intracoded; this picture has an associated PTS; and the presentation time of succeeding pictures, if any, is later than of the still picture by at least two picture periods
System target decoder (STD) [System]	A hypothetical reference model of a decoding process used to describe the semantics of an ISO/IEC 13818 multiplexed bit stream
Time stamp [System]	A term that indicates the time of an event
Transport packet header [System]	A data structure used to convey information about the transport stream payload
Variable bitrate	Operation where the bitrate varies with time during the decoding of a compressed bit stream

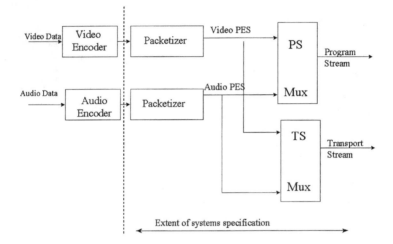

FIGURE 4.32   Simplified system overview.

do not overflow or underflow. Such information is coded in the syntax using time stamps concerning the decoding and the presentation of coded audio and visual data and time stamps concerning the delivery of the data stream itself. Both stream definitions are packet-oriented multiplexes.

The basic multiplexing approach for simple video and audio elementary streams is illustrated in Figure 4.32. The video and audio data is encoded as described in Parts 2 and 3 of the MPEG-2 standard. The resulting compressed elementary streams are packetized to produce Packetized Elementary Steams (PES), which are shown in Figure 4.33.* Information needed to use PES packets independently of either transport stream or program stream may be added when PES packets are formed (this information is not needed and need not be added when PES packets are further combined with system-level information to form transport streams or program streams). The systems standard (Part 1 of MPEG-2) covers those processes to the right of the vertical dashed line in Figure 4.32.

The program stream is analogous and similar to MPEG-1 Systems Multiplex (ISO 11172-1 MPEG Systems). It results from combining one or more packetized elementary streams, that have a common time base, into a single stream. The program stream definition can also be used to encode multiple audio and video elementary streams into multiple program streams, all of which have a common time base. Like the single program stream, all elementary streams can be decoded with synchronization. The

---

*To understand the full functionality of the PES, including the exact meaning of all fields, the reader is referred directly to the MPEG-2 standard.

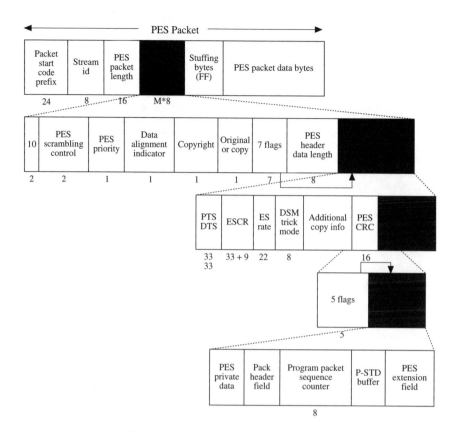

FIGURE 4.33 Packetized elementary stream.

program stream is designed for use in relatively error-free environments and is suitable for applications that may involve software processing of system information such as interactive multimedia applications. Program Stream packets may be of variable and relatively great length.

The transport stream combines one or more programs with one or more independent time bases into a single stream. PES packets, made up of elementary streams that form a program, share a common timebase. The transport stream is designed for use in environments where errors are likely, such as storage or transmission in lossy or noisy media. Transport stream packets are 188 bytes in length.

Program and transport streams are designed for different applications and their definitions do not strictly follow a layered model. It is possible and reasonable to convert from one to the other; however, one is not a subset or superset of the other. In particular, extracting the contents of a program from a transport stream and creating a valid program stream is possible and is accomplished through the common interchange format of packetized elementary streams; but not all of the fields needed in a program stream are contained within the transport stream — some must be derived. The transport stream may be used to span a range of layers in a layers model, and is designed for efficiency and ease of implementation in high-bandwidth applications.

The scope of syntactic and semantic rules set forth in the systems specification differ: the syntactic rules apply to systems layer coding only, and do not extend to the compression layer coding of the video and audio specifications; by contrast, the semantic rules apply to the combined stream in its entirety.

The systems specification does not specify the architecture or implementation of encoders or decoders, nor those of multiplexers or demultiplexers. However, bit stream properties do impose functional and performance requirements on encoders, decoders, multiplexers, and demultiplexers. For instance, encoders must meet minimum clock tolerance requirements. Notwithstanding this and other requirements, a

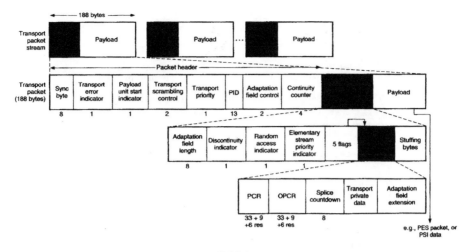

FIGURE 4.34 MPEG-2 transport stream.

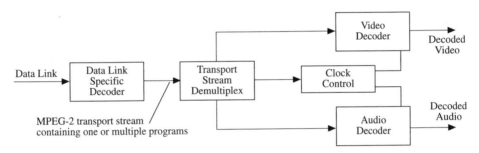

FIGURE 4.35 Prototypical transport demultiplexing and decoding example.

considerable degree of freedom exists in the design and implementation of encoders, decoders, multiplexers, and demultiplexers.

4.4.2 Transport Stream

The transport stream is a stream definition that is tailored for communicating or storing one or more programs of MPEG-coded data and other data in environments in which significant errors may occur. Such errors may be manifested as bit value errors or loss of packets.

The MPEG-2 transport stream may be constructed by any method that results in a valid stream. It is possible to construct a transport stream containing one or more programs from elementary coded data streams, from program streams, or from other transport streams that may themselves contain one or more programs (Figure 4.34).*

The transport stream is designed in such a way that several operations on a transport stream are possible with minimum effort. Among these are the ability to:

1. Retrieve the coded data from one program within the transport stream, decode it, and present the decoded results, as shown in Figure 4.35.
2. Extract the transport packets from one program within the transport stream and produce as output a different transport stream with only that one program, as shown in Figure 4.36.

---

*To understand the full functionality of the transport stream, including the exact meaning of all fields, the reader is referred directly to the MPEG-2 standard.

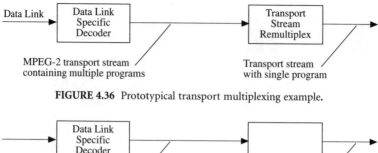

FIGURE 4.36 Prototypical transport multiplexing example.

FIGURE 4.37 Prototypical transport to program stream conversion.

3. Extract the transport packets of one or more programs from one or more transport streams and produce as output a different transport stream.
4. Extract the contents of one program from the transport stream and produce as output a program stream containing that one program as shown in Figure 4.37.
5. Take a program stream, convert it into a transport stream to carry it over a lossy environment, and then recover a valid, and in certain cases, identical program stream.

Figures 4.35 and 4.36 illustrate prototypical demultiplexing and decoding systems that take as input an MPEG-2 transport stream. Figure 4.35 illustrates the first case, where a transport stream is directly demultiplexed and decoded. MPEG-2 transport streams are constructed in two layers: a system layer and a compression layer. The input stream to the transport stream decoder has a system layer wrapped about a compression layer. Input streams to the video and audio decoders have only the compression layer.

Operations performed by the transport stream decoder either apply to the entire MPEG-2 Transport stream ("multiplex-wide operations") or to individual elementary streams ("stream-specific operations"). The MPEG-2 transport system layer is divided into two sublayers, one for multiplex-wide operations (the transport packet layer) and one for stream-specific operations (the PES packet layer).

A prototypical audio/video transport stream decoder is depicted in Figure 4.35 to illustrate the function of a decoder. The architecture is not unique — some transport stream system decoder functions such as decoder timing control might equally well be distributed among elementary stream decoders and the data link specific decoder — but this figure is useful for discussion. Likewise, indication of errors detected by the data link specific decoder to the individual audio and video decoders may be performed in various ways and such communications paths are not shown in the diagram. The prototypical decoder design does not imply any normative requirement for the design of an MPEG-2 transport stream decoder. Nonaudio/video data is also allowed, but not shown.

Figure 4.36 illustrates the second case, where a transport stream containing multiple programs is converted into a transport stream containing a single program: in the case shown in Figure 4.36 the remultiplexing operation will include the correction of PCR time stamps to account for the change in arrival of time stamps.

Figure 4.37 illustrates a third case in which an MPEG-2 multiprogram transport stream is first demultiplexed then converted into an MPEG-2 program stream.

Figures 4.36 and 4.37 indicate that it is possible and reasonable to convert between different types and configurations of MPEG-2 transport stream. There are specific fields defined in the transport stream and program stream syntaxes that facilitate the conversions illustrated. There is no requirements that specific implementations of demultiplexers or decoders include all of these functions.

4.4.2.1   Transport Stream Coding Structure

The transport stream coding layer allows one or more groups of one or more elementary streams to be combined into a single stream. A group of elementary streams, with a common system_clock_frequency time base, is called a program. Data from each elementary stream are encoded and multiplexed together with information that allows elementary streams within a program to be replayed in synchronism.

An MPEG-2 transport stream consists of one or more programs each containing one or more elementary streams and other streams multiplexed together. Each elementary stream consists of access units, which are the coded representation of presentation units. The presentation unit for a video elementary stream is a picture. The corresponding access unit includes all the coded data for the picture. The access unit containing the first coded picture of a group of pictures also includes any preceding data from the group of pictures, as defined in Part 2 of the MPEG-2 standard, starting the group_start_code. The access unit containing the first coded picture after a sequence header, as defined in Part 2 of MPEG-2, also includes a sequence header. The sequence_end_code is included in access unit containing the last coded picture of a sequence. The presentation unit for an audio elementary stream is the set of samples that correspond to samples from an audio frame (see Part 3 of the MPEG-2 standard for the definition of an audio frame).

Elementary stream data is carried in PES packets. A PES packet consists of a PES packet header followed by packet data, as seen in Figure 4.33. PES packets are inserted into transport packets, as seen in Figure 4.34. The first byte of each PES packet header is located at the first available payload location of a transport packet.

The PES packet header begins with a 32-bit start code that also identifies the stream to which the packet data belongs. The PES packet header may contain decoding and/or presentation time stamps (DTS and PTS) that refer to the first access unit that commences in the packet. The PES packet header also contains a number of flags opening up a range of optical fields. The packet data contains variable number of contiguous bytes from one elementary stream.

Transport packets begin with a 4 byte prefix, that contains a 13 bit packet ID (PID). The PID identifies, via the Program Specific Information (PSI) tables, the contents of the data contained in the transport packet (see list below). Transport packets of one PID value carry data of one and only one elementary stream. The PSI tables are carried in the transport stream. There are four PSI tables:

1. Program association table
2. Program map table
3. Network information table
4. Conditional access table

These tables contain the necessary and sufficient information to demultiplex and present programs. The program map table specifies, among other information, which PIDs, and therefore which elementary streams are associated to form each program. This table indicates the PID of the transport packets which carry the PCR for each program.

Transport stream packets may be null packets. Null packets are intended for padding of transport stream. They may be inserted or deleted by remultiplexing processes and, therefore, the delivery of the payload of null packets to the decoder cannot be assumed.

PID. The payload_unit_start_unit_indicator contained in the transport steam is a one bit flag that has normative meaning for transport packet that carry PES or PSI data. When the payload of the transport packet contains PES data, the payload_unit_start_indicator has the following signification: A "1" indicates that the payload of this transport packet will commence with a PES packet header and a "0" indicates there is no PES header in this transport packet payload. When the payload of the transport packet contains PSI data, the payload_unit_start_indicator has the following signification: If the transport packet carries the first byte of a PSI section, the payload_unit_start_indicator value is "1," indicating that the first byte of the payload of this transport packet carries the pointer_field. If the transport packet does not carry

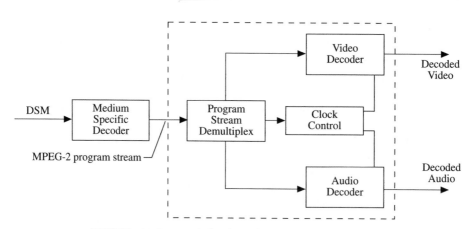

FIGURE 4.38 Prototypical audio/video program stream decoder.

the first byte of a PSI section, the payload_unit_start_indicator value is "0," indicating that there is no pointer_field in the payload. (The meaning of this bit for private-data-carrying transport packets is not defined in this international standard.)

4.4.3   Program Stream

As in the case of the transport stream, a prototypical audio/video program stream decoder system is depicted in Figure 4.38 to illustrate the function of a decoder. The architecture is not unique — system decoder functions including decoder timing might equally well be distributed among elementary stream decoders and the medium specific decoder — but this figure is useful for discussion. The prototypical decoder design does not imply any normative requirement for the design of an MPEG-2 program stream decoder. Nonaudio/video data are also allowed, but not shown.

The prototypical MPEG-2 program stream decoder shown in Figure 4.32 is composed of system, video, and audio decoders conforming to Parts 1, 2 and 3, respectively, of the MPEG-2 standard. In this decoder the multiplexed coded representation of one or more audio and/or video streams is assumed to be stored on a digital storage medium (DSM), or network, in some medium-specific format. The medium specific format is not governed by the MPEG-2 standard, nor is the medium-specific decoding part of the prototypical MPEG-2 program stream decoder. Figure 4.39 depicts the program stream Syntax.[*]

The prototypical decoder accepts as input an MPEG-2 program stream and relies on a program stream decoder to exact timing information from the stream. The program stream decoder demultiplexes the stream, and the elementary stream so produced serves as input to video and audio decoders, whose outputs are decoder video and audio signals. Included in the design, but not shown in the figure, is the flow of timing information among the program stream decoder, the video and audio decoders, and the medium specific decoder. The video and audio decoders are synchronized with each other and with the DSM using this timing information.

MPEG-2 program streams are constructed in two layers: a system layer and a compression layer. The input stream to the program stream decoder has a system layer wrapped about a compression layer. Input streams to the video and audio decoders have only the compression layer.

Operations performed by the program stream decoder either apply to the entire MPEG-2 program stream ("multiplexer-wide operations") or to individual elementary streams ("stream-specific operations"). The MPEG-2 program system layer is divided into two sublayers, one for multiplexer-wide operations (the pack layer) and one for stream-specific operations (the PES packet layer).

---

[*]To understand the full functionality of the program stream, including the exact meaning of all fields, the reader is referred directly to the MPEG-2 standard.

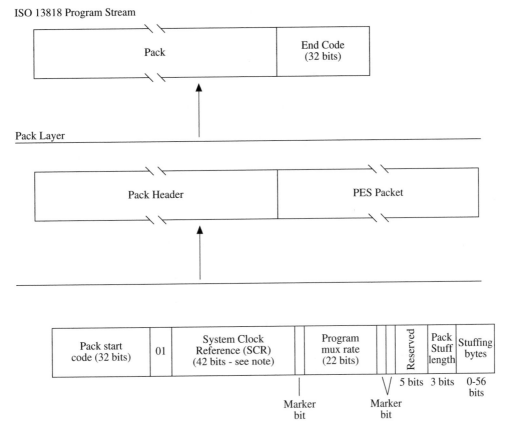

FIGURE 4.39 Program stream pack. *Note:* SCR is a 42-bit number coded in four segments as follows: 3 bits (bits 32 → 30) system clock reference base; 1 marker bit (in addition to the 42 bits); 15 bits (bits 29 → 15) system clock reference base; 1 marker bit (in addition to the 42 bits); 15 bits (bits 14 → 0) system clock reference base; 1 marker bit (in addition to the 42 bits); 9 bits system clock reference extension.

4.4.4 Conversion between Transport Stream and Program Stream

It is possible and reasonable to convert between transport stream and program streams by means of packetized elementary streams (PES). This results from the specification of transport stream and program stream. PESs may, with some constraints, be mapped directly from the payload of one multiplexed bit stream into the payload of another multiplexed bit stream. It is possible to identify the correct order of PES packets in a program to assist with this. Certain other information necessary for conversion, e.g., the relationship between elementary streams, is available in tables and headers in both streams. Such data must be available and correct in any stream before and after conversion. Not all transport streams will have been formed from program streams, but it is possible to create a valid program stream from a valid transport stream.

4.4.5 Packetized Elementary Stream

Transport streams and program streams are each logically constructed from packetized elementary stream packets, as indicated in the figures above. PES packets are to be used to convert between transport streams and program streams; in some cases the PES packets need not be modified when performing such conversions. PES packets may be much larger than the size of a transport packet.

A continuous sequence of PES packets of one elementary stream with one stream ID may be used to construct a packetized elementary stream. PES streams include elementary stream clock reference (ESCR) fields and elementary stream rate (ES-Rate) fields, and the elementary stream data must be contiguous bytes from the elementary stream in their original order. PES streams do not contain some necessary systems information that is contained in program streams and transport streams. Examples include the information in the pack header, program stream map, program stream directory, program map table, and elements of the transport packet syntax.

The PES stream is a logical construct that may be useful within implementations of this standard; however, it is not defined as a stream for interchange and interoperability. Applications requiring streams containing only one elementary stream can use program streams or transport streams which each contain only one elementary stream. These streams contain all of the necessary system information. Multiple program streams or transport streams, each containing a single elementary stream, can be constructed with a common time base and therefore carry a complete program, i.e., with audio and video.

4.4.6   Timing Model

MPEG systems, video, and audio all have a timing in which the end-to-end delay from the signal input to an encoder to the signal output from a decoder is a constant. This delay is the sum of encoding, encoder buffering, multiplexing, communication or storage, demultiplexing, decoder buffering, decoding, and presentation. As part of this timing model all video pictures and audio samples are presented exactly once, unless specifically coded to the contrary, and the interpicture interval and audio sample rate are the same at the decoder as at the encoder. The system stream coding contains timing information that can be used to implement systems that embody constant end-to-end delay. It is possible to implement decoders that do not follow this model exactly; however, in such cases it is the decoder's responsibility to perform in an acceptable manner. The timing is embodied in the normative specifications of this standard, which must be adhered to by all valid bit stream, regardless of the means of creating them.

All timing is defined in terms of a common system clock, referred to as a system time clock. In the program stream this clock may have an exactly specified ratio to the video or audio sample clocks, or it may have an operating frequency that differs slightly from the exact ratio while still providing precise end-to-end timing and clock recovery. In the transport stream the system clock frequency is constrained to have the exactly specified ratio to the audio and video sample clocks at all times; the effect of this constraint is to simplify sample rate recovery in decoders.

4.4.7   Conditional Access

Encryption and scrambling for conditional access to programs encoded in the program and transport streams is supported by the system data stream definitions, but Conditional Access (CA) mechanisms are not specified in the MPEG-2 standard. The stream definitions are designed so that implementation of practical conditional access systems is reasonable, and there are some syntactical elements specified which provide specific support for such systems.

4.4.8   Multiplex-Wide Operations

Multiplex-wide operations include the coordination of data retrieval from the DSM or data link, the adjustment of clocks, and the management of buffers. The tasks are intimately related. If the rate of data delivery off the data link or DSM is controllable, then data delivery may be adjusted so that decoder buffers do not reach illegal states of overflow or underflow; but if the data rate is not controllable, then elementary stream decoders must slave their timing to the data source to avoid overflow or underflow.

Program streams are composed of packs whose headers facilitate the above tasks. Pack headers specify intended times at which each byte is to enter the program stream decoder and buffer management. The schedule need not be followed exactly by decoders, but they must compensate for deviations about it.

Similarly, transport streams are composed of transport packets with headers containing information that specifies the times at which each byte is intended to enter program stream decoder from the data source. This schedule provides exactly the same functions as that which is specified in the program stream.

An additional multiplex-wide operation is a decoder's ability to establish what resources are required to decode a transport stream or program stream. The first pack of each program stream conveys parameters to assist decoders in this task. Included, for example, are the stream's maximum data rate and the highest number of simulation video channels. The transport stream likewise contains globally useful information.

The transport stream and program stream each contain information that identifies the pertinent characteristics of and between the elementary streams that constitute each program. Such information may include the language in audio channels, as well as the relationship between video stream when multilayer video coding is implemented.

4.4.9   Individual Stream Operations (PES Packet Layer)

The principal stream-specific operations are (1) demultiplexing and (2) synchronizing playback of multiple elementary streams.

4.4.9.1   Demultiplexing

On encoding, program streams are formed by multiplexing elementary streams, and transport streams are formed by multiplexing elementary streams, program streams, or the contents of other transport streams. Elementary streams may include private, reserved, and padding streams in addition to MPEG-2 audio and video streams. The streams are temporally subdivided into packets, and the packets are serialized. A PES packet contains codes from one and only one elementary stream.

In the program stream both fixed and variable lengths are allowed. For transport streams the packet length is 188 bytes. Both fixed and variable PES packet lengths are allowed, but will be relatively long in most applications.

On decoding, demultiplexing is required to reconstitute elementary streams from the multiplexed program stream. Stream_ID codes in program stream packet headers, and packet ID codes in the transport stream make it possible.

4.4.9.2   Synchronization

Synchronization among multiple elementary streams is effected with Presentation Time Stamps (PTS) in the program and transport bit stream. Time stamps are generally in units of 90 kHz, but the System Clock Reference (SCR), the Program Clock Reference (PCR), and the optional elementary stream clock reference have extensions with a resolution of 27 MHz. Decoding of N elementary streams is synchronized by adjusting the decoding of streams to a common master time base rather than by adjusting the decoding of one stream to match that of another. The master time base may be one of the N decoder's clocks, the DSM or channel clock, or it may be some external clock. Each program in a transport stream that contains multiple programs has its own time base.

Because presentation time stamps apply to the decoding of individual elementary streams, they reside in the PES layer of both the transport streams and program streams. End-to-end synchronization occurs when encoders have time stamps at capture time, when the time stamps propagate with associated coded data to decoders, and when decoders use those time stamps to schedule presentations.

Synchronization of a decoding system with a data source is achieved through the use of the SCR in the program stream and by the equivalent PCR in the transport stream. The SCR and PCR are time stamps encoding the timing of the bit stream itself in terms of the same time base as is used for the audio and video PTS values from the same program. Since each program may have its own time base, there are separate PCR fields for each program in a transport stream containing multiple programs. It is possible to have only one PCR for some or all programs in a multiprogram transport.

4.4.9.3 Relation to Compression Layer

The PES packet layer is independent of the compression layer in some senses, but not in all senses. It is independent in the sense that PES packet payload need not start at compression layer start codes, as defined in Parts 2 and 3 of MPEG-2. For example, a video packet may start at any byte in the video stream. However, time stamps encoded in PES packet headers apply to presentation times of compression layer construct (namely, presentation units).

4.5 QoS Issues and Video over the Internet

This section returns to the concepts introduced in Section 4.1 in reference to the evolving desire to deliver business video over enterprise networks. It provides some basic QoS-focused background material that needs to be understood by any developer addressing this aspect of video distribution.

4.5.1 IP/Internet Background

Network communications can de categorized into two basic types: *circuit-switched* (sometimes called connection-oriented) and *packet/fastpacket-switched* (these can be connectionless or connection-oriented) networks. Circuit-switched networks operate by forming a dedicated connection (circuit) between two points. In packet-switched network, data to be transferred across a network is segmented into small blocks called *packets* (also known as datagrams or protocol data units) that are multiplexed onto high-capacity interswitch trunks. A packet, which usually contains few hundred bytes of data, carries routing information that enables the network hardware to know how to send it forward to the specified destination. In frame relay, the basic transfer unit is the (data link layer) *frame*; in cell relay this basic unit is the (data link layer) *cell*. Services such as frame relay and ATM use circuit-switching principles; namely, they use the call setup mechanism similar to that of a circuit-switched (ISDN) call. IP has become the *de facto* standard connectionless packet network layer protocol for both LANs and WANs. In a connectionless environment there is no call setup; each packet finds its way across the network independently of the previous one.

4.5.2 Internet Protocol Suite

TCP/IP is the name for a family of over 100 data communications protocols used in the Internet and in intranets. In addition to the communication functions supported by TCP (end-to-end reliability over a connection-oriented session) and IP (subnetwork-level routing and forwarding in a connectionless manner), the other protocols in the suite support specific application-oriented tasks, e.g., transferring files between computers, sending mail, or logging in to a remote host. TCP/IP protocols support layered communication, with each layer responsible for a different facet of the communications (as seen in Table 4.10). Some of the voice-over-IP or video-over-IP applications will utilize TCP, while others utilize RTCP and UDP.

4.5.3 The Internet

The same IP technology now used extensively in corporate intranets is used in (in fact, originated from) the Internet. The Internet is a collection of interconnected government, education, and business computer networks — in effect, a network of networks. Recently there has been a near-total commercialization of the Internet, allowing it to be used for pure business applications. (The original roots of the Internet were in the research and education arena.) A person at a computer terminal or personal computer with the proper software communicates across the Internet by having the driver place the data in an IP packet and "addressing" the packet to a particular destination on the Internet. Communications software in routers in the intervening networks between the source and destination networks "read" the addresses on packets moving through the Internet and forward the packets toward their destinations. TCP guarantees end-to-end integrity.

TABLE 4.10 Functionality of the TCP/IP Suite Layers

Network interface layer	This layer is responsible of accepting and transmitting IP datagrams. This layer may consist of a device driver (e.g., when the network is a local network to which the machine attaches directly) or a complex subsystem that uses its own data link protocol.
Network layer (Internet layer)	This layer handles communication from one machine to the other. It accepts a request to send data from the transport layer along with the identification of the destination. It encapsulates the transport layer data unit in an IP datagram, uses the datagram routing algorithm to determine whether to send the datagram directly onto a router. The Internet layer also handles the incoming datagrams, and uses the routing algorithm to determine whether the datagram is to be processed locally or is to be forwarded.
Transport layer	In this layer the software segments the stream of data being transmitted into small data units and passes each packet along with a destination address to the next layer for transmission. The software adds information to the packets including codes that identify which application program sent it, as well as a checksum. This layer also regulates the flow of information and provides reliable transport, ensuring that data arrives in sequence and with no errors.
Application layer	At this level, users invoke application programs to access available services across the TCP/IP Internet. The application program chooses the kind of transport needed, which can either be messages or stream of bytes, and passes it to the transport level.

From a thousand or so networks in the mid-1980s, the Internet has grown to an estimated 1 million connected networks (by 1997) with about 100 million people having access to it (by 1997). The majority of these Internet users currently live in the U.S. or Europe, but the Internet is expected to have ubiquitous global reach over the next few years.

Starting in 1973, ARPA initiated a research program to investigate techniques and technologies for interlinking packet networks of various kinds. The objective was to develop communication protocols that would allow networked computers to communicate transparently across multiple packet networks. The project became very successful and there was increasing demand to use the network, so the government separated military traffic from civilian research traffic, bridging the two by using common protocols to form "an internetwork" or "internet." The term *Internet* is defined as "a mechanism for connecting or bridging different networks so that two communities can mutually interconnect." So, in the mid-1970s ARPA became interested in establishing a packet-switched network to provide communications between research institutions in the U.S. With the goal of heterogeneous connectivity in mind, ARPA funded research by Stanford University and Bolt, Beranek, and Newman to create an explicit series of communications protocols. The ARPA-developed technology included a set of network standards that specified the details of the computers that would be able to communicate, as well as a set of conventions for interconnecting networks and routing traffic. The result of this development effort, completed in the late-1970s was the *Internet suite of protocols*. Soon thereafter, there were a large number of computers and thousands of networks using TCP/IP, and it is from their interconnections that the modern Internet has emerged. ARPA was also interested in integrated voice and data.

While the ARPANet was growing into a national network, researchers at the Xerox Corporation Palo Alto Research Center were developing one of the technologies that would be used in local-area networking, namely, the ethernet. Ethernet became one of the important standards for how to implement building and campus data communications networks. At about the same time, ARPA funded the integration of TCP/IP support into the version of the UNIX operating system that the University of California at Berkeley was developing. It follows that when companies began marketing non-host-dependent workstations that ran UNIX, TCP/IP was already built into the operating system software, and vendors such as Sun Microsystems included an ethernet port on the device. Consequently, TCP/IP over ethernet became a common way for workstations to interconnect.

The same technology that made PCs and workstations possible made it possible for vendors to offer relatively inexpensive add-on cards to allow a variety of PCs to connect to ethernet LANs. Software vendors took the TCP/IP software from Berkeley UNIX and ported it to the PC, making it possible for PCs and UNIX machines to use the same protocol on the same network.

In 1986, the U.S. National Science Foundation (NSF) initiated the development of the NSFNet. NSFNet has provided a backbone communication service for the Internet in the U.S. It should be noted that the NSFNet operated utilizing a service Acceptable User Policy (AUP). The policy stated that the NSFNet is to support open research and education in and among U.S. research and intellectual institutions, plus research arms of for-profit firms when engaged in open scholarly communication and research. Use for other purposes was not acceptable. The commercialization of the Internet that we are experiencing is not based on the AUP, in effect abrogating it. By the end of 1991, the Internet had grown to include some 5000 networks in over three dozen countries, serving over half-a-million host computers. These numbers have continued to grow at geometric rates throughout the 1990s. There are now several thousand Internet Access Provides (ISPs), although the number is expected to decrease greatly in the next 5 years. Table 4.11 depicts highlights of the history of the Internet over a 30-year span.

TCP and IP were developed for basic control of information delivery across the Internet. Application layer protocols, such as TELNET (Network Terminal), FTP (File Transfer Protocol), SMTP (Simple Mail Transfer Protocol), HTTP (HyperText Transfer Protocol), have been added to the TCP/IP suite of protocols to provide specific network services. Access and backbone speeds have increased from 56 kbps, to 1.5 Mbps (most common now), to 45 Mbps and beyond, for most of the backbones. Voice applications over IP have to ride over the Internet systems developed for traditional data services. Most problematic is the lack of QoS support; this, however, is expected to slowly change. Nonetheless, in spite of the emergence of new technologies, such as RSVP and RTP, a retarding factor to true QoS support is the very success of the Internet: the number of people using it is increasing at such a rapid rate that it is difficult to add enough resource and protocol improvements to keep up with the demand.

Intranets use the same WWW/HTML/HTTP and TCP/IP technology used for the Internet. When the Internet caught on in the early-to-mid-1990s, planners were not looking at it as a way to run their businesses. But just as the action of putting millions of computers around the world on the same protocol suite fomented the Internet revolution, so connecting islands of information in a corporation via intranets is now sparking a corporate-based information revolution. Thousands of corporations now have intranets. Across the business world, employees from engineers to office workers are creating their own home pages and sharing details of their projects with the rest of the company.

4.5.4   QoS — Problems and Solutions

Voice and video over IP is impacted by network congestion. QoS encompasses various levels of bandwidth reservation and traffic prioritization for multimedia and other bandwidth-intensive applications.

The specific QoS solutions depend on the applications and circumstances at hand. QoS is generally not required for batch applications; it is needed for most if not all real-time applications. See Table 4.12.[48] For nonmultimedia applications, QoS in enterprise networks is useful for allocating and prioritizing bandwidth to specific users. For example, accounting departments may need more bandwidth when they are closing the books each month and a CEO needs more bandwidth during an extensive videoconferencing session. QoS is also important to supply streams of data that continuously move across the user's computer screen, such as stock tickers, real-time news, or viable data.

Various QoS solutions for intranets and the Internet are available, beginning at the low end with more bandwidth to the LAN desktop via Layer 2 switching. New protocols and standards offer the next level of QoS for enterprise network environments, including 802.1p, 802.1q, and RSVP.[48] Using ATM as a backbone improves bandwidth between subnetworks, and Layer 3 switching adds performance improvements in environments where IP dominates. Finally, end-to-end ATM provides many levels of built-in QoS.

Besides the capability for bandwidth reservation, QoS is affected by abilities of switches to perform real-time IP routing. Advances in silicon integration are being brought to bear for optimizing the performance of third-wave switches and paving the way for wire-speed IP routing capabilities.[49] Third-wave switches are optimized for switching at Gbps speeds. This is a possible thanks to advancements in

TABLE 4.11 A Snapshot of Internet-Related Activities over the Years

- Late 1960s: ARPA (think tank of DoD) introduces ARPANet
- During 1970s ARPAnet expands geographically and functionally (allows nonmilitary traffic, e.g., universities and defense contractors)
- By late 1970s realization takes hold that ARPAnet cannot scale
- TCP/IP is developed for heterogeneous networking, interenterprise connectivity; protocols to support global addressing and scalability
- Early 1980s (1983) TCP/IP is a standard operating environment for all attached systems
- Network splits into a military component (MILNET) and a civilian component (ARPANet)
- 1986: Six supercomputer centers established by NSF
- Interagency dynamics and funding considerations lead to creation of NSFNet by the NSF; IP protocol and newer equipment utilized in NSFNet; NSFNet and ARPANet intersected at Carnegie Mellon University
- Late 1980s: ARPANet absorbed into NSFNet
- "Phase 1": Three-tiered architecture developed
  1. NSF to undertake overall management; fund the backbone operationally and in terms of technology upgrades
  2. Regional and state network providers; supply Internet services between universities and the backbone; to become self-supportive on service fees
  3. Campus networks, organizations, colleges, and universities use TCP/IP-based system to provide widespread access to researchers and students
- Six supercomputer sites interconnected in 1987 using DEC routers and 56 kbps links
- Traffic congestion begins to be experienced
- "Phase 2": Merit partnership formed with IBM and MCI, to upgrade network
- By mid-1988 a DS1-line (1.544-Mbps) network connects over a dozen sites, IBM-based switches used
- Re-engineering in 1989 due to fast growth (15%/month); new routers and additional T1 links (MCI) installed
- "Phase 3": Third redesign of NSFNet: Outsourcing approach, whereby NSFNet is "overlaid" over a public Internet (NSF relieved from responsibility of upgrading the network on an ongoing basis); lines upgraded to DS-3 rates (45 Mbps)
- Merit, IBM, and MCI form Advanced Network Services, Inc. (ANS): the not-for-profit organization is to build/manage a commercial Internet
- DS3 lines provided by MCI; routers by IBM (RS/6000-based); network also called ANSNet; NSFNet is now a virtual network in the ANSNet (migration accomplished in 2 years).
- 1992: Original NSFNet dismantled
- ANS launches for-profit subsidiary (ANS CORE) to face costs
- Debates sparked by commercial Internet providers
  - PSINet, CERFNet, and AlterNet formed Commercial Internet Exchange (CIX) as a backbone and bypass to the NSFNet; 155 other members join, including NEARNet, JvNCNet, SprintLink, and InfoLAN
  - (Based on CIX approach) CICNet, NEARNet, BARRNet, North WestNet, NYSERNet WestNet, and MIDNet form the Corporation for Regional & Enterprise Networking (CoREN)
  - Regional commercial providers (not in CoREN) compete against CoREN
- "Phase 4": Rapid increase requires NSF to redesign the backbone
- Two years of bidding and planning leads to two awards to replace current NSFNet: (1) MCI to deploy Very High Speed Backbone Network Service (vBNS) based on 155 Mbps, SONET/ATM to connect NFS supercomputing centers; (2) Merit and USC Information Sciences Institute to do routing coordination
- Network Access Providers (NAPs) to provide access to the vBNS; NAP functions went to Ameritech, Sprint, MFS, and PacTel
- NFS instituted the Routing Arbitrer for fair treatment among various Internet service providers with regard to routing administration; database of route information, network topology, routing path preferences, interconnection information; deployment of routing which supports type of service, precedence routing, bandwidth on demand, and multicasting (accomplished by "route servers" using Border Gateway Protocol and Interdomain Routing Protocol)
- Fund established to support Network Information Center (NIC)
  - Registration Services (by Network Solutions Inc.): IP and Domain Names; who is and white pages
  - Directory Services (by AT&T): Directory of directories, white/yellow pages
  - Information Services (by General Atomics): coordination services, clearinghouse for information, training, workshops, reference desk, education (General Atmonics operated CERFNet — now owned by TCG — and San Diego Supercomputer Center)
- Complete commercialization of the Internet completed in mid-1990s; now about two dozen national Tier 1 Network Service Providers
- Thousands of ISPs emerge by late 1990s, but shake out of ISP was predicted by the turn of the century

TABLE 4.12 Applications and QoS

	QoS Required	Applications
Non-real-time data	Little or none	Data file transfer, imaging, simulation, and modeling
Non-real-time multimedia	Little or none	Exchange text E-mail, exchange audio/video E-mail, Internet browsing with voice and video, intranet browsing with voice and video
Real-time one-way	Various QoS levels	Multimedia playback from server, broadcast video, distance learning, surveillance video, animation playback
Real-time interactive	Various medium or high QoS levels	Videoconferencing, audioconferencing, process control

high-performance custom ASICs that can process packets simultaneously and in real-time across multiple ports in a switch. Furthermore, the design of ultrawide data paths and multigigabit switching backplanes enable third-wave switches to perform at gigabit speeds through full-duplex connections on all ports without blocking.[51]

4.5.5 Protocols for QoS Support for Audio and Video Applications

4.5.5.1 RSVP Applications

RSVP, along with available network bandwidth, is required to ameliorate the overall *quality* in IP networks. New applications are now emerging that require such capabilities. For example, some companies are adding Web telephone access to their call centers, letting customers reach the carrier's customer service agent by clicking an icon at their Web site that reads "speak to the agent." But in order to scale this on a broad scale, standards are required so that QoS can be supported and made available as a network service.

RSVP is based on receiver-controlled reservation requests for unicast or multicast communication. RSVP carries a specific QoS through the network, visiting each node the network uses to carry the stream. At each node, RSVP attempts to make a resource reservation for the stream. To make a resource reservation at a node, the RSVP daemon communicates with two local decision modules, *admission control* and *policy control*. Admission control determines whether the node has sufficient available resources to supply the requested QoS. Policy control determines whether the user has administrative permission to make the reservation. If either check fails, the RSVP program returns an error notification to the application process that originated the request. If both checks succeed, the RSVP daemon sets parameters in a packet classifier and packet scheduler to obtain the desired QoS. The packet classifier determines the QoS class for each packet and the scheduler orders packet transmission to achieve the promised QoS for each stream (Figure 4.40).

A receiver-controlled reservation allows scaling of RSVP for large multicast groups. This support is based on the ability of RSVP to merge reservation requests as they progress up the multicast tree. The reservation for a single receiver does not need to travel to the source of a multicast tree; rather, it travels only until it reaches a reserved branch of the tree.

RSVP does not perform its own routing; instead, it uses underlying routing protocols. There is vendor interest in delivering RSVP on routers. A draft version of RSVP was approved by the IETF in 1996, and by 1997 vendors such as Cisco and Bay Networks were expressing interest, although they were being quoted as stating that "there is little demand for RSVP from applications at the moment."

To ensure delivery through the network, RSVP allows listeners to request a specific QoS for a particular data flow. Listeners can specify how much bandwidth they will need and what maximum delay they can tolerate; internetworking devices then set aside the bandwidth for that flow. Users are either granted the channel they have requested or are given a "busy signal." RSVP hosts and networks interact to achieve a guaranteed end-to-end QoS transmission. All the hosts, routers, and other network infrastructure elements between the receiver and sender must support RSVP. They each reserve system resources such as bandwidth, CPU, and memory buffers to satisfy a request.

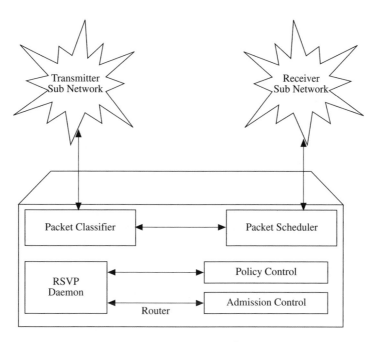

FIGURE 4.40 RSVP routing.

RSVP operates on top of IP (either IPv4 or IPv6), occupying the place of a transport protocol in the protocol stack, but provides session-layer services (it does not transport any data). The RSVP protocol is used by routers to deliver control requests to all nodes along the paths of the flows.

Vendors have implemented RSVP both above and below Winsock. RSVP-aware applications can be developed with Winsock 2, which has a QoS-sensitive API. Another approach is to use an RSVP proxy that runs independently of the real application, making RSVP reservations.

RSVP raises questions about billing for Internet bandwidth. In the current model, ISPs oversell their available capacity, and customers accept slowdowns. Since resource reservation puts a specified demand on bandwidth, overselling would result in unacceptable performance (by the admission control module). ISPs will probably offer different service levels, and premiums will be charged for RSVP reservations. Billing across multiple carriers will also have to be resolved, as will the allocation of computational resources to routers to inspect and handle packets on a prioritized basis. It is unclear whether existing routers would be able to handle widescale implementation of RSVP across the whole Internet.

Developers now see the use of RSVP at the edges, as a signaling protocol; MPLS is likely to emerge in the core of a large IP-based network. Typically such network would be powered by gigarouters.

4.5.5.2 Real-Time Streaming Protocol Applications and Active Streaming Format

The Internet provided the impetus for the development of streaming technologies. The growth of real-time media on the Internet has stretched the HTTP capabilities used for downloading files to their maximum. The IETF is now attempting to standardize functions such as starting and stopping data streams, synchronizing multiple media elements, and implementing other controls. To this end, the main IETF work is embodied in RTSP, which was jointly proposed by Progressive Networks, Netscape, and Columbia University toward the end of 1996.

The latest version of RTSP essentially provides HTTP-level services to real-time streaming data types. However, RTSP differs from HTTP in that data delivery takes place out-of-band utilizing a distinct protocol. RTSP establishes and controls either single or several time-synchronized streams of continuous media. RTSP is expected to use TCP as the transport layer (for control only), but UDP may be optionally supported. Although in draft specification, RTSP implementations are available today.

In parallel with RTSP development, Microsoft has implemented a proprietary protocol called Active Streaming Format (ASF) in its new Netshow server platform. While it offers capabilities similar to RTSP, Microsoft documents refer to ASF as a "file format" and describe it as a component of the Microsoft overall ActiveX strategy. It is a kind of metafile that packages multiple "media objects" into a unified framework. Like RTSP, it may be used to synchronize a number of multimedia objects including audio, video, still images, events, URLs, HTML pages, script commands, and executable programs. Unlike RTSP, an ASF stream includes both control and content elements.

ASF content may either be constructed off line or captured in real time. This multimedia content is stored into ASF as objects. These elements may be combined into a single ASF file. ASF retains all forms of media (e.g., audio and video compression) and (optional) synchronization information so that when the file is played over a network, the user sees and hears the file exactly as the file creator intended.[50]

ASF data objects are stored within an ASF file as "packets." Each packet is designed to be directly inserted "as is" into the data field of a data communication transport protocol. These packets are designed to be streamed across a network at a specific bit rate. The packet structure contains one or more payloads (i.e., distinct media streams) of data. Each packet may contain the data from a single media stream, or interleaved data from several media streams. A "packet" is a collection of multimedia data that is ready to be streamed "as is" over the Internet/intranet. Ideally, the packet has been correctly sized so that all that needs to be done to ship it "over the wire" is to append the appropriate data communication protocol headers. ASF does not impose a packet size limitation; however, in practice, the packet sizes generally run from 512 bytes to the data communication's maximum transmission unit (MTU) size. Each packet may contain interleaved data (i.e., composed of data from multiple multimedia streams). The format of the packet data is fairly complex in order to ensure that the packet data is as dense as possible for efficient transmission over a network.[50]

ASF data can be tailored to satisfy a variety of network requirements. For example, the data in each ASF file have been designed to stream at a distinct bit rate. The actual streaming bit rate is determined by the file's creator. The file's content creator has a range of streaming bit rates to choose between (e.g., 14.4 kbps to 6 Mbps). ASF content can thus be flexibly targeted to specific network environments with distinct capacity requirements. Similarly, there are no data communications dependencies within ASF: ASF data can be carried over a wide variety of differing transport protocols. ASF multimedia streams can be stored on traditional file servers, HTTP servers, or specialized media servers, and can be transmitted efficiently over a variety of different network transports. These transports include TCP/IP, UDP/IP, RTP, and IPX/SPX. This data may be sent as either unicast- (point-to-point) or multicast streams.[50,51]

4.5.5.3 Internet Stream Protocol Version 2

The first version of the Stream Protocol (ST) was published in the late 1970s and was used throughout the 1980s for experimental transmission of voice, video, and distributed simulation. The experience gained in these applications led to the development of the revised protocol version ST2. The revision extends the original protocol to make it more complete and more applicable to emerging multimedia environments. The specification of this protocol version is contained in RFC 1190 that was published in October 1990.

With more and more developments of commercial distributed multimedia applications under way, and with a growing dissatisfaction at the network-secured QoS for audio and video over IP (particularly in the MBONE context), interest in ST2 has grown over the last few years. Companies have products available incorporating the protocol. Implementations of ST2 for Digital Equipment, IBM, NeXT, Macintosh, PC, Silicon Graphics, and Sun platforms are available.[52]

ST2 is an experimental resource reservation protocol intended to provide end-to-end real-time guarantees over the Internet or intranet. It allows applications to build multidestination simplex data streams with a desired QoS. The ST2 is an connection-oriented internetworking protocol that operates at the same layer as IP. It has been developed to support the efficient delivery of data streams to single or multiple destinations in applications that require guaranteed quality of service. ST2 is part of the IP protocol family and serves as an adjunct to, not a replacement for, IP. The main application areas of the protocol are the real-time transport of multimedia data, e.g., digital audio and video packet streams.

ST2 can be used to reserve bandwidth for real-time streams. This reservation, together with appropriate network access and packet scheduling mechanisms in all nodes running the protocol, guarantees a well-defined QoS to ST2 applications. It ensures that real-time packets are delivered within their performance targets, that is, at the time where they need to be presented. This facilitates a smooth delivery of data that is essential for time-critical applications, but cannot typically be provided by best-effort IP communication.[52]

ST2 consists of two protocols: ST (stream transport) for the data transport and SCMP (stream control message protocol), for all control functions. ST is simple and contains only a single PDU format that is designed for fast and efficient data forwarding in order to achieve low communication delays. SCMP packets are transferred within ST packets. For comparison, SCMP is more complex than ICMP.

ST2 is designed to coexist with IP on each node. A typical distributed multimedia application would use both protocols: IP for the transfer of traditional data and control information, and ST2 for the transfer of real-time data. Whereas IP typically will be accessed from TCP or UDP, ST2 will be accessed via new end-to-end real-time protocols. The position of ST2 with respect to the other protocols of the Internet family is represented in Figure 4.41.[52]

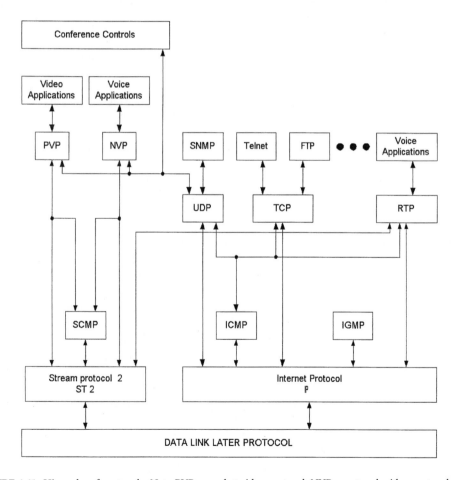

FIGURE 4.41 Hierarchy of protocols. *Note:* PVP = pocket video protocol; NVP = network video protocol; SCMP = stream control message protocol.

Both ST2 and IP apply the same addressing schemes to identify different hosts. ST2 and IP packets differ in the first four bits, which contain the internetwork protocol version number: number 5 is reserved for ST2 (IP itself has version number 4). As a network layer protocol, like IP, ST2 operates independently of its underlying subnets.

As a special function, ST2 messages can be encapsulated in IP packets. This link allows ST2 messages to pass through routers, which do not run ST2. Resource management is typically not available for these IP route segments. IP encapsulation is, therefore, suggested only for portions of the network that do not constitute a system bottleneck.[52] In Figure 4.41, the RTP protocol is shown as an example of transport layer on top of ST2. Others include the packet video protocol (PVP), and the network voice protocol (NVP).

ST2 proposes a two-step communication model. In the first step, the real-time channels for the subsequent data transfer are built. This is called stream setup; it includes selecting the routes to the destinations and reserving the correspondent resources. In the second step, the data are transmitted over the previously established streams; this is called data transfer. While stream setup does not have to be completed in real time, data transfer has stringent real-time requirements. The architecture used to describe the ST2 communication model includes[52]:

- Data transfer protocol for the transmission of real-time data over the established streams
- Setup protocol to establish real-time streams based on the flow specification
- Flow specification to express user real-time requirements
- Routing function to select routes in the Internet
- Local resource manager to handle resources involved in the communication appropriately

4.5.5.4 IP Multicast

For the Internet to be a viable real-time audio/video medium, it needs a method for serving a community of users. IP multicast is a suite of tools that addresses the bandwidth cost, availability, and service-quality problems facing real-time, large-scale Webcasting.

Rather than duplicating data, multicast sends the same information just once to multiple users. When a listener requests a stream, the Internet routers find the closest node that has the signal and replicates it, making the model scalable. IP multicast can run over just about any network that can carry IP, including ATM, frame relay, dial-up, and even satellite links. Originally developed in the late 1980s, it is now supported by virtually all major internetworking vendors, and its implementation and usage is picking up speed.

Reliability is a challenge with multicast because there is not necessarily a bidirectional path from the server to the user to support retransmission of lost packets. A string of lost packets could create enough return traffic to negate multicast bandwidth savings. For this reason, TCP/IP cannot be used. Among the transport protocols developed for IP multicast, RTP and RTCP are the main ones for real-time multimedia delivery. RTP adds to each packet header the timing information necessary for data sequencing and synchronization. It does not provide mechanisms to ensure timely delivery or provide QoS guarantees; it does not guarantee delivery, nor does it assume that the underlying network is reliable. RTP and RTCP are currently in draft status; both were expected to be final in 1998.

Uninterrupted audio requires a reliable transport layer; nevertheless, existing basic concealment techniques such as frequency domain repetition combined with packet interleaving work reasonably well if packet loss is minimal and occasional departures from perfection can be tolerated. One approach is to use FEC. Adding some redundant data improves performance considerably; combined with interleaving, this can be a good strategy, but it requires more bandwidth for a given quality level. This can be a challenge on a 28.8 kbps modem connection.

Reliable multicast can be used to increase the performance of many applications that deliver information or live events to large numbers of users, such as financial data or video streaming. Reliable multicast creates higher-value application services for today's IP-based networks. According to a study recently conducted by the IP Multicast Initiative (IPMI), 54% of information systems managers stated that IP multicast had created new business opportunities for their companies and these numbers are likely to grow from year to year.[51]

4.5.5.5 Additional Information

QoS is a relatively new topic. The reader interested in additional information in this topic should consult the book, D. Minoli, and A. Schmidt, *Internet Architecture*, John Wiley, New York, 1999. About two thirds of the book deals with the QoS topic from a video-over-Internet point of view.

References

1. U. Reimers, Digital video broadcasting, *IEEE Commun. Mag.*, June 1998, p. 104.
2. N. Farber and J. Villasenor, Extensions of ITU-T recommendations for error-resilient video transmission, *IEEE Commun. Mag.*, June 1998, p. 120.
3. L. J. Thorpe, Digital television is coming, *TV Broadcast*, March 1997, p. 26.
4. B. Bhatt, D. Birks, and D. Hermreck, Digital television: making it work, *IEEE Spectrum*, October 1997, p. 19.
5. D. Minoli, Packetized speech networks. Part 1: Overview, *Austr. Electr. Eng.*, April 1979, pp. 38–52.
6. D. Minoli, Packetized speech networks. Part 2: Queuing model, *Austr. Electr. Eng.*, July 1979, pp. 68–76.
7. D. Minoli, Packetized speech network. Part 3: Delay behavior and performance characteristics, *Austr. Electr. Eng.*, August 1979, pp. 59–68.
8. D. Minoli and E. Minoli, *Voice over Packet Networks*, John Wiley & Sons, New York, 1998.
9. D. Minoli, Supporting multimedia and other evolving applications using broadband, *Annu. Rev. Commun.*, 50, 729, 1997.
10. *Desktop Video Commun.*, July/August 1997, p. 45.
11. A. Davis, Cameras for desktop videoconferencing, *Desktop Video Commun.*, March/April 1997, p. 18.
12. M. Pihlman, H.323 videoconferencing over packet networks, *Desktop Video Commun.*, March/April 1997, p. 13.
13. S. Borthic, Turning up the heat at DVC west, *Desktop Video Commun.*, July/August 1997, p. 7.
14. D. Minoli, *Video Dialtone Technology, Approaches, and Services: Digital Video over ADSL, HFC, FTTC, and ATM*, McGraw-Hill, New York, 1995.
15. D. Minoli, Digital video compression: getting images across a net, *Network Comput.*, July 1993, p. 146.
16. D. Minoli, Distributed multimedia, bringing the infrastructure up to the challenge, *WAN Connections–Commun. Wk.*, August 1993, p. 60.
17. D. Minoli, Multimedia: Opportunities for Carriers and Service Providers, Market Report, Probe Research Corporation, June 1993.
18. D. Minoli, Distance Learning Applications, Broadband Networking, DataPro Report 1015BBN, November 1993.
19. D. Minoli, Concocting a recipe for the right multimedia mix, *Network World*, September 12, 1994, p. L10.
20. D. Minoli, Imaging Communications, Broadband Networking, DataPro Market Report, June 1994.
21. D. Minoli, Videoconferencing, Broadband Networking, DataPro Market Report, April 1994.
22. D. Minoli, Designing Scalable Networks, Network World, Collaboration, January 10, 1994, p. 17.
23. D. Minoli, Communications-Based Imaging, Market Report, Probe Research Corporation, September 1994.
24. D. Minoli, An Assessment of Digital Video and Video Dialtone Technology, Regulation, Services, and Competitive Markets, DataPro Market Report on Convergence Strategies & Technologies, April 1995.
25. D. Minoli, 1995: the year of video in enterprise nets, *Network World*, December 5, 1994, p. 21.

26. D. Minoli, Video Dialtone (VDT): Overview, Datapro Report 1090CNS, May 1995.

27. D. Minoli, Video Compression Schemes, DataPro Market Report on Convergence Strategies & Technologies, May 1995.

28. D. Minoli and B. Keinath, *Distributed Multimedia through Broadband Communication Services*, Artech House, Norwood, MA, 1994.

29. P. Jerram, Videoconferencing gets in sync, *Newmedia*, July 1995, p. 48.

30. FAQ for the alt.video.dvd Usenet newsgroup, available at http://www.videodiscovery.com/vdyweb/dvd/dvdfaq.html.

31. G. Pensinger, Sarnoff Assumes Role in Broadcasting's "Third Revolution," *TV Broadcast*, June 1995, p. 19.

32. GDC Corporation promotional material.

33. R. Eggers, On-line, Off-line Editing Systems Converging, *TV Broadcast*, June 1995, p. 19.

34. J. Van Pelt, Objective Testing for Video Compression, *TV Broadcast*, June 1995, p. 86.

35. Picturetel promotional literature.

36. B. Doyle, Crunch time for digital video, *Newmedia*, March 1994, p. 47.

37. L. W. Lockwood, MPEG-2: a wide ranging standard, *Commun. Technol.*, October 1993, p. 16.

38. M. Barezzani et al., Compression Codecs for Contribution Applications, *Electr. Commun.*, 3rd Quarter, 220, 1993.

39. T. Wechselberger, Conditional access and encryption options for digital systems, *Commun. Technol.*, November 1993, p. 20.

40. O. Eldib and D. Minoli, *Telecommuting*, Artech House, Norwood, MA, 1995.

41. P.E. Walker, Squeezing the picture: video compression, *Broadcast Eng.*, February 1994, p. 54.

42. R. Talluri, Error-resilient video coding in the ISO MPEG-4 standard, *IEEE Commun. Mag.*, June 1998, p. 112.

43. *Multimedia Week*, March 14, 1994, p. 8.

44. G. K. Wallace, The JPEG still picture compression standard, *Commun. ACM*, 34(4), 30, 1991.

45. A. G. MacInnis, The MPEG systems coding specification, *Signal Proc. Image Commun.*, 4, 153–159, 1992.

46. ISO/IEC JTC1/SC29/WG11 CD1-11172, Coding of Moving Pictures and Associated Audio for Digital Storage Media at up to 1.5 Mbits/s, Part 1 (Systems), November 1991.

47. *Video Technology News*, December 20, 1993, p. 2.

48. Infonetics Research, Inc., Quality of Service White Paper, available at www.atminc.com.

49. Extreme Networks, Inc., Promotional material.

50. Microsoft ASF White Paper, available at www.microsoft.com.

51. S. Pizzi and S. Church, Audio Webcasting demystified, *Web Techn. Mag.*, 2(8), 1997.

52. L. Delgrossi and L. Berger, Category: Experimental, ST2 Working Group, August 1995.

53. PictureTel, available at www.picturetel.com.

Index